D0773116

Sound Design in the Theatre

John L. Bracewell
Ithaca College

Prentice Hall, Englewood Cliffs, New Jersey 07632

Library of Congress Cataloging-in-Publication Data

Bracewell John L.
Sound design in the theatre / John L. Bracewell.
p. cm.
Includes index.
ISBN 0-13-825167-3
1. Theaters--Electronic sound control. I. Title.
TK7881.9.B73 1992
792'.024--dc20 92-30762
CIP

Acquisitions editor: Steve Dalphin
Production editor: Elaine Lynch
Copy editor: Mary Louise Byrd
Editorial assistant: Caffie Risher
Cover design: Bruce Kenselaar
Illustrations: John Bracewell
Pre-press buyer: Kelly Behr
Manufacturing buyer: Mary Ann Gloriande

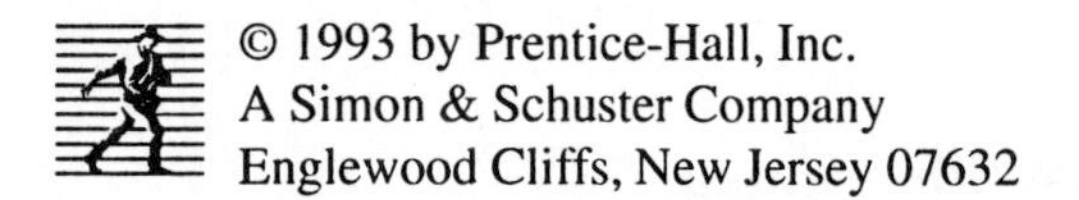

A Simon & Schuster Company
Englewood Cliffs, New Jersey 07632

Printed in the United States of America

10 9 8 7 6 5 4 3 2 1

ISBN 0-13-825167-3

Prentice-Hall International (UK) Limited, *London*
Prentice-Hall of Australia Pty. Limited, *Sydney*
Prentice-Hall Canada Inc., *Toronto*
Prentice-Hall Hispanoamericana, S.A., *Mexico*
Prentice-Hall of India Private Limited, *New Delhi*
Prentice-Hall of Japan, Inc., *Tokyo*
Simon & Schuster Asia Pte. Ltd., *Singapore*
Editora Prentice-Hall do Brasil, Ltda., *Rio de Janeiro*

Contents

CHAPTER 12: Synthesizers and Samplers 139

CHAPTER 13: Survey of Audio Systems 153

CHAPTER 14: Assembling Audio Components into Audio Systems 162

CHAPTER 15: Basic Characteristics of Human Hearing 175

CHAPTER 16: Binaural Perception and Psychoacoustics 187

CHAPTER 17: The Psychological Basis of Auditory Aesthetics 197

Preface

Work on this text started almost 20 years ago, shortly after I had completed my doctorate. Theatre and theatre sound were different than they are today. Theatre was less mechanized and not at all computerized. Sets were more illusionistic and less realistic. Sound was, if not in its infancy, still struggling through a difficult childhood. The only sound text of any merit at the time was *Sound in the Theatre*, by Harold Burris-Meyer and Vincent Mallory. As a graduate student, first at the State University of Iowa, then at Florida State University, *Sound in the Theatre* had been my holy writ, but one that, like any self-respecting *meisterschrift*, contained some elements of puzzling content that never became clear to me until I actually met Burris-Meyer himself.

Sound in the Theatre was intended mainly as a thumbnail outline of what Burris-Meyer and Mallory had learned in their early experiments with sound as a theatrical design element. Working at Stevens Institute of Technology during the 1930s, they had begun to break the ground that would lead to the development of audio-generated and -controlled sound as an adjunct of dramatic production. In order to use and control sound adequately within a theatrical environment, Burris-Meyer and Mallory had to design and build an adequate audio control system. It embodied both mix-down and fan-out stages, and it was capable of what seemed at the time to be amazing auditory gymnastics, making sound appear to come from specific directions, to move around the stage, and to produce unearthly voices for characters such as the Great Boyg in *Peer Gynt*.

As their work became known, Burris-Meyer and Mallory were invited to take their Stevens Mark II sound system into New York professional theatres and even into the Metropolitan Opera House to try out the new audio art in a professional setting. Unfortunately, the American entry into World War II interrupted the theatrical activities of both men. They were drafted into wartime research, and when they were discharged from military service, after almost a decade, the dominant names in New York theatre had changed. No one remembered Burris-Meyer and Mallory or their initial efforts in applying audio to theatre.

Sound in the Theatre, written in 1952, was meant only to summarize and preserve their work and to point the way for those of us in the next generation who would pick up the idea of audio as a theatrical art. Unfortunately, most of us who developed an interest in audio in the high-fidelity fad of the 1950s were addicts of technology and of the amazing loudspeakers and amplifiers that were pushing the limits of sound to what we thought were the extremes. I never quite understood why Burris-Meyer bothered with things like equal loudness contours and psychophysical properties of hearing. I really didn't see what it had to do with theatre sound—until I met Burris-Meyer.

During my doctoral studies, I was the resident authority in theatre sound at Florida State University, and I called Harold Burris-Meyer (who was then at Florida Atlantic University) to ask if he would serve as an unofficial major professor for my dissertation. My first meeting with him occurred on a rather warm south Florida day in the fall of 1969. When we sat down in his office, the first thing he asked was, "What do you know about the ear?" Rather startled, I admitted that I knew virtually nothing about the ear or hearing. Burris-Meyer immediately asked, "How can you design for something you don't understand?" He really said very little more to me that day, other than to suggest that I add a minor in psychology of perception to my doctoral curriculum. With that one question, however, he set my entire perspective about sound and theatre on its head. As I began to take classes in psychophysics, I began to unravel the puzzles of those misunderstood sections of *Sound in the Theatre*. The result, some 20 years later, is that this text goes much deeper into the nature of perception and the relationship between perception and aesthetics than was possible for Burris-Meyer in 1952.

Fortunately for me, my association with Harold did not end with that one encounter. He did supervise my dissertation, even though he could not do so officially, and he opened my mind to a new way of thinking about sound—as an art supported by a marvelous technology rather than as a technological playground that could also do theatre sound effects.

I projected the need for a comprehensive text in theatre sound in my dissertation, but for me to try my hand at creating that book took a long time. The initial problem was, essentially, one of what a theatre sound person needed to know. With little time to write and a lot to figure out about how to deal with sound as an art, progress was slow. Over the years, a number of other theatre sound texts have appeared, almost all very good and useful contributions to the field. Of those, most concern audio technology only. Only two have attempted any significant effort to discuss the art of sound. It isn't an easy subject to discuss. Lighting, at about the time Burris-Meyer and Mallory were involved with their experiments at Stevens, was in a situation similar to that of sound in the early 1970s—the inheritor of a new technology with great artistic potential, but lacking an artistic tradition on which to base its principles. Scenic art had a long history of painting and architecture to build on. Costume had the entire tradition of fashion as its base. Electric lighting and audio-based sound entered theatre as developments of electrical technology and engineering. The earliest instruments in both disciplines were too primitive to do much more than generate illumination or amplify a voice.

Lighting overcame its artistic deficit mainly because of the thought and work of Stanley McCandless, who gave us the four functions and controllable properties of light as anchor points and the idea of emulating the light of Flemish painting as a method of application. The only artistic area closely related to sound was music. Music offered very little that could serve as a launching point for an art of theatre sound. Burris-Meyer attempted to use musical scoring as a method of notating intricate cues for sound operators. However, most theatre people don't read music, and musical notation isn't really suitable for most forms of cuing anyway. Sound as a theatrical design art really needs a base of functions and an alphabet of controllable properties. Once those are available, then musical concepts (though not musical notation and techniques) are usable as methods of applying sound.

About eight years ago, when the section on the artistic elements of sound design was just taking shape, one of my students said to me, "Doc, you'd better hurry up and publish that book before the whole world goes digital!" That statement points to another problem in writing a text such as this one. The field changes so very rapidly. For most of its legitimate existence, theatre sound as an art based in electronic audio has struggled to keep up with the pace of growth in the robust professions of recording, broadcast audio, and, recently, acoustic reinforcement. The world hasn't quite gone all digital, but it certainly is on the brink of doing so. The last major effort in putting this text together was to try to bring the information it contains into the computer age. If a second edition of this book ever materializes, a number of chapters will have disappeared altogether. Others will have metamorphosed into discussions of a very different kind. Even the basic staple of the sound booth, the tape recorder, may be obsolete by then.

One reason that this text contains what is admittedly a considerable amount of older information is that a number of schools and theatres that could serve as training ground for sound designers still have basic analog equipment and are not likely to have anything else within the next few years. Another reason is that a record of the way theatre sound was done in the 1970s and early 1980s has some historical value. Most of our efforts were invested in the consoles, tape recorders, and signal processors that we used at the time. In the future, those efforts will center on the use of disc recording technology and waveform editing, but the principles of shaping and processing sound will remain the same as we move into an age of digital process. A filter is a filter, whether the technology that implements it is a hardware-based configuration of capacitors and inductors or a mathematical algorithm carried out by software and a computer. The artistic process of applying sound to serve the mood and the environment of drama does not change.

This text is roughly divided, into two portions. The first covers the technical information on which the craft of sound design rests. The second deals with the art of sound as a design element of the theatre and includes a massive design example.

Chapter 1 provides a basic discussion of the physical nature of sound. The coverage includes the way sound is produced and propagates through the atmosphere, the properties of sound, the way sound behaves in open and closed spaces, and how sound behaves in complex circumstances.

Chapter 2 provides the reader with an elementary understanding of the electrical principles underlying audio equipment on the theory that, to some extent, audio is an engineer's game. Beginning with an analogy to explain the behavior of current flow in a circuit, the chapter covers basic electrical laws, circuit configurations, magnetism, and the behavior of circuit elements (resistance, capacitance, and inductance) in audio devices. The chapter also covers the basics of audio systems including transmission lines and connectors. Chapter 3 introduces the reader to principles of audio measurement; and Chapter 4 introduces the families of audio systems that exist and the functions of different classes of equipment within those families.

Chapters 5 through 8 provide detailed examination of the various kinds of equipment used in audio systems and examples of how each kind is used for theatre sound purposes. Chapter 5 covers loudspeakers and explains how each type of loudspeaker may be used in the theatre. Chapter 6 discusses microphones and their uses for recording and reinforcement. Chapter 7 provides a basic discussion of audio amplifiers, concentrating on what one needs to know in order to interconnect equipment properly and in order to understand the operation of control consoles and signal processors. Chapter 8 explains the various families of equalizers, filters, and automatic intensity control devices.

Chapter 9 provides a very thorough discussion of audio consoles, explaining several different kinds of consoles that may be found in theatre sound booths. The chapter covers mix-down consoles designed for public address, reinforcement, broadcast audio, and studio recording. The final portion of the chapter covers a form of control designed primarily for theatre sound performance.

Chapter 10 explains principles of audio storage and retrieval, including tape recording, phono disc recording, compact discs, and digital tape systems. The chapter also introduces the newer computer mediated, direct-to-disc recording and editing systems. Time code synchronizing systems for multiple tape recorders and for automated consoles are also explained.

Chapter 11 explains time-based signal processing devices including delay and reverberation generators, harmonizers, pitch shifters, and phasers.

Chapter 12 takes up the subject of synthesizers and the associated equipment that makes so much of present-day music possible. Especially important is the discussion of samplers and the explanation of the Musical Instrument Digital Interface, including MIDI Show Control, the recently developed protocol particularly for interfacing elements of theatrical production such as lighting, pyrotechnics, sound, and motors.

Chapters 13 and 14 are designed to be a synthesis of the sections on technical processes. Together they discuss how the audio components discussed in the previous chapters, are assembled into systems. Chapter 13 reexamines the kinds of systems used in the audio industry, taking a more complete look at the interface of components and the organization of each system. Chapter 14 discusses how systems are organized technically, providing three examples related especially to theatre.

Chapters 15 through 27 make up the second portion of the text, covering the aesthetics of sound and sound design as an art of the theatre.

Chapters 15 and 16 explain the nature of hearing and the characteristics of hearing, focusing on the aspects of hearing and psychoacoustics that are of special importance to the sound designer in creating effective and appropriate sounds to support a dramatic environment.

Chapter 17 sets out the basis of the sound designer's art in terms of the aesthetics of sound and the interrelationship of vision and audition as applied to the shaping of meaningful emotional experience. The chapter examines the potential that sound offers as a means of leading and shaping audience reaction to dramatic events. Chapter 18 examines the relationship among craftsmanship, creativity, and design. The discussion includes an explanation of phases of maturation in design and in mastery of craft.

Chapter 19 is a particularly important chapter that sets out the relationship of sound to staging theatrical events. An exposition of the functions of sound and its controllable properties is included.

Chapters 20 through 25 contain a large design example, taking a specific playscript (*The Rose Tattoo* by Tennessee Williams) and fitting a substantial sound design to it. Chapters 20 and 21 discuss the creative activity involved in generating a sound design. The chapters include an exhaustive analysis of the playscript in terms of general production values and then in terms of sound as a supporting element. Chapters 22 and 23 provide examples of the effort necessary to obtain a usable framework on which to base the development of the design, of the research necessary to find appropriate sounds, and of the scheduling and planning required to execute the design within the available time between the beginning of the design phase and the start of technical rehearsal.

Chapters 24 and 25 track the progress of construction of the design from the first attempts to put sounds on tape to the end of the rehearsal period.

Chapter 26 reviews the design example to extract additional information related to design activities and processes.

Chapter 27 provides a brief sketch of how the design for *The Rose Tattoo* might be accomplished using state-of-the-art equipment, substituting sampling synthesizers, sequencers, a digital waveform editing station, and a MIDI-capable sound scoring control system.

Appendix I provides a decibel table, for reference.

Appendix II examines theatrical intercommunication systems. Because the business of assembling and installing the production intercom usually falls to sound, this addendum to the text attempts to explain the basics of intercom system equipment and the ways in which it can be usefully assembled to serve a variety of purposes in both rehearsal and performance.

Although reinforcement equipment is examined in some detail, acoustical reinforcement itself is not given anything more than a cursory treatment in this book. Reinforcement is a separate and very important area of sound—one that deserves separate discussion in its own book. This text is intended as a work explaining the design of auditory enhancement and auditory environments for dramatic production. Because full discussion of reinforcement would have doubled the length, that area of theatre sound is all but omitted.

Acknowledgment is made to those who served as reviewers for Prentice Hall: Thomas A. Beagle, Antioch, California City School District; David G. Flemming, Southwest Texas State University; and Richard K. Thomas, Zounds Productions.

Finally, my hope is that, had Harold Burris-Meyer lived to see the publication of this book, he would have approved. On one occasion I saw him pick up a recently published text on theatre sound, page through it, growl, and put it down complaining that it was just another book on hardware, not on design and the relationship of psychophysics of perception to theatrical aesthetics. I believe that he would have found the treatment contained here more to his liking. At least, that is my fervent hope.

John L. Bracewell
Ithaca, New York

Introduction

On the Use of Sound as an Art of the Theatre

Theatre relies, essentially, on the two primary modes of human perception: sight and hearing. The object of theatre is for the audience to see and hear the staged event. The words "staged event" are important: Theatre, regardless of content (mime, dance, dramatic narrative), is the visual and auditory realization of some form of dramatic action on the stage.

For most of its history, theatrical staging has been a predominantly visual art. Dialogue excepted, the auditory element served a minor role furnished mainly by property noisemakers. As a result, the visual aspects of theatre are both artistically and technologically mature. Sound, by contrast, is a young art in the theatre, one that began to develop only after the advent of sophisticated electronic audio equipment and after sound as an adjunct of dramatic art had been fully established in such areas as broadcasting and film.

The concept of sound as a design art for theatre still needs definition. For many people, theatre sound means acoustic reinforcement; for others, the term refers to music and environmental sounds used to enhance dramatic impact. Both are part of the uses of sound in the theatre, but neither alone is the sum of the art.

Sound in the theatre has seven functions:

Audibility
Motivation
Music
Vocal alteration
Vocal substitution
Mood
Extension of dramatic space/time

These functions are mentioned frequently, directly or indirectly, in the course of the design example given in the latter part of this book. They are fundamental to the art of sound design, and each is explained in detail. These functions are the basis of the sound designer's work and play a role in all decisions the designer makes about virtually every cue in a production.

Implementation of the functions is possible because of the controllable properties of sound. Some of these properties can be manipulated directly by means of audio control and processing equipment (e.g., regulation of intensity by a fader on a control console); others are matters of the designer's choice of sounds (e.g., choosing bright instead of dark sounds to suit the mood of the scene or moment). The controllable properties are

Intensity
Frequency
Duration
Timbre
Envelope
Directionality

Again, like the functions of sound, the controllable properties are mentioned throughout the book, and each is explained in its turn.

One consideration in particular is central to the understanding of sound as an element of theatrical design: Of all the theatrical arts, sound offers one of the most powerful means to affect the response of an audience to a dramatic event. Sound is such a commonplace of human daily existence that we seldom realize how extremely potent it is as an afferent force in our emotional lives. But consider how empty a film seems when we see only the picture without the sound track. Even if we can hear the dialogue, a great deal of the emotional impact of the film is lost without the environmental sounds and underscore music. Because of the difficulty of blending live and recorded sound, and because of the problems of keeping recorded sound synchronized with live acting (which can vary in pacing from performance to performance), theatre has been conservative in the use of sound. Recent developments in audio technology, however, have alleviated most, though not all, of these difficulties, and use of sound is now limited mainly by the imagination of directors and sound designers.

The designer's imagination is critical. In any application of an artistic medium to the stage, whether visual or auditory, the job of the designer is to provide a metaphor for the meaning of the staged event. The metaphor must appeal to the senses and the emotions through the qualities of the medium within which the artist works. Consequently, a thorough familiarity with the expressive qualities and capabilities of the medium is essential to the development of the designer's art.

Like the visual aspects of theatre, theories of visual art are well developed. Aesthetic principles for theatre sound do not rest on any such long tradition, but they do exist. The basis of an art of sound is most closely related to the principles of musical composition. Although the sound designer need not use musical notation to set down design ideas, the process of developing and shaping ideas in sound for dramatic production is similar to the process of composition. Both musical composition and sound design are processes of building structural relationships through time and delineating the marker events that the human mind needs to recognize patterns in time. Indeed, if the sound designer is also a musician, so much the better.

The technical craft of any art is important, especially for sound. Traditional audio is almost inescapably an "engineering" art. Sound itself is a complex phenomenon. Control of sound requires a sophisticated technology, and audio equipment is, in general, designed by engineers for engineers. Equipment customarily found in recording studio control rooms bears labeling terminology drawn from the field of audio engineering. Therefore, the theatre sound artist needs a fundamental grounding in the technology of audio, and must understand the various kinds of equipment and know how each is used. More important, the sound designer must be able to imagine novel and untried uses for audio equipment in order to fulfill the strange and unusual demands that theatre can often impose.

All spaces affect quality of sound. The behavior of sound within particular spatial configurations (the science of acoustics) is another area with which the theatre sound designer must be conversant. Understanding the characteristics of acoustic spaces and how to interface audio equipment with those spaces can mean the difference between a successful sound design and one that fails to achieve its purpose. Besides an understanding of acoustics, a general understanding of the fundamentals of the physical behavior of sound is important to the sound designer.

The theatre sound designer not only deals with sound, acoustics, and electronic audio but must also understand the characteristics of human hearing and its peculiarities and limits. Familiarity with the principles of psychoacoustics (the psychophysics of sound) enables the designer to use sound to create impact, to add sound to dramatic performance without masking dialogue or detracting from the actors' lines, and to lead and channel emotional response. A knowledge of the behavioral characteristics of the ear–brain process is essential to the full realization of the possibilities of sound as an art of the theatre.

All of these various aspects of the sound designer's art are covered in this book. In general, the first part of the book deals with the technology and the craft of audio; the second part concentrates on the art of sound design, although the principles of the art are never very far from consideration in any portion of the book. The sections on hardware always include examples of potential uses of each piece of equipment, uses drawn from real design problems in staging dramatic productions.

In all portions of the text sound is treated both historically and practically. That is, some background on the development of the art and the technology is given for each aspect of the subject. Current practices and methods are also given as complete an exposition and explanation as space allows. Every attempt has been made to bring the coverage as near to state of the art as possible. Success in this last attempt, however, is hardly possible, as almost any aspect of audio is in the process of obsolescence as soon as it becomes a viable technique.

Throughout this book one aspect of sound design must remain firmly in view: that the art of sound in theatre rests on an aesthetic grounded in the psychology of perception and the development of meaning—an aesthetic that deals with the special importance of space and time within human symbolization of perceived reality, with the importance of sound to the division and articulation of time, and to the intuitive perception of the nature of space. The psychological element of this aesthetic pays special attention to the dual, parallel nature of the human information processing system and to the ways in which sound affects each half of that system.

Chapter 1

Basic Characteristics of Sound

The word *sound* is the name by which we designate our subjective experience of a particular kind of physical event: acoustic energy carried in the air around us. Sound, especially in the contemporary world, is almost continuous. Almost everything in our experience generates noise—automobile engines, whistles, musical instruments, animals, wind moving through tree leaves, and, of course, our own bodies, especially our vocal folds.

Sound is actually a sequence of pressure variations traveling through the air to our ears. The variations occur so rapidly that we do not sense them individually, just as we do not see the blades of an electric fan individually. **Acoustics** is the study of the behavior of vibrating bodies and of the behavior of vibrations within various configurations of space and conducting media. **Psychoacoustics**, the study of how humans hear sound, is a related discipline, but one that deals with sound from a somewhat different point of view. Later in this book we shall take a close look at psychoacoustics, because it is important to sound design. The purpose of this chapter is to provide an introduction to the physical nature of sound. We need to understand the behavior of sound, first, in order to understand the way in which audio systems record, process, and reproduce sound, and second, to construct sound designs that integrate fully and believably with the dramatic environments that they must support.

The Physical Nature of Sound

Sound of any sort begins with a vibrating body, which sets a surrounding medium into motion. Air is the medium in which we normally experience sound. Air is an excellent acoustic conductor because it is **elastic**. An elastic medium can be distorted momentarily but tends to snap back to its basic shape whenever the distorting force is removed. Elasticity is essential to any sound-conducting medium because acoustic energy travels by pressing and stretching the substance of a medium. A vibrating object acts as a force that pushes and pulls the air. Pressing creates a small momentary region of high density, called **compression**. Stretching creates a small momentary vacuum, usually called a **rarefaction**. (See Fig. 1-1A.) As the air is compressed and stretched, energy pulses radiate away from the vibrating source. The regions of air that are compressed and stretched do not travel outward from the generator; the compression effect does, however.

Sound energy (the compression effect) travels at a speed of 1128 feet per second in warm air, which works out to approximately two tenths of a mile in a second, or 13 miles per minute, or 770 miles per hour. One pulse of energy can travel only a limited distance before the source generates another pulse. The distance between the beginning of one energy pulse and the beginning of the next is determined by two things: the speed of sound and how rapidly the source vibrates. That distance is called the **wavelength** of the sound.

One period of compression plus one period of rarefaction are called a **cycle** of energy. The number of cycles that occur during 1 second is called the **frequency** of the sound. (See Fig. 1-1B.) **Cycles per second** are measured in **hertz** (abbreviated **Hz**). Saying that a sound has a frequency of 440 Hz is equivalent to saying that the energy source **oscillates** (changes direction of motion) at a rate of 440 cycles per second. The strength of compression and rarefaction is called the **amplitude** (or **intensity**) of the sound. Intensity and frequency are two of the controllable properties of sound. Wavelength and frequency are inversely proportional. The higher the frequency the shorter the wavelength; the lower the frequency the longer the wavelength. (Note illustration of wavelength in Fig. 1-1B.)

Sound tends to radiate spherically (nondirectionally) away from any generator. Each spherical compression wave contains a certain amount of energy. As the surface of the sphere expands, the energy it contains pushes against a larger and larger body of air. In consequence, sound diminishes in intensity as it travels away from its source. This phenomenon is called the **inverse square law**, because the sound pressure varies as the inverse of the square of the radius of the sphere. (See Fig. 1-1C.)

FIG-1-1A: COMPRESSION WAVES

CONCENTRIC SPHERES OF PRESSURE EXPANDING AWAY FROM SOUND SOURCE, AT CENTER

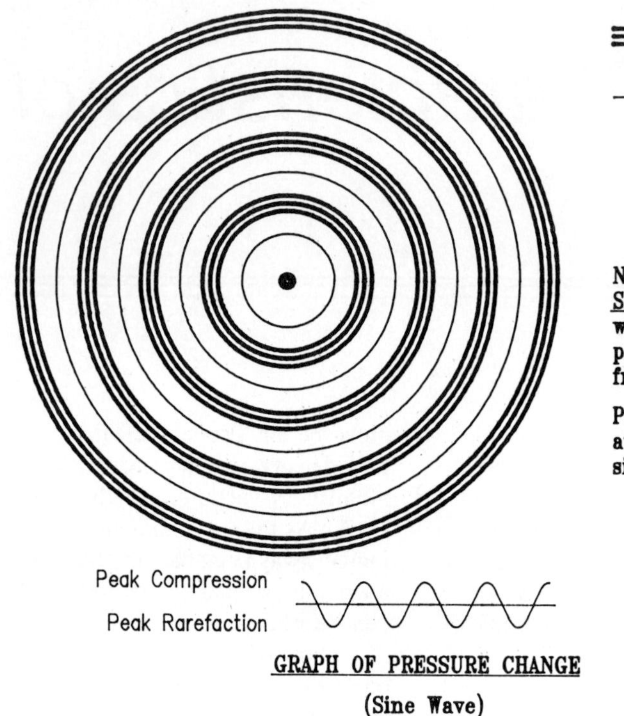

= Compression

= Rarefaction

NOTE: Pressure waves are SPHERICAL. Surface of wave constantly expands as pressure wave radiates away from sound source.

Pressure variation in atmosphere describes a sinusoidal curve.

Peak Compression

Peak Rarefaction

GRAPH OF PRESSURE CHANGE

(Sine Wave)

FIG 1-1B: FREQUENCY

SOUND SOURCE EMITTING A 10 Hz. TONE

NOTE: Pressure waves shown only as a segment of the total spherical radiation pattern emitted by the sound source.

= Compression

= Rarefaction

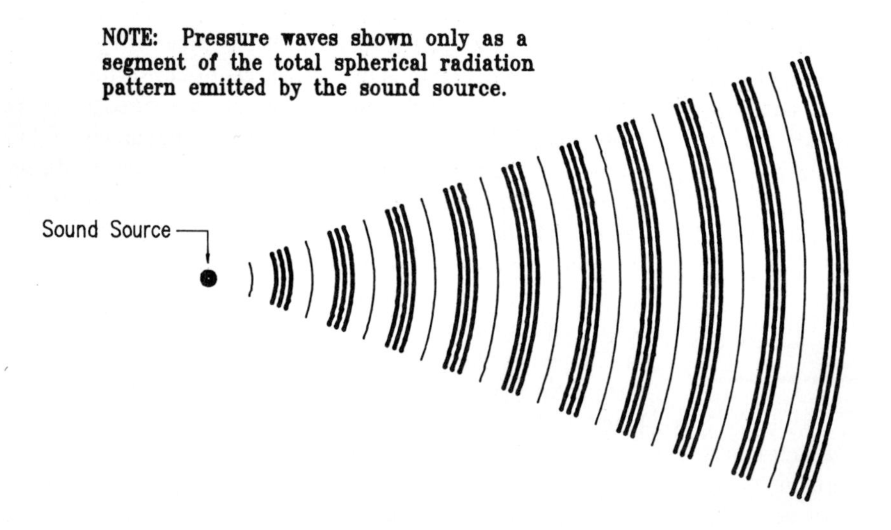

Compression wave #1 travels 1/10 second before succeeding compression wave leaves sound source. Compression wave #1 travels 10 wavelengths in 1 second. Source emits ten compression waves in one second; thus, frequency is 10 cycles per second, or 10 Hertz.

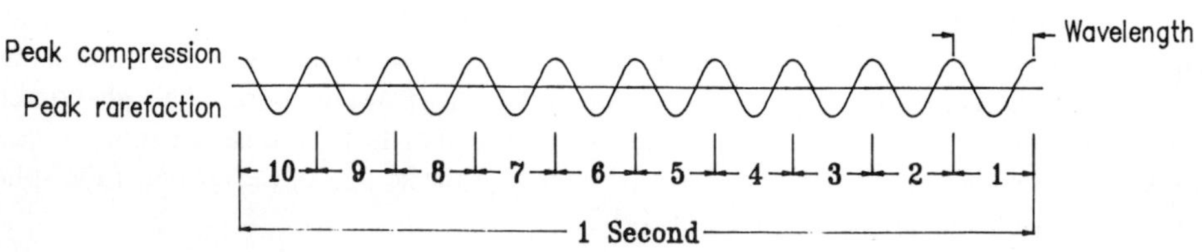

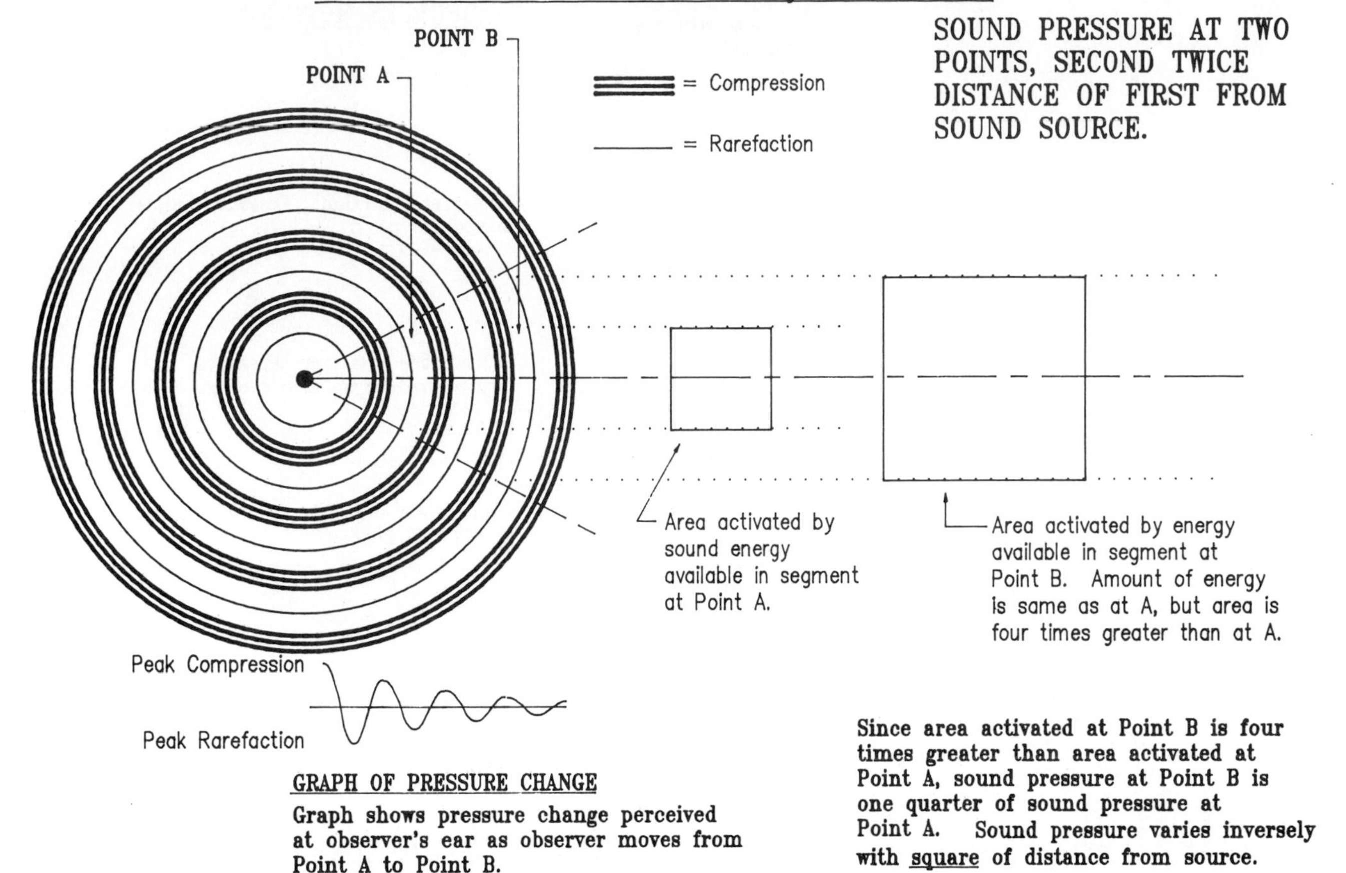

Every sound has a particular waveform that develops as the energy of the source produces compression and rarefaction in the air. The shape of a wave determines the exact quality of a sound. The simplest waveform is a **sine wave**. The pressure wave illustrated in Figs. 1-1A through 1-1C is a sine wave. A sine wave represents a pure tone, one that contains only a single frequency of vibration.

Very few sounds are pure sinusoids. Most sounds are **complex**—that is, they contain internal variations in magnitude and direction of pressure change. Over the wavelength of the basic frequency of a sound, smaller changes may occur. The changes may be analyzed as many higher frequencies sounding at the same time as the basic frequency. The waveshape describing air pressure variation when many frequencies sound simultaneously is much more intricate than the simplicity of a sine wave. Of course, we do not hear the many individual frequencies in complex sounds; we hear only a single sound. Complexity gives individual sounds their characteristic qualities. The particular quality of a sound is called its **timbre**. The various frequency components that make up timbre are called **harmonics**.

Harmonics may be either **linear** or **nonlinear**. Linear harmonics are integral multiples of the fundamental frequency (i.e., f_0 x 1, f_0 x 2, f_0 x 3, . . .) and are often called **partials** or **overtones.** Non-linear harmonics are fractional multiples of the fundamental. Musical instruments generally produce linear harmonics; random noise (wind turbulence, a table leg scraping over a wood floor) contains nonlinear harmonics. **Usually, the perceived pitch of a sound is that of the fundamental**. In a sense, all sounds are perceived as their fundamental frequency **modulated by the subordinate components**.

The sine curvature of a simple tone gives us some useful descriptive terminology used both in acoustics and audio. A sine wave gets its name from the fact that the height of the curve at any given point is proportional to the sine of the *angle* of rotation. The beginning of the waveform represents 0 degrees; the end represents 360 degrees. In fact, if one placed a moving chart beneath a wheel and fitted the wheel with some form of scribing device (a pencil or pen to draw a trace on the chart as the wheel turns), a sine curve would be the result traced on the chart. Because the position of a pressure wave in time can be expressed in degrees of rotation, all waveforms, whether sinusoidal or not, are described in terms of rotation through a complete circle. The position of a waveform in time is called **phase**. (See Fig. 1-2A.)

Two or more waveforms can be compared by their position in time. If waveforms rise and fall in synchrony, they are **in phase**; if not in synchrony, they are **out of phase**. Figure 1-2B shows sound sources physically displaced from each other as a cause for the out-of-phase condition. That is, the pressure wave from one of the sources arrives at an observer's

FIG. 1-2A: PHASE AND ROTATION.

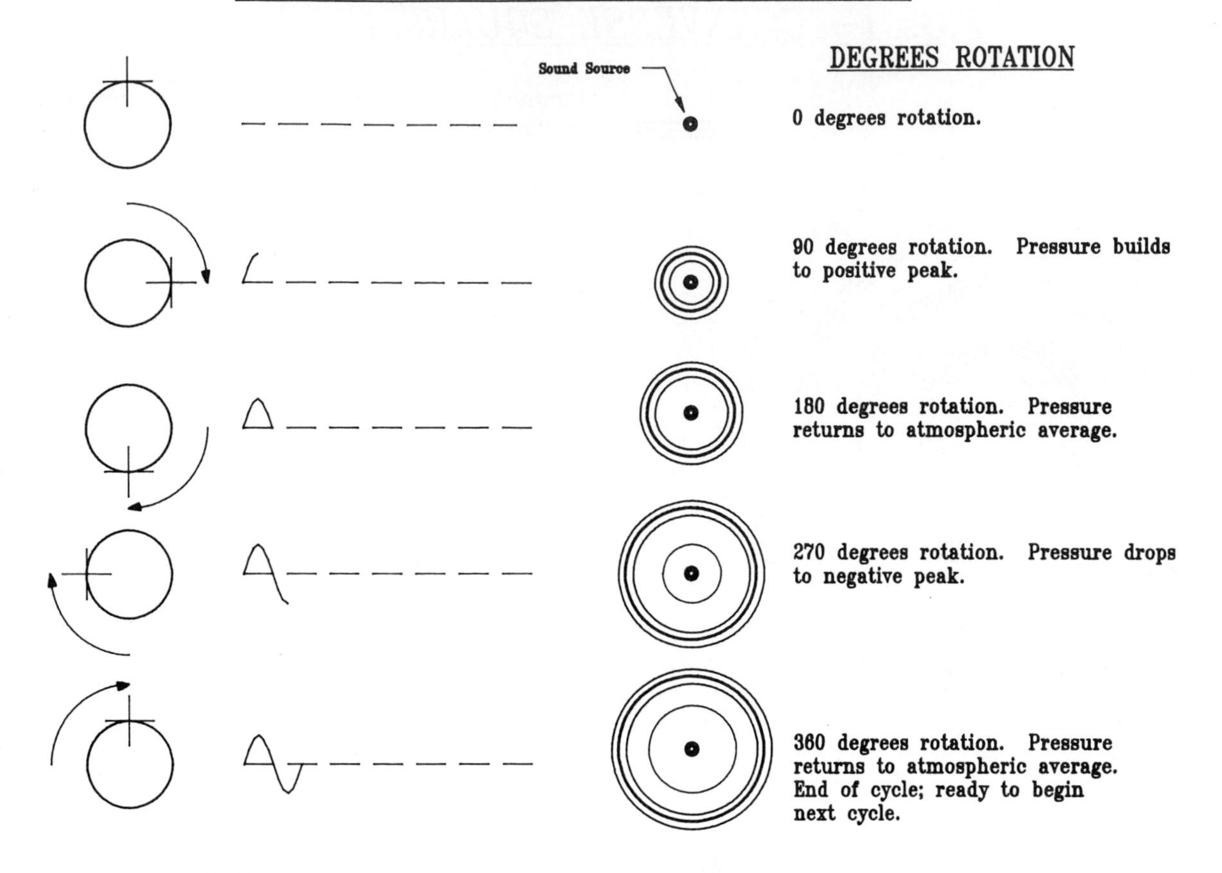

FIG. 1-2B: PHASING BETWEEN SOURCES.

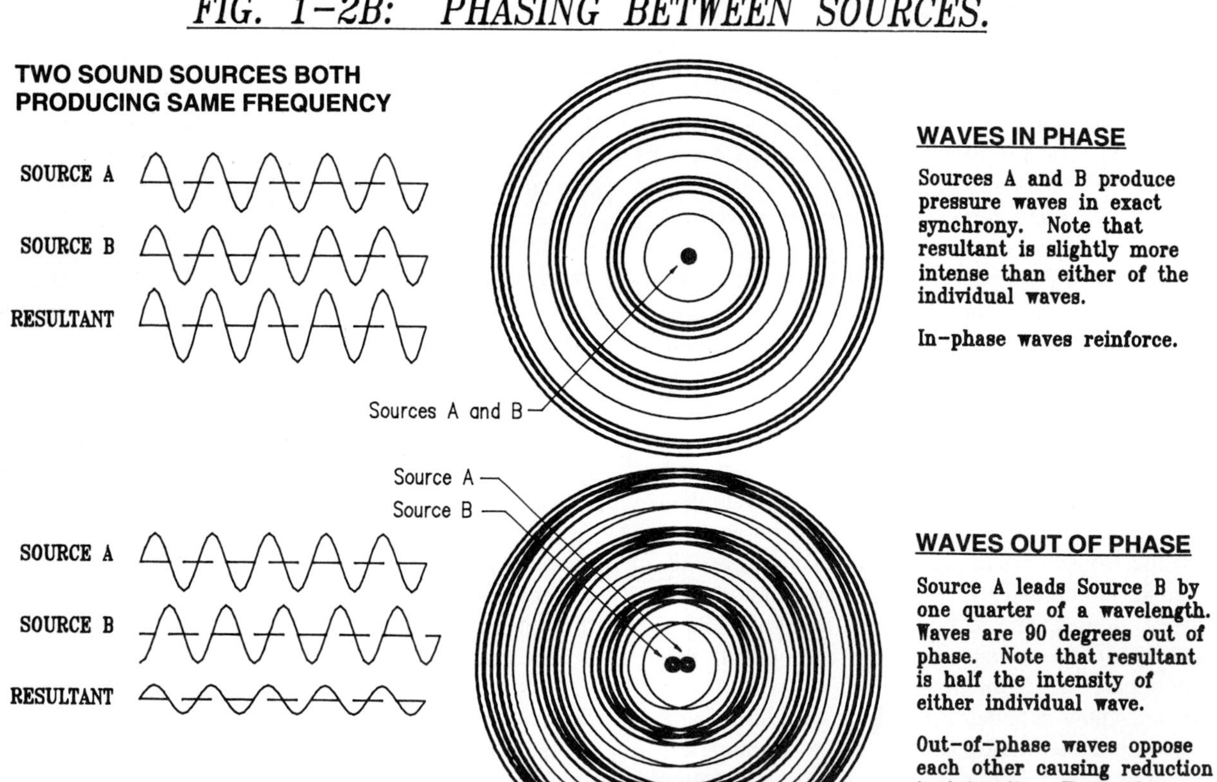

position before the pressure wave from the other. Phasing problems may also result from a situation in which one source begins to produce sound earlier or later than another. Phase is important in both acoustics and audio. In audio especially, phasing errors can destroy signal quality.

All sounds appear to us as individual events or groups of events. The human brain tends to group a large number of elements of sound into a synthetic entity that we hear as a particular sound. The component structure of a sound remains largely concealed in our perception of the sound as an event. Part of the hidden structure is the harmonic complexity of the sound's timbre. Another part is the sound's **envelope**. Envelope is the variation of the amplitude of the sound energy plotted against the duration of the sound. Every envelope is made up of four parts, each part consisting of a portion of the overall duration. The graph in the Fig. 1-3 represents an average envelope that describes all elements of intensity variation within a sound. Individual sounds may vary from the average.

All sounds begin with a rise in intensity, from silence to some finite peak level. The rise is known as the **attack**—the time required for the sound energy to develop from the inaudible to peak intensity. The top of the attack curve is the most intense part of the sound, though the duration of peak energy may be so short that the peak has little effect in the perceived loudness of the sound. The second part of the intensity envelope is called **decay** (or initial decay)—the time required for the energy to drop from attack peak to the **sustaining level** of the sound. Sustaining level is defined as the average intensity maintained while the sound source continues to supply energy to generate vibrations in the surrounding air. Varying the difference between peak level and sustain level contributes to the percussive quality of the sound. Short attack with a large drop from peak to sustaining levels will give the effect of a very percussive attack. The third part of the envelope is called the **sustaining period** (or, usually, just **sustain**). Length of the sustain period constitutes most of the perceived duration of the sound and most of the perceived loudness of the sound. The final part of the envelope is the **release period**, which is the time required for the generator to cease producing audible vibrations after the sound source ceases to supply energy. Envelope is often called **ADSR**, after the initials of the names of each subperiod.

Every sound has an envelope. A note sung by the human voice tends to have a moderate attack, quick decay, usually a relatively long sustain, and a quick release. Percussive sounds often have a short attack, no decay, no sustain, and a short release. By varying the value of ADSR, all qualities of envelope can be obtained.

Envelope is one of the controllable properties of sound. Choice of envelope quality is an effective tool in control of the emotional properties of sound used in the context of a dramatic

FIG. 1-3: ENVELOPE.

VARIATION OF INTENSITY WITH TIME IN AN INDIVIDUAL SOUND.

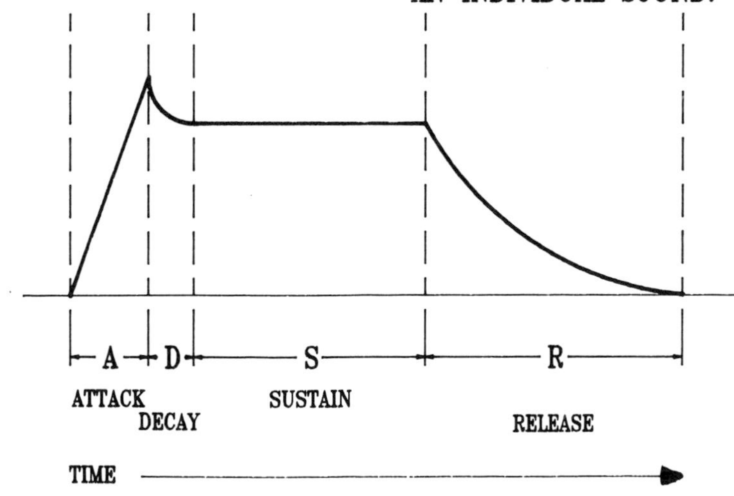

ATTACK (A): Rise time of the sound.
Typically short compared to Sustain and Release times.

DECAY (D): Initial decay.
Typically very short compared to Sustain and Release times.

SUSTAIN (S): Sustaining time.
Period during which source supplies energy to maintain sound. Typically, very long -- 1 second to as much as several minutes.

RELEASE (R): Release time.
Time required for sound to drop from sustaining level to inaudibility. May be very short or very long. Affected by reverberance of environment.

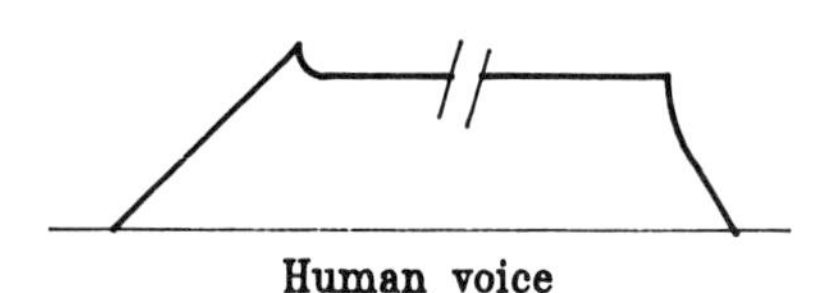

Human voice

A = Long (250 milliseconds typical)
D = Short (10 milliseconds typical)
S = Long (1 or more seconds)
R = Short (10 milliseconds or less)

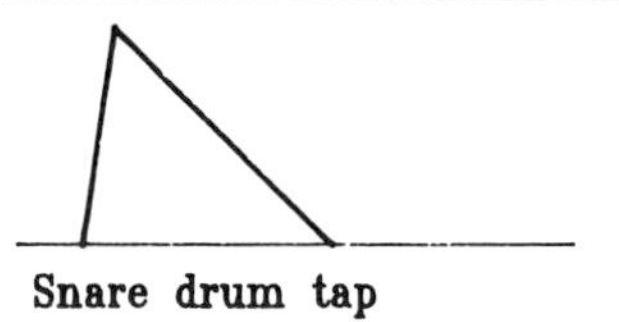

Snare drum tap

A = Short (5 milliseconds)
D = 0
S = 0
R = Short (100 milliseconds or less)

work. Film and operatic scores have made use of envelope as a controller of emotional response for many years. For example, a *sforzando* chord (sharp attack with quick drop lower sustaining level) is frequently used to set the mood for a catastrophic development in a drama. A sound with a slow rise may be used to enhance a statement of love or to reinforce the emotion of a gradual understanding on the part of a character.

Practical sound generators consist of an energy source, a generator, and a means of coupling the action of the generator to the medium. Coupling devices improve the efficiency of the generator. Sound sources use two kinds of coupling devices: **radiators** and **resonators**. (See Fig. 1-4.) Both radiators and resonators contact large amounts of air, and both tend to support some frequency or band of frequencies while suppressing others. Loudspeaker systems are built on the use of radiators and resonators.

Practical sound generators, as we have noted, produce complex waveforms—waveforms that are formed by multiple sine waves vibrating simultaneously. The waveforms that complex tones produce vary from the sinuous in proportion to the number of component sines that go into their construction. Depending on their constituents, complex sounds resemble one of three generalized waveforms: **sawtooth, triangular,** or **rectangular**. (See Fig. 1-5.) Analog synthesizers and electronic test generators produce sawtooth, triangular, rectangular, and a pseudo sine wave.

The behavior of sound is always affected by the boundary surfaces that surround a space. A space with no boundaries is called an **open field**. A space that is partly or completely limited by boundary surfaces is called a **closed field**. Sound is either transmitted, absorbed, or reflected by boundaries. Open surfaces transmit sound; soft surfaces tend to absorb sound, hard, closed surfaces reflect sound. In an open field an observer would hear the sound source directly, but no reflections would occur. (See Fig. 1-6.) A field surrounded with transmissive boundaries is almost equivalent to an open field. A closed field bounded by soft surfaces creates almost no reflections. In a field closed by hard surfaces, sound energy produces reflections. (See Fig. 1-7.) Multiple reflections create **reverberation**, and the duration of these reflections is called **reverberation time**. Reverberation time is measured as the length of time required for reflected sound intensity to drop to one millionth of the original level of the source sound. A space with a long reverberation time is said to be **live**; spaces with little or no reverberation time are said to be **dead**.

The kind of activity that will occur most frequently in a particular location determines the necessary acoustic treatment of the space. Spaces designed primarily for speech need a relatively short reverberation time, otherwise the intensity of the reflected sound would tend to mask subsequent speech sounds. Music, on the other hand, requires a relatively long reverberation time. Spaces for speech, therefore, are usually

FIG. 1-4: RADIATOR AND RESONATOR

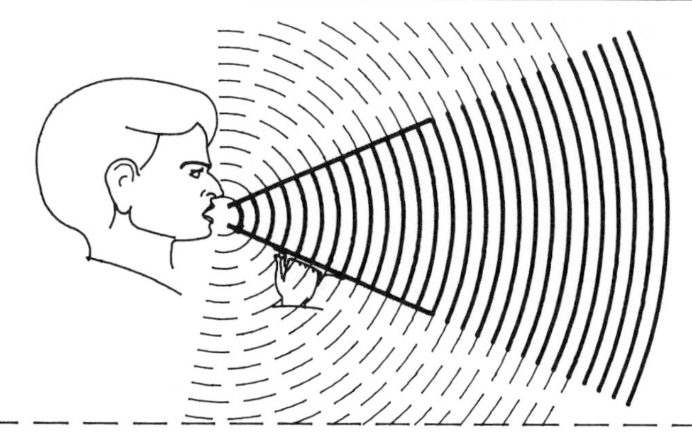

RADIATOR. Megaphone acts as radiator to project the voice forward. Weaker pressure waves not enhanced by effect of radiator dissipate rapidly.

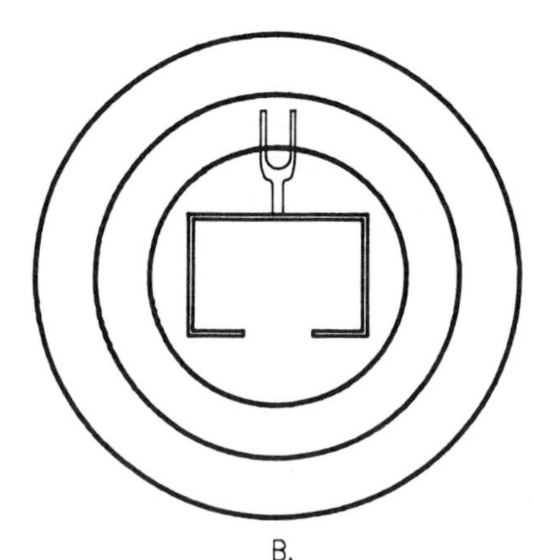

A. B.

RESONATOR. The tuning fork at A. produces very little sound. It's energy level is too small to move the surrounding air.

The tuning fork at B. in contact with the resonator is clearly audible. Note that the focus of sound energy is the resonator, not the fork. The resonator contacts a large amount of air, increasing the intensity of the tuning fork.

FIG. 1-5: COMPLEX WAVEFORMS

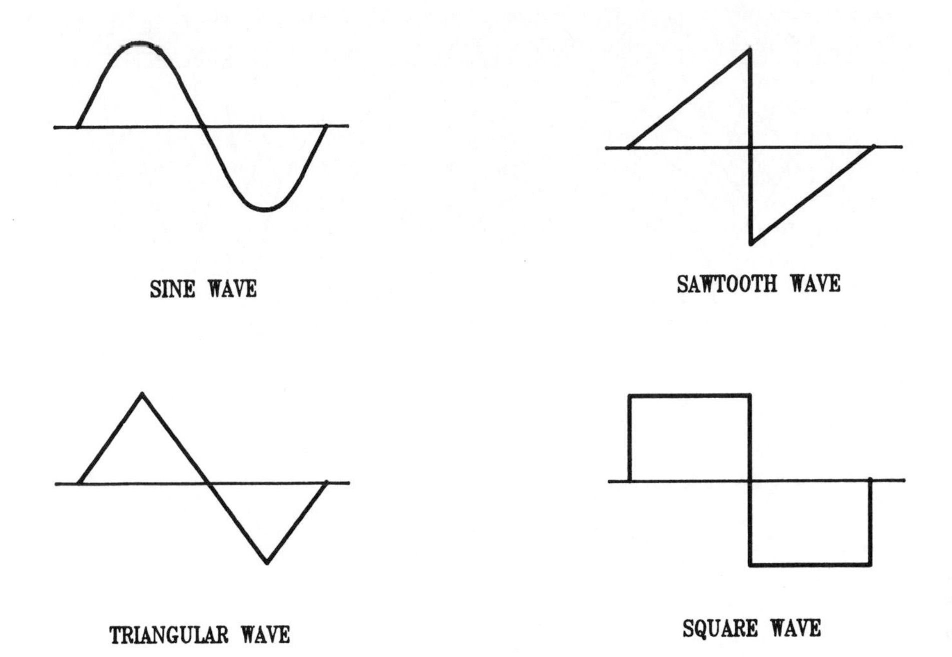

FIG. 1-6: OPEN FIELD

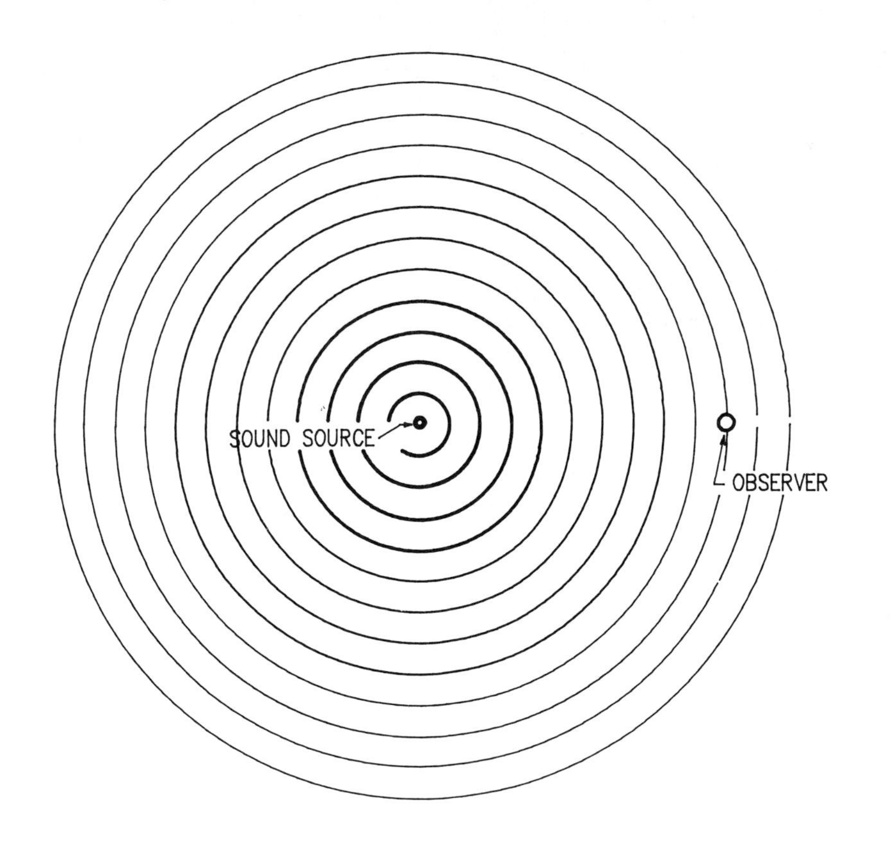

No occluding surfaces in sound field. Observer hears sound directly with no echoes.

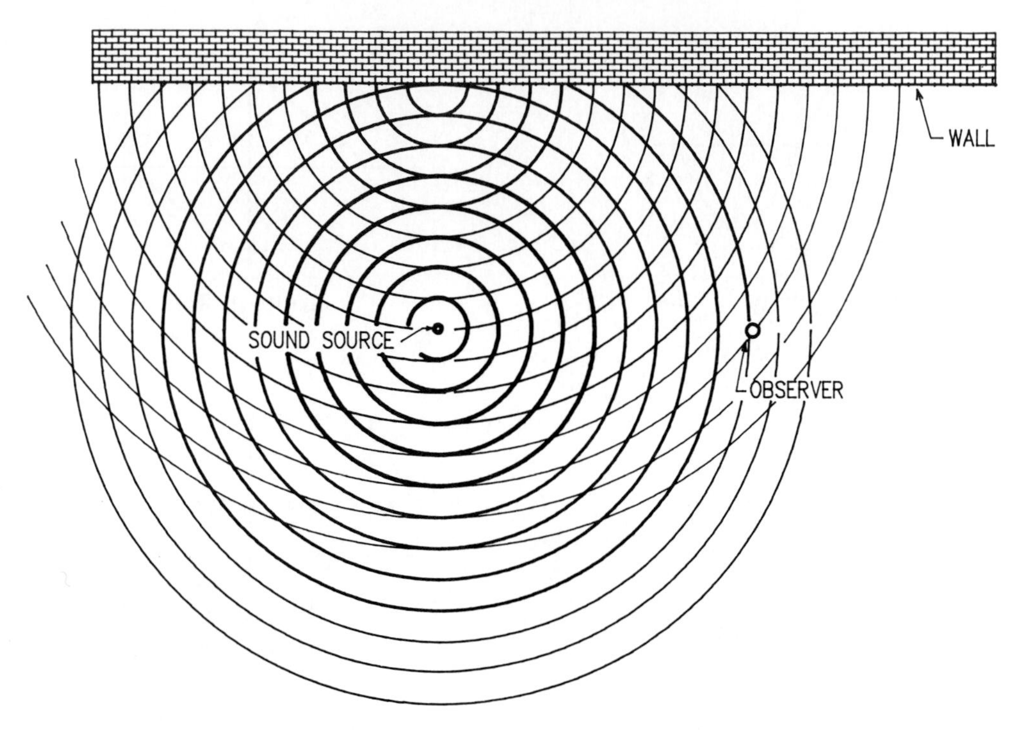

Sound hits wall and is reflected. Observer hears direct sound followed by single echo.

incompatible with spaces for concerts—one reason why the design of an auditorium as a multipurpose hall is extremely difficult. In general, however, one can say that, whatever the purpose, the larger the space the longer the reverberation time needed.

Any enclosed space tends to act as a resonator of some kind, a fact that is as true of auditoriums as of any other acoustic space. Like all resonators, enclosed rooms tend to sustain some frequencies and suppress others. The frequencies at which a room tends to resonate are called **ring modes**. Good acoustic design attempts to minimize significant ring modes; if such modes do persist within a space, special measures are usually required to overcome the problems. When electronic sound reinforcement systems are used in an auditorium, the system must be tuned to suppress all energy at any significant ring modes that exist within the space.

Sound, Space, Time, and Motion

The speed of light energy (186,202 miles per second) makes possible the perception of events at a distance virtually as they occur. Compared to the speed of light energy, sound energy travels slowly. Consequently, if an event takes place at a considerable distance from our particular vantage point, we experience a noticeable delay in perception of any sound the event may produce. We are all familiar with the experience of seeing someone at a distance strike a nail with a hammer or kick a football, then hearing the noise of the action slightly afterward.

Propagation of sound through the air also produces one significant effect related to distance and motion. Consider what happens when a car races toward you with the horn blaring. As the car nears and then passes you, the pitch of the horn seems to decrease, slowly at first, and then to drop rapidly to a lower pitch. This phenomenon is known as **Doppler effect** (Fig. 1-8) and clearly demonstrates the relationship between pitch and wavelength. As the car moves toward you, its horn produces a pulse of energy that radiates away from the horn at 1128 feet per second.

If the frequency of the horn is 200 Hz, then the first pulse of sound energy will travel 5.64 feet (the wavelength for the frequency) before the next pulse of energy is generated by the horn. Assume that the automobile is traveling at 50 miles per hour, or 73 feet per second. In the time between pulses of sound energy the car will have traveled 0.365 foot (approximately 4 1/3 inches), reducing the effective wavelength to 5.28 feet. A wavelength of 5.28 feet is equivalent to a frequency of 213.6 cycles per second—a higher pitch than the actual frequency of the horn.

The closer the automobile comes to the observer (assuming that the observer is not directly in the path of the vehicle) the less effect the forward motion of the car has on the apparent frequency of the sound. This is because the angle

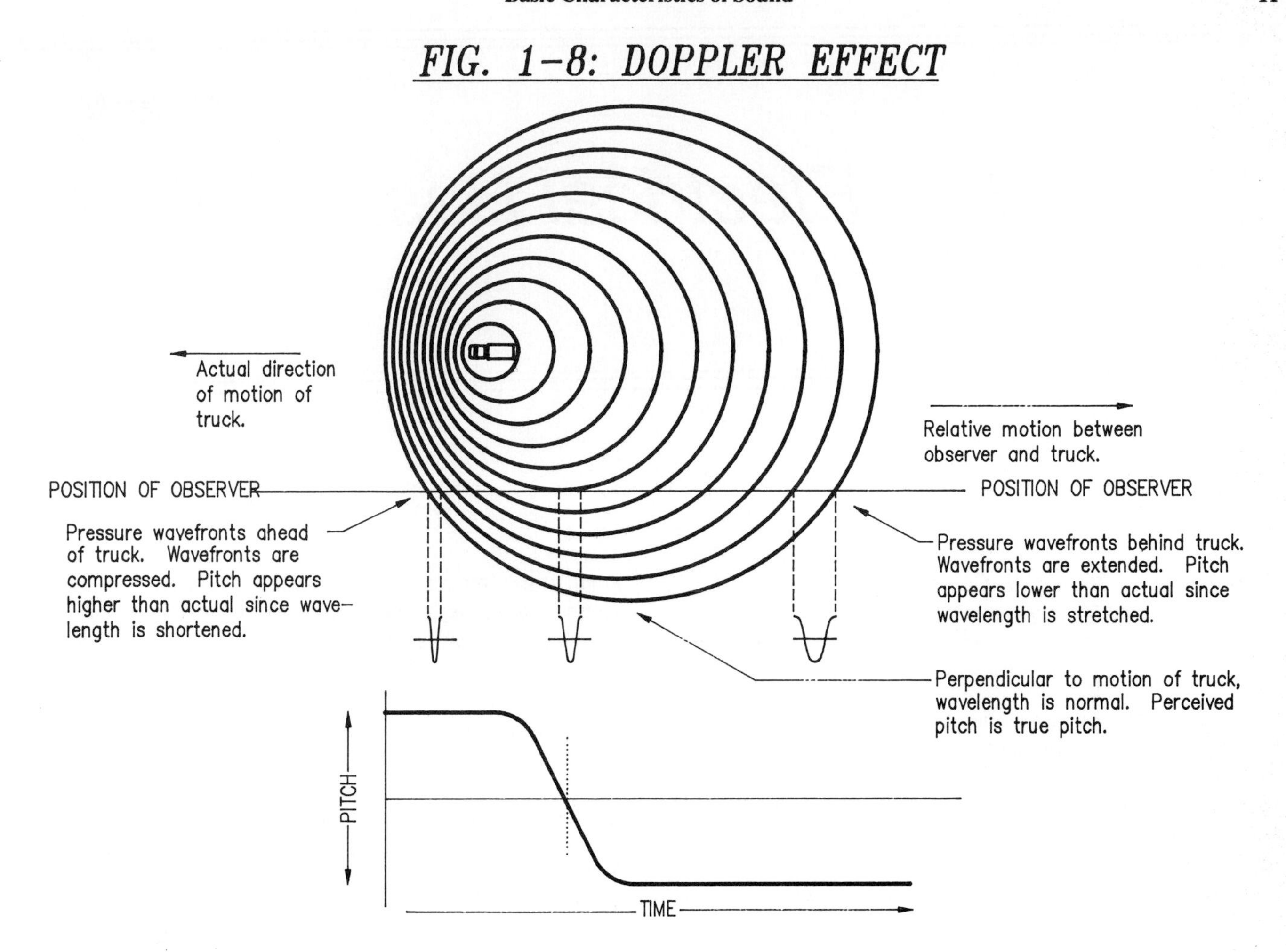

between the car's line of travel and the line of direction from horn to listener becomes increasingly wider. When the line of direction from horn to observer is exactly 90 degrees to the direction of travel, the motion of the car has no effect at all on the apparent frequency of the sound.

As the car passes and begins to recede, the reverse process occurs. The angle between line of travel and line of direction again begins to decrease and the motion of the car begins to take on a significant effect, except that now the apparent wavelength is increasing. As before, a pulse of energy from the horn will travel 5.64 feet before the next wave is generated; however, in the time required for the pulse to travel that distance the car will have traveled 0.365 foot, which now increases the apparent wavelength to 6.005 feet, equivalent to a frequency of 187.84 cycles per second—a lower pitch than the actual frequency of the car horn.

This simple description of the behavior of sound together with the terms introduced in this chapter should facilitate understanding of matters discussed throughout this book, especially the use of audio equipment. Audio, after all, is a means of capturing, processing, storing, and reproducing sound through an electronic medium. Note the use of the term *medium* again. Sound travels only in a medium. As acoustic energy, sound travels by compressing and rarefying air. Converted to audio, sound travels as a greater or smaller number of electrons flowing through a conductor in a given period of time. Terms such as phase, frequency, intensity, timbre, complexity, and wavelength all apply to audio as they do to acoustics. Behavior such as echo and reverberation can be simulated by audio equipment. The equipment and basic characteristics of audio systems form the subject matter of the first part of this text.

Chapter 2

The Electrical Basis of Audio

Audio and the Development of Electrical Technology

Considered from the viewpoint of technology in general, the most important development of the twentieth century was probably the evolution of a practical way to harness mechanical energy to generate electricity. Readily available electric energy to provide light and do work radically changed the way we live.

The first mechanically driven electric generators were based on ideas developed when the only known practical method of producing electric power was chemical. Chemical piles (more commonly known as batteries) produced fixed electrical pressure and polarity. Whenever some kind of external circuit was connected to the battery, electric current would flow in a single direction, from the negative terminal of the battery to its positive terminal. This form of current flow is known as **direct current**, or **DC**. The first mechanical generators were, therefore, designed to produce DC. Somewhat later came the realization that a cyclical alternation in direction of current flow offered useful advantages. This form of electricity is called **alternating current**, or **AC**. Devices that produce this form of power are called **alternators**, and are designed to reverse polarity a given number of times in a second (called the frequency of alternation).

Except for the development of electricity itself, no event in the history of audio is more significant than the discovery of alternating current. Alternating the direction of current flow in a circuit mimics vibrating energy in an elastic medium. Acoustic energy is, essentially, alternating pressure in the elastic medium of a gaseous atmosphere. Alternating current in an electric circuit, therefore, can serve as an analog of alternating pressure in air. If the frequency of alternation in electrical pressure is set equal to the frequency of alternation in acoustic pressure, the electrical waveform will be a replica of the acoustic waveform. The capability to make an electrical signal resemble an acoustic signal is what makes audio possible.

Audio is the technology of using electric circuits to transmit, store, and reproduce sound. Electricity is the basis of audio. In order to understand how audio equipment operates, we need at least minimal information about the nature and principles of electricity. Fortunately, basic electricity is not difficult to learn. Its fundamental laws are simple and are based on a very few essential facts. Let's begin with a simple analogy to provide a way to visualize what happens in an electric circuit.

An Analogy to Get Started

Figure 2-1A shows a tank of water, elevated above the ground so that the water will flow out under gravitational pressure. A large pipe with a valve in it regulates the flow of water from the tank. When the valve is opened just a bit, a small amount of water will flow out of the tank; when the valve is fully opened, a large quantity of water will flow out. The valve regulates the rate of flow by constricting the size of the passage through the pipe.

The tank of water, the pipe, and the valve provide a close parallel to the action of an electric circuit. The tank, elevated so that gravity provides pressure to cause the water to fall, represents the force that moves electricity through a circuit. The water itself, as a material substance that can flow through the pipe, parallels the flow of electric current. The pipe with its valve, creating a limit to the amount of water that can flow in a given period of time, simulates electrical resistance that limits the flow of current. As you read the following explanation of electric circuit behavior, refer to Fig. 2-1B.

FIG. 2-1A: WATER ANALOG OF ELECTRICITY

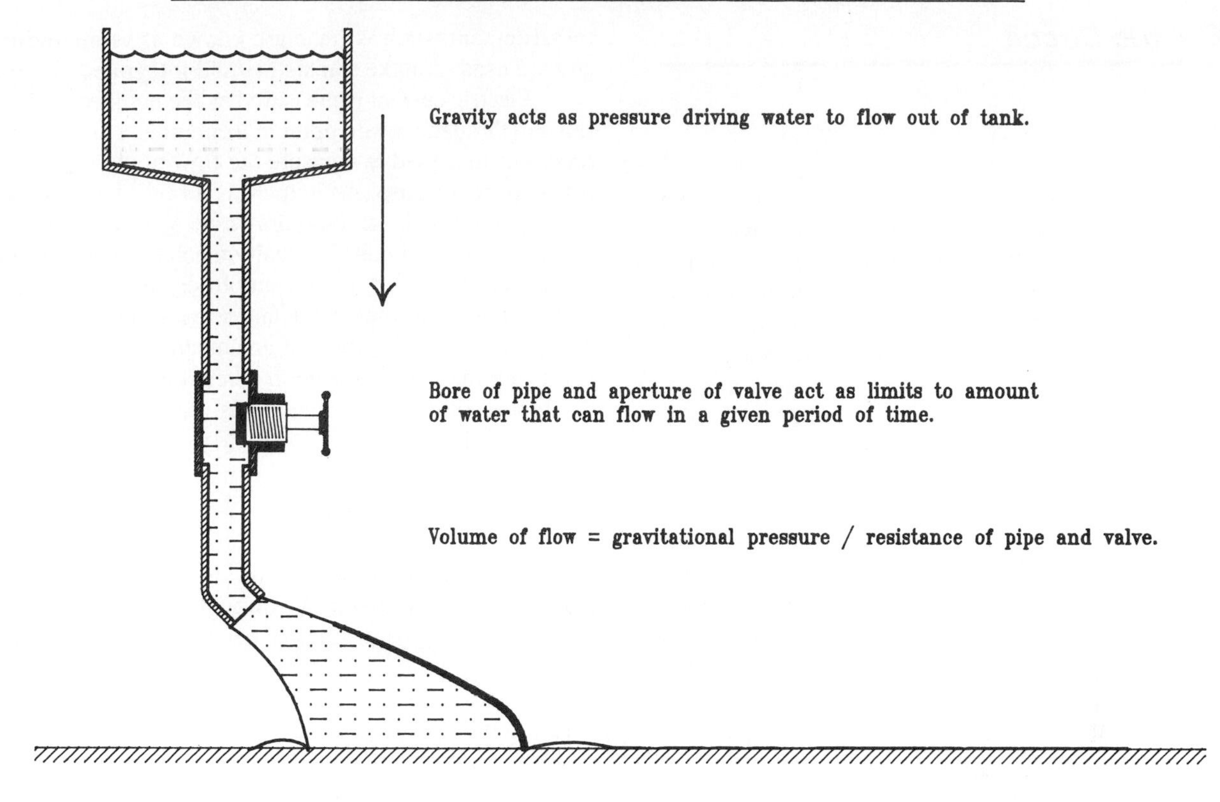

FIG. 2-1B: BASIC ELECTRICAL CIRCUIT.

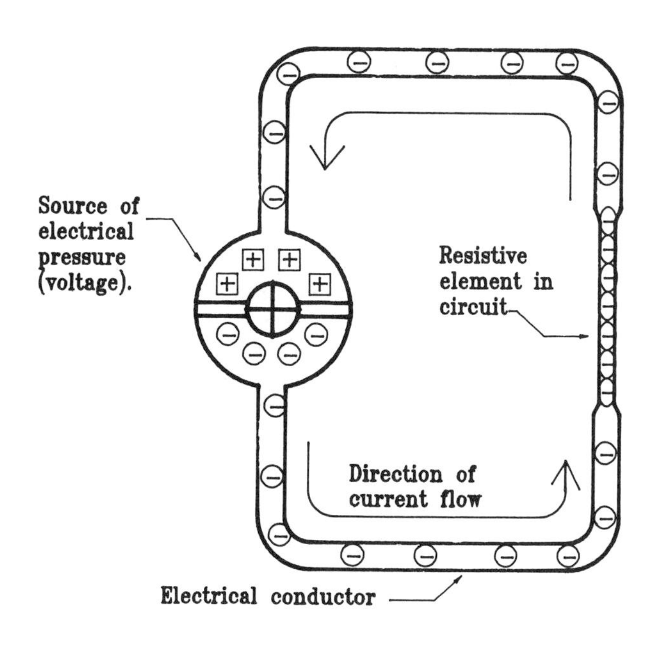

[+] = positive ion

(−) = electron

CURRENT FLOW: Free electrons move through circuit from negative terminal of voltage source, through circuit, returning to positive terminal of source.

SOURCE OF ELECTRICAL PRESSURE:
Source of pressure is force which frees electrons from atoms, causing concentration of positive ions at one terminal of source and concentration of electrons at opposite terminal. Force used to separate electrons and ions is usually chemical (battery) or magnetic (generator).

RESISTANCE: Any element which restricts passage of electrons thereby limiting the flow of current through the circuit.

OHM'S LAW: In Fig. 2-1A the water flow is proportional to the gravitational pressure divided by the restriction of the pipe bore and the valve opening. In an electric circuit, electron flow is proportional to the voltage pressure divided by the resistance of current-limiting elements in the circuit. Current flow is symbolized by I (for Intensity), pressure by E (for Electromotive force), and current-limiting by R (for Resistance). Stated as a formula:

$$I = E/R$$

The Electric Circuit

Current

The first basic fact about electricity is that when a current flows in a circuit, something actually moves. The things that move are called **electrons**. (See Fig. 2-1C.) Electrons are components of atoms, and they pass through an electric conductor much like water flows through a pipe. In order for electrons to flow, they must break free of their atoms. The atoms of certain metals, such as gold, silver, and copper, release electrons quite easily. Thus, these metals are called **conductors**. Other substances, such as rubber, glass, and some plastics, do not release electrons except under tremendous force. These substances are called **insulators**. A few substances, such as carbon, germanium, and silicon, release electrons but require moderate levels of force to do so. Thus, they are neither good conductors nor good insulators. Elements such as these are known as **semiconductors** and are used to make transistors and integrated circuits.

Electric current is measured as the number of electrons flowing through a given point in a circuit in a given period of time. The unit used to measure the flow of electric current is the **ampere**, or **amp**. An ampere is a fairly large amount of current, roughly the amount drawn by a 100-watt light bulb. Current in audio circuits is usually much less than an ampere. Signal current in a solid-state amplifier rarely exceeds a few thousandths of an ampere; signal current from microphones is often only a few millionths of an ampere. Such small current values are measured in milliamperes (the prefix *milli* means one thousandth) and microamperes (the prefix *micro* means one millionth).

Electrical Force

A second basic fact about electricity is that force is required to move electrons through a conductor, just as gravitational pressure, acting on a reservoir of water, is

FIG. 2-1C: CURRENT FLOW.

CURRENT FLOW: Free electrons move through circuit from negative terminal of voltage source, through circuit, returning to positive terminal of source.

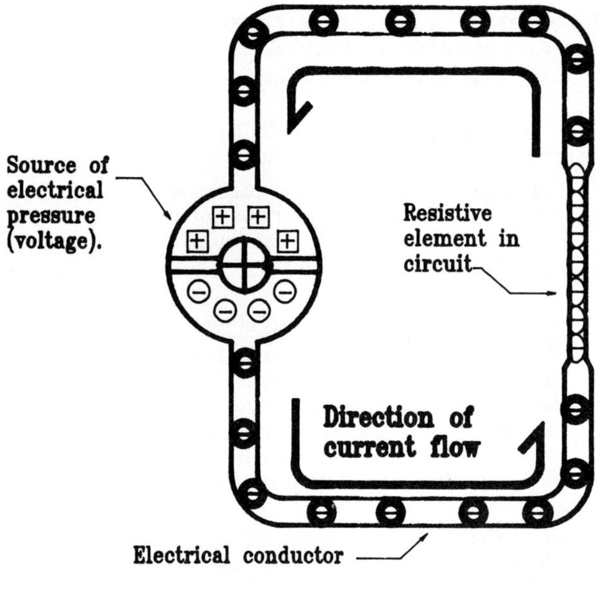

SOURCE OF ELECTRICAL PRESSURE: Source of pressure is force which frees electrons from atoms, causing concentration of positive ions at one terminal of source and concentration of electrons at opposite terminal. Force used to separate electrons and ions is usually chemical (battery) or magnetic (generator).

RESISTANCE: Any element which restricts passage of electrons thereby limiting the flow of current through the circuit.

OHM'S LAW: In Fig. 2-1A the water flow is proportional to the gravitational pressure divided by the restriction of the pipe bore and the valve opening. In an electric circuit, electron flow is proportional to the voltage pressure divided by the resistance of current-limiting elements in the circuit. Current flow is symbolized by I (for Intensity), pressure by E (for Electromotive force), and current-limiting by R (for Resistance). Stated as a formula:

$$I = E/R$$

FIG. 2-1D: ELECTROMOTIVE FORCE.

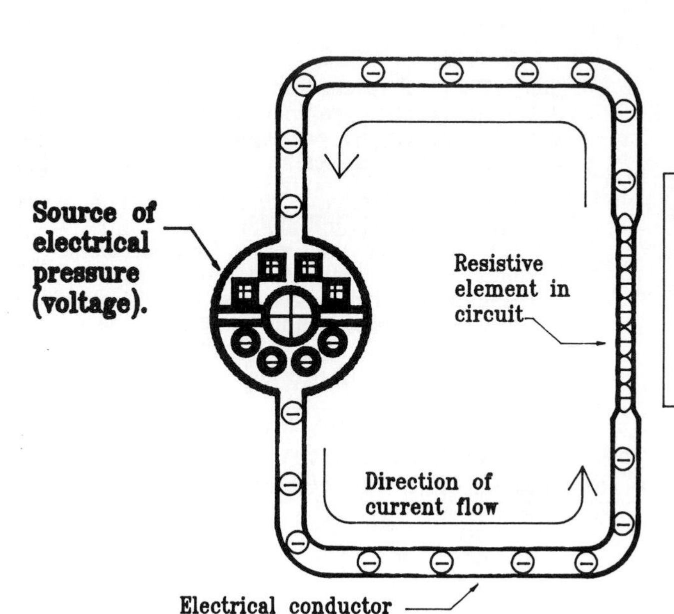

⊞ = positive ion

⊖ = electron

CURRENT FLOW: Free electrons move through circuit from negative terminal of voltage source, through circuit, returning to positive terminal of source.

SOURCE OF ELECTRICAL PRESSURE:
Source of pressure is force which frees electrons from atoms, causing concentration of positive ions at one terminal of source and concentration of electrons at opposite terminal. Force used to separate electrons and ions is usually chemical (battery) or magnetic (generator).

RESISTANCE: Any element which restricts passage of electrons thereby limiting the flow of current through the circuit.

OHM'S LAW: In Fig. 2-1A the water flow is proportional to the gravitational pressure divided by the restriction of the pipe bore and the valve opening. In an electric circuit, electron flow is proportional to the voltage pressure divided by the resistance of current-limiting elements in the circuit. Current flow is symbolized by I (for Intensity), pressure by E (for Electromotive force), and current-limiting by R (for Resistance). Stated as a formula:

$$I = E/R$$

required for water to flow through a pipe. Electrical pressure is created by such devices as generators and batteries that force atoms to release electrons. Electrical pressure creates **regions of charge** inside a generator or battery—a **negative** region and a **positive** region. (See Fig. 2-1D.) The oppositely charged regions are known as **poles**. Connections from these regions to the outside world are called **terminals**. The attraction between regions of opposite charge creates a force known as **difference in potential** or **electromotive force** (**emf**) between the terminals of the battery or generator.

Without an external path through which electrons can flow, the energy stored in the attractive force between the charge regions is **potential energy**. When an external circuit is available, the potential energy becomes **kinetic energy**. Electrical pressure is measured in **volts**; consequently, electrical force is also known as **voltage**. Voltage levels under 1 volt are measured in millivolts (thousandths of a volt) or microvolts (millionths of a volt). The external circuits we deal with in audio are amplifiers, signal processors, and other audio equipment used for theatre sound.

Resistance

Another basic fact about electricity is that there are always limits to the number of electrons that can flow at any given time, both in the generator and in the external circuit. (See Fig. 2-1E.)

The pipe and valve in the water tank analogy restrict the amount of water that can drain out of the tank in a given amount of time. The size of the opening determines the volume of water that can pass at any instant, and friction between the water and the walls of the pipe acts to retard the flow. Together, these factors limit the flow of the water.

The factors limiting the flow of electrons through a conductor are not unlike those controlling the flow of water through the pipe. The size of the conductor determines how many electrons can travel through the conductor at a given time. A large wire conducts many electrons; a small wire conducts few electrons. Also, as we have mentioned, some kinds of substances release electrons more easily than others, and this is analogous to the friction of a pipe. Together, the two current-limiting characteristics constitute a property called **resistance**, which is measured in **ohms**.

FIG. 2-1E: RESISTANCE.

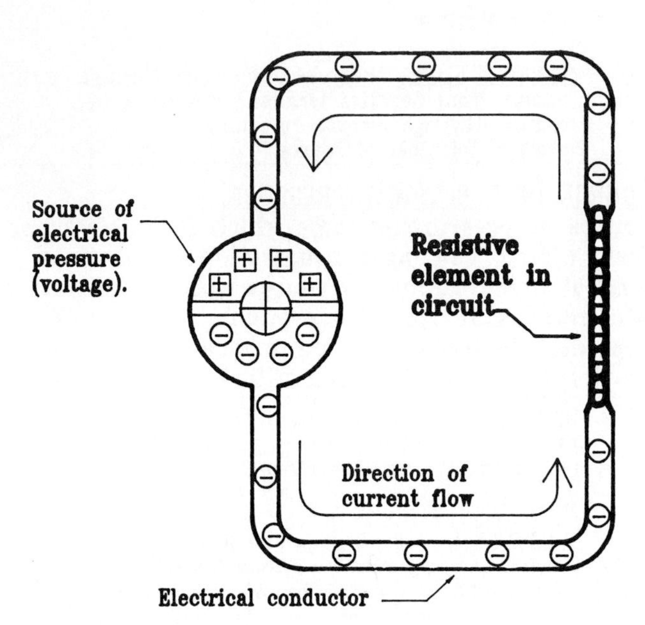

⊞ = positive ion

⊖ = electron

CURRENT FLOW: Free electrons move through circuit from negative terminal of voltage source, through circuit, returning to positive terminal of source.

SOURCE OF ELECTRICAL PRESSURE: Source of pressure is force which frees electrons from atoms, causing concentration of positive ions at one terminal of source and concentration of electrons at opposite terminal. Force used to separate electrons and ions is usually chemical (battery) or magnetic (generator).

RESISTANCE: Any element which restricts passage of electrons thereby limiting the flow of current through the circuit.

OHM'S LAW: In Fig. 2-1A the water flow is proportional to the gravitational pressure divided by the restriction of the pipe bore and the valve opening. In an electric circuit, electron flow is proportional to the voltage pressure divided by the resistance of current-limiting elements in the circuit. Current flow is symbolized by I (for Intensity), pressure by E (for Electromotive force), and current-limiting by R (for Resistance). Stated as a formula:

$$I = E/R$$

Basic Circuit Relationships

To put these three basic facts together, let's use the water example again. If we express the relationship between volume of flow, gravitational pressure, and the limits imposed by the pipe as a formula, we derive the following equation:

Volume of water flow = gravitational pressure / resistance of pipe.

We can set up a similar equation to describe the flow of current through an electric circuit:

Volume of current = electrical pressure / circuit resistance

Using the units of measurement for each of these quantities, we have

Amperes = volts / ohms

The number of electrons that flow through the external circuit around a battery or generator is called the **intensity** of current; thus amperage is symbolized by I, voltage by E (for electromotive force), and resistance by R. With these letter symbols, the formula becomes

$$I = E / R$$

This equation is the fundamental relationship applying to all electric circuits, no matter how simple or complex, and is called **Ohm's law**.

Note the similarities between an electric circuit and an acoustic medium. Voltage in an electric circuit is equivalent to pressure in the atmosphere; current is equivalent to the motion of air molecules; and resistance is similar to the inertia of air.

Electric Power

We use electricity to accomplish work. Accomplishing work means expending power. In mechanical systems the object to be moved or lifted is called a load. To move a heavy steel beam from the ground onto a rack, for example, we might use a forklift. The forklift burns gasoline or diesel fuel to get its energy to lift the steel beam. Lifting the beam means providing more force in an upward direction than gravity exerts in the downward direction. The engine that burns the fuel can use only a fraction of the combustion energy to develop force. The rest of the energy is given off as heat. Accomplishing work is a process of energy exchange.

FIG. 2-2A: BASIC POWER RELATIONSHIPS

⊞ = positive ion

⊖ = electron

→ = heat

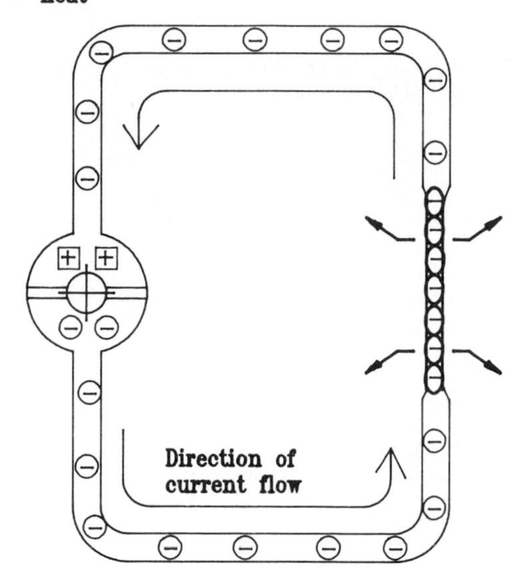

SMALL RESISTANCE

Small resistance in circuit requires low power and relatively small source of electromotive force to drive current through the circuit.

Small amount of heat dissipated as current is forced through resistance.

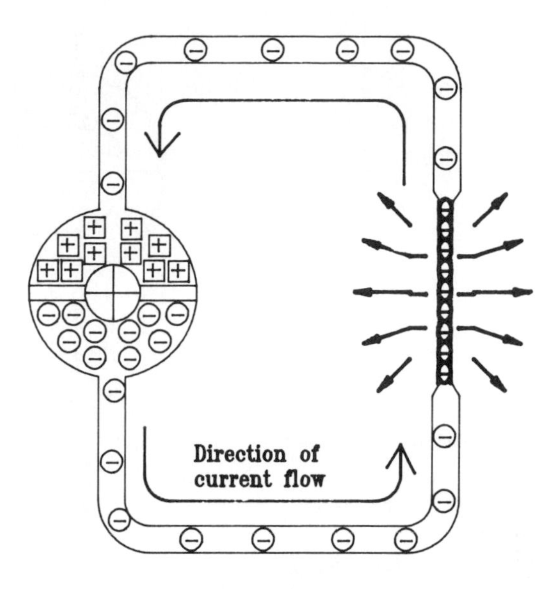

LARGE RESISTANCE

Large resistance in circuit requires large amount of power and relatively large source of electromotive force to drive same amount of current as with small resistance.

Large amount of heat dissipated as current is forced through resistance.

Driving electrons through a circuit is also a matter of energy exchange. Electrons have to acquire energy in order to flow. Subsequently, they lose their acquired energy (as heat) when they recombine with an atom. The amount of heat dissipated by a circuit element thus becomes an indicator of the amount of power consumed in driving electrons through that part of the circuit.

Power must be expended to drive electrons through resistance. To cause a given level of current to flow in a circuit, a small amount of power is required if resistance is small; a proportionally larger amount of power is necessary as resistance increases. (See Fig. 2-2A.) Resistance determines the amount of energy required to make electrons flow in a particular substance. Resistance is the inverse of conductance. Good conductors (those that release electrons easily) have low resistance; insulators have high resistance. Resistance is the electrical equivalent of something to be moved or lifted; therefore, resistance in a circuit is called a *load*.

Anything that requires power to overcome opposition to current flow in a circuit constitutes an electrical load. Agents other than resistance can limit the number of electrons that will flow in a circuit, including mechanical loads to which an electric circuit is coupled. An electric motor running with no load (other than its own mass) requires relatively little current and power. When an electric motor is used to lift a heavy object, however, the weight (gravitational attraction acting on mass) of the object acting against the pull of the motor tends to stall the motor, stopping the flow of electrons. Like the forklift hoisting a steel beam, enough energy must be put into the system to maintain current flow through the motor to keep it turning against the opposing pull of the weight. (See Figs. 2-2B and 2-2C.)

Electric power may be defined as the capability to supply sufficient force to move a given quantity of electrons in a fixed period of time against the resistance of the load. Increase in electrical pressure is required to drive electrons through a circuit under load. Electric power is the product of electrical pressure and current. Thus

$$\text{Electric power} = \text{current} \times \text{emf}$$

The unit of power is the watt. Therefore

$$\text{Watts} = \text{amperes} \times \text{volts}$$

The usual symbol for power is P; the power formula is most commonly expressed as

$$P = I \times E$$

FIG. 2-2B: POWER AND LOAD

MOTOR RUNNING WITH NO LOAD

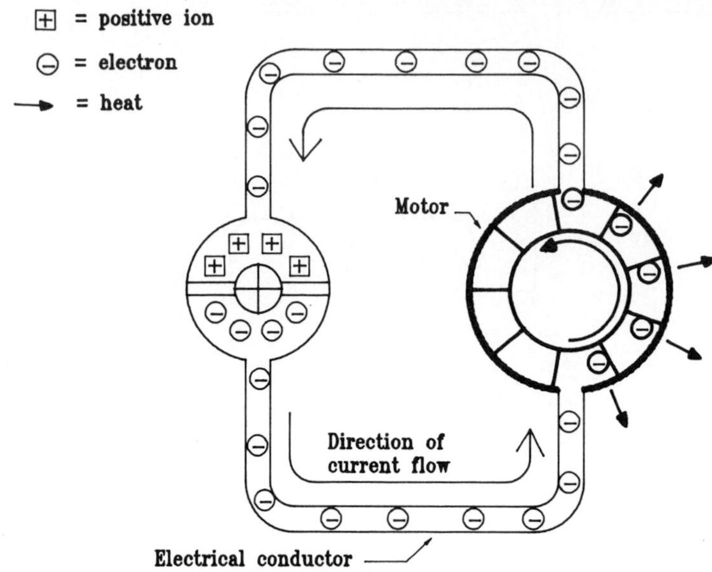

Motor running under no-load conditions draws minimal current. Little power is required to turn motor mass alone.

Energy used in causing motor to rotate dissipated as heat.

$$P = IE \quad \text{(Equation 1)}$$

$$P = I^2R \quad \text{(Equation 2)}$$

WATT'S LAW: Power is the product of the volume of current flowing in the circuit and the electromotive force required to drive it (Equation 1). Power may also be expressed as the square of the current times the resistance of the load (Equation 2). Both forms of the power law are useful. The first clearly shows that adequate force must be available to supply the power to overcome resistance. The second that in order to double the volume of current, power must be increased by a factor of four—in other words, power requirement varies as the square of the current.
The power formula is known as Watt's Law.

FIG. 2-2C: POWER AND LOAD

MOTOR RUNNING UNDER LOAD

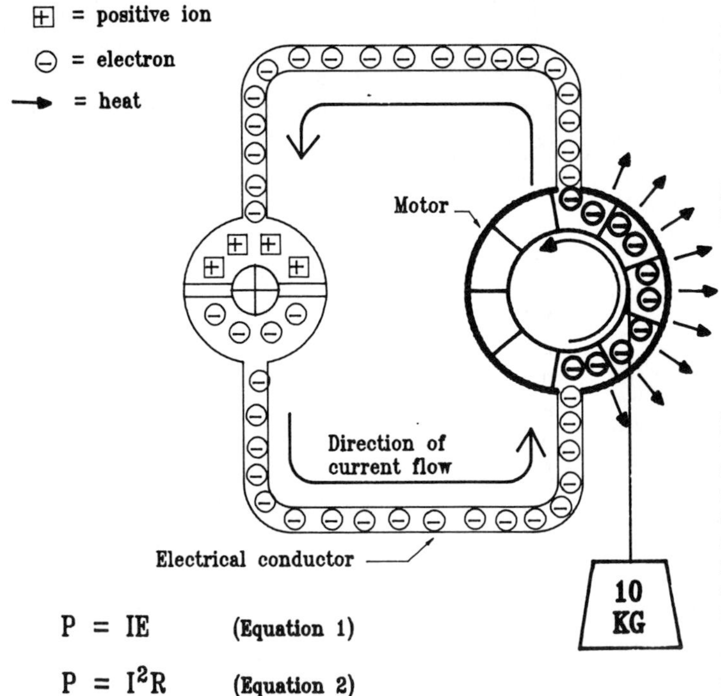

With load, more power is required to keep motor turning in order to lift load. Current volume increases accordingly.

Energy used in causing motor to rotate dissipated as heat.
Large amount of heat dissipated as motor acts to lift load.

$$P = IE \quad \text{(Equation 1)}$$

$$P = I^2R \quad \text{(Equation 2)}$$

WATT'S LAW: Power is the product of the volume of current flowing in the circuit and the electromotive force required to drive it (Equation 1). Power may also be expressed as the square of the current times the resistance of the load (Equation 2). Both forms of the power law are useful. The first clearly shows that adequate force must be available to supply the power to overcome resistance. The second that in order to double the volume of current, power must be increased by a factor of four—in other words, power requirement varies as the square of the current.
The power formula is known as Watt's Law.

where P = power in watts, I = current in amperes, and E = electromotive force in volts. The power equation is known as **Watt's law**.

Acoustic systems do work just as electrical systems do. To overcome the inertia of a volume of air requires power from whatever the source of vibration may be. Each time we speak, we release chemical energy in order to make our muscles do the work necessary to set air into motion. That chemical energy is derived from the food we consume. Pipes in a pipe organ draw energy from compressed air, which, in turn, is energized by electricity driving an electric motor.

Audio systems also do work and act on loads. The volume of air in a large room serves as load for a loudspeaker. When the loudspeaker has to set this volume of air into vibration, the resistance of the air to change in pressure acts against the motion of the loudspeaker just as a weight pulls against the rotation of an electric motor. Energy must be consumed by the loudspeaker to overcome the inertia of the volume of air.

Polarity

The negativity or positivity of each terminal of a voltage source is known as its **polarity**. Concentration of charge makes a terminal either positive or negative. A negative terminal is a concentration of electrons; a positive terminal represents a concentration of positive ions (atoms from which an electron has been released). Current always flows from the negative to the positive terminal of a source, but the polarity of a circuit can be either positive or negative, depending entirely on how we measure. If the source supplies a difference in potential of 10 volts, for example, then the positive terminal will measure +10 volts with respect to the negative terminal. However, we could just as easily measure -10 volts at the negative terminal with respect to the positive terminal.

In all circuits there is one point to which all circuit elements return. This one point is called **common**. Common will always appear to be **neutral**—that is, a point of zero voltage. As long as the only frame of reference is the voltage source, polarity in the circuit is governed exclusively by our choice of which terminal to use as common.[1] If, for example, we measure with a voltmeter, connecting its so-called "common" probe to the negative terminal and the "hot" probe to the positive terminal of the voltage source, the meter will read all of the electromotive force applied to the circuit as a positive voltage. However, if we reverse the connections (common to positive and hot to negative), the meter will read a negative voltage. (See Fig. 2-3A.)

Grounding

For all electric circuits, the planet—the Earth itself—constitutes the ultimate common. The planet effectively forms a stable mass having a net electric charge of zero; thus, any part of a circuit that is directly connected to the earth is also forced to a voltage level of zero. Returning a circuit to earth is called **grounding**. Objects that are clamped to a zero-voltage point are said to be neutral—hence, the frequent reference to common as neutral. Objects that are large compared to the size of a circuit can simulate ground. Thus, the chassis of an amplifier is normally used as the ground for an individual piece of electronic equipment. External grounding is necessary when several pieces of equipment must be connected together to form an assemblage, such as an audio system.

The term *common* is sometimes used interchangeably with the term *ground*, but the two are not the same. Common is the point to which all circuits of a system are returned. Ground is an absolute zero-voltage point referred to the electrical mass of the planet. Normally, common will appear to be neutral, but its zero-voltage level applies exclusively to the local circuit. The common for any piece of equipment may be at a nonzero level with respect to some external point. Unless common is connected to true ground, a difference in potential can, and often does, exist. This difference in potential can become a source of noise.

Audio systems are highly susceptible to problems in grounding. Audio systems are usually composed of several individual pieces of equipment, some of which may be separated from the main body of the system. In such cases, grounding conductors, if not properly connected, can become unplanned paths for the signal or, worse, paths for induced hum and noise. Such paths are called **ground loops**.

Alternating Current

An understanding of the nature of electrical polarity enables us to acquire a basic understanding of alternating current, which, as we have stated, is so very important to audio. A DC circuit has fixed polarity. Its hot terminal is always positive or negative. In an AC circuit, polarity is not fixed but undergoes a cyclical alternation where the hot terminal is positive during one half of the cycle and negative during the other half. (See Fig. 2-3B.)

Negativity and positivity imply concentrations of charge carriers—free electrons in the case of a negative region and positive ions (atoms from which an electron has been released)

[1] Circuit polarity is an engineering decision and, for users of equipment, is critical. Solid-state audio amplifiers and digital devices are good examples of such critical circuits. Power supply connections are generally marked as clearly as possible in such devices. If we connect a power supply to an audio amplifier with the terminals reversed, the amplifier could easily be damaged.

FIG. 2-3A: DC POLARITY

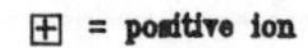

= electron

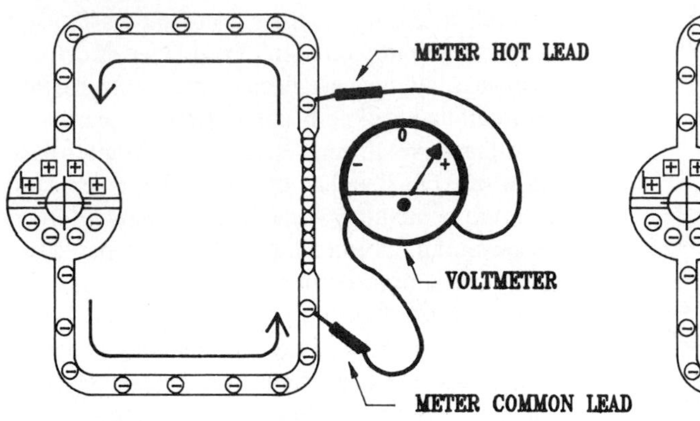

COMMON TO – ; HOT TO +

With common probe connected to negative side of voltage source, and "hot" probe connected to positive side, meter displays positive voltage reading.

Negative side of circuit becomes common.

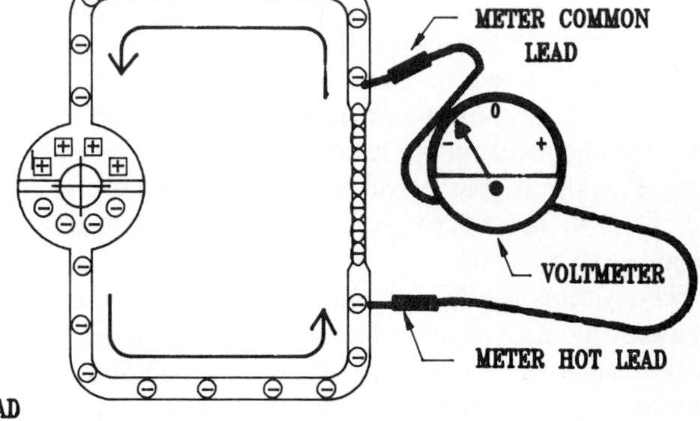

COMMON TO + ; HOT TO –

With common probe connected to positive side of voltage source, and "hot" probe connected to negative side, meter displays negative voltage reading.

Positive side of circuit becomes common.

FIG. 2-3B: AC POLARITY

= positive ion

= electron

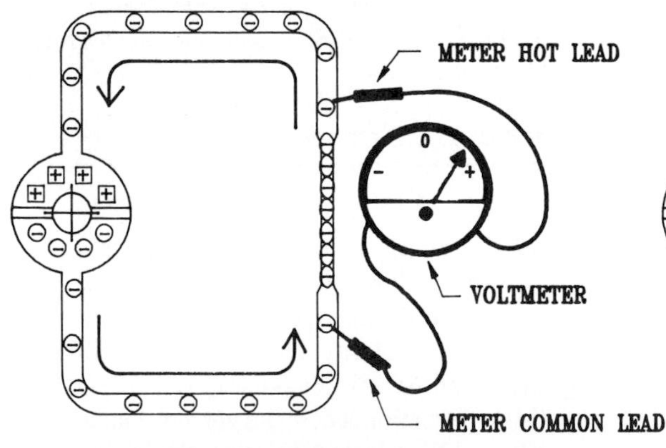

POSITIVE HALF-CYCLE

Meter is connected exactly as in DC, with common lead to negative side of voltage source and hot lead to positive side. Meter reads positive voltage, but voltage will rise to peak, then decline to zero.

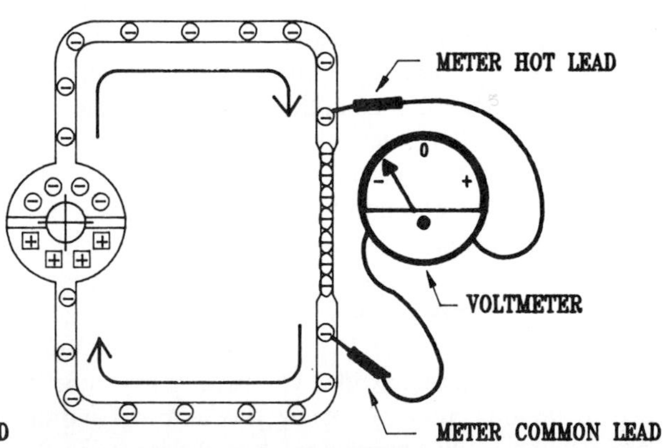

NEGATIVE HALF-CYCLE

Note that meter remains connected as at left, but polarity of voltage source has changed. Now the meter reads negative voltage. The voltage will rise from zero to negative peak, then decline to zero.

In a DC circuit we can change polarity by reversing the meter or by reversing the voltage source, both of which require physical intervention, rearranging the circuit in some way. In an AC circuit, the voltage source changes polarity naturally with each half-cycle. The physical arrangement of the circuit remains constant (i.e., neutral is always neutral; hot is always hot.)

in a positive region. Because current flow is always from negative to positive (i.e., from a concentration of electrons to a concentration of positive ions), the direction of current flow reverses with each change in polarity. The common terminal of an AC circuit appears to have a neutral charge with respect to the system, just as in a DC circuit. The opposite terminal in an AC system, however, will alternate between a positive and a negative polarity.

The basic Ohm's law and Watt's law calculations are essentially the same for AC as for DC. Of course, any physical system that is in constant change is more complex than a fixed system; therefore, AC has its own laws, which go considerably beyond the simple calculations necessary for DC circuits. Fortunately, for a basic understanding of audio circuits, only two characteristics of AC need be considered. One is the close relationship between AC current and magnetism; the other is the way in which the behavior of electric circuits varies with frequency of alternation.

Alternation in polarity corresponds to compression and rarefaction in acoustics. A voltage swing in the direction of positive potential simulates compression; a voltage swing in the direction of negative potential simulates rarefaction. Inside most audio circuits there is always a flow of current. That normal flow of current is equivalent to average atmospheric pressure. A greater than normal flow of electrons simulates compression; a smaller than normal flow simulates rarefaction. Changes in potential and current flow also give audio the equivalent of wavelength in acoustics inasmuch as a lapse in time exists between one positive peak and the next. The term *wavelength* is used just as though there really were some distance traveled between the onset of one energy pulse and the next.

Because the term *wavelength* applies to audio, so, also, does the term *phase*. Phase and all its characteristics of reinforcement and cancellation affect audio as they affect acoustics. If two waves of identical frequency are slightly out of phase, overall intensity will be reduced. If the two waves are completely opposite in phase, the frequency will be canceled. If the two waves are in phase, the overall intensity will increase.

Magnetism

Electricity and magnetism are inseparable phenomena. If a conductor is moved through the field of a magnet, a flow of electrons will be **induced** in the conductor. (See Fig. 2-4A.) Current will flow so long as either the conductor or the field is moving. When either one stops moving, the current flow ceases. Conversely, whenever a current of electrons moves through a conductor, a magnetic field surrounds that conductor.

FIG. 2-4A: BASIC ELECTRICITY AND MAGNETISM

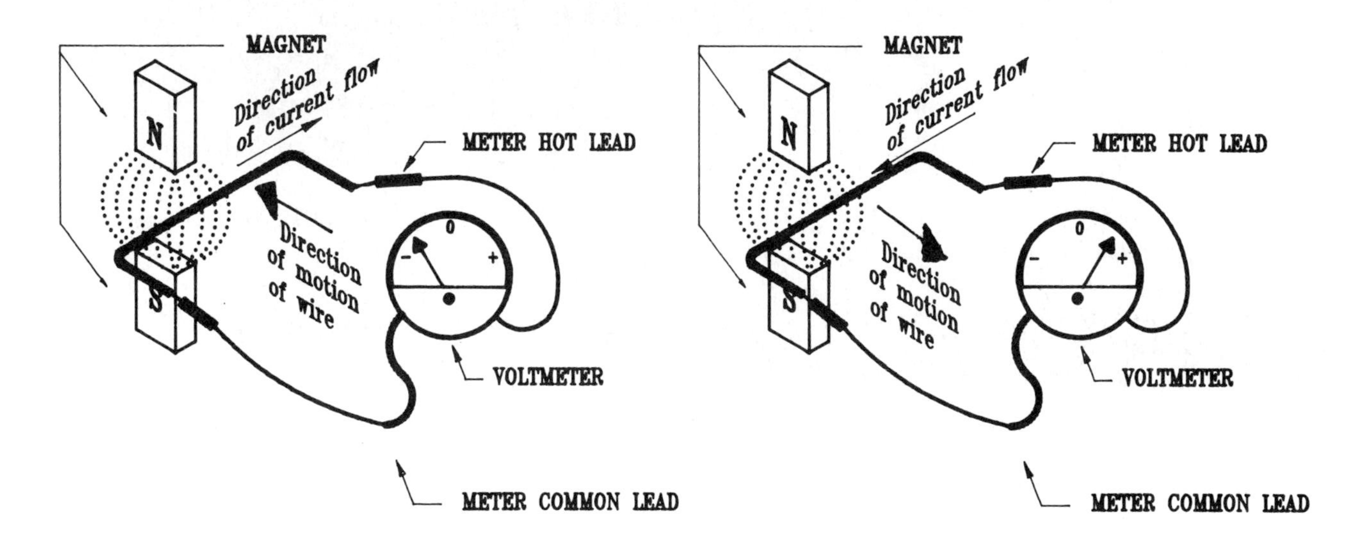

Conductor moves through magnetic field. Current is generated in conductor. Voltmeter registers negative polarity.

Conductor moves in opposite direction through magnetic field. Current reverses direction of flow. Voltmeter registers positive polarity.

When a conductor and a magnetic field move relative to each other, current is generated in the conductor. **Magnetic generation of current is the operational principle of dynamic microphones and of most phonograph pickups. Magnetic generation of current is also the source of almost all household and industrial electric power.**

(See Fig. 2-4B.) Like the flow of current in an electric circuit, magnetism has properties that can be used to simulate the behavior of acoustic energy. Unlike acoustic energy and electrical energy, magnetism can be recorded and stored. That is what makes magnetism an important aspect of audio systems.

Two forms of sound recording use electromagnetism: the phonograph and the tape recorder. Phonograph records are cut using a stylus driven by small coils of wire. The coils generate a magnetic field that pulls the stylus from side to side to cut the trace in the master disc. Phonograph pickup cartridges move tiny coils of wire through the field of a small magnet to generate audio from the grooves in a record. Tape recording is based on the fact that electric currents produce magnetic fields, and vice versa. In a tape recorder, the recording head is really a small electromagnet in which currents flowing at audio frequencies produce a variable magnetic field. As it passes by the record head, the tape becomes magnetized and holds an image of the changes in size and direction of magnetic field strength. When the tape is replayed, the stored magnetism induces a corresponding current in the playback head. That current is then amplified and converted back into the acoustic energy that we hear as the recording.

Magnetism also constitutes the operational principle of dynamic microphones and loudspeakers. In each, a coil of wire is suspended between the poles of a magnet. In the microphone, acoustic pressure change causes the coil to move back and forth in the magnetic field, generating current at the frequencies of the acoustic energy. In the loudspeaker, current from an amplifier flows through the coil, generating a magnetic field that changes at the frequencies of the audio signal. The interaction between the magnetic field generated around the coil and the field of the magnet causes the loudspeaker to move, setting the surrounding air into motion.

Transformers

Transformers are devices based on the interaction between electricity and magnetism. They may not be as familiar as are tape recorders and phonographs, but we've all heard of them (usually when we experience a power blackout and the electric company blames it on a transformer that failed for some reason or other.) Transformers are quite simple in principle: Because current flow in a wire induces a magnetic field, and a changing magnetic field induces current flow in a wire, we simply place two wires side by side. When a current flows in one wire, the magnetic field that it creates forces a current to flow in the other wire. Of course, placing two straight wires side by side accomplishes nothing useful. A transformer is made of two *coils* of wire. (See Fig. 2-5A.) Transformers for audio utilise a core of soft iron. (See Fig. 2-5B.) The core conducts the magnetism like a wire conducts electrons. The

FIG. 2-4B: BASIC ELECTRICITY AND MAGNETISM

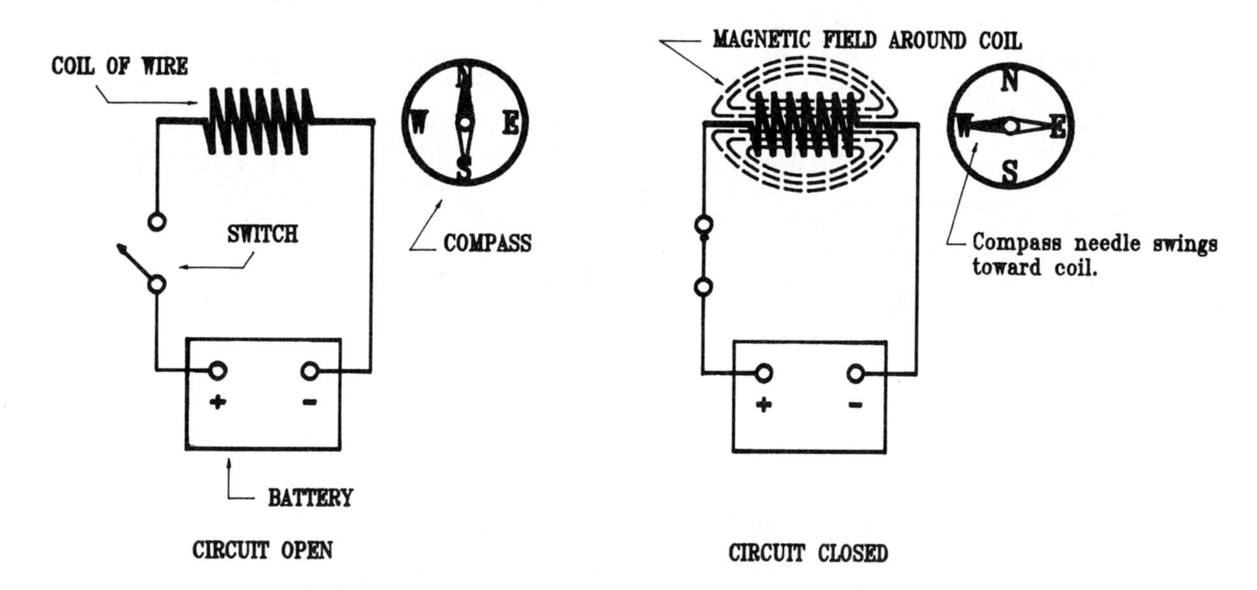

Whenever current flows through a conductor a magnetic field builds up around the conductor. Wrapping the conductor into a coil strengthens the field. We can prove the existence of the field by placing a compass near the coil. When the circuit is closed, the compass needle swings away from magnetic north and toward the magnetic field surrounding the coil.

FIG. 2-5A-D: TRANSFORMERS

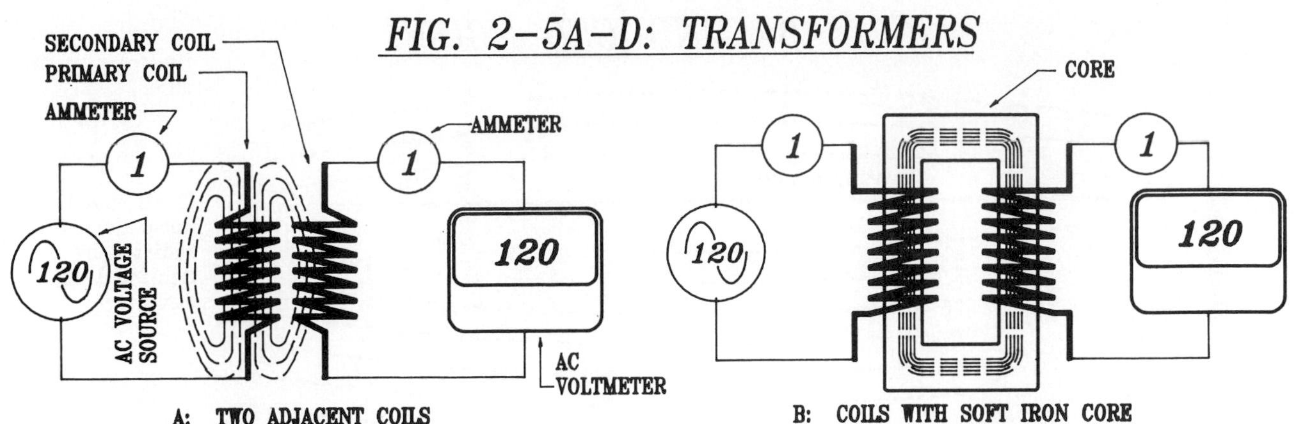

A: TWO ADJACENT COILS
Magnetic flux from primary coil induces current into secondary coil. Note that power in each circuit equals 120 watts.

B: COILS WITH SOFT IRON CORE
Core channels flux from primary to secondary. Note that current and voltage in secondary circuit equal those in primary. Power in both = 120 watts.

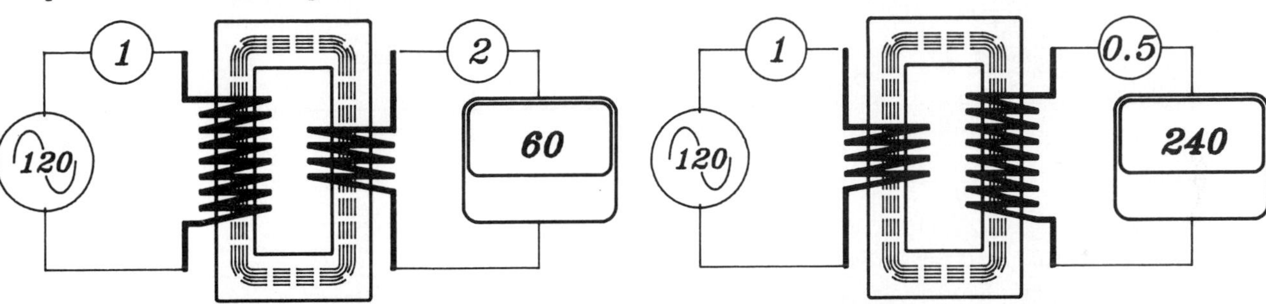

C: STEP-DOWN TRANSFORMER
Secondary coil has half the turns of primary coil. Voltage is half that of primary circuit but current is doubled. Power in both circuits equals 120 watts.

D: STEP-UP TRANSFORMER
Secondary coil has twice the turns of primary coil. Voltage is twice that of primary circuit but current is reduced to half that of primary. Power in both circuits equals 120 watts.

relative size of each coil determines the amount of current and voltage that will be transferred by the magnetic field. If the primary coil (the one connected to the source of voltage and current) is large and the secondary coil (the one to which the magnetic field transfers power) is small, the voltage from the secondary will be lower than that of the primary, but the secondary's current will be much larger. (See Fig. 2-5C.) Conversely, if the primary coil is smaller than the secondary, the secondary's voltage will be greater than the primary voltage, but the current will be smaller. (See Fig. 2-5D.)

Transformers are often used in audio circuits to adjust current and voltage levels, as for example, in microphones. Microphones produce extremely small amounts of voltage and current, usually not enough to overcome the resistance of a long cable connecting the microphone to the control console. The transformer steps up the voltage enough so that the weak signal of the microphone can travel through the cable without significant loss.

Magnetism and Noise

The interrelationship of current and magnetism is useful in audio, but it also causes problems. If an audio signal current can induce a magnetic field, which can, in turn, induce a secondary signal current, nothing keeps a standard electric power line from producing a magnetic field which an audio circuit subsequently picks up as induced hum and noise. (Lighting circuits are troublesome sources of induced noise, especially in theatre, where hundreds of thousands of watts may be involved. Dimmers tend to produce an insidious and pervasive kind of induced noise.) To overcome the problem of induced noise, almost all audio circuits (and especially microphone lines) are **shielded**. (See Fig. 2-6.) The shield may be fine wire woven into a tubular casing that encloses the signal-carrying conductors, or it may be a sheet of metal foil wrapped around the conductors. In either case, the shield is a protective layer of conductive metal that is clamped to a ground point.

Circuit Configuration: Balanced and Unbalanced

A source of voltage and current always has two poles. Normally, we force one of the poles to appear neutral and let the other pole alternate between negative and positive potential.

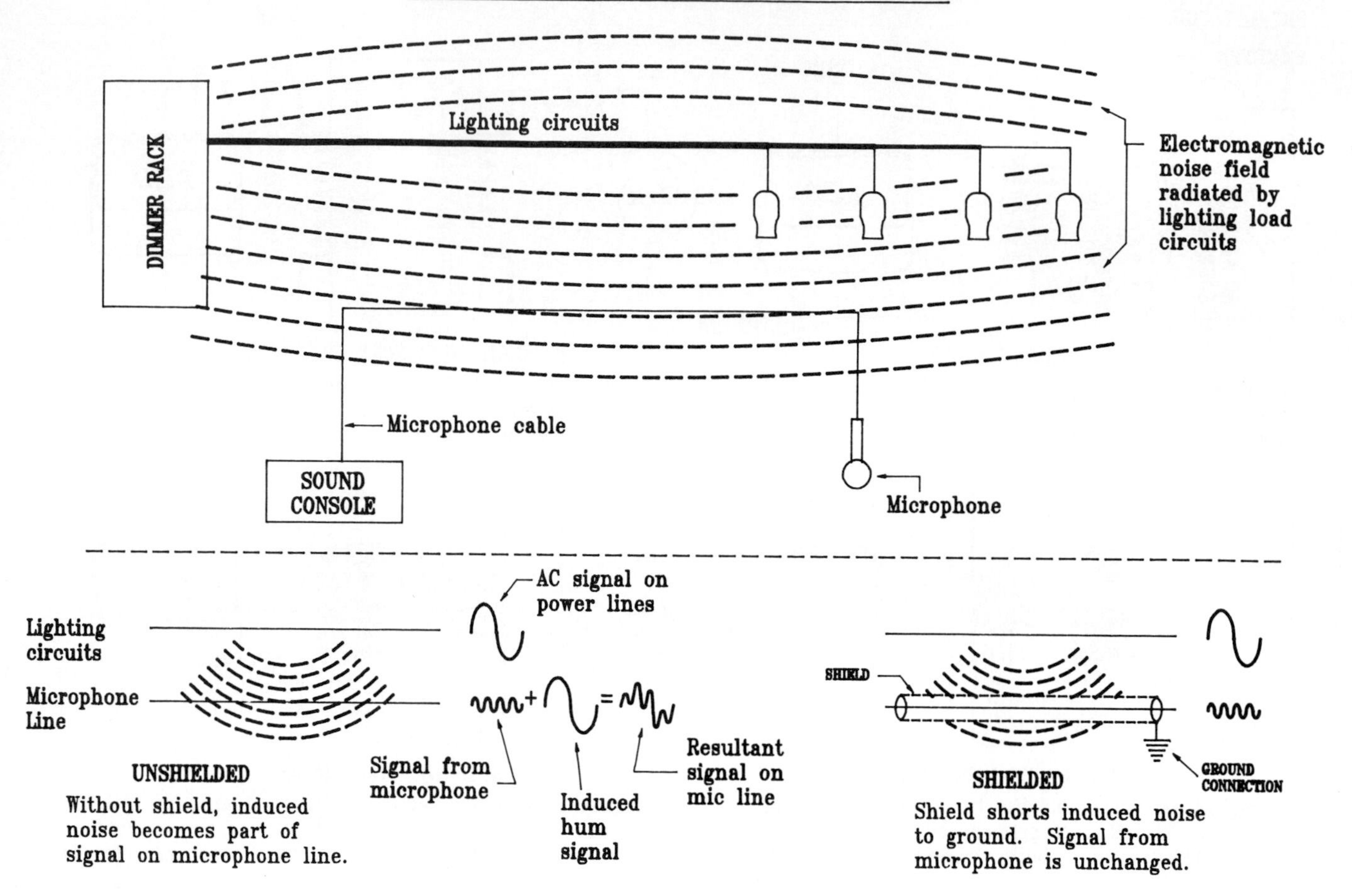

If we construct the external circuit so that the circuit resistance—the load—is divided into two equal parts and then ground the middle of the circuit, as in Fig. 2-7, the circuit will have two hot terminals in addition to a common. With signal flowing in the circuit, one terminal is always positive with respect to common, and the other is negative. The source voltage is divided, half to the positive side of the circuit and half to the negative side. Such a circuit is said to be **balanced**.[2] Circuits that are not divided by a center-neutral are called **unbalanced**.

The audio world uses both balanced and unbalanced circuits. The system user deals with balanced and unbalanced circuits when connecting the signal between one piece of equipment and another. Using the right kind of transmission line is essential. Balanced transmission lines are standard in professional audio installations, because balanced circuits provide natural immunity to external sources of noise. Because of this noise rejection characteristic, professional microphone circuits are always balanced.

Balanced circuits require two of everything and are, therefore, costly. In semiprofessional and consumer-grade audio, unbalanced transmission lines are standard, mainly because they are much less expensive. Unbalanced circuits, however, require great care in shielding to make sure that induced noise does not become a problem. Any transmission line that must run more than 25 to 30 feet should be balanced.

Kinds of Circuits

No matter how complex a circuit may look, only two fundamental configurations of circuit components are possible: **series** and **parallel**. Most audio equipment contains both series and parallel arrangements of components.

[2]A variant of the balanced circuit,called a differential circuit, is often used in modern audio equipment. See discussion of differential circuits in Chapter 7.

FIG. 2-7: BALANCED CIRCUIT

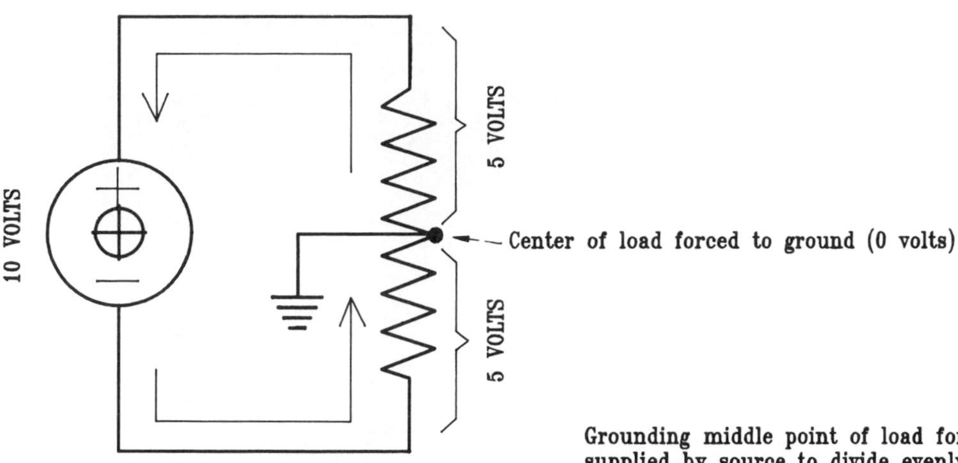

Grounding middle point of load forces voltage supplied by source to divide evenly across load. Voltage to positive side of source assumes positive potential; voltage to negative side of source assumes negative potential.

$$I_{LOAD} = (E_{POS} + E_{NEG})/R_{LOAD}$$

SYMBOL KEY

= Resistance

= Ground

Series Circuits

The arrangement in Fig. 2-8A illustrates a very simple circuit. It contains a voltage source, a resistive element, and a complete conducting path leading from one pole of the battery through the resistor to the other pole of the battery. The circuit illustrated is called a ***series circuit*** because all of its elements are strung together end to end. The current flows equally through all parts of the circuit. The entire difference in potential created by the battery is impressed across the resistor (i.e., the end of the resistor nearest the positive terminal assumes the voltage level of that terminal, and the end of the resistor nearest the negative terminal assumes that voltage level).

If we could take measurements inside the resistor in the circuit of Fig. 2-8A, we would find that the voltage level changes gradually through the body of the resistor. (A level control in an audio amplifier is usually a resistor made so that we can slide a contactor up and down the body of the resistor, tapping off the percentage of the total voltage that we need.) The change in level of emf across the resistor is called a **voltage drop**. All of the emf applied to a circuit must be used by the circuit. (We cannot get to the other end and still have force left over.)

The circuit in Fig. 2-8A is seldom referred to as a series circuit. Normally, when people talk about a series circuit, they mean one where all the *load* elements are in series, like links in a chain. If a circuit contains a chain of several resistors connected end to end, as in Fig. 2-8B, the total resistance in the circuit is the sum of all of the resistances. Like the single resistance in Fig. 2-8A, the full voltage of the source is distributed across the string of resistors, and each resistor will produce a voltage drop proportional to its resistance.

A good illustration of a practical series circuit is an old-fashioned string of Christmas tree lights—the kind where losing one bulb knocks out the whole string. Now that we know what a series circuit is, the reason the string goes off when one bulb burns out should be obvious: The faulty bulb breaks the circuit, just like shutting off a switch. Until the **continuity** through that lamp socket is restored, the whole circuit is out.

Parallel Circuits

Again, consider the water analogy with which we began our discussion of electricity. We had set up a large pipe with a valve in it to change the resistance and control the outflow of water from the tank. We could just as easily have replaced the large pipe with several smaller pipes.

FIG. 2-8A-B: SERIES CIRCUITS

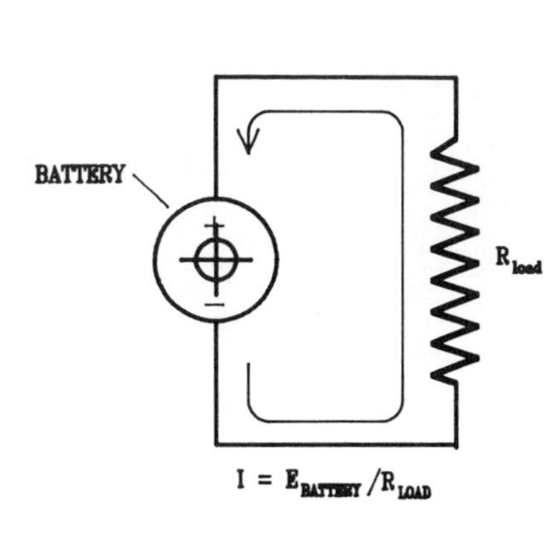

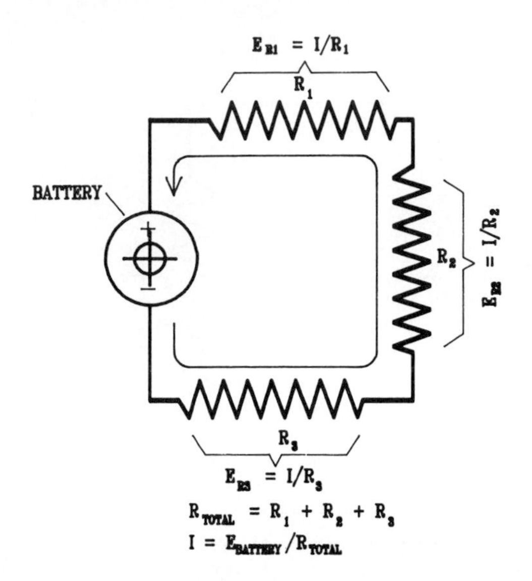

A. BASIC CIRCUIT

EMF supplied by battery is applied across circuit. Current flows through circuit wiring and resistance and returns to battery. Components are connected end to end. Current through circuit equals applied voltage divided by load resistance.

SYMBOL KEY
—/\/\/\— = Resistance

B. MULTIPLE LOAD ELEMENTS IN SERIES

The circuit above is effectively identical to the circuit shown in A, except that the load resistance has been divided into three parts. All components are connected end to end, and the current flowing through the circuit is the quotient of battery voltage divided by the total resistance of the circuit. Each element of the load drops a portion of the voltage. The amount of voltage required to drive current through a given load element is proportional to the magnitude of its resistance.

(See Fig. 2-9A.) The pressure forcing water to flow out of the reservoir affects each pipe equally, so each will carry no more or less than it would were it the only outflow in the system, but the total volume of water flowing out through all of the pipes together, however, is equal to the outflow from a single large pipe. Using a number of smaller pipes to drain water from the tank is like placing resistors in parallel across a source of voltage.

Figure 2-9B shows a circuit with three resistances placed across a battery in parallel with each other—a **parallel circuit**. In this configuration the emf across each resistor is the full supply voltage, and each resistance draws current as though it were the only element in the circuit. The total current drawn from the supply, however, will be the sum of all the currents drawn by the individual resistive elements. As the total current drawn is greater than for any of the individual resistances, the effect of a group of resistors in parallel is to *lower* the total resistance that the circuit presents to the voltage source.

Parallel circuits are probably the most common kind of multiple load circuit that everybody uses daily. Plug a lamp or a CD player or a hair dryer into a wall outlet, and you've probably connected the device in parallel with something else. When we need to connect more than one loudspeaker to a power amplifier, we usually connect them in parallel, but therein lies a problem. How low can we make the load on a power amplifier before we run the risk of doing damage? After all, too low a load will try to draw more current from the amplifier than it can provide, and the amplifier will burn out. Assuming that all the loudspeakers that we want to connect have equal resistance, the total load will be the resistance of one loudspeaker divided by the number of units to be connected to the one amplifier. For example, if we want to connect two 8-ohm loudspeakers to a power amplifier, we divide 8 (the resistance of one loudspeaker) by 2 (the number of units to be used) to get a value of 4 ohms load. (Four ohms load, by the way, is about the minimum impedance that most power amplifiers can stand.)

Frequency-Dependent Characteristics of AC Circuits

Resistance is resistance in any circuit. Nonresistive loads, however, are not so simple. These elements behave very differently in AC circuits than they do in DC circuits and have

FIG. 2-9A: WATER ANALOGY FOR PARALLEL RESISTANCES

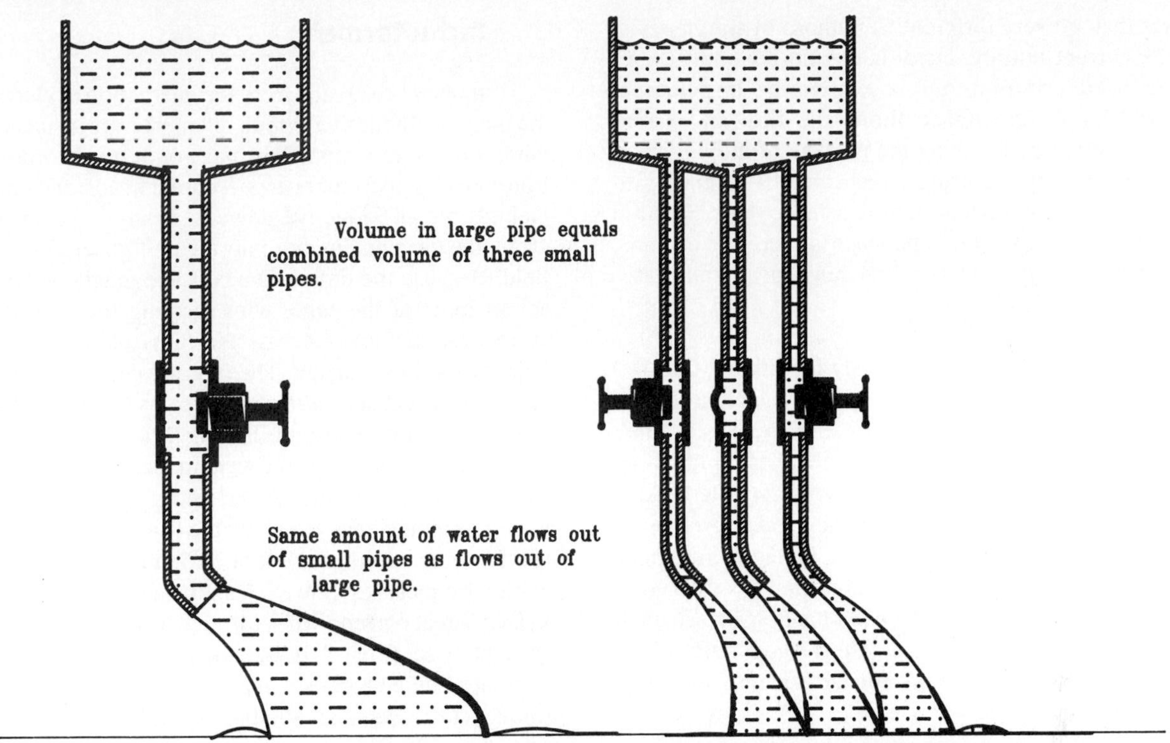

Each small pipe in the tank at right has more resistance to water flow than the large pipe in the tank at left. With three small pipes releasing water, however, the combined flow equals that through the larger pipe. The illustration shows that the combination of resistances in parallel produces an effective resistance that is <u>lower</u> than the smallest resistance of the combination.

FIG. 2-9B: PARALLEL CIRCUIT

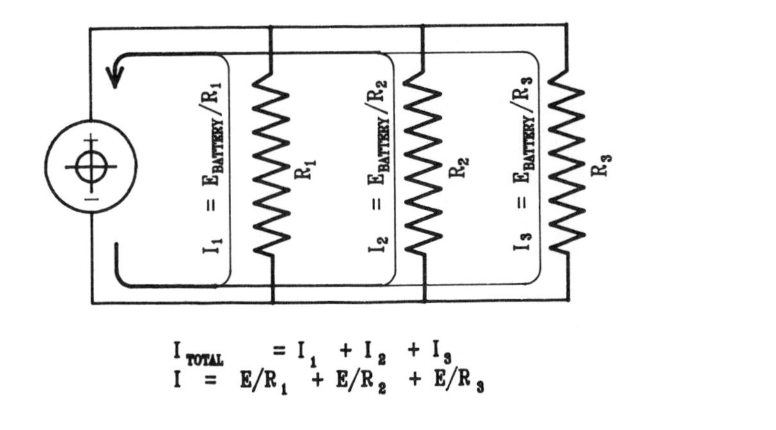

$$I_{TOTAL} = I_1 + I_2 + I_3$$
$$I = E/R_1 + E/R_2 + E/R_3$$

THREE PARALLEL LOAD ELEMENTS

Three load resistors in parallel configuration. Full voltage is applied across each resistor, and each resistor draws current as if it were the only load element in the circuit. The total current is equal to the sum of the currents drawn by the individual resistors.

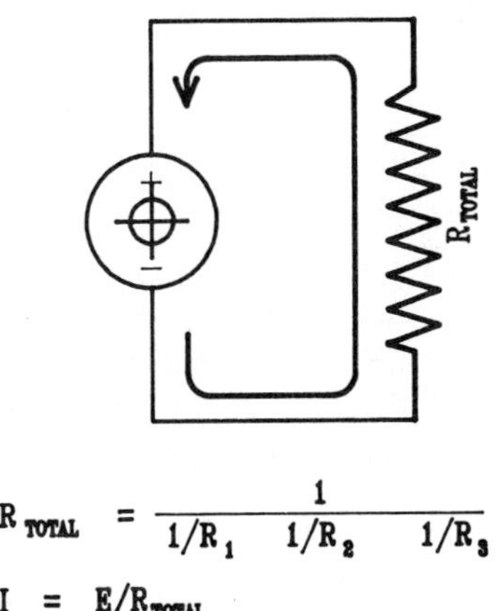

$$R_{TOTAL} = \frac{1}{1/R_1 \quad 1/R_2 \quad 1/R_3}$$
$$I = E/R_{TOTAL}$$

EQUIVALENT CIRCUIT

The voltage source sees a load element equivalent to a single resistor that would draw the total amount of current flowing in the circuit. That single resistor would be smaller than the least of the three parallel resistors.

properties that are very different from those of simple resistance. The current-limiting factor in circuits containing nonresistive load components is known as **impedance**, symbolized by Z. Impedance limits the flow of current through the circuit by blocking the passage of moving electrons, rather than by requiring large amounts of energy to force electrons into conduction, as resistance does. Because it is a current-limiting factor, impedance is measured in ohms, so, for impedance, we can write the Ohm's law formula as

$$I = E/Z$$

Resistance behaves the same way for either AC or DC circuits because it is not sensitive to frequency. Nonresistive circuit elements, however, are sensitive to frequency. They behave quite differently when polarity changes rapidly than when polarity changes slowly or is constant (DC). Such components are said to react to frequency and, therefore, to have the property of **reactance**. Two kinds of reactance exist: **inductance** and **capacitance**. Impedance is defined as *current limiting produced by the combined effects of reactance and resistance in a circuit.* The behavior of an AC circuit depends on whether its impedance is mainly inductive or capacitive.

Inductance

Inductance results from the interaction of electricity and magnetism. Inductive circuit elements are usually coils of wire, sometimes wrapped on paper or plastic or wound on soft iron cores. Transformers, tape recorder heads, and phonograph pickups are all inductive devices. In an inductance (Fig. 2-10A), the current flowing into the coil generates a magnetic field. Because the device is a coil, the magnetic field moves across turns of the same wire carrying the field-generating current. As it moves across the turns of the coil, the field induces a second current. The induced current flows *opposite* the main current and *subtracts* electrons from the main current flow, thus limiting current flowing through the coil.

If inductive reactance were the same for all frequencies, from DC to the limits of the audio spectrum, inductance would be almost indistinguishable from resistance. Inductive reactance is much different at DC and low frequencies than at high frequencies, however. Inductances offer no reactance to DC; direct current flows through a coil as though it were a straight wire (Fig. 2-10B). As soon as current begins to alternate, some reactance appears, although reactance to low frequencies is very small (Fig. 2-10C). The higher the fre-

FIG. 2-10A-D: INDUCTIVE REACTANCE

A. EXPLANATION OF INDUCTANCE

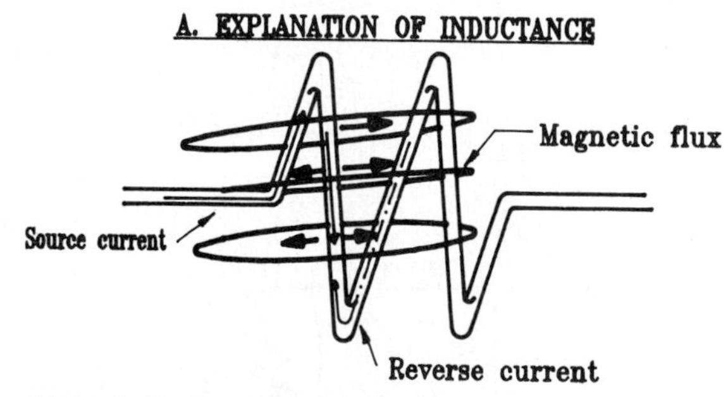

Current flowing into coil from source generates magnetic field around conductor. Field radiating out from each turn of coil cuts across adjacent turns of coil and generates a reverse current. Reverse current limits (subtracts from) original current. Reverse current limits main current flow and constitutes basis of inductive reactance.

B. INDUCTANCE AND DIRECT CURRENT

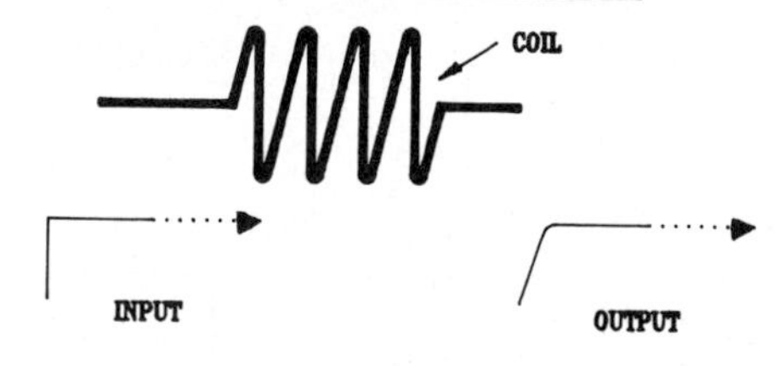

Direct current rises instantaneously to full voltage level and continues without change until shut off. Inductive reactance rises momentarily as DC rises to level; then reactance disappears as magnetic field stabilizes. Output voltage rise is slightly retarded, then settles at same voltage level as input.

C. INDUCTANCE AND LOW FREQUENCY

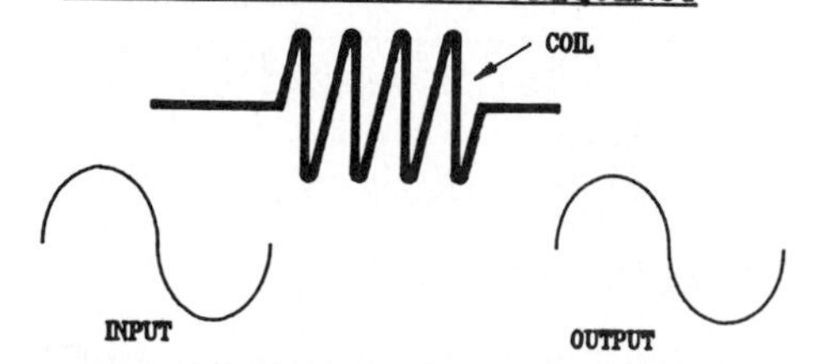

With low frequency (slow rise and fall of input voltage and current) magnetic field builds quickly, then stabilizes for most of waveform period. Output is slightly less intense than input, almost like DC.

D. INDUCTANCE AND HIGH FREQUENCY

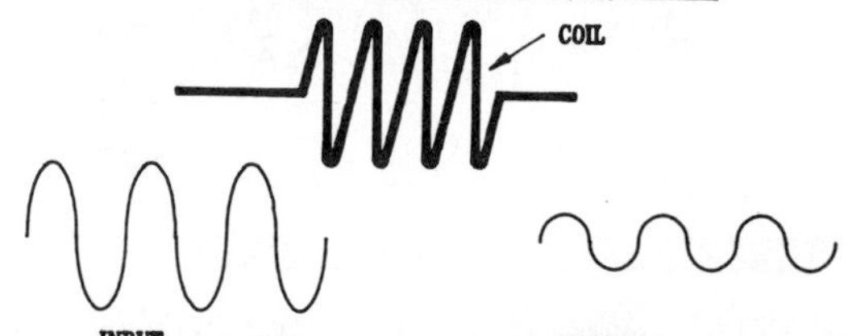

At higher frequency, input waveform changes quickly. Magnetic field cannot stabilize before waveform changes. Reactance is significant. Output is less intense than input.

quency, the greater the inductive reactance (Fig. 2-10D). In the upper range of the audio spectrum, inductive reactance becomes quite large. Simply stated, *inductive reactance passes low frequencies and impedes high frequencies.*

Capacitance

Capacitance stores charge. For capacitance, we can create another water analogy. Figure 2-11A shows two small tanks connected by a pipe with a reversible pump in the middle of the pipe. The system contains enough water for one tank and part of the other, but not enough to fill both. The pump (equivalent to the voltage source in an electric circuit) starts moving water in one direction. Eventually it will fill one tank; but when the tank reaches its *capacity*, no more water will flow, no matter how long the pump continues to run. A unidirectional pump is equivalent to direct current.

If we make the pump change directions so that it moves water, first in one direction, then in the other, it will fill first one tank, then the other. If the pump alternates directions slowly, one tank will fill and flow will stop until the pump changes directions again. Then water will flow until the opposite tank is filled. If, however, we make the pump alternate directions rapidly, one tank cannot fill before the direction changes. Water will flow back and forth constantly. Thus, at a low frequency of alternation, water flow will be minimal during the time the pump moves water in any given direction. At high frequency, however, a relatively large volume of water will flow constantly.

A capacitor consists of two conductive plates separated by the thinnest possible insulating layer. (See Fig. 2-11B.) When voltage is applied to the circuit, one plate goes negative, the other positive. The negative plate fills up with electrons as the opposite plate is drained of electrons. As the plates reach their capacity to hold charge carriers, current flow diminishes and finally stops. The capacitor blocks the flow of current when its plates can no longer store charge.

For DC, capacitance is like an open circuit: The plates charge quickly, and then no more current flows at all. (See Fig. 2-11C, part 1.) At low frequencies of alternation, small blips of current flow during each half-cycle, but during most of the voltage waveform the plates are charged and no current flows at all (Fig. 2-11C, part 2). At high frequencies, current direction changes long before the plates can charge or discharge, so current flows easily (Fig. 2-11C, part 3). Capacitive reactance, therefore, is the opposite of inductive reactance: *Capacitive reactance blocks DC and low frequencies but passes high frequencies.*

FIG. 2-11A: WATER ANALOGY FOR CAPACITANCE

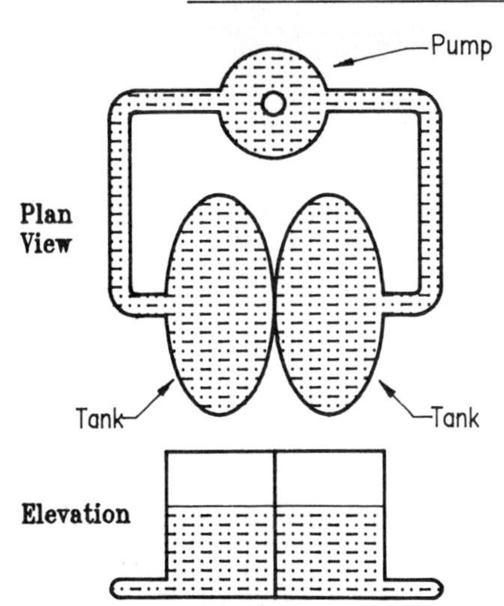

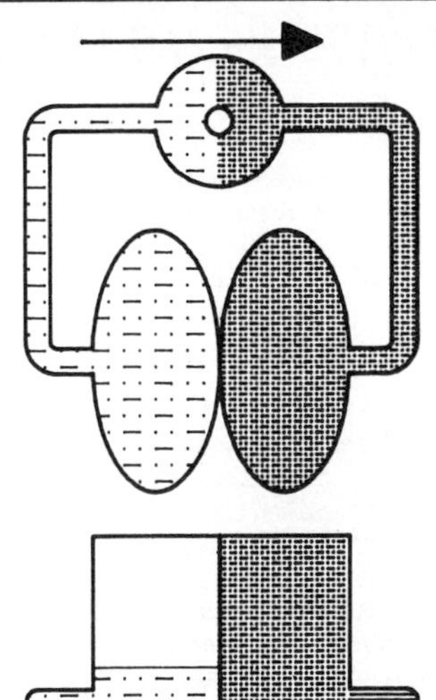

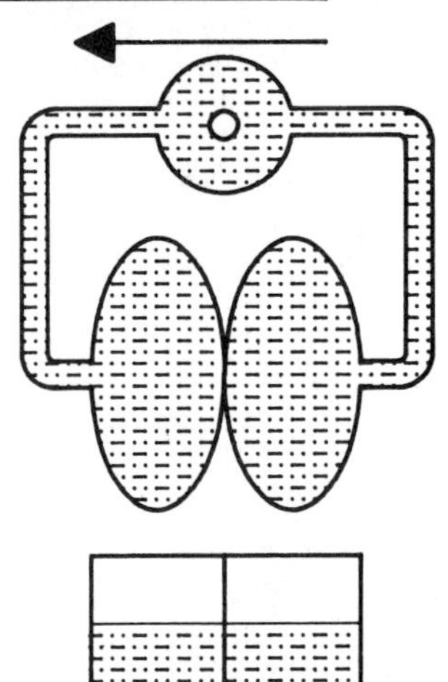

1. Pump not running; both tanks at normal level. Pressure is balanced.
2. Pump runs. Tank at right is pumped to full capacity; tank at left is mostly emptied. Pressure is unbalanced.
3. Pump shuts off; pressure in tank at right pushes water back through pipes and brings tank at left to level. Pressure rebalanced.

If pump runs in only one direction, one tank will fill to capacity and water will stop flowing, even if pump keeps running. Pump will simply maintain pressure. Pump can run in either direction and can fill either tank. If the pump alternates direction tanks will fill and discharge alternately. If frequency of alternation is slow, one tank will fill to capacity long before pump changes direction. Consequently, water flow (current) will stop. If frequency of alternation is rapid, direction will change long before tank can fill, and water flow (current) will be continuous and constantly changing direction. Average flow will be small at low frequencies and large at high frequencies.

FIG. 2-11B: CAPACITIVE REACTANCE

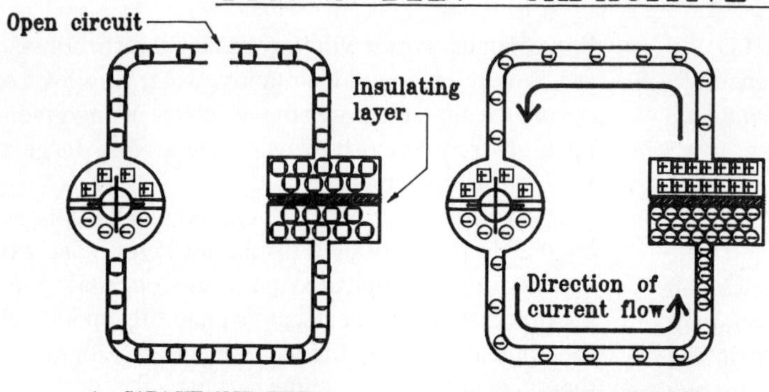

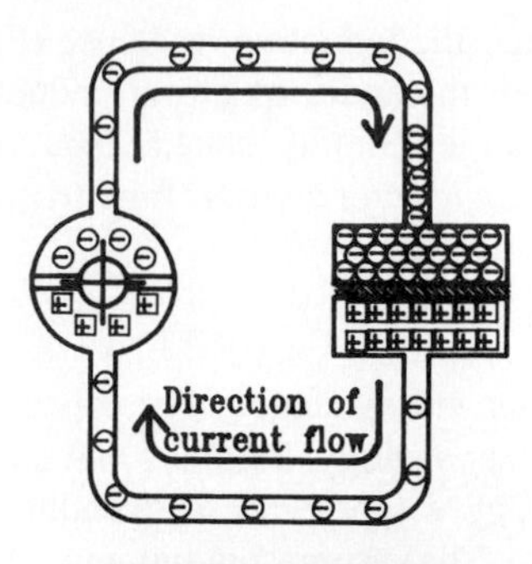

1. CAPACITANCE WITH NO APPLIED VOLTAGE

With circuit open, voltage has no effect on circuit. Circuit wiring and capacitor are all balanced.

2. CAPACITOR CHARGED

Circuit closed. Voltage applied to capacitor draws electrons out of top plate and forces electrons into bottom plate. As soon as capacity is reached, no more electrons can flow, and current stops.

3. POLARITY ALTERNATES TO OPPOSITE CHARGE

Polarity reverses. Current flows in opposite direction, drawing electrons out of bottom plate and filling top plate. Again, when capacity is reached, no more electrons can flow, and current stops.

The current limiting properties of a capacitor depend on the volume of electrons and positive ions (amount of charge) that can be forced onto its plates. Current will flow rapidly as plates first begin to charge. Then current flow will decrease as plates near full charge. When full charge is reached, current flow will stop. Opposition to further current flow is the capacitive form of reactance.

SYMBOL KEY
⊞ = Positive ion
⊖ = Electron
O = Balanced atom

FIG. 2-11C: CAPACITIVE REACTANCE AND FREQUENCY

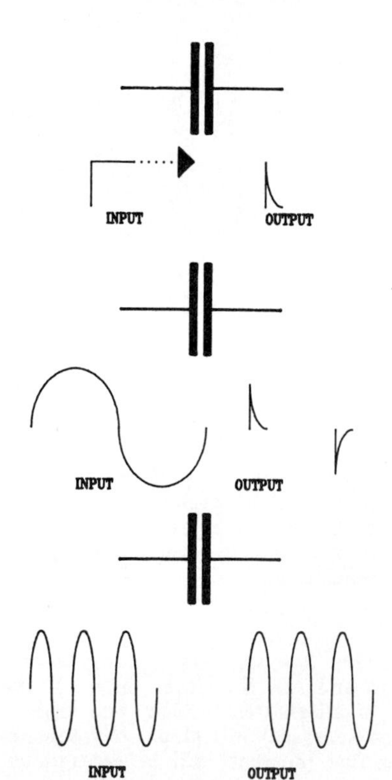

1. Direct current applied to capacitor. Voltage rises and stabilizes at maximum level. Output from capacitor is small pulse of voltage and current. Following pulse, no voltage or current appears on output side of capacitor. Capacitor blocks DC.

2. Low frequency applied to capacitor. Voltage rises slowly, but capacitor charges before waveform can reach peak. Output from capacitor is a positive pulse. Output side of capacitor remains at zero voltage and current for remainder of positive half-cycle. As waveform enters the negative half-cycle the waveform drops toward negative peak. The capacitor reverses polarity and reaches full reverse charge before the waveform can reach negative peak. Output from capacitor is a negative pulse. Output side of capacitor remains at zero voltage and current for remainder of half-cycle. Capacitor offers high reactance to low frequencies.

3. High frequency applied to capacitor. Voltage rises to peak <u>before</u> capacitor can charge. Output voltage and current follow input voltage and current. Capacitor offers low reactance to high frequencies.

Reactance and Power

Reactive components behave like resistance for Ohm's law but not for Watt's law, because, in subtracting from current flow (as an inductor does) or in ceasing to store charge (as a capacitor does), *reactances do not consume power.* Reactances do affect power calculations, however, because they affect the phase of voltage and current. Further, phase shift changes as external loads affect the amount of current drawn through the circuit, and external loads do consume power.

In an AC circuit containing reactance, power calculations must take into account both the phase angle of the load current and the frequency of the AC signal. Reactance is not equal at all frequencies. Therefore, the current drawn through the circuit, even under load, is not the same at all frequencies. AC power calculations are complex, and space does not permit an adequate discussion here. The reader should realize, however, that when a loudspeaker draws power from an amplifier, that power is expended mainly to overcome the inertia of the air—to set all the air in a room into motion.

Practical Applications of Reactive Circuit Elements

The response of reactive components to frequency makes possible devices like the crossover networks that are used in loudspeaker systems to direct the appropriate frequency bands to the high- and low-frequency elements of the system. Figure 2-12 shows a simple crossover network. The entire audio bandwidth is fed to the input of the network at the junction point between the series capacitor and the series inductor. The capacitor, however, passes only frequencies above 500 Hz easily, whereas the inductor passes only those below 1000 Hz. Following the series capacitor is an inductor placed in parallel across the circuit. This inductor readily conducts any frequency less that 800 Hz. Energy at frequencies less than 800 Hz is, therefore, shorted to ground between the series capacitor and its output to the high-frequency element of the loudspeaker system. Following the series inductor is a capacitor placed in parallel across the circuit. This capacitor readily conducts any frequency greater than 800 Hz, shorting energy above that frequency to ground between the series inductor

FIG. 2-12: CROSSOVER NETWORK

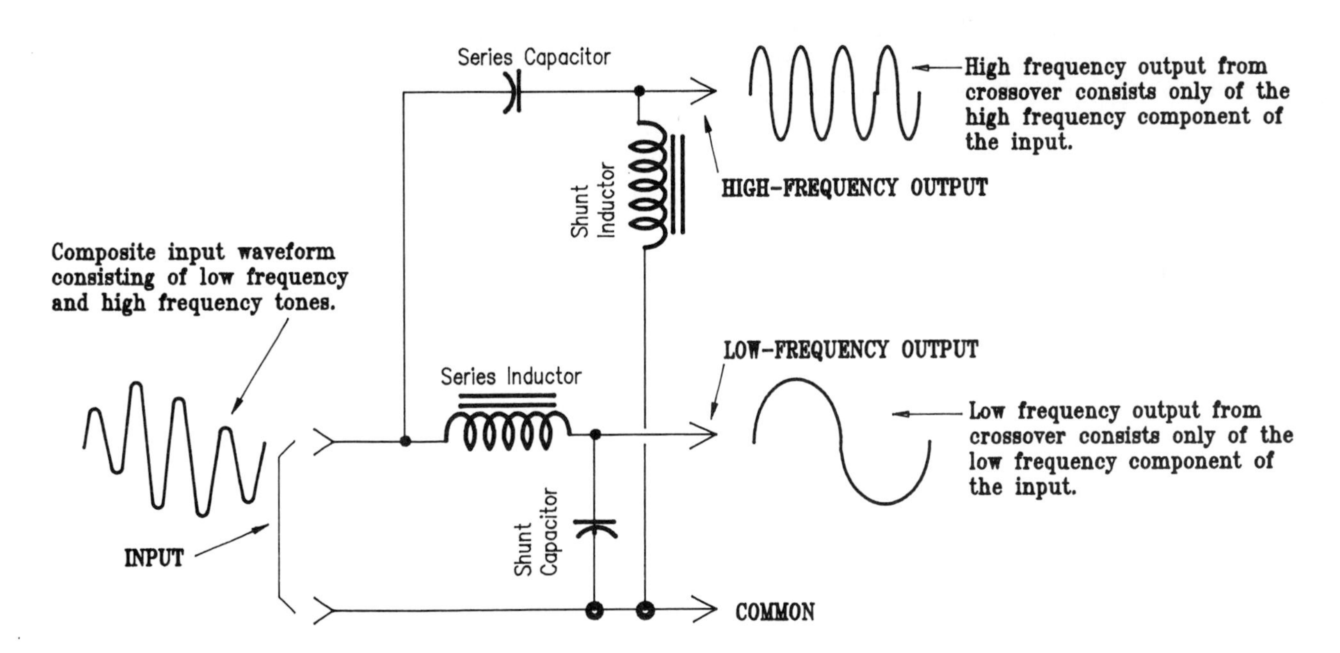

Composite signal is applied equally to both the series capacitor and the series inductor. The capacitor blocks low frequency components of the input signal but passes high frequency components. Any residual low frequencies are passed to ground (common) through the shunt inductor. High frequencies cannot pass through the shunt inductor. The series inductor blocks high frequency components of the input signal but passes low frequency components. Any residual high frequencies are passed to ground (common) through the shunt capacitor. Low frequencies cannot pass through the shunt capacitor.

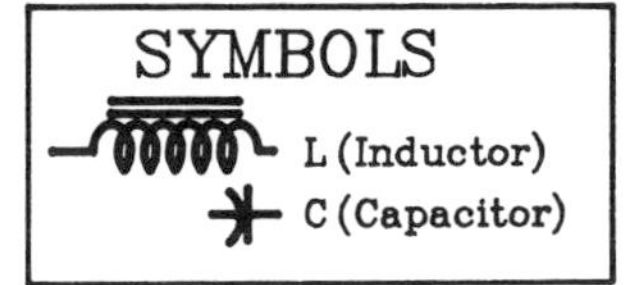

and its output to the low-frequency element of the loudspeaker system. The two branches of the crossover network form a circuit that smoothly divides the audio spectrum into two parts, making the transition from one band to the other at 800 Hz. The high-frequency section of the system reproduces only frequencies higher than 800 Hz, and the low-frequency section handles only those below 800 Hz.

The Nature of an Audio Signal

As mentioned earlier, alternating current can be made to mimic alternating pressure to provide an electrical analog of an acoustic waveform. Acoustic waveforms, however, are complex variations in pressure, and therefore, audio signals must be complex variations in voltage and current.

The human ear can perceive frequencies from roughly 20 cycles per second to 20,000 cycles per second—a spectrum of 11 octaves. Therefore, audio must be capable of handling a frequency range of 20 to 20,000 cycles of alternating current per second. (Most other electronic systems handle little more than one octave.) Moreover, many different frequencies will be present simultaneously in almost any audio signal. Because of the wide bandwidth of audio, distortion tends to become a problem. **Distortion** is any difference between the input waveform and the output waveform. Current passing through reactive components tends to produce some amount of distortion. Because audio involves many reactive components, audio systems are designed to keep current as low as possible. Audio signals, therefore, are amplified, processed, and transferred from unit to unit mainly as voltage. Only in the final stage of power amplification is current increased. Through most of an audio system, the signal will appear as a voltage tracing a complex pattern of alternation in direction and polarity. Because current is low and must be kept low, the signal is especially vulnerable when it must be transferred from one piece of equipment to another. Any mistake made by the system operator in interconnection can result in distortion or in a weakened or lost signal.

Electrical Concepts and Practical Audio Systems

Generally, sound technicians and designers probably don't have to think about electrical considerations, certainly not to the extent that lighting designers and technicians do; yet audio equipment is much more complex technically than most lighting equipment. As long as an audio system is assembled, installed, and working correctly, little reason exists to spend time thinking about how, where, or why electrons are flowing in the equipment. People who have to assemble equipment or keep a system going so that the show isn't interrupted, however, must know at least a few basic principles. The main technical concepts that almost all sound operatives—technicians and designers—need to understand are four: impedance, level, line configuration, and direction of signal flow.

Impedance and Audio Equipment

Impedance, as we have seen, is the combination of resistance and reactance that makes up the current-limiting properties of a circuit. We meet impedance at the input and output of audio equipment. The impedance characteristics of an output must be compatible with the input to which we connect it. In order to understand this process, we need to apply what we've learned about Ohm's law and Watt's law.

When we pass a signal from one piece of equipment to another, we are transferring power. That is, we are using the voltage and current producing capability of the first piece of equipment in the chain to drive the input of the second piece of equipment. The power level involved may be extremely small, but we are still transferring power. Generally, one very important rule applies: *For optimum power transfer, load impedance must equal source impedance.* When the load imposed by the second piece of equipment is large, as a loudspeaker presents a large load to a power amplifier, impedance match is critical. If we want to use a loudspeaker that has an input impedance of 8 ohms, we must have a power amplifier that has an output impedance of 8 ohms. Older power amplifiers used transformers to adapt the impedance of the power circuits to the impedance of a loudspeaker. The transformer had **taps** marked in ohms, so that it was easy to connect the loudspeaker to the right set of terminals. As we shall see in a later chapter, newer power amplifiers only provide a pair of terminals to which any loudspeaker within a prescribed impedance range may be connected. Knowing the prescribed range of a power amplifier and the impedance rating of loudspeakers to be used is the responsibility of the system user.

Impedance is significant whenever a **transducer** must be connected to an audio system. A transducer converts energy from one form to another: acoustic energy to electrical energy, for example. Microphones, phono pickups, and loudspeakers are all transducers. The rule that impedances must match for optimum power transfer always applies when interfacing transducers to electrical systems.

We also meet impedance in connecting amplifier to amplifier, as when we plug a tape recorder into a mixing console. In order to understand the requirements of interconnecting amplifiers, however, we need to look at the second of the technical concerns, level.

Level

Level refers to the magnitude of voltage and current to be passed from one unit to another. Essentially, we are talking about power, but, except at the output of a power amplifier, matters of level usually concern voltage. Audio systems pass signals at three different levels: low level, line level, and power level. The breakdown represents a generalization, and other names exist for the various levels; but this classification works well enough for purposes of theatre sound.

Low level means signal voltage on the order of a few millivolts and currents of a few microamperes—the signal strength of microphone and phono pickup outputs. *Line level* means emf of approximately 1/2 to 1 1/2 volts and currents of a milliampere or less. Most consoles, processing devices, tape recorders, and synthesizers operate at line level, at least at their outputs. Power level refers to voltages ranging from 10 to near about 100 volts and current of 1 to 10 amps. *Power level* applies almost exclusively to the output of power amplifiers and the inputs to loudspeakers.

Because impedance is a current-limiting factor, and level is concerned with power, each different level has a characteristic range of impedances. Impedance at low level is 50 to 250 ohms in microphone circuits and approximately 50,000 ohms for phono pickups. Impedance at line level is usually either 600 ohms or 10,000 ohms. Impedance at power level is typically 4, 8, or 16 ohms.

Line Configuration

Line configuration refers to the setup of inputs and outputs and the kind of transmission line needed to interconnect pieces of equipment. Inputs, outputs, and transmission lines are either balanced or unbalanced circuits. A special kind of balanced circuit, called *differential*, also exists. A differential circuit is one that has the properties of a balanced circuit, but creates the negative and positive terminals through the use of a device called a differential amplifier rather than through the use of a balanced transformer. As noted earlier, balanced lines (including differential inputs) have a high immunity to induced hum and noise. Unbalanced lines are easily affected by hum and noise. Usually, a particular output configuration must be connected to a similar input configuration.

Audio connecting cables are invariably shielded. Cables for balanced lines usually have two internal conductors—one for each hot lead—and use the shield as the neutral conductor. Cables for unbalanced lines usually have only one hot conductor and use the shield as neutral. The usual arrangement of conductors and shield in audio cables works well as long as we connect balanced to balanced and unbalanced to unbalanced. Attempting to interconnect unlike inputs and outputs is almost always disastrous. (Interconnecting unlike circuit configurations is possible, but it requires specialized treatment, as we shall learn in a later chapter.)

Signal Direction

Signal direction refers to the normal flow of an audio signal from the output of one piece of equipment to the input of the next. Proper signal routing requires that one understand the function of each connector on each piece of equipment and whether that connector represents an input to the device or an output from the device.

Specification of Connectors

Each connection to a piece of audio equipment can be specified in terms of impedance, level, configuration, and direction. A microphone connector on a professional audio console is a balanced (configuration), low-level input (direction) having an impedance of 150 ohms. An effect-send connector on a console is usually an unbalanced (configuration), line-level output (direction) having an impedance of about 10,000 ohms. The main output channels of a console might be differential (configuration), line-level outputs (direction) having an impedance of 600 ohms.

Naturally, all of these specifications will seem confusing to the novice. As we study the various kinds of audio equipment and learn the basic principles of audio systems, though, most of the terms and the associated standard numbers will become quite familiar. The concepts stated here are to help the reader establish a general understanding for the electrical basis of audio equipment. No one will become an audio engineer or even a qualified technician based on the concepts set out in this chapter, but these concepts will help the reader to gain a better understanding of the nature and function of the kinds of audio equipment used in theatre sound as we discuss them in the succeeding chapters.

Chapter 3

Audio Measurements

Chapter 2 introduced electrical terms such as volts, amperes, ohms, watts, impedance, inductance, and capacitance. Except for those involved in Ohm's and Watt's laws, none of these terms were introduced as mathematical entities. Throughout this study of theatre sound, most of the technical concepts can be handled without treatment in terms of numbers and equations. Usually, one can work with audio equipment and never have to worry about the engineering math associated with the design and development of electronic devices—but then the subject of signal measurement turns up.

Signal measurement is a very important element of good audio practice. Measurement is inescapably quantitative, and some understanding of the underlying mathematics is essential. To understand about such matters as signal level, signal-to-noise ratio, and the way in which the meters and controls on audio equipment are indexed, one must deal with a little bit of math. Usually, once the basic principles are well understood, one can simply read meters and know what is going on without having to have a calculator handy. Perhaps the easiest way to start is to go back to the physical nature of sound, to elements that we did not cover in Chapter 1.

Normal human hearing covers a large dynamic range. The loudest sound that we can tolerate is one trillion times more intense than the softest sound we can detect. One trillion is a very large number, and working with large numbers is cumbersome. Because audio is meant to reproduce sounds that humans can hear, audio equipment must be able to approximate a power range similar to that of human hearing. How would you like it if you had to glance at a meter and figure out the difference between a trillion and a billion quickly? Or what about representing numbers up to a trillion on the scale of a fader? Worse yet, what happens when you're dealing with low power at the console and high power at a power amplifier, but meters on both devices need to reflect the range of loudness as those levels will seem to your ear?

The way that audio engineers deal with large numbers and with shifting the dynamic range through different levels is by use of logarithms. Okay, almost everyone groans on hearing that word, logarithms. But stay with us; use of logarithms really isn't that difficult. (Mathematical wizards, bear with us while we take a few baby steps.)

Logarithmic Notation

A logarithm is simply the power to which a number must be raised to equal some other number. When we speak of raising a number to a power, we mean how many times the number must be multiplied by itself—how many times must a must be multiplied by itself in order to equal b. For example, if we multiply 10 by 10 we get 100. Because two 10s are required for the operation, the power is 2. Therefore 10 to a power of 2 equals 100. The number to be multiplied by itself is called the *base;* in this example the base is 10. The number of times the base is used in the operation is called the *exponent.* In the example, 2 is the exponent. The words *logarithm* and *power* are often used to mean the same thing as *exponent.*

As we learn to use logarithmic notation, we will need some easy way to determine just what the logarithm of any given number is. Some engineering and scientific calculators provide a log function. Reference books also provide tables of logarithms, and a table of logarithms in base 10 is provided in Appendix A. As we go through the explanations of decibels and other measurements, you should refer to the table of logarithms in the appendix, or else use a calculator that provides a log function. Each time we introduce an example, try to work it out for yourself to gain an understanding of the procedure and a feeling for how the system of logarithmic notation operates.

Logarithmic notation expresses numbers in terms of powers in the following manner:

$$\log_{\text{base}} X = y$$

where X is the number to be expressed as a power of the base and y is the exponent. Thus

$$\log_{10} 100 = 2$$

In words, this is "The logarithm of 100 in base 10 is 2."

One hundred is a larger number than 10. What if the number to be expressed in logarithmic notation were (a) not an integral power of the base or (b) smaller than the base or a fraction? For example, what is the logarithm of 50 in base 10?

$$\log_{10} 50 = 1.69897$$

In this case, the exponent is not an integer—that is, not a whole number. Rather it is a whole number plus a fraction. Ten cannot be multiplied by itself a whole number of times to equal 50.

What is the logarithm of 3 in base 10?

$$\log_{10} 3 = 0.477$$

Here the exponent is simply a fraction. In other words, 10 cannot be multiplied by itself even once to equal 3.

And, for another example, what is the logarithm of 0.25 in base 10?

$$\log_{10} 0.25 = -0.60206$$

The exponent in this case is a negative number. When the exponent is negative, the number to be expressed is a fraction, that is, less than 1.

Logarithmic notation includes fractional exponents and negative exponents. Fractional exponents represent numbers for which the exponent is not a whole number, meaning that the base cannot be multiplied by itself an integral number of times. Negative exponents represent numbers that are smaller than 1—fractions.

0.25, 3, 50, and 100 are not particularly large numbers, so expressing them as powers of 10 may seem more complicated that simply saying 0.25, 3, 50, or 100. But what about a number like one trillion, which is a 1 followed by 12 zeroes? Well, how many times must 10 be multiplied by itself to equal one trillion? The answer is 12. Therefore

$$\log_{10} 1{,}000{,}000{,}000{,}000 = 12$$

Twelve is an easier number to handle than one trillion, but not if we have to write the number one trillion plus a formula every time we want to use 12 to mean one trillion. That implies that we need a system that defines the one number in terms of the other, so that whenever we see the exponent we already know that it is an exponent and what base it serves. The system used by audio engineers is called *decibel notation.*

Everyone who works with audio equipment has seen the term *decibel* and its abbreviation *dB*. Most of us will have even gotten the impression that a decibel is some kind of unit of power. So what kind of a unit is a decibel and how do we use it? Let's briefly trace the history of the unit in order to find some answers to these questions.

Origin of Decibel Notation

Investigation of human hearing started with questions like the following:

"What is the softest sound that the average ear can detect?"

"What is the loudest sound that the average ear can tolerate?"

"How much louder is the loudest sound than the softest sound?"

The third question is the most critical for our purposes. "How much louder" suggests the relation of one quantity to another. We have to ask that kind of question many times in dealing with sound and audio. How much louder is a sound when one is 2 feet from its source than when one is 50 feet from the source? How much louder is the signal coming out of an amplifier than the signal that went into the amplifier? These questions all concern matters of *relative level*—that is, power at one point relative to power at another.

We need the answers to the first two questions in order to answer the third, of course. Questions one and two deal with specific quantities of power. When we specify some amount of power, that is called an *absolute level.* Absolute level is specified in watts. Although the term *watt* is used mainly as a unit of electric power, it could represent any power in any form. When we measure acoustic systems, we speak of power in terms of *acoustical watts.*

Back to the first question: "What is the softest sound that the average ear can detect?" The answer is one trillionth of an acoustical watt, which is about the amount of sound power that will reach your ear when you first detect the buzz of a mosquito somewhere nearby. This power level is called the *threshold of hearing.*

We've already said that the loudest sound that most of us can tolerate is one trillion times greater than the softest sound we can detect, so if the threshold of hearing is one trillionth of a watt, then the loudest sound we can tolerate must be 1 acoustical watt. One acoustical watt represents the *threshold of pain.* Relative to the threshold of hearing, the threshold of pain is one trillion times greater. Notice that, stated as a relative value, no specific units are involved. The wattage levels could be anything, so long as the larger is one trillion times more than the smaller. But now we're back to the problem of size. One trillion and one trillionth are still cumbersome numbers.

In order to deal with the very large ratios and the very small fractions that were evidently going to be involved in comparisons of power levels, given that the dynamic range of hearing is so large, engineers and scientists set up a logarithmic system of notation in base 10. Expressed as an equation, the notation looks like this:

$$\text{Relative power} = \log_{10} (P_1 / P_0)$$

where P_1 is the larger level and P_0 is the smaller level. They named the unit of relative power the *bel*, after Alexander Graham bell, the inventor of the telephone.

Once they began using the bel, researchers found it too coarse. One was always dealing with fractions of bels. A unit about one tenth the size of the bel seemed more appropriate, so the bel was divided into 10 parts, called *decibels*. The prefix *deci* means one-tenth. The abbreviation of decibel is *dB*. Because the decibel sets 10 units in the place of 1, the equation must now read

$$\text{decibel} = 10 \log_{10}(P_1 / P_0)$$

We can begin to quantify the answer to "How much greater" questions in more meaningful terms, now that we have a unit of relative power. Earlier we said that the threshold of pain was one trillion times greater in intensity than the threshold of hearing. Let's use the bel first, because the equation is a bit simpler. We would write

$$\text{Bels} = \log_{10} (1 / 0.000000000001)$$

as threshold of pain equals 1 acoustical watt and threshold of hearing equals one trillionth of an acoustical watt. Dividing, we get

$$\text{Bels} = \log_{10} 1{,}000{,}000{,}000{,}000 = 12 \text{ bels}$$

In words, the number of bels is equal to the log in base 10 of the number one trillion. We've already stated that one trillion is equal to 10 raised to the twelfth power. Therefore the threshold of pain is 12 bels greater than the threshold of hearing.

Now, let's turn bels into decibels. Expressed in decibels, the formula reads

$$\text{decibels} = 10 \log_{10} (1 / 0.000000000001) = 120 \text{ dB}$$

Thus, the number of decibels will be 10 x 12, or 120. So the threshold of pain is 120 dB greater than the threshold of hearing. We now have our system that tells us, without having to write the number we are trying to simplify, that the number written stands for another value. When we say that signal level at one point is 120 dB greater than at another point, we mean that the signal has increased in power by one trillion times.

So far, P_1 (the dividend in the ratio) has been greater than P_0 (the divisor) in all of the cases. The opposite can, and does, happen. Dividends are often smaller than divisors, yielding fractional quotients. For example, if we reverse the question and compare the threshold of hearing to the threshold of pain, we write

$$0.000000000001 / 1 = 0.000000000001$$

What is the log in base 10 of one trillionth? The answer is

$$\log_{10} 0.000000000001 = -12$$

Therefore, relative to threshold of pain, the threshold of hearing is 120 dB less in intensity. Signed exponents give us a way to represent both relative increase and relative decrease in power.

Let's examine how relative specification works in some other instances. Say that a power amplifier delivers an average power of 1 watt to a loudspeaker; then we double the level to 2 watts. How much greater is 2 watts than 1 watt in decibels?

$$\text{dB} = 10 \log_{10} (2 / 1)$$

Two divided by 1 equals 2, and the log in base 10 of 2 is 0.301. Ten times 0.301 equals 3.01; therefore 2 watts is approximately 3 dB greater than 1 watt.

Now let's try a different set of numbers for the ratio. Suppose that, from an average power of 30 watts, we increase the output of a power amp to 60 watts. We write

$$\text{dB} = 10 \log_{10} (60 / 30)$$

Sixty divided by 30 equals 2, so we get the same answer: 3 dB. Herein lies one of the advantages of relative power specification. Equal amounts of increase or decrease are represented by the same number of units, no matter whether the actual power levels involved are fractional or in the thousands of watts.

Reference Levels

Relative specification of power levels is convenient, partly because we can deal with more manageable numbers, partly because equal amounts of change are specified as equal units, and partly because we usually want to know how much louder or softer one sound is than another, rather than the specific power of each sound. Nevertheless, we always need to know what range of power we're talking about. Are we dealing with the fractional wattage levels found in consoles and signal processing equipment, or are we talking about the tens or hundreds of watts delivered by power amplifiers to loudspeakers?

In order to know the power range, some sort of qualifier needs to be added to our terminology, and we need a standard power within that range to which to refer our measurements. People like to think of starting points as unity. Example: We tend to start counting from 1. Logarithmic notation has a convention for representing unity: Any base to its zero power equals 1. Thus

$$\log_{10} 1 = 0$$

A little thought about this convention should tell us that 0 dB doesn't equal no sound; it means no *change* in power level. Remember, the decibel is a unit of *relative* power. So 0 dB means that the power level hasn't changed compared to the most recent previous measurement. Zero can also be used as a marker, to signify some important point within a range—a maximum or a minimum level, for instance.

Going back to the dynamic range of human hearing, the custom is to use the threshold of hearing as a reference level. Because one trillionth of an acoustical watt becomes our starting point, that power level is designated as *0 dB SPL*. The letters *SPL* stand for *sound pressure level*. They indicate that we are talking about acoustic power, and that all measurements will be referred to threshold of hearing. Thus, threshold of pain is 120 dB SPL; average conversational level is approximately 60 dB SPL.

In audio, most parts of a system operate at relatively low power levels. Engineers define a power level of one one-thousandth of a watt (0.001 watt) as the reference level. The prefix *milli-* indicates one one-thousandth, so the reference level is commonly known as 1 *milliwatt*. The term that tells us that we are operating in a range of power based on 1 milliwatt is *dBm*. The abbreviation stands for "*d*eci*b*els referred to a power of one 1 *m*illiwatt." One milliwatt is designated as 0 dBm. A power level of one milliwatt is called *standard operating level*.

If a power amplifier delivers 5 watts of power to a loudspeaker, then its level with respect to standard operating level is

$$\text{dBm} = 10 \log_{10} (5/0.001) = 37 \text{ dBm}$$

5 divided by 0.001 equals 5000. The log in base 10 of 5000 is 3.7; therefore, the relative power level to the loudspeaker is 10 x 3.7, or 37 dBm.

Having stated that a power level of 1 milliwatt is the standard of reference for audio power level measurements, we must note one further qualification. Recall that in Chapter 2 we mentioned four conditions that specify every input and output to and from units of audio equipment: impedance, level, line configuration, and direction of signal flow. Impedance usually figures into the specification of operating level in the following manner. Operating level is defined as a voltage of 0.775 volt across an impedance of 600 ohms. If one substitutes these values into Watt's law, the result is

$$P = E^2 / Z = 0.775^2 / 600 = 0.001$$

that is, 1 milliwatt. Clearly, if the impedance of a circuit is not 600 ohms, then a different level of voltage is required to produce a power level of 1 milliwatt. At a nominal level of 10,000 ohms, for example, voltage must be more than 3 volts in order for power to reach a level of one milliwatt.

One milliwatt is not an inconsequential power level, even though one thousandth of a watt may sound quite small. One milliwatt is not, like the threshold of hearing, a minimum level marker. On the contrary, 1 milliwatt of power is very near the maximum safe level on many pieces for audio equipment.

Volume Units

A reference level is very important for one aspect of audio that every designer and operator needs to understand thoroughly. That aspect is the use of the meters on consoles, tape recorders, and often on power amplifiers. Audio meters are calibrated in a unit called the *volume unit*, abbreviated *VU*, which is why the meters are called *VU meters*. Figure 3-1 shows a typical audio VU meter. Notice several important characteristics of the meter. First, its scale is logarithmic, meaning that the divisions are not equal. At the left-hand end of the scale the divisions are compressed, but they expand as the scale rises toward the right-hand end. (Compare the distances between -2 and 0 and between -20 and -10. They are approximately the same, but that distance incorporates only two units near the top of the scale and 10 units near the bottom of the scale.) Second, the scale is divided into two sections, one black and one red. The black portion of the scale is toward the left, the red portion to the right. Third, the numbers rise from -20 to 0 on the black portion, then from +1 to +3 on the red portion. Some VU meters read even higher, to +4 or +6.

Notice, also, that the VU meter has a linear scale from 0 to 100, with 0 at the low (left hand) end of the scale, adjacent to -20 on the logarithmic scale and 100 directly under the 0 on the logarithmic scale. The 0 to 100 scale is used to indicate percentage modulation in broadcast operations. An alternate form of the VU meter scale places the modulation scale above in bold characters and the logarithmic scale below in smaller characters.

Visually, the VU meter behaves as one might expect: Staying in the black indicates safe levels; breaking over into the red suggests approaching dangerous power levels. Typically, one sets faders so that the loudest levels to be handled read just at or slightly over 0 VU, meaning that the indicator will swing just to the top of the black portion of the scale or slightly into the red. You should realize, however, that a VU meter tells the status of level for one piece of equipment only. The VU meter bears no fixed relationship to how loud sound will be on the control room monitor or how much power loudspeakers in the theatre will project into the space.

The VU meter is a sliding scale: It measures the same amount of change in relative power no matter where in the audio chain it may be placed. In a well-calibrated system, a VU meter on the console should read exactly the same as the VU meter on the tape recorder feeding the console input.

FIG. 3-1: VU (VOLUME UNITS) METER

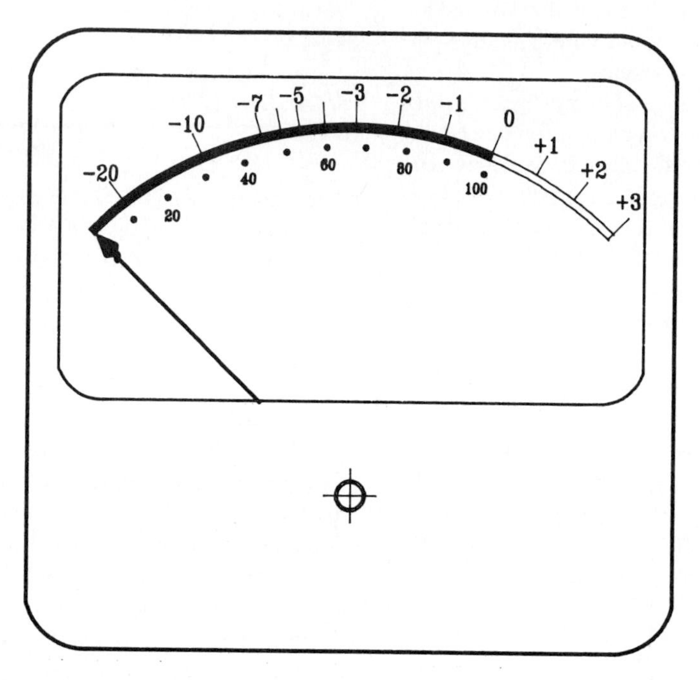

Theoretically, a VU meter on a power amplifier taking signal from the console should also show the same reading. The critical piece of information is the reference: To what power level is 0 VU calibrated? Usually the level reference is expressed in dBm, which means that the ultimate reference is the standard milliwatt operating level. 0 VU, however, may be set to any number of dBm.

For many years, 0 VU on most consoles and tape recorders was set at 0 dBM, but most engineers like a slightly more powerful signal at the console output. Present practice, therefore, is to set 0 VU equal to +4 dBm on consoles and tape recorders. +4 dBm serves as *operating level* in newer systems, meaning the level at which signal is passed between most pieces of equipment in the system. Tape recorders, CD players, synthesizers, most processing equipment, consoles, and the input to power amplifiers all work at operating level. Manufacturers' specifications for the various pieces of audio equipment in an installation almost always specify the sensitivity of the input and the signal level at the output for each piece of equipment.

Typically, a VU meter reads the output level from a device. By knowing the reference level, we can estimate the power level involved. What is the reference when a VU meter appears on the face of a power amplifier, since the output of the power amp is going to be well above standard operating level? No industry standard comparable to console operating level (+4 dBm) applies here. Typically, 0 VU will be referenced to an arbitrary value determined by the manufacturer, reflecting safe maximum output. For example, if the manufacturer rates the amplifier at 50 watts per channel, then 0 VU for one channel might represent the number of dB gain between 1 milliwatt and 50 watts. Thus

$$\text{dBm} = 10 \log_{10} (50 / 0.001) = 47 \text{ dBm}$$

Ideally, VU meters on power amplifiers should be calibrated to reflect a desired maximum sound pressure level. In a permanent installation, the system could be calibrated and never changed. In a portable installation, the reference for 0 VU would have to be set with each installation. In either case, a variable-zero VU meter would assume the knowledge and competence on the part of the system installer and operator to make the appropriate measurements and calibrations of the system, and to use the information correctly.

Fader Indexing

Look at the index scale printed beside a fader on an audio console. (See Fig. 3-2.) If the console is fairly recent, the scale will show infinity ∞ at the bottom, reading through a series of negative numbers up to 0, then on to +3 at the top. The arrangement of values corresponds to the scale on the VU meter, and you should be able to treat it as such—if you calibrate your console properly. We'll discuss setting up a

FIG. 3-2: FADER SCALE

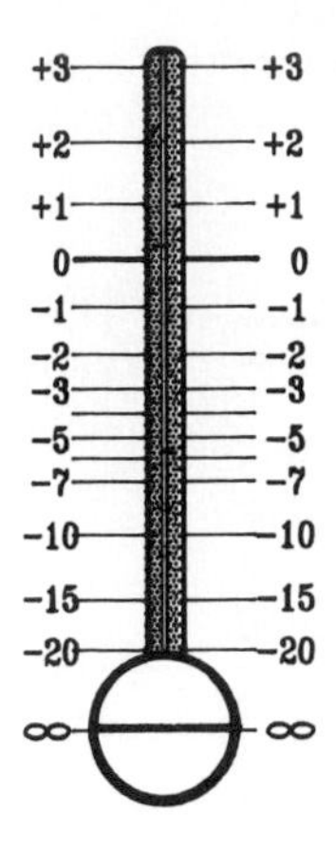

console for proper operation later on, but, in simple terms, you should be able to make zero on a fader produce zero on the VU meter.

The component that lies behind a fader is a network of resistances called an *attenuator*. The word *attenuate* means "to weaken or reduce." A fader reduces signal strength. At the top of its scale, the fader passes the output of an amplifier unaltered. Bring the fader down toward infinity and the signal strength will be reduced. The scale will indicate the number of dB reduction in level.

Gain Control

The question of how to make zero on a fader equal zero on the VU meter may have occurred to you by this time. Anyone who has had experience with audio and especially with recording will know that one cannot always predict how loud a sound source will be. The question brings us to the matter of *gain*. Gain is the difference between the level of the input to an amplifier and the level of its output. Gain is expressed in decibels. Most modern audio consoles provide gain controls for each input module, and we shall examine their function and use more closely in Chapter 9. For now, we need only point out that by setting the fader at 0—remember that 0 means no attenuation, not no sound—then regulating the amount of gain, we usually can set the loudest signal to be recorded so that it measures 0 VU. If we're using a tape recorder or some other device that has its own VU meter, then we set the output level from the recorder so that the loudest passage reads 0 VU on its own meter, then adjust the gain control on the console to match. In either case, an understanding of decibels, VU, and reference levels will make the business of using audio equipment a great deal easier.

Review of Audio Measurement Principles

Because of the importance of audio measurements, let's check over the information we've learned to this point. First, logarithmic notation is used to simplify dealing with the large numbers involved in sound and audio power measurement. Logarithmic notation expresses numbers as a base value raised to a power. In the case of audio, we use 10 as the base. We compare one value to another by making a ratio—that is, by dividing one power level into another. Usually, we divide output by input to determine the amount of gain contributed by an amplifier stage. The quotient of the operation is the number to be expressed as a logarithm. The log of the number becomes a value expressed in a unit called the bel. The bel turns out to be too coarse a unit, and so we divide it into tenths. A tenth of a bel is a decibel. In order to get the number of decibels, we multiply the logarithm of the ratio by 10.

No matter what the specific power values, equal ratios will always produce equal decibel readings. If two amplifiers each double the level of the input, the gain will be 3 dB, even if one turns an input of one tenth of a watt into an output of two tenths of a watt while the other raises an input of 5 watts to an output of 10 watts. Because decibel notation is comparative, expressing the relative relationship of two quantities, we need reference values to tell us just what range of power a decibel reading represents. In audio, the standard reference value is a power level of 1 milliwatt. One milliwatt is called operating level. A decibel notation referenced to operating level is written as dBm.

Because an audio system handles a range of power levels from the very small to the very large, we need a sliding scale of measurement that can present a similar indication at any point in the system. That scale is built on the volume unit, an arbitrary unit that can be assigned to any range of power by associating 0 VU with some particular number of dBm. For most consoles and tape recorders, 0 VU is set at +4 dBm. For power amplifiers, the value is a function of maximum safe output.

dB Applied to Voltage and Current

The explanation of decibels, dBm, and volume units that we've discussed to this point represents the main information that all audio operators need to know. For those who are interested in a more detailed understanding of electronic technology, the following explanation of dB as a means of stating change in the level of voltage and current is provided.

Decibel notation can be used to represent change in the voltage and current levels that produce change in power, *but only if impedance is the same at both points of measurement.* Recall from Chapter 2 that

$$P = I \times E$$

The Watt's law formula can be expressed in several ways, one of which is the preceding equation. Others are

$$P = I^2 \times R$$

and

$$P = E^2 / R$$

Notice that the resistance of the circuit plays a significant role. As we stated in Chapter 2, resistance is not the only current-limiting element that may be present in an electric circuit. When current limiting is either nonresistive or only partly resistive, we use the term *impedance*, symbolized by the letter Z. We can substitute Z for R in any of the Ohm's or Watt's law equations.

Notice, also, that in both of these permutations of Watt's law, power is proportional to the *square* of the voltage or current. In logarithmic notation, the square of a number is two times its exponent. Thus

$$\log_{base} X^2 = 2y$$

Therefore, for bel notation expressing change in voltage or current

$$\text{bel} = 2 \times \log_{10} (P_1 / P_0)$$

and for decibels

$$\text{dB} = 10 \times 2 \times \log_{10} (P_1 / P_0) = 20 \log_{10} (P_1 / P_0)$$

Be sure that you understand, however, that decibel notation ultimately refers to *power levels only.* We can represent gain as a function of voltage and current only so long as each point of measurement has the same impedance. If the impedance is different at each point of measurement, then a ratio of voltages or currents is meaningless.

Chapter 4

Overview of Audio Systems

In today's world, almost everyone owns an audio system, and most of us own more than one. We run with them; we ride with them; we use them as functional furniture in our homes. We've become reasonably sophisticated in our knowledge of audio controls. The knobs, buttons, sliders, meters, and LED indicators on modern home music systems would scare the wits out of the typical stereo owner of a decade ago. Still most of us would be totally lost in a professional audio control room.

Professional audio is a technical craft based on electronic engineering. Although technological advances continually make equipment easier to use, professional audio will always be an engineer's art. Creative use of audio requires both the artist's intuition and the technician's expertise. The purpose of Chapters 2 and 3 was to provide an understanding of the electronic basis of audio and of the necessary audio measurements. The purpose of this chapter is to present a preliminary survey of types of audio equipment and an introduction to the use of that equipment to make up complete audio systems. Following this chapter, we begin a detailed examination of the various kinds of audio equipment.

Audio is the use of electrical devices to pick up, transmit, amplify, process, store, and reproduce sound. Increasingly, audio equipment in the form of new musical instruments is used to generate sound. Audio systems are made up of individual component devices. All systems contain a combination of amplifiers and signal modification equipment bracketed by input and output elements. The number of practical system configurations can be generalized to a very few. The categories of systems are based primarily on the most common applications of audio.

Basic Audio System Principles

At the most abstract level any audio system consists of a device to initiate an electrical signal, an amplifier to process the signal in some way, and an output device to which the signal is delivered. Every practical audio system, however complex, merely expands this basic schema.

Audio systems consist of **strings of components**, the output of one unit connected to the input of the next until the end of the signal path is reached. At the end of the signal path, the audio signal is either stored for later replay or restored to an acoustic medium for immediate use. Some **subsystems** in audio processing involve side by side arrangements, where the signal path is divided to pass through two separate units at the same time. An audio signal can also be branched to more than one destination (e.g., sending the signal to more than one tape recorder or more than one loudspeaker at a time). Mainly, however, audio systems are chains of equipment, and the signal passes through each link in the chain in sequence. Understanding how to use an audio system means thinking in terms of steps in the path of the signal from one end of the system to the other.[1] (See Fig. 4-1.)

Basic Audio System and System Terminology

Refer to Fig. 4-2 as we go through this discussion. An audio system must have sources of sounds, a control center that receives the sounds and distributes them, and some means to reproduce the sounds. Along with the control center and the playback components, we frequently need ways to modify the sound or to execute some kind of automatic control over some aspect of the sound. We also need ways to store the sound. All of these functions are illustrated in Fig. 4-2.

The sound sources for the system are several: microphone, tape recorder, compact disc player, and synthesizer. All of these sources connect into the central mixing console. Mixing consoles are the main control elements in most professional systems. The console outputs run in several directions: main outputs to power amplifiers and loudspeakers and secondary outputs to effect devices.

[1] End-to-end and side-by-sidearrangements of audio components are not the same as series and parallel at the electric circuit level. Audio components are connected in parallel electrically, even if only two units are involved.

FIG. 4-1: BASIC AUDIO SYSTEM CONFIGURATION

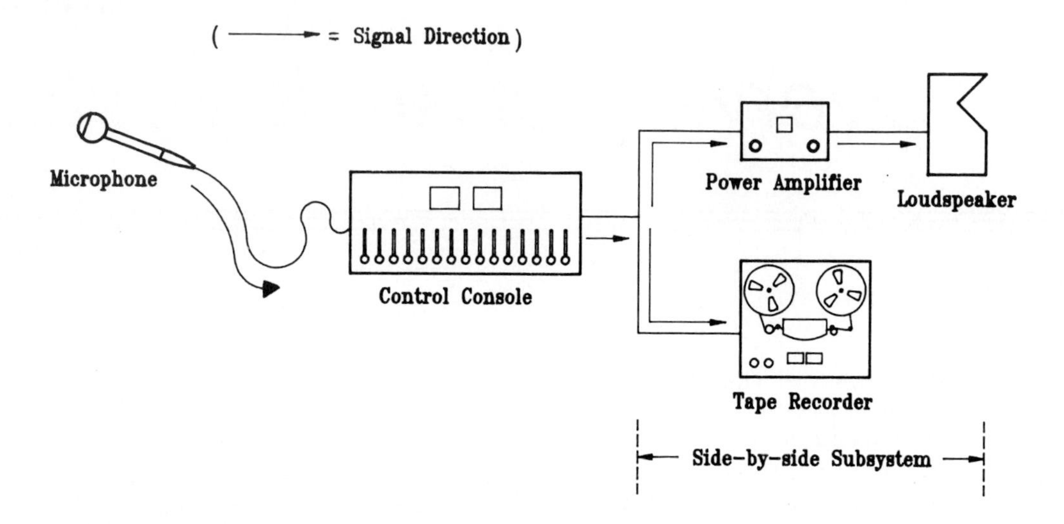

Note that components are connected end-to-end. The output of one element drives the input of the following element.

After the console, the signal path branches to two different units. One branch goes to a power amplifier and loudspeaker and the other branch to a tape recorder. The signal paths are side-by-side, but within each path the components are still placed end-to-end.

FIG. 4-2: BASIC AUDIO SYSTEM ELEMENTS

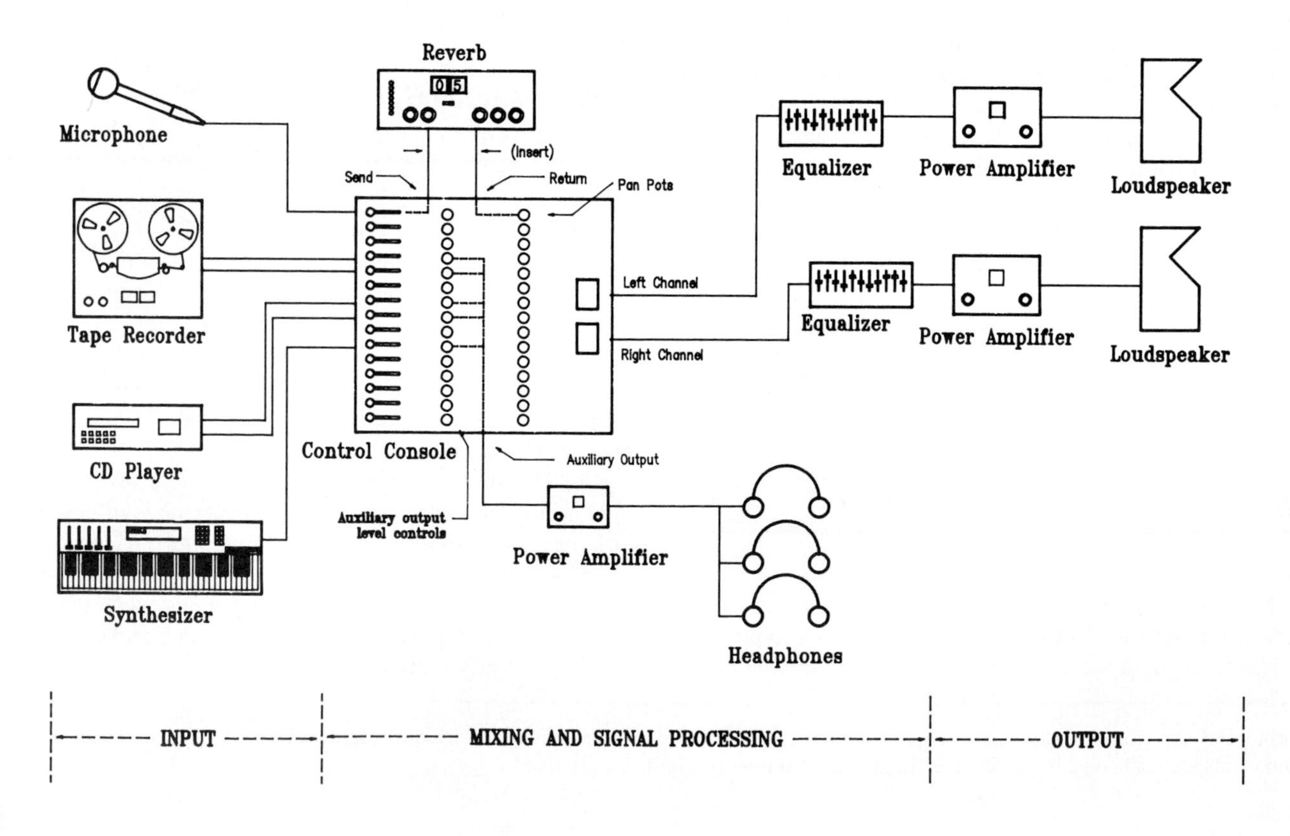

The main function of the console, the heart of a professional audio system, is **mixing**. Mixing, as the term suggests, means combining two or more signals into a composite output. For example, we might have actors speaking into the microphone, background music synthesized, and sound effects on the CD and the tape recorder. The console provides level controls that vary the amount of each signal that the final output contains. That output signal is referred to as a **mix**. Because most present-day audio systems are stereophonic, the mix is usually divided into two parts: the left channel and the right channel. Consoles provide controls that permit input signals from each of the sources to be assigned proportionally to the left and right channels. The proportional distribution may range from exclusive distribution to left or right to equal distribution to both sides. Placement of the signal between left and right extremes is called **panning**.

The outputs of the system go to power amplifiers and loudspeakers as a final destination. Between each console output and the power amplifier it drives is a piece of **signal processing** equipment. In Fig. 4-2, the device is an **equalizer**. The equalizer is something like a tone control on a home stereo system, except that the equalizer is capable of regulating any part of the audio spectrum, not just treble and bass. The equalizer, in this application, is to compensate for acoustic deficiencies in the listening space.

The console also provides secondary connection points called **inserts** and **sends**. Inserts permit an effect device to be inserted in the signal path of an individual part of the console. In Fig. 4-2, we have placed an echo device in the section that regulates the level of the microphone, which gives us the ability to echo the actors' voices without affecting the background music or the effects. The console provides a secondary output called the **auxiliary send**. The auxiliary send can collect the signals of some or all of the input channels and provide a mix of them to the auxiliary output. We use the send for **foldback** through headphones to permit the actors to hear their own voices and the background music and sound effects.

The system in Fig. 4-2 provides a good illustration of the ways in which audio equipment may be interconnected to form a system. The kinds of equipment and the arrangement of that equipment depend entirely on the purpose and function of the system. A number of standard component configurations exist that make up audio systems for particular applications. The applications and their corresponding systems include the following:

Public address
Acoustic reinforcement
Stage monitor
Broadcast audio
Studio recording
Background music/paging
Domestic music and entertainment
Film sound mixing
Theatrical sound scoring

Naturally, discussions of the various kinds of systems will be easier after we have a basis for understanding the role of each component within a system, so we shall hold the consideration of these various system configurations for later.

Basic Families of Audio Systems

All forms of audio equipment belong to one of two basic operational families: **analog** or **digital**. Analog equipment generates an electrical replica of an acoustic waveform and essentially keeps that replica of the original signal intact throughout the entire system path. (See Fig. 4-3.) Digital equipment produces a stream of numbers that represent encoded samples of the acoustic waveform. Enough samples are taken during each cycle of the signal for the system to be able to reconstruct a satisfactory replica of the signal waveform at the system output. (See Fig. 4-4.) Each family has its advantages and disadvantages.

Analog Equipment

The term *analog* suggests that one thing is like another. In the case of audio, the term means that an electrical waveform is like an acoustic waveform. An acoustic waveform, as we have seen, is variation in air pressure that changes in proportion to the energy level of a vibrating body. The air pressure builds to a peak of compression, then falls back to normal or continues to a peak of rarefaction. Similarly, an electrical wave representing that acoustic energy variation would build to a peak of positive potential and fall back to zero potential or continue to a peak of negative potential.

Analog audio is technically simple and relatively easy to process, and because of their simplicity, analog systems are cost-effective. Analog equipment can produce excellent audio quality, but its signal is always subject to some amount of **noise** and **distortion**. Noise is defined as extraneous energy that becomes attached to a signal. Distortion is an unwanted change in signal waveshape somewhere in the amplification process.

Noise can come from either inside or outside a piece of analog equipment. Internal noise, the result of random current flow in the circuits of an audio device, produces white noise, normally called **hiss**. Hiss is present in all amplifiers, but in well-designed equipment, hiss should be practically inaudible. External noise results from electric power lines and sometimes from radio frequency electromagnetic radiation, adding a noise component called **hum** to the original signal. Good analog equipment reduces hum by means of well-designed shielding and grounding. Usually, when hum appears in the output of a system, a shield connection is open at the input or

FIG. 4-3: ANALOG SOUND REPRODUCTION

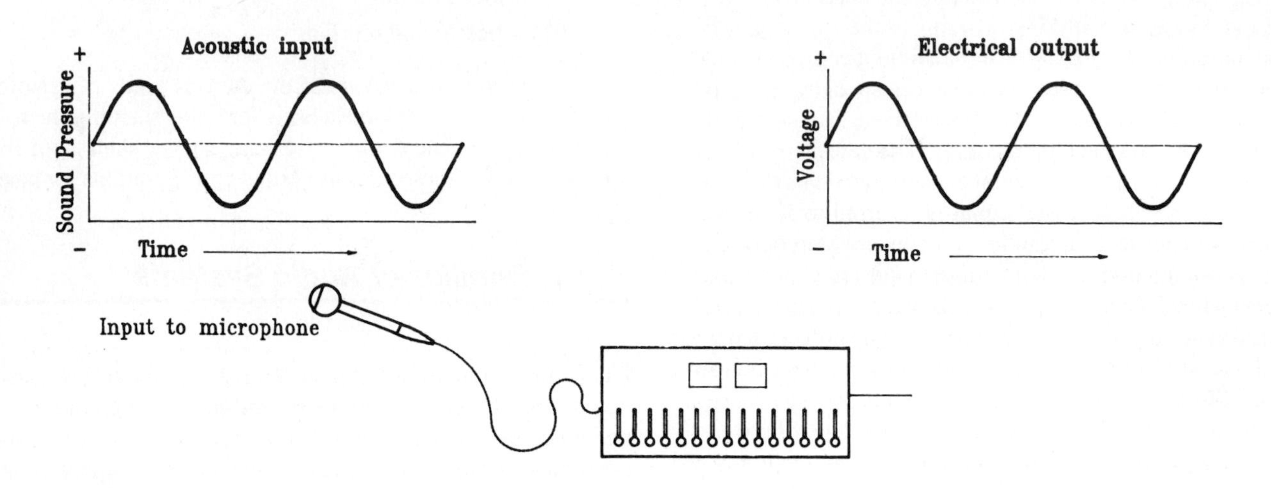

Pressure change in response to acoustic energy is sensed by microphone and converted into change in voltage and current. Amplifiers in console pick up output of microphone and amplify the signal. Ouptut of console remains an electrical replica of the acoustic input.

FIG. 4-4: DIGITAL SIGNAL

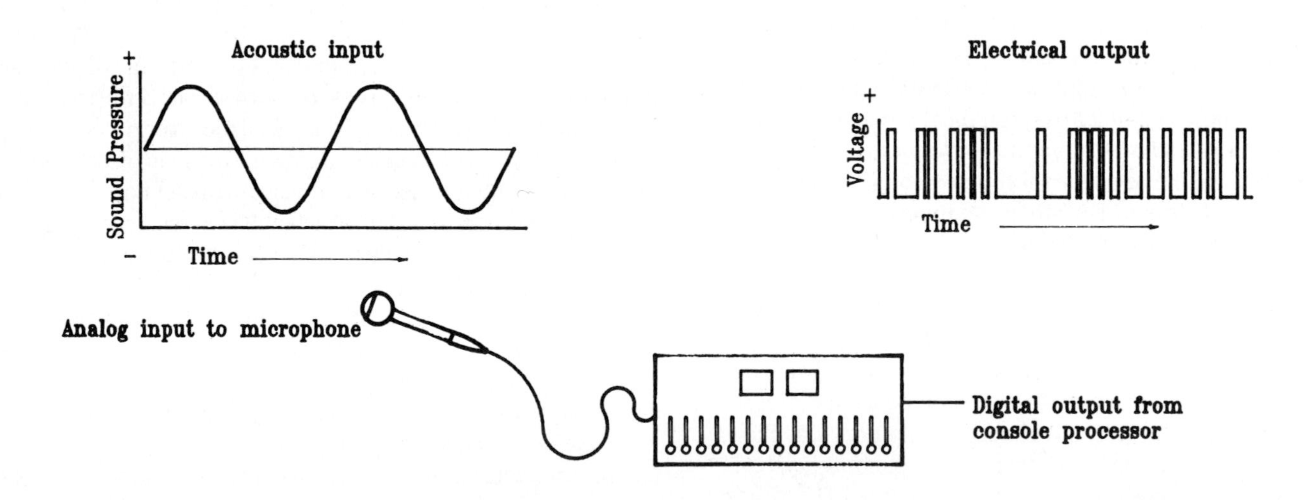

As in an analog system, the varying acoustical pressure is sensed by the microphone and delivered to the console input as an electrical replica of the acoustic signal. Inside the console, the signal is converted to an encoded representation of the signal. The encoded signal becomes the output of the digital device and is the input to the next device in the audio chain.

output of some component within the string of devices that make up the system.

Like noise, distortion can arise from internal or external causes. Internal signal distortion is caused by **nonlinear**[2] elements in the signal path and by interaction of high-frequency and low-frequency elements of the signal itself. The former (nonlinearity) is called **harmonic distortion**. Harmonic distortion is produced when some part of the signal handling system alters the waveshape from its original form. The latter (interaction of signal components) is called **intermodulation distortion**. Intermodulation distortion means that a high frequency tone and a low frequency tone interact to produce a third tone, usually the difference between the two frequencies. Because this third tone is not part of the original signal, it constitutes distortion.

Externally produced distortion occurs when radio frequency energy **modulates** the audio signal. In this form of distortion, a high-frequency signal rides the lower frequency audio signal, changing the instantaneous value of the audio signal but not its overall waveshape. Radio frequency distortion does not necessarily result from radio or television transmission. It can result from a variety of sources. One such source that is particularly troublesome in theatre is noise produced by electronic dimming systems. Electronic dimmers control current by switching on and off rapidly. As the dimmer changes state (switches from off to on) it produces a burst of radio frequency energy.

Digital Equipment

Digital audio equipment is considerably more complex than analog equipment. Digital processing requires that the input signal be converted into a series of "snapshots" of the waveform at tiny but successive intervals of time. The snapshots, called **samples** (see Fig. 4-5), must be converted into numbers, which are passed as data through the system. The process is similar to the way in which a computer handles information.

FIG. 4–5: SAMPLING AN ANALOG SIGNAL

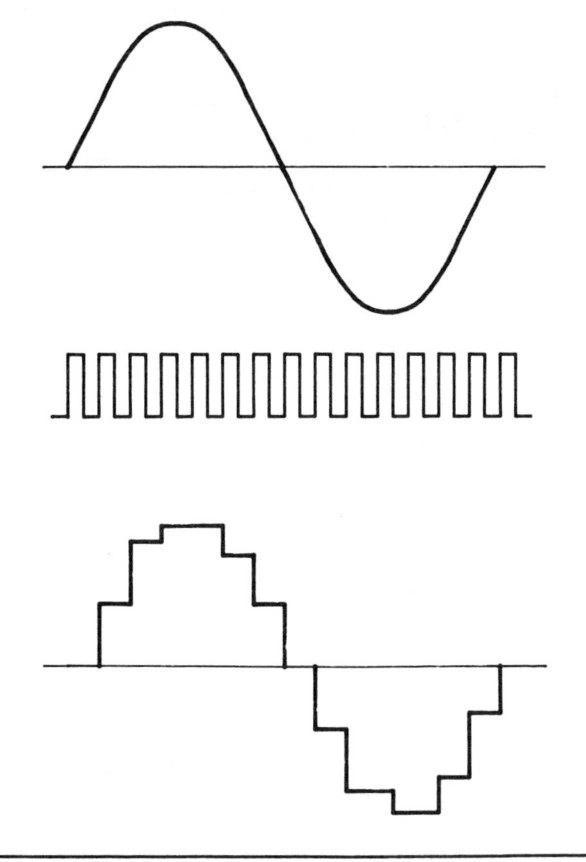

1. **The original signal is a sine wave. The waveform varies smoothly from peak to peak.**

2. **The strobe signal is a pulse that turns on and off the sampling circuitry. When the sampling circuitry turns on it grabs the immediate voltage level and holds it. The voltage remains at the last sample level until the strobe turns the sampling circuitry on again.**

3. **The sampler output is a stepped waveform that approximates the shape of the analog original.**

[2] The term *linear* implies that all portions of the output signal maintain a constant relationship to the input signal. For example, if an amplifier increases signal level by a factor of 10, all portions of the output waveform should be exactly 10 times greater in intensity than the input waveform. Non-linearity implies that some portions of the signal do not maintain a constant relationship to the input signal. Some parts of the output waveform could be 12 times the magnitude of the input signal; other parts might be only 3 times the intensity of the input signal. Obviously, the output waveform would not be identical to the input waveform.

In order for a digital audio system to reproduce sound accurately, two conditions must be met: first, the number of samples per second must be adequate to permit reasonably complete reconstruction of the original waveform, and second, the resolution (size of the largest encoding number) of the system must be adequate to capture the smallest significant variation in signal magnitude and polarity. Let's consider each of these conditions separately.

The number of samples per cycle necessary to recreate a good approximation of the original waveform is a function of the highest signal frequency to be reproduced. If too much of a waveform goes by without being sampled, the output will cease to resemble the input—a form of distortion called **sampling error**. The possibility of sampling error increases as the frequency of the audio signal increases. The maximum signal frequency at which a digital system can maintain any semblance of the original waveform is half the sampling frequency. At that rate, the audio signal will be sampled only once per half-cycle. This maximum reproducible signal frequency is called the **Nyquist frequency**. One sample per half-cycle returns information adequate to construct only the crudest reproduction of a sine wave and fails to provide any useful information at all about the partials of a complex waveform. Consequently, sampling rate should exceed 40 kHz in order to reproduce audio frequencies approaching 20 kHz.

Resolution refers to the ability of a system to capture minute changes in signal behavior. The ideal resolution would enable a system to reproduce everything that the human ear can hear. The resolution of a digital system is a function of the number of **bits** used to represent the numbers into which a signal sample is converted. Each sample of an audio signal has two primary *attributes*: **intensity** and **polarity**. Intensity is the magnitude of the instantaneous signal voltage; polarity reflects the positive or negative state of the instantaneous signal voltage.

The difference between one sample and the next may be large in some circumstances and small in others. A very small change may require the ability to encode a minute fraction of a volt. The smallest fraction that a digital system can record is the reciprocal of the largest number that the system can express. To understand the limits of a digital system, we must consider the way in which computers encode data.

Computers recognize only two electrical states: 0 or 1, encoded as "off" or "on." A two-value counting system is called a **binary** or base-2 system. Numbers greater than 1 can be expressed in a binary system in the same way that numbers greater than 9 can be expressed in a decimal (base-10) system—by shifting one place to the left. Shifting one place to the left raises the base number by one order of magnitude. The binary number 1 signifies 2^0. (Any base raised to the 0 power is always 1.) The binary number 10 (one-zero) signifies 2^1 (two to the first power, i.e., shifted one place left), or decimal two. Early computers used an 8-bit number, which meant that the largest number that the computer could handle was

$$2^7 + 2^6 + 2^5 + 2^4 + 2^3 + 2^2 + 2^1 + 2^0 = 11111111_{(binary)} = 255_{(decimal)}$$

An 8-bit number is called a **byte**. 0 through 255 equals 256 values.

Early attempts to digitize audio used 8-bit numbers, which meant that the smallest variation in signal intensity that could be recorded was 1/256 of the maximum signal voltage level. A smaller change was simply not recognized by the digitizing system. Many components of complex waveforms represent changes smaller than 1/256 of maximum intensity, so reproduction of timbre suffered somewhat in 8-bit digital systems. Originally, 8-bit systems were tolerated because higher resolution was simply too expensive, requiring a large investment in memory and logic integrated circuit chips. Chip prices have since fallen dramatically, and digital systems now represent signals using 16-bit numbers. A 16-bit number is called a **word**. Sixteen binary bits encode the decimal number 65,535. Thus, the smallest change in signal magnitude that can be recorded by a 16-bit digitizing system is 1/65536 of maximum signal amplitude, which is considered adequate to represent any significant attribute of a signal.

Data words representing encoded signal attributes are passed through the system. Various standard audio operations are accomplished by mathematical processes. Amplification, for example, becomes a matter of multiplying each data word by some specified factor. Fades become a matter of proportional scaling. At the output of the system, so long as nothing has corrupted the data, the system should be able to construct an acoustic waveform that is functionally identical to the signal picked up by the microphone. Induced noise and hum, hiss, and amplifier noise should not be a factor in the output of a digital system. A digital system should be completely immune to all the forms of distortion and noise to which analog audio is subject.

Digital audio is not completely free of problems, though. The fundamental problem is that of adequately describing the analog input. Sampling rate determines the ability of a digital system to describe attributes of an analog waveform; resolution determines its ability to encode those descriptions as data. Sampling error and inadequate resolution distort the signal in a digital system. Another form of distortion, also related to sampling, is known as **aliasing**. If any signal component exceeds the Nyquist frequency, not only does the waveform for that component fail to be sampled accurately, but it tends to generate the appearance of an incorrect frequency component within the signal. Signals above the Nyquist frequency are sampled irregularly, meaning less than two samples per cycle. Irregular sampling produces the effect of a spurious, low-frequency component. (See Fig. 4-6.) The spurious frequency is an **alias** of the true frequency. Aliasing is particularly insidious because significant harmonics of high signal frequencies often exceed sampling frequency.

FIG. 4-6: ALIAS DISTORTION

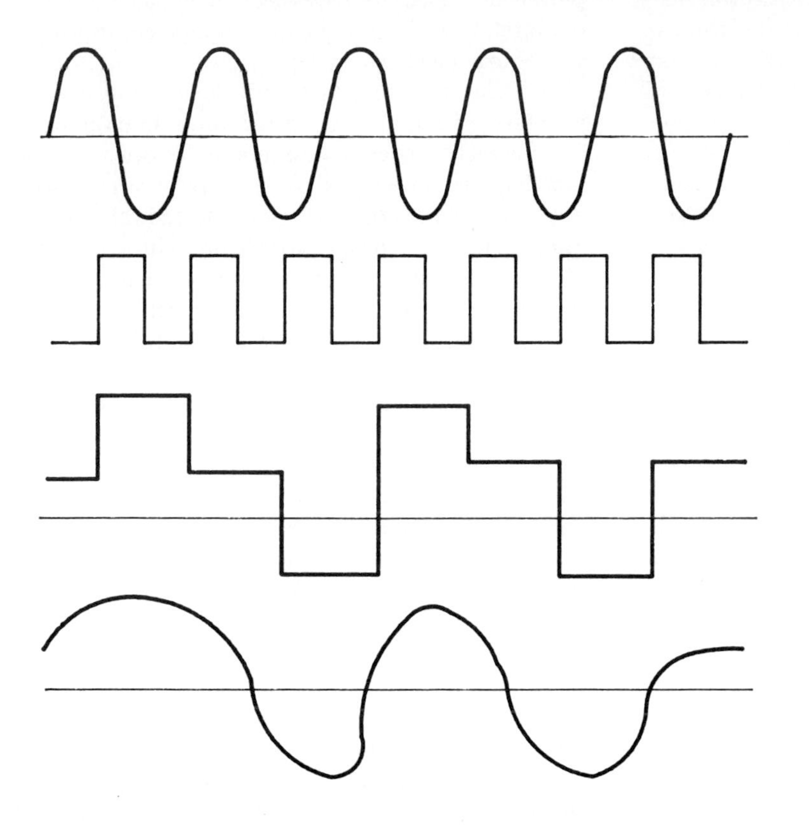

1. **High frequency signal component above Nyquist frequency.**
2. **Strobe signal. Note that strobe samples signal frequency less than twice per cycle on average.**
3. **Resultant sample.**
4. **Approximated output waveform from device that converts digital signal back to analog. Note that waveform is approximately 1/2 the frequency of the original signal frequency (i.e., an alias of the original signal component.)**

The solution to the problems of sampling error and aliasing might appear obvious: Increase the sampling rate so that no signal component could ever be higher. Unfortunately, high sampling frequency makes equipment outrageously expensive. As with resolution, the cost of hardware sometimes makes less than desirable alternatives tolerable. The cheaper alternative in the case of sampling rate is to limit the highest frequency the system is permitted to encode. Most digital systems include a filter designed to remove any frequencies higher than a specified maximum. Thus, aliasing is eliminated and some minimum acceptable number of samples per cycle is maintained.

Additional discussion of digital signal handling is found in Chapter 10 as part of an examination of the process of digital signal storage.

Functions of Audio Equipment

Both families of equipment, analog and digital, may be divided by function into four fundamental categories: transducers, signal processors, storage devices, and signal generators. The various pieces of individual equipment that make up audio systems fall into one of these categories. Sometimes subcomponents of equipment will belong in various categories. For example, tape recorders are signal storage devices, but the heads that write to and read from the tape are transducers. As we explain each function, we shall give a brief, preliminary description of the kinds of equipment that fall into that category. If the precise nature of the kinds of equipment is not immediately clear, recall that a detailed explanation of each kind of equipment is provided in the chapters that follow.

Transducers

Transducers are devices that convert energy from one form to another—from acoustic to electrical, for example. Audio uses a number of different kinds of transducers. Transducers constitute the input and the output of most audio systems. Microphones receive acoustic energy as input and in turn put out an electrical analog of that energy. Loudspeakers receive electrical energy and convert it to acoustic energy. Transducers are used in the storage (recording) of audio signals. Phonograph disc cutters receive electrical input and use it to cut a groove into the plastic surface of the disc, converting the signal into potential mechanical energy. On playback, the phono pickup in a tonearm converts mechanical energy back into electrical energy. Tape heads convert electrical energy into magnetic fields to be stored on and retrieved from audio tape. Small light sources convert electrical energy into light energy to expose the sound track on motion-picture film; a light source and a photoelectric cell reconvert the light energy into electrical energy when the film is run through a projector.

A transducer is generally analog. The output of a transducer is variation in energy corresponding to energy variation at its input. A transducer that must provide digital information to a system requires a converting device. Also, at the output of a digital system, the signal must be converted to analog format before a transducer can convert the signal back into acoustic energy. Analog-to-digital converters (ADCs) are used at the input to a digital system and digital-to-analog converters (DACs) are used at its output.

Signal Processors

Signal processors alter one or more parameters[3] of a signal. The controllable audio signal parameters include intensity, frequency, bandwidth, complexity, reverberance, and duration. Almost any piece of audio equipment could be classified as a signal processor, as even basic amplification meets the definition of the term. In common practice, however, the term designates those kinds of equipment that alter the signal in some way other than overall increase in signal level. The main forms of signal processing are equalization, filtering, compression, delay, and frequency shifting.

Equalization (EQ) and filtering alter the spectral quality of a signal. Equalizers and filters are very similar devices. Both target a band of frequencies and affect the intensity of that band. EQ can either **boost** (increase intensity) or **cut** (reduce intensity) the band of frequencies; filtering only cuts the band. The purpose of a filter is to *remove* the band as completely as possible from the signal output; EQ is meant only to *adjust* the intensity of the band with respect to the rest of the audio spectrum. EQ does not remove the band of frequencies.

Compression affects the **dynamic** range (maximum and minimum loudness level) of an audio signal. Devices in this category—compressors, limiters, and expanders—serve as automatic level controllers. Compressors reduce the dynamic range of an audio signal; limiters impose an absolute maximum output level on the signal; and expanders increase the dynamic range of an audio signal.

All compression devices alter the *rate of change* in output signal amplitude relative to input signal amplitude. Chapter 8 will explain more completely, but the essential idea is that, up to a predetermined *threshold*, output amplitude exactly follows input amplitude. Beyond threshold, output amplitude holds some relationship other than unity with respect to input amplitude. Compressors reduce output amplitude *above* threshold; expanders reduce output amplitude *below* threshold. Limiters *clamp* output at threshold—that is, prevent the signal from ever becoming louder than the threshold level. Uses for compression equipment in theatre sound will be examined in Chapter 8.

Delay and frequency-shift devices fall into a general category of equipment called **time-domain processors**. Time-domain processors essentially lengthen the amount of time required for a signal to pass from one point to another. The most natural effect of such time control is signal delay. The simplest use of delay is to produce echo and reverberation. Most of the delay devices can produce pitch-shift effects ranging from a slight vibrato to psychedelic, frequency-twisting distortion of a signal. Devices specifically designed as pitch shifters (often called **harmonizers**) take advantage of the natural relationship between frequency and time to effect a constant alteration of the pitch of a signal. For example, a pitch shifter can be used to bring an out of tune instrument into accord with the rest of a group (if the instrument is recorded on its own track of a tape, of course.) Some pitch shifters can also work with a keyboard to provide multiple pitches from the same signal, for example, turning one singer's voice into a trio or quartet.

Signal processors come in both analog and digital versions, although most of the time-domain processors are digital. Analog processors operate directly on the shape of the signal waveform; digital processors alter the numbers that constitute the signal samples. Analog devices are comparatively simple and reliable. Controls for analog equipment are usually in the form of sliders and knobs, and the audible effect on the signal usually bears a direct relationship to the amount of change in a control setting. Digital devices, because they process numbers, operate purely by mathematical operations on the data samples. The user does not have to perform the math, but simply selects a function and a value of rate, frequency, or level. The circuitry of the device handles the math. Controls on digital processors may be sliders and knobs, but may also be numerical keypads and command buttons. The relationship between control settings and audible effect tends to be more abstract.

Signal Storage

Signal storage devices exist in both digital and analog formats, and provide a means to preserve a relatively permanent record of a signal. Storage devices include tape recorders, phonodisc cutters and players, compact disc writers and readers, and (with the introduction of the computer into the realm of audio) any of the standard computer data storage devices.[4] Most storage devices, whether digital or analog, require

[3]The term *parameter* comes from linguistic roots that mean, in essence,"something to be measured." Modern usage has adapted the term to signify a variable that will assume a certain value, dependent on immediate circumstances. All of the measurements with which we describe the characteristics of an audio signal are parameters.

[4]Computer storage formats can record and preserve digitized sound. Because digitized sound requires immense amounts of storage space to hold even a few seconds of sound, only the largest data storage devices are practical for recording digital sound.

transducers since the form in which energy must be stored is usually magnetic, mechanical, or optical rather than electrical.

Signal Generators

Signal generators do not require an original acoustic signal. They create an electrical waveform that has characteristics similar to those that might be encountered in an acoustic waveform; however, signal generators may also be designed to create either pure tones or signals containing dominantly nonlinear elements such as no natural sound generator could ever produce. Examples of signal generators include electronic organs, electric keyboard instruments, and the whole range of music synthesizers, as well as laboratory and engineering equipment that is designed to produce very precise waveforms for the testing and measurement of audio and other electronic devices. The most recent versions of any of these devices may be partly or completely digital—especially electronic keyboards and synthesizers. Older devices were invariably analog.

Chapter 5
Loudspeakers

All practical audio equipment is based on the theoretical considerations we have discussed in previous chapters. To understand the various kinds of equipment, we begin by studying individual components and work our way to the point where we can examine components as they combine to form systems. The study begins with loudspeakers—the output components of an audio system. Because loudspeakers utilize basic acoustic and electrical principles, we can begin our study of audio equipment on familiar ground.

Basic Principles of Loudspeakers

With a few exceptions, loudspeakers are similar to electric motors. Both the loudspeaker and the motor both have coils through which current flows, setting up a magnetic field that forces some other part of the device into motion. The motor's motion is rotary; the loudspeaker's is reciprocal (i.e., a limited back-and-forth motion).

The purpose of a loudspeaker is to set the volume of air in its immediate neighborhood into motion at frequencies identical to those contained in the electrical audio signal. Virtually any loudspeaker, no matter how small, can project sound over a surprisingly large space. A tiny transistor radio with the volume set nearly full can be heard a long distance away, but one hears mainly the frequencies nearest 3000 Hz, which is the reason a small unit sounds so "tinny." A loudspeaker that small simply cannot move very much air. In order to project a full frequency range over a large space (especially out of doors), large loudspeakers, sometimes arrayed in clusters, are required. Most loudspeakers used for professional purposes must move huge volumes of air and require large amounts of power in order to do so.

Any loudspeaker consists of the following basic elements: a membrane designed to excite the surrounding air at acoustic frequencies, a motor element to drive the membrane, and the mechanical framework and structural supporting elements required to house and position the membrane and motor. Depending on the particular loudspeaker design, ancillary elements such as radiators and/or resonators may also be required. Loudspeakers are mainly of three kinds: cone, horn/driver, and electrostatic. Closely related are such variants as ribbons, dome radiators, and piezoelectric devices.

Cone Loudspeakers

Figure 5-1 shows a basic **cone** loudspeaker. The device is made of a metal frame that supports a cone of rigid, treated paper or synthetic material connected to a movable diaphragm. The diaphragm is driven by a coil of wire suspended in the field of a permanent magnet. The cone has concentric corrugations at the outer edge to allow it to flex easily. Dependent upon the lowest frequency to be reproduced and the volume of air to be moved, the cone must present a certain amount of surface area to the surrounding air (the larger the surface area, the lower the minimum frequency for a given volume of air). Signal current from an amplifier flows through the coil producing a magnetic field. The coil field reacts with the field of the permanent magnet (either being attracted by it or repelled from it). Because the coil is fixed to the diaphragm of the loudspeaker, the diaphragm is set into motion, in turn moving the cone. The cone compresses and rarefies the surrounding air, delivering a pulse of acoustic energy. This type of unit is known as a **dynamic** loudspeaker.

Cone loudspeakers are made in a full range of sizes, from 30-inch bass units to 2- and 3-inch devices for subminiature radios and tape players. Most cone speakers are round, but oval units (mainly used in automotive audio systems) also exist. Large speakers designed to handle low frequencies are collectively designated **woofers.** Woofers generally range in size from 10 to 18 inches with 12 inches and 15 inches being the most commonly used size. The family of cone loudspeakers also includes full-range units (loudspeakers that reproduce most of the audible spectrum reasonably well).

Early in the history of audio all loudspeakers were paper cone. Low-frequency units were called woofers, midrange elements were "honkers," and small, high-frequency

FIG. 5-1: CONE LOUDSPEAKER

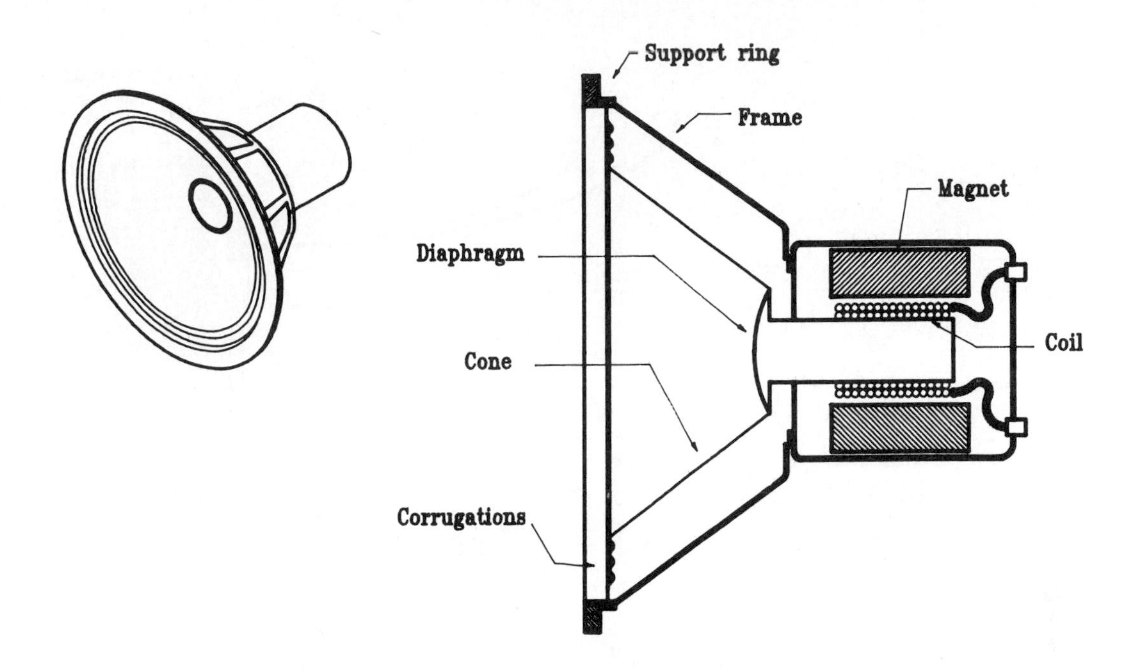

speakers were known as "tweeters." Cone speakers survive today mainly as low-frequency reproducers. The mid- and high-range functions have been taken over by more sophisticated types of transducers. (For these later units only the expression "tweeter" has remained in use to designate any form of high-frequency reproducer.)

Horn/Driver Loudspeakers

Reproduction of mid- and high-frequency energy is usually accomplished by devices called **horn-loaded drivers**. (See Fig. 5-2.) The driver uses a diaphragm to compress air within a small, short tube. The diaphragm, made of lightweight metal or plastic, is driven electromagnetically using a strong, permanent magnet and a coil of fine wire (as in a cone loudspeaker.) By itself the driver is extremely inefficient and produces only a weak, thin sound. Drivers are usually fitted with **horns**, flared radiators, that dramatically increase the efficiency of coupling to the surrounding air. The length and shape of the horn determine the **dispersion angle** through which the loudspeaker radiates. The most common horn shapes are exponential and conical.

Horns make use of the natural fact that high frequencies assume directional characteristics more easily than do low frequencies. When mid- to high-frequency sound energy is pushed through a tube or throat equal in length to a significant fraction of the longest wavelength to be reproduced, the radiational field of the loudspeaker will be essentially **unidirectional**. The dispersion angle will be determined primarily by the length of the throat and secondarily by the flare of the horn. Horns are designed to create radiational angles from as little as 15 degrees (relatively long throat and little flare) to as much as 120 degrees (relatively short throat and wide flare). Throat length is always relative to the frequency band to be reproduced, however. Honkers must have long horns in order to create a narrow field pattern whereas tweeters need only short throats. In order to smooth the field pattern horns are sometimes designed with cellular structures filling the mouth of the flare. Such a unit is called a **sectoral** horn, and the individual sectors form horn subsystems within the overall horn.

Wide-Angle, High-Frequency Radiators

Synthetic materials technology has benefited the loudspeaker industry by developing artificial substances that can reproduce short wavelengths at high levels of energy. Tweeters have been made in the form of domes or ribbons (rather than as horns or cones). (See Fig. 5-3.) These units produce excellent high-frequency energy through an angle approaching 180 degrees. Dome or ribbon radiators tend to be used in smaller systems (monitor speakers and home entertainment units) rather than in large concert reinforcement systems.

Electrostatic Loudspeakers

Although little used in professional systems (and, indeed, not used much at all at the present), a completely different form of loudspeaker is the so-called **electrostatic** loudspeaker. Electrostatic units are based on the principle of

FIG. 5-2: HORN-LOADED DRIVER

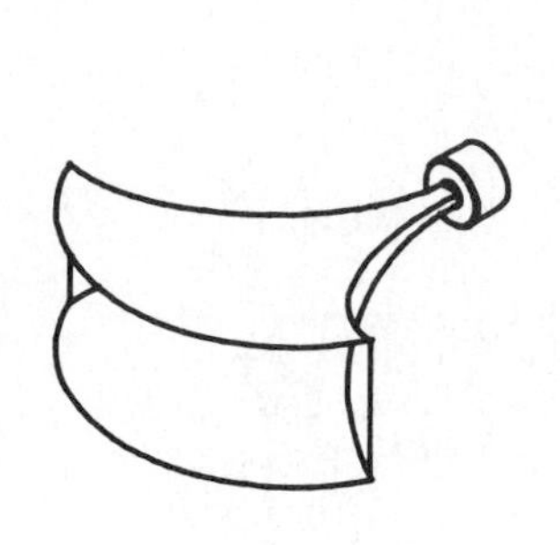

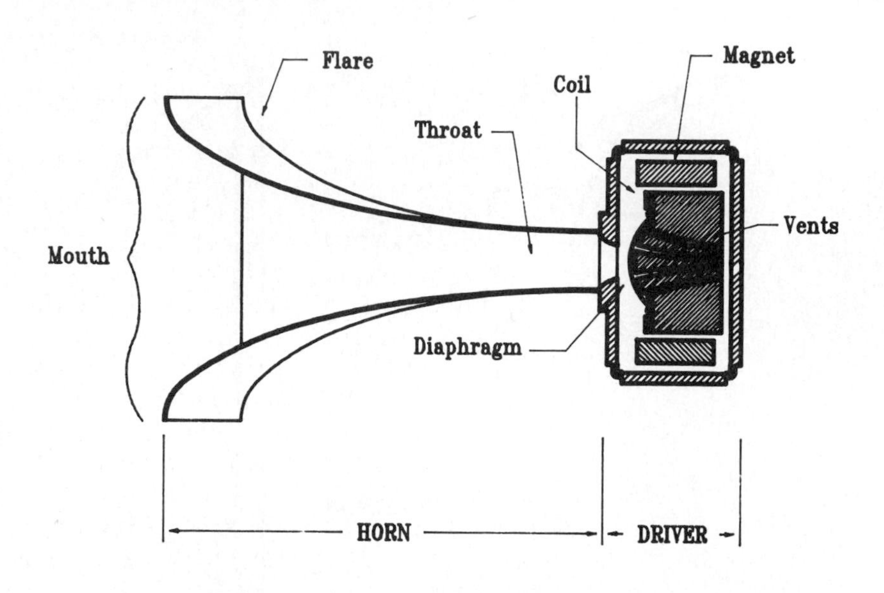

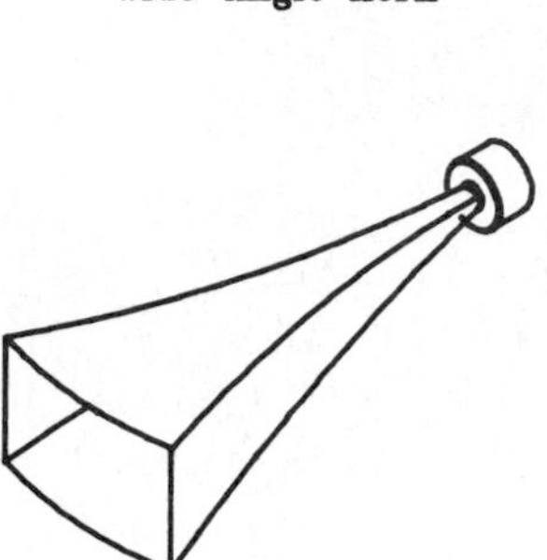

The driver compresses air in the throat of the horn. The pressure wave expands and excites the atmosphere directly in front of the mouth of the horn. The flare of the horn acts as a radiator to couple the energy produced by the driver to the surrounding air.

The width of the flare determines the coverage angle of the horn. A wide mouth and a short throat produce a wide beam; a narrow mouth and a long throat produce a narrow beam. Choosing the **appropriate coverage angle is similar to choosing the proper focal length for the throw and coverage of an ellipsoidal reflector spotlight.**

the mutual attraction of unlike charges and repulsion of like charges. In effect, these devices are large capacitors in which one plate is held at a constant charge while the charge of at least one other plate is varied in proportion to the audio waveform. As the relative charge between the two plates changes, the plates move toward or away from each other. If the plates are made sufficiently large, their movements will set air into motion. A diagram of an electrostatic speaker is shown in Fig. 5-4. Electrostatic loudspeakers are generally best as high-frequency reproducers.

Loudspeaker Enclosures

In order to achieve the most efficient transfer of energy to the air, low-frequency loudspeakers need additional support from a well designed **enclosure**. The enclosure serves two purposes: to modify the **resonant frequency** of the loudspeaker; and to suppress or vent the **backwave** (the acoustic energy from the rear of the cone).

All loudspeakers tend to respond to one frequency, known as the resonant frequency, more easily than to any other. A loudspeaker's frequency response curve displays a noticeable peak at resonance while lower frequencies drop in intensity, quickly falling below useful level. When a loudspeaker is used as the driver element in an enclosed system, the enclosure is tuned to force resonance to a lower frequency, thereby extending usable response. Tuning is a function of size, shape, and the way in which the enclosure treats the backwave.

An enclosure which completely suppresses the back wave is called an **infinite baffle**. One form of infinite baffle is known as an **acoustic suspension system**. Acoustic suspension systems seal the enclosure and use the trapped air as a tuning element. **Vented** enclosures redirect the backwave to the front of the cabinet, using it as a tuning element. The most common type of vented enclosure is the **folded horn**, which uses interior baffles to lengthen the backwave path, thereby bringing the backwave into phase with the frontwave. The size and shape of the **port** that vents the enclosure determines the tuning of the system.

A loudspeaker enclosure must distribute the energy of the loudspeaker into a desirable pattern of radiation. All sound sources tend to radiate spherically; therefore, the first priority in the design of an enclosure usually is to cut down the angle

FIG. 5-3: HIGH FREQUENCY RADIATOR

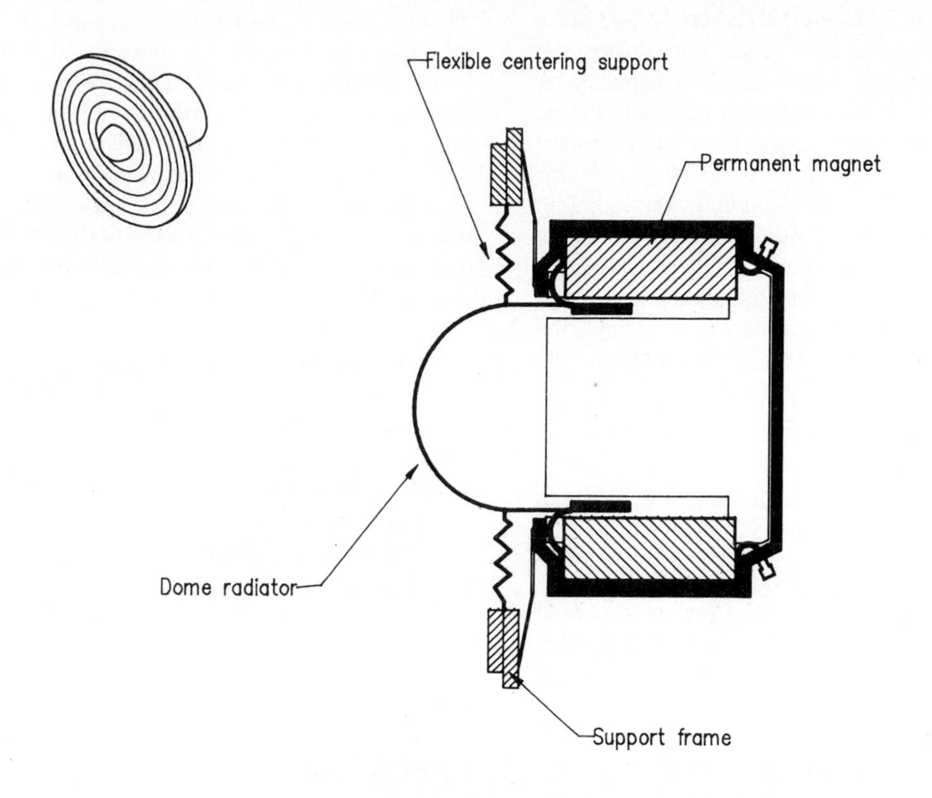

FIG. 5-4: ELECTROSTATIC LOUDSPEAKER

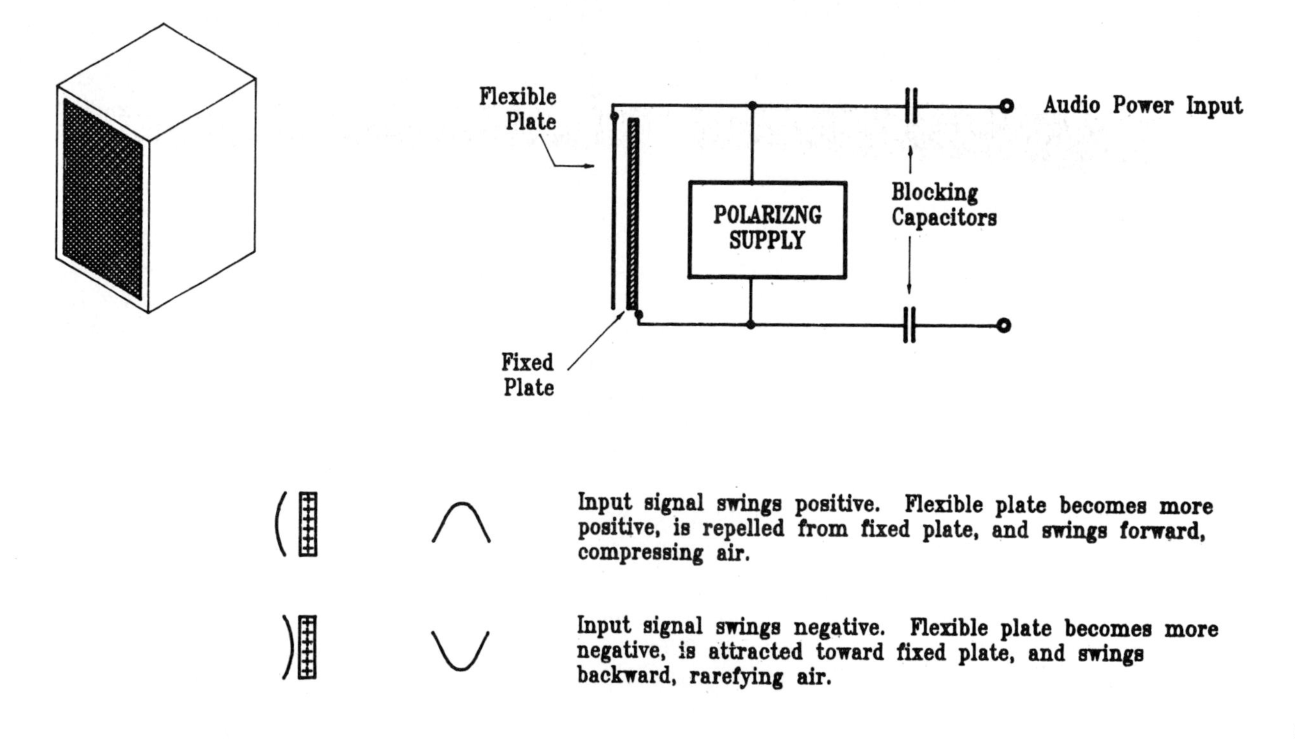

into which the unit radiates, and this can be done in any of several ways, some more efficient than others. If we place the loudspeaker at the center of a large, flat surface, the effective radiating area of the unit becomes a hemisphere. (The flat surface blocks radiation to the rear side of the loudspeaker and redirects that energy toward the forward hemisphere.) Placing the speaker in the perpendicular intersection of two large, flat planes reduces the radiating angle to one quarter of a sphere. Adding a third flat plane oriented at 90 degrees to each of the other two surfaces restricts the loudspeaker's energy to a radiating angle of one eighth of a sphere. With each decrease in radiating angle the sound pressure level increases because the energy of the speaker is concentrated into a smaller field angle.[1]

Loudspeaker Systems

Loudspeaker systems are made up of various individual components, assembled to fit the requirements of a particular kind of application. Most systems consist of one or more bass units and a group of mid- and high-frequency elements combined. Some more elaborate systems separate the mid from the high frequencies. Some kinds of systems are made up exclusively of cone loudspeakers; others contain both cone and horn/driver elements.

Crossover Networks

Loudspeakers perform best, as the preceding descriptions of low-, medium-, and high-frequency units suggests, when required to handle only a small band of frequencies. Devices called **crossover networks** divide the audio spectrum into low-, mid-, and high-frequency bands. A crossover network contains **band-pass filters**. Given an input containing energy across the full audible spectrum, the output of a band-pass filter will contain only some desired range of frequencies. Each element of a loudspeaker system must have its own filter designed to pass exactly the part of the spectrum that the speaker is to reproduce. A crossover network is the assemblage of filters required to direct energy, as needed, to the elements of a loudspeaker system.

Crossover networks may be either *active* or *passive*. Passive crossover networks contain only the inductors, capacitors, and resistors needed to make a band-pass filter. An active crossover network uses amplifiers (transistors or integrated circuits) in combination with filters. Passive crossovers are placed between the power amplifier and the loudspeaker system, and the components of the network must be capable of carrying high current. An active crossover, on the other hand, is normally placed *before* the power amplification section. Because the frequency division occurs before power amplification, each loudspeaker element in the system must be driven by a separate power amplifier. Systems using active crossovers and multiple power amplifiers are bi-amplified (if the system is divided into two frequency bands) or tri-amplified (if the system is divided into three frequency bands.) Passive crossovers usually divide the spectrum into two bands only, but three-way systems are possible using passive crossovers.

System Configurations

Generally, loudspeaker systems are assembled in one of three ways: as an integrated package contained in an optimized enclosure, as a particular kind of integrated unit called a **column radiator**, or as a **cluster** of separate units.

Integrated units may be two-way or three-way systems. All integrated systems contain one woofer as the low-frequency transducer. For two-way loudspeakers, the tweeter may be a cone, a dome, or a small horn/driver unit. The tweeter element in a two-way system handles all of the mid and high frequencies. In three-way systems the mid frequencies are reproduced by a cone or a horn/driver unit, and high frequencies are handled by a small horn/driver or by a dome tweeter. Integrated units normally use passive crossover networks that are built into the enclosure and optimized for the system. (Some manufacturers make the individual components accessible for bi- or tri-amplification.) Integrated systems are commonly used as stage speakers for theatrical productions, as loudspeakers for electronic musical instruments, and as studio monitor loudspeakers. Some larger systems (like the classic Altec A7) may use horn/driver elements outboarded on top of the bass bin but are still essentially integrated units.

Column radiators are two- or three-way loudspeaker systems that utilize an array of elements to form a beamlike dispersion pattern. The array is always linear. For example, such a system might contain six woofers and six tweeters. The six woofers are arrayed one above the other. The tweeters, also placed in a vertical array, are set to one side of the woofer array. The arrangement of the elements produces a radiation pattern that is wide in one dimension horizontally and narrow vertically. Column loudspeaker elements are usually relatively small, physically. An array of small elements can set as much air into motion as can a single large loudspeaker. Column radiators are a relatively inexpensive way to provide a beam of sound that can be focused toward a particular section of a room.

Loudspeaker clusters are usually the most flexible and versatile of all the ways in which loudspeaker systems can be

[1]Reduction xof the field angle into which a loudspeaker radiates is roughly analogous to using longer lenses in ellipsoidal reflector spotlights. A 500 watt 6 x 9 at 50 feet from the stage will produce a circle of light nearly 38 feet wide, and the light will be extremely dim. A 500 watt 6 x 22 placed at the same location will produce a circle of light 10 feet wide, and the light will be bright. Both instruments radiate the same amount of light energy, but the longer lens concentrates the energy into a smaller field.

assembled, but they are also the most expensive. The components are purchased separately, each selected for its particular characteristics with respect to the space in which it is to be used. Crossovers, mounting hardware, and everything else needed to make the system operational must also be specified and purchased separately. Clusters are most often used in fixed reinforcement situations and in reinforcement for large concert performances. For fixed installations, the clusters are mounted in specially designed parts of the auditorium; for concerts, the usual mounting method is on towers made of scaffolding. Clusters for large spaces (huge sports arenas or outdoor locations) may well involve batteries of elements—that is, a bank of several woofers and multiple horn/driver units for each channel of the reinforcement system. Cluster systems used in demanding situations are almost always bi- or tri-amplified. Electronic crossovers used for large clusters may also contain phase correction, level setting controls, and dynamic limiting and signal attenuation to linearize and optimize performance of the cluster.

Cluster Installation and Balancing

Once the basic elements of a cluster are acquired, then a crew of specialist technicians is needed to install all of the units and align them correctly. Output of the cluster is carefully checked using special testing devices to ensure that signal level is balanced throughout the full audio spectrum, that all components of the signal are in phase, and that the directional characteristics of the cluster are as specified in the system design.

Instruments called **real-time analyzers** are used to check the spectral balance. The RTA reads acoustic energy in bands centered on a group of frequencies specified by the International Standards Organization (ISO). The ISO frequencies are set for octave, half-octave, or third-octave measurements. Audio input to the cluster must be equalized so that energy is approximately the same at all ISO centers.

Other instruments called **sound pressure level meters** are used to check the dispersion pattern of the cluster. The SPL meter is carried to various different parts of the space, reading sound intensity at each location. Noting where intensity begins to decline and plotting the angle to the loudspeaker reveals the actual dispersion pattern of the cluster. An RTA can also be used toward the sides of the field pattern to detect aberrations in frequency response.

Time Alignment

One aspect of the cluster balancing process deserves particular consideration, especially as the practice is useful for integrated units as well (though not practical to carry out in the field since for integrated units the process must generally be done at time of manufacture). That aspect is the correction of phase misalignment among elements of the loudspeaker system. The process is known as **time alignment**.

Whenever multiple loudspeakers are combined, the fact that a composite sound will be produced by two or more vibrating elements is inescapable. If any one of the vibrating elements of the system is not precisely in phase with the other elements, some degree of distortion will result. Ideally, the distance from each individual vibrating element to the listener's ear should be identical. When disparities in distance occur, small phasing errors can result. Time alignment is the process of correcting such problems. Time alignment may be carried out, either by physically aligning the vibrating elements so that point of origin in space is the same for all bands of the system output (difficult, to say the least) or by introducing small correcting delays into the various crossover bands so that pressure variations generated by each element of the loudspeaker system are synchronized.

Cables and Connectors

Connectors and cables used to connect power amplifiers to loudspeakers should, ideally, be uniquely identifiable as loudspeaker-related items. Such identification may be accomplished either by use of unique connectors in each different part of an overall audio installation or by clear and highly visible labeling. In the first method connectors for loudspeakers would be physically different than those for microphones or line-level devices. In the second method the same kinds of connectors might be used throughout the audio system, but the connectors would be clearly labeled. The first method is more characteristic of multiuse facilities that keep a stock of in-house equipment that is reused in a different way for each staged performance. The second method probably works better for companies that carry one show for a long period of time or that travel with a production. (If a microphone cable breaks, for instance, a cable from some other part of the system could be substituted with merely a change in labeling, whereas in the first method, an entire stock of replacement cables would have to be carried as backups.)

Cables for loudspeakers must be sized appropriately to guarantee adequate current delivery over the distance between power amplifier and loudspeaker. In other words, low resistance is essential. Because cable resistance is a function of cross-sectional area and length, a (relatively) large diameter conductor is advisable, even for moderate cable runs. For stage purposes, loudspeaker cables should be extremely flexible (i.e., stranded conductors), jacketed with a tough exterior casing (usually one of the newer synthetics), and buffered inside with string fibers. At the entry of the cable into any connector, a strain-relief device should be clamped to the jacketing so that no tension is placed directly on the conductors.

Connectors in common use for loudspeaker cabling include Twistlok devices, XLR-type connectors, and phone plugs. Any of these are satisfactory for the purpose, but all are used, either in other parts of the audio system or in other forms of stage electrical systems. Where idiot-proofing is essential, connectors should be planned so that attachment of one kind of device into the circuit for some other sort of signal is virtually impossible.

Chapter 6

Microphones

Loudspeakers convert the audio signal from electrical to acoustic. At the other end of the audio chain, microphones serve to transduce acoustic energy into the electrical form required by the audio system. For many years microphones were almost the only means of input to an audio system; they, or their close relatives, are still the only means of acquiring signals from acoustic events. Microphones are, essentially, the inverse of loudspeakers, and they operate on the same principles. (In fact, inexpensive intercommunications systems use the loudspeaker as a microphone.)

Basic Types of Microphones

The earliest microphones were either crystal or carbon devices, both of which worked on the principle that acoustic pressure against the substance created a change in electrical behavior. For the crystal, pressure creates stress in the crystalline structure, which forces a difference in potential to appear across one dimension of the substance. The resulting variation in potential can be amplified as an audio signal. Carbon is a substance that changes its resistance under variation in pressure; thus, a current flowing through the carbon element of a microphone varies in magnitude following variations in acoustic pressure. Neither carbon nor crystal can reproduce acoustic input with fidelity. Both substances respond better to low frequencies than to high frequencies. The frequency response typical of carbon and crystal microphones was a range of 200 to 8000 Hz.

Dynamic Microphones

The first microphones to achieve reasonably high quality sound pickup were **dynamic microphones**. (See Fig. 6-1.) Dynamic microphones, like dynamic loudspeakers, use a coil suspended in the field of a permanent magnet. In the case of the microphone, the fluctuations of a tensile membrane in response to acoustic pressure variations drives the coil back and forth in the field of the magnet. In effect, a dynamic microphone is a small electric generator. As the coil moves within the magnetic field, a current is induced in the coil. The direction and magnitude of the current are proportional to the direction and rate of movement of the diaphragm (membrane) in response to the fluctuations in the acoustic field.

Until recently, dynamic microphones were not capable of reproducing very low or very high signal frequencies, although they were a great improvement over carbon and crystal microphones. Early in the development of microphones, metals that could respond to high frequency were not strong enough to carry the mass of the coil and coil support required to generate a sufficient current; nor were they able to follow extremely slow changing (low-frequency) pressure variations. The first dynamic microphones had a frequency response range of approximately 100 to 12,000 Hz. The development of newer technology in plastics and metals extended the range to near the limits of the audio spectrum. Dynamic microphones at present have an average frequency response of 50 to 18,000 Hz.

Ribbon Microphones

Before wide response dynamic microphones became available, other forms of acoustic pickup were developed in an effort to improve frequency response. One of the most successful products used a thin ribbon of metal as the transducer element. **Ribbon microphones** possessed excellent sensitivity over a wide range of frequencies and sound pressure levels. The ribbon, which served both as diaphragm and field coil, was suspended vertically between the two poles of a permanent magnet, and, when vibrated, produced a current. Where the dynamic microphone responds to change in pressure applied to the surface of the diaphragm, a ribbon element responds to **pressure gradient**, that is, to the difference in pressure between the air on either side of the element. Because the ribbon did not have to move the mass of a coil of wire and the form supporting the coil, it could respond to the rapid changes in pressure gradient accompanying high-frequency energy.

FIG. 6-1: DYNAMIC MICROPHONE

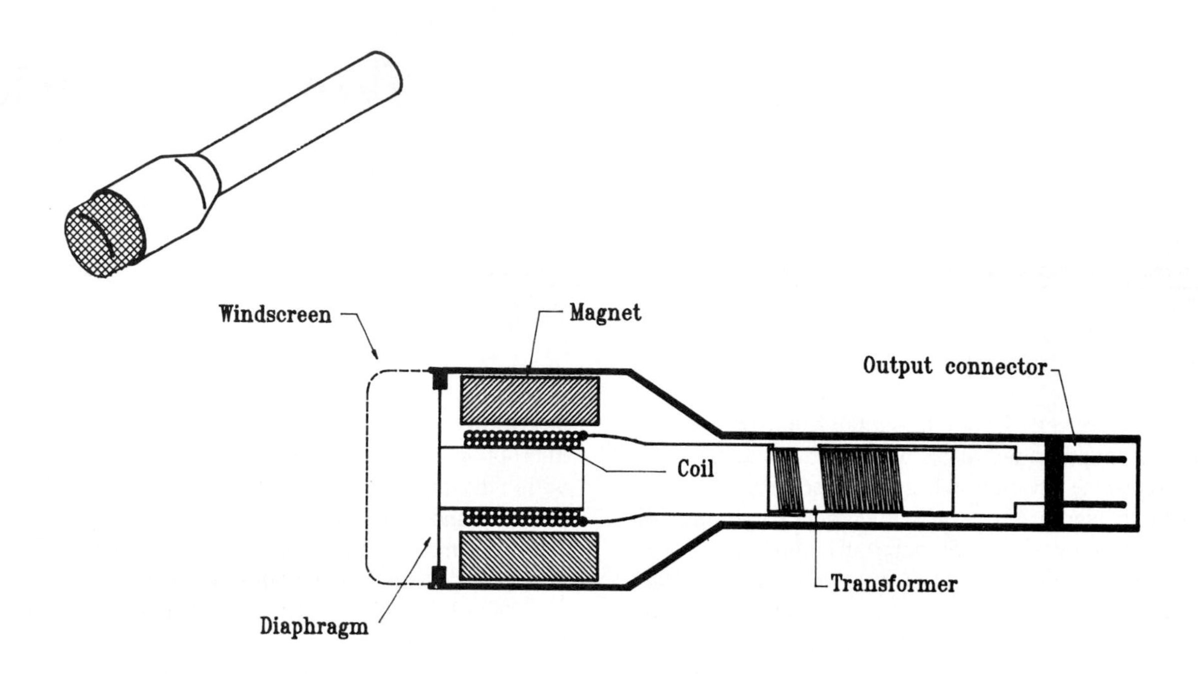

Condenser and Electret Microphones

Possibly the most significant development in microphone technology came with the advent of amplifiers small enough to place directly in the case of the microphone itself. Such small amplifiers made possible the use of capacitive plates as acoustic pickup elements. (See Fig. 6-7.) Like the electrostatic loudspeaker, operation of a condenser (another name for a capacitor) microphone depends on the balance of charge between two or more plates, one rigid and the other able to move in response to variation in acoustic pressure. The motion of the variable plate changes the distance between the elements, altering the capacitance and varying the amount of charge that can be stored by the plates. Because the variable plate need not carry the mass of a coil (as with a dynamic microphone), it can be extremely compliant. Condenser microphones, therefore, are very sensitive to high frequency. The signal output from the capacitive element is extremely small, so that even a short length of cable has too high a capacitance for the device to overcome. Preamplifiers placed only millimeters away from the pickup element are essential. Power for the preamplifier as well as the plates must be provided, so that a condenser microphone is not as simple a device as is a dynamic microphone. One cannot simply plug in a condenser device to an amplifier and start recording. On the other hand, because the microphone has its own preamplifier, its output can be higher in level than the output of a dynamic microphone.

The earliest condenser microphones were fairly large, cumbersome affairs with massive power supplies and multiconductor cables. Condenser microphones quickly evolved into much more practical packages, especially after the main developments in solid-state electronics. Another discovery that aided in the miniaturization of such microphones was the **electret** element, a pressure-sensitive, self-polarizing structure that requires no external polarizing voltage. The preamplifier of the microphone still requires a power supply, but the voltage can be quite low. (Power supplies for condenser and electret microphones can pose a significant problem, as we shall see later on.)

Microphone Signal Standards

Standards for microphone sensitivity, impedance, and level have been developed by the audio industry. Microphone impedance is, nominally, 150 ohms. Signal level from most dynamic microphones is, typically, about -50 dBm at an input pressure of 10 microbar per square centimeter. Ten microbar (10 dynes acting on an area of one square centimeter) represents a sound pressure level of approximately 100 dB, near the top of the dynamic range of human hearing. Most microphones can tolerate somewhat higher levels and must be able to pick up much smaller signals.

Current and voltage generated by a dynamic microphone pickup element are small. To make the signal usable, a low-loss transmission cable must connect the microphone to the mixer, and the signal must be boosted to a level suitable for processing. This boost is known as **preamplification**. A microphone preamplifier usually raises the signal level from about -60 dBm to something near operating level (0 to +4 dBm). All consoles and mixers incorporate one microphone preamplifier for each input channel.

Directivity

One very important characteristic of a microphone is its ability to discriminate the direction from which signal arrives. Directivity is measured in terms of signal strength with respect to the central axis of the microphone and is expressed as a **polar graph**. If the microphone picks up signal uniformly from all directions, it is called an **omnidirectional** microphone. Figure 6-2 shows a polar graph of an omnidirectional pickup pattern. A microphone that picks up signal only from one hemisphere of the surrounding environment is a **unidirectional** microphone. (Fig. 6-3.) Pickup that rejects signal from the sides of the microphone but accepts signals from front and rear is called **bidirectional**. (See Fig. 6-4.) A very useful variant of the directional pattern is shown in Fig. 6-5. This pattern picks up signal that is in front and on axis quite well, but sensitivity falls off as the signal moves to either side. Signal rejection is almost complete just to either side of the rear axis with a slight loop of sensitivity directly on the rear axis. The pattern, somewhat heart-shaped, is called **cardioid**.

Dynamic microphones are inherently omnidirectional. Omnidirectional pickup is not always desirable, however. In fact, most recording or reinforcement tasks require directional pickup characteristics. Making the microphone reject a portion of the acoustic field is accomplished by changing the signal phase relationships within the microphone. As shown in Fig. 6-6, dynamic microphones have a small port at the back of the microphone. Sound from in front of the microphone bypasses the port. Sound from behind the microphone enters both the port and the front of the microphone. The port ducts the rear-field signal into an acoustic labyrinth. The labyrinth shifts the phase of the signal by 180 degrees. At the rear of the diaphragm the phase-shifted signal tends to produce membrane motion opposing the portion of the rear-field signal reaching the front of the microphone. The rear-field signal cancels itself out, making the microphone sensitive only to its front-field information. In practice, most microphones do not achieve complete rear-field rejection, usually by design. Some

FIG. 6-2: OMNIDIRECTIONAL MICROPHONE PATTERN

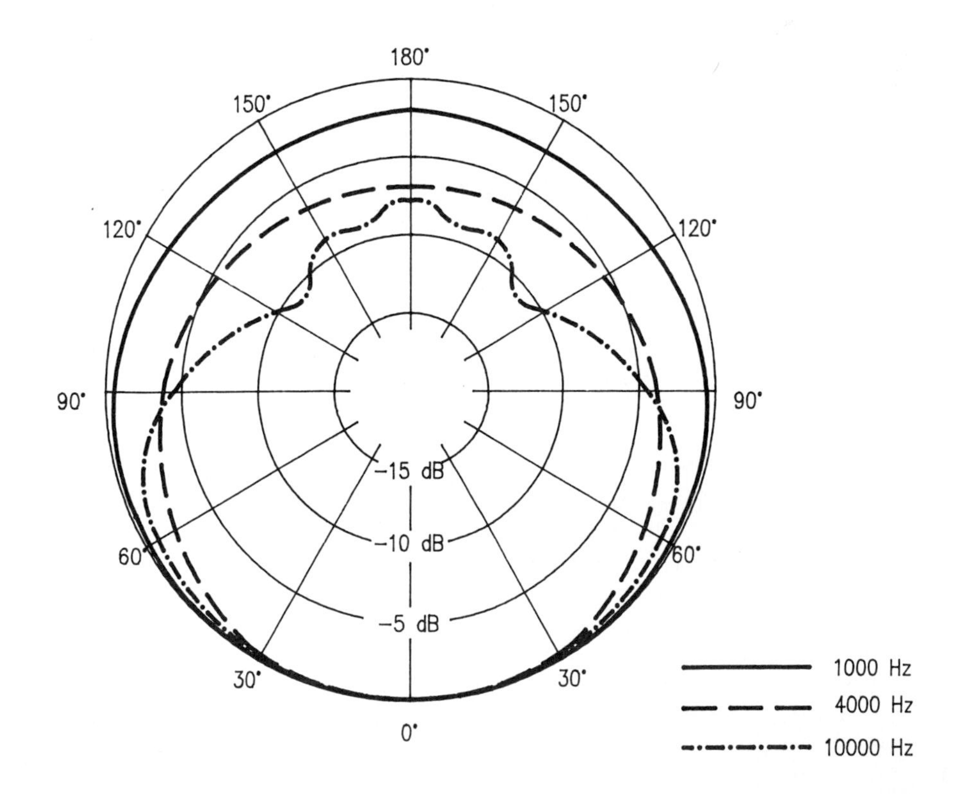

FIG. 6-3: UNIDIRECTIONAL MICROPHONE PATTERN

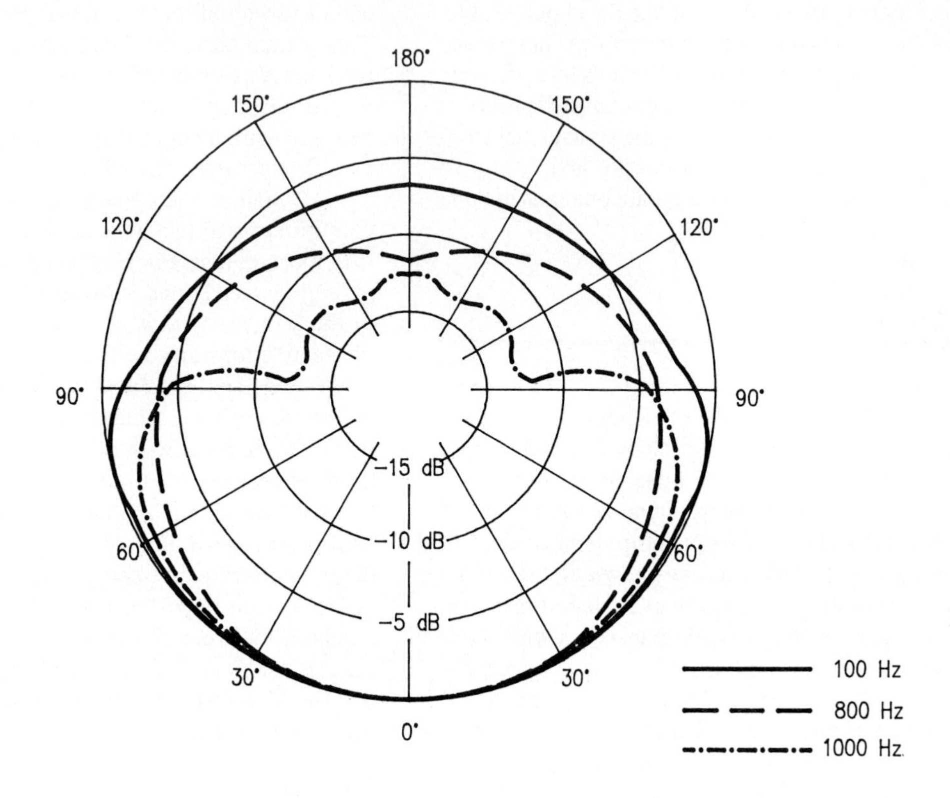

FIG. 6-4: BIDIRECTIONAL MICROPHONE PATTERN

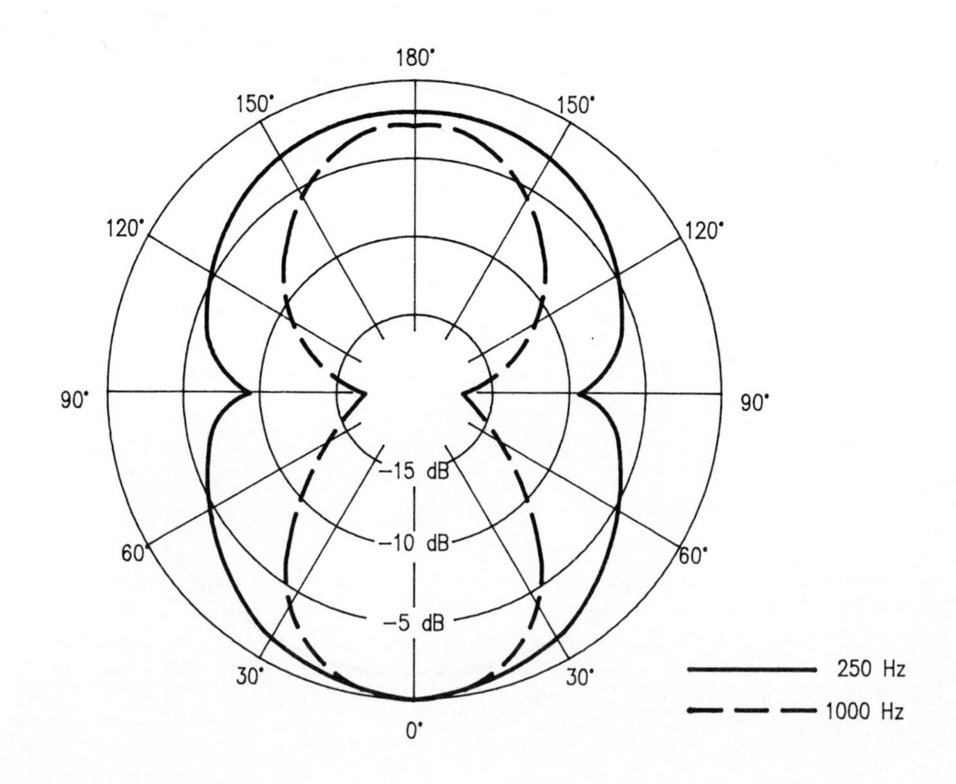

FIG. 6-5: CARDIOID MICROPHONE PATTERN

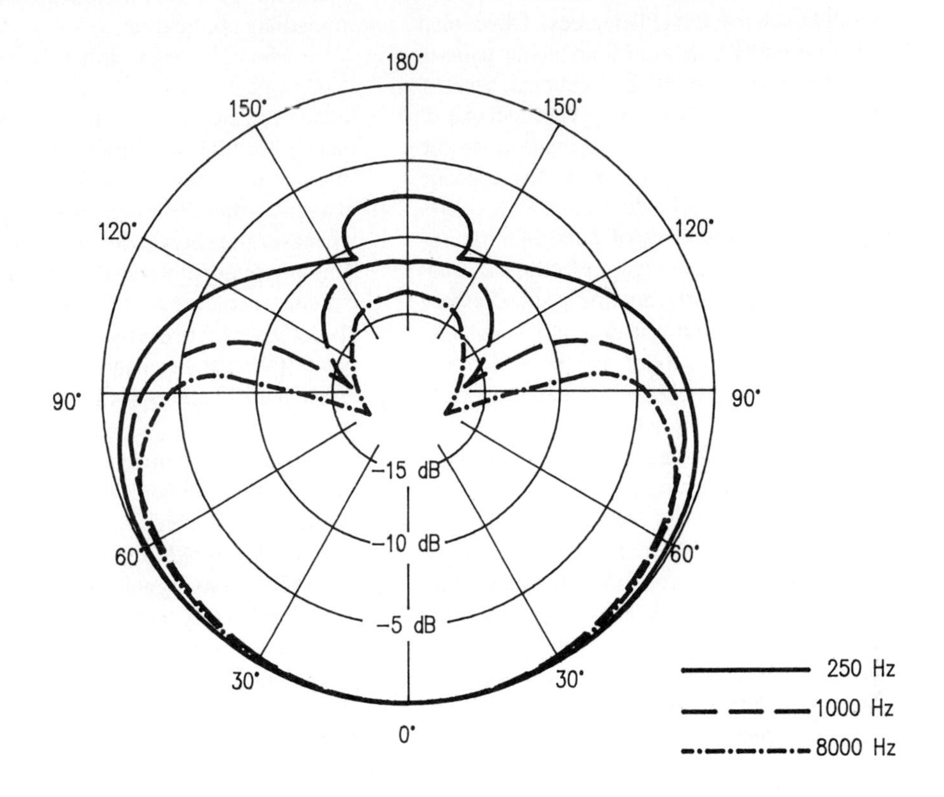

FIG. 6-6: CARDIOID DYNAMIC MICROPHONE

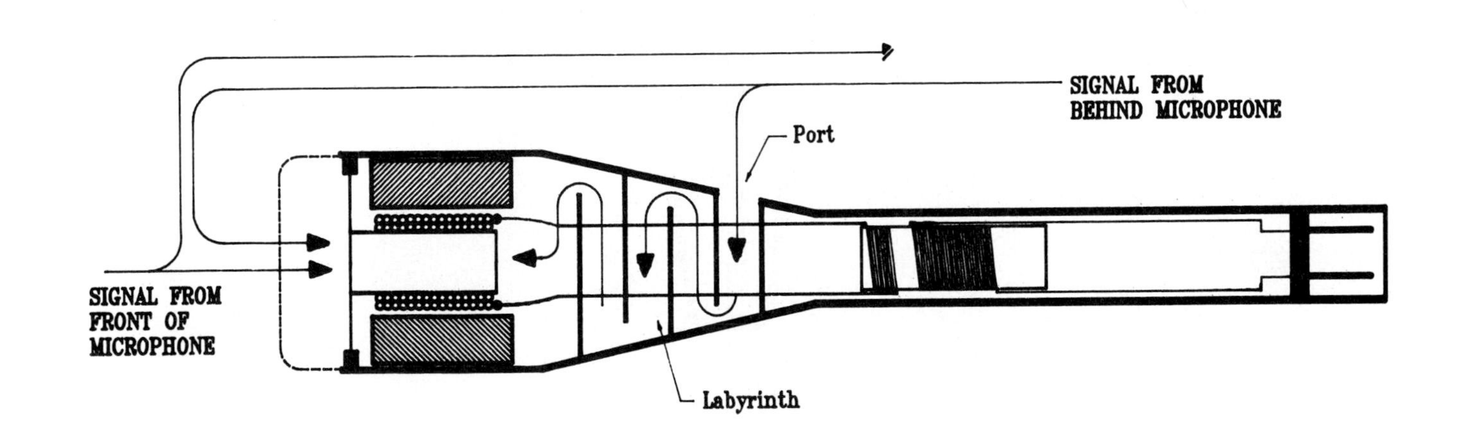

Signal from front of microphone enters windscreen and activates diaphragm. As signal passes around microphone, frontal signal does NOT enter port.

Signal from rear of microphone enters port and labyrinth. Rear signal also enters windscreen, but rear signal reaching diaphragm through labyrinth is 180 degrees out of phase with signal rear signal reaching diaphragm through windscreen. Result is cancellation of effect of rear signal. Therefore, only signal from sources in front of microphone are picked up.

small degree of rear-field information is deliberately included (by making phase shift less than a full 180 degrees). Directional dynamic microphones normally have a cardioid pickup pattern.

Ribbon microphones are naturally bidirectional, because the ribbon presents a large, flat area to both the front and rear of the microphone. Ribbon microphones could be made to assume omnidirectional or cardioid patterns by use of variable acoustic labyrinths placed within the body of the device.

Directional pickup characteristics of condenser microphones may be varied simply by changing the electrical characteristics of the plates. Elaborate acoustic ducting is not necessary as it is with dynamics; however the principle is similar. As shown in Fig. 6-7, a condenser microphone may have one fixed and two movable plates. The movable plates face in opposite directions. With both movable plates active and in phase, the microphone is bidirectional. Introducing a phase shift between the front and rear plates reduces microphone sensitivity to the rear-field signal. At 180 degrees out of phase the microphone becomes completely unidirectional. Switching off the voltage supply to the rear plate makes the microphone omnidirectional. The selection of directional characteristic is made by an adjustment on the microphone itself, except in the more expensive units, where directionality may be varied by the operator during use. Some condenser microphones permit continuous variability, from omnidirectionality to unidirectionality, so that the operator may select any pickup characteristic best suited to the immediate application.

Microphones, whether dynamic or capacitive, come in a wide range of types and sizes. In additional to directional characteristics, microphones may be optimized for music or for vocal pickup, for high sensitivity and wide frequency response or for minimal size and weight. Dynamic cardioid microphones, for example, naturally have an increased low-frequency response whenever the sound source is within 2 feet of the microphone. This low-frequency boost is called **proximity effect**. Under some circumstances, proximity effect may be useful. Many announcers, for instance, use proximity effect to help them achieve a deep, powerful vocal resonance. Under other conditions, proximity effect is not at all desirable, as, for example, in theatre when an actor's voice must be recorded, but the auditory perspective required must be similar to the quality of the actor's natural voice as the audience would hear it. Many cardioid dynamic microphones provide a bass-rolloff switch that attenuates low frequencies, helping to reduce proximity effect.

Optimum signal pickup occurs when a microphone is placed relatively near the sound source, but such optimal placement is not always possible. For the times when a mic cannot be placed near the sound source, unidirectional microphones that pick up a very narrow frontal field are used. Microphones of this type usually have a long train of

FIG. 6-7: CONDENSER MICROPHONE

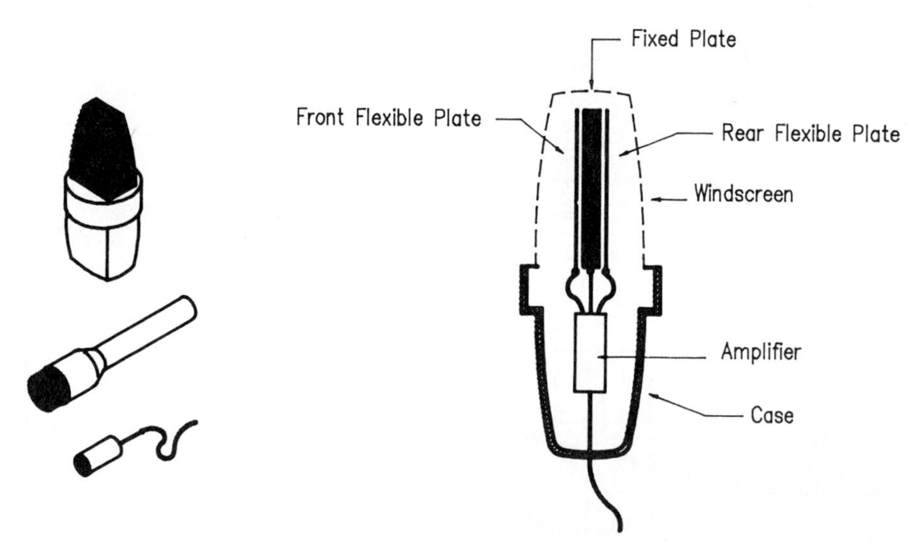

Condenser microphones come in several forms. The three shapes shown are common.

The condenser microphone has either two or three plates, one of which is fixed. Acoustic pressure variation makes the flexible plates move, which changes the charge level on the plates. As the charge level changes, electrical current replicating the acoustic waveform flows to the amplifier.

Varying the polarizing charge supplied from the power supply to the plates will change the directional characteristics of the microphone.

waveguides in front of the pickup element. This long waveguide looks much like a shotgun barrel; therefore, this kind of microphone has come to be known as a "shotgun" microphone. The narrower pickup patterns of shotgun microphones are known as **hypercardioid** (medium narrow) and **supercardioid** (very narrow). The gain in narrow forward pickup angle is offset, to some extent, by a sacrifice in exclusion of sounds from the rear field.

For times when they must be highly inconspicuous but still in view, microphones are made in very small packages, to be worn on a tie clip or on a lanyard around a person's neck. These are called **lavaliere** microphones. Lavaliere microphones are also useful as small suspended pickups over otherwise inaccessible parts of a stage, or as mics to hide in small parts of the set. Lavaliere mics are generally omnidirectional, so that wherever they are used, the exposed field must be considered.

One fairly recent development in microphones is a device that only receives reflected sounds, called a pressure-zone microphone (PZM). The pickup element itself is housed in a rigid arm suspended a fraction of an inch above a large, flat footplate. Acoustic energy is reflected from the footplate into the pickup element. Because they are placed at a major boundary in the acoustic space, PZMs usually manage to avoid the sound quality common to most microphones placed relatively far from a sound source. PZMs are naturally omnidirectional, and a basic pressure-zone device cannot be made truly directional. The best one can do is to shield the mic from unwanted sound sources. A version of the PZM called a phase-coherent microphone (PCM) uses phase cancellation techniques to achieve a reasonably directional pickup characteristic without sacrificing the advantages of pressure-zone technology.

Microphone Transmission Lines and Powering

Generally, for professional equipment, microphone signals are transmitted over **balanced lines**. The primary purpose of using a balanced line is to reduce susceptibility to induced noise. A balanced line uses two signal conductors and a grounded shield. Both signal leads are electrically symmetrical but take opposite polarity with respect to ground. Therefore, the signal on one of the conductors will assume a negative polarity, and the signal on the opposite conductor will assume the positive polarity. The signal condition is shown in Fig. 6-8. Note that,

FIG. 6-8: SIGNALS ON BALANCED LINES

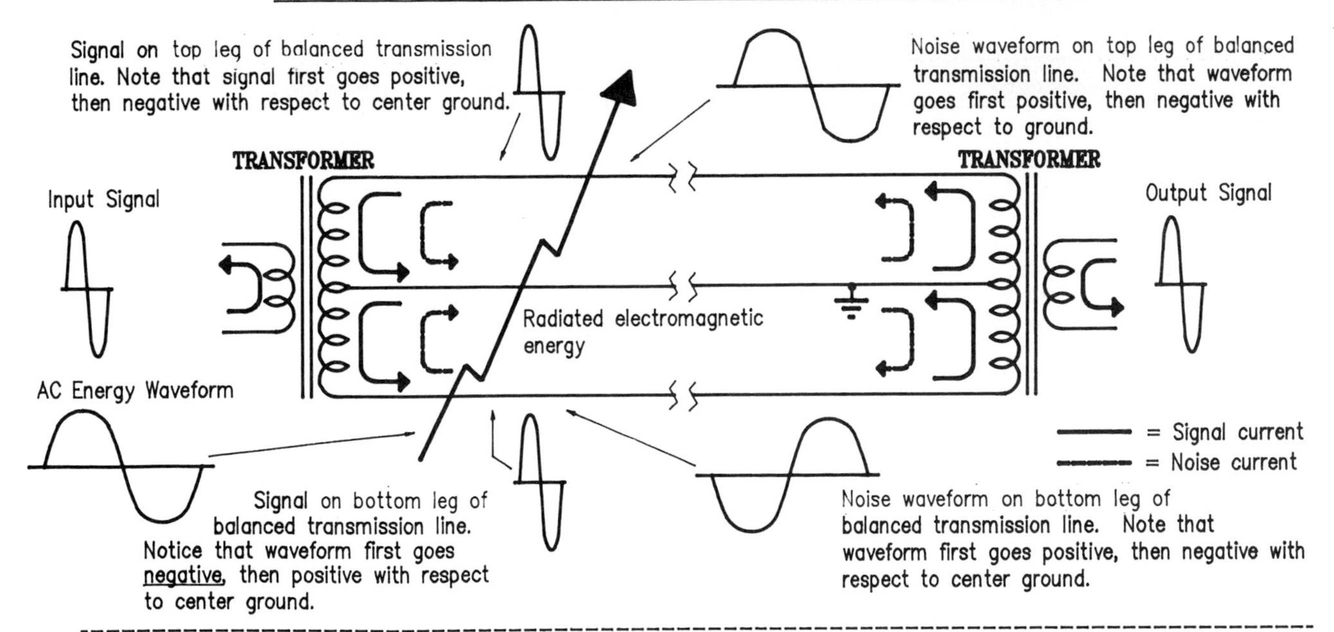

Audio signal is passed from the source device to the balanced transmission line through a transformer. The secondary of the transformer is forced to ground at its center, making the coils above the center tap and the coils below the center tap behave as if they were two separate transformers connected in **series. For signal, however, current in the primary coil generates a unidirectional current in the secondary coil. Therefore, for signal, the coil essentially behaves as a single coil. At the other end of the line, the signal is passed on to the next stage in the audio chain.**

Noise waveforms induced in the line by external source treat the two halves of the line as separate, unbalanced lines. The noise waveform takes the same polarity with respect to ground in each line. As a result, the noise-induced current in the line is 180 degrees out of phase in the transformer coils, and so is canceled out.

although the polarity in each half of the line is opposite, the signal is *in phase* between the two sides of the line. Noise picked up by the line is induced equally into both signal leads and with *similar polarity*. That is, the noise signal goes positive or negative with respect to ground in both signal leads at the same time. As shown in the figure, noise voltages are *out of phase*. Because the input to a microphone preamplifier is balanced, the noise signal is effectively 180 degrees out of phase on each signal line and becomes self-canceling.

Because condenser and electret microphones universally require internal preamps[1], powering condenser microphones is always of concern. Amplifiers need a DC power source. The earliest condenser mics used multiconductor cables in which some conductors were for signal and others supplied amplifier and polarizing power; but the cables were large and clumsy, and they were always breaking. The ideal was to be able to use standard microphone cable for condenser mics. Then both condensers and dynamics could be used interchangeably without the requirement of special lines or special console inputs. Dynamic microphones, however, respond very poorly if DC gets into their signal lines. Any DC voltage finding its way to a dynamic microphone's coil element would react with the field of the mic's permanent magnet to lock the diaphragm into immobility. The problem was solved by a technique called **phantom powering.** Phantom power derives its name from the fact that it is "invisible" to standard microphones but is available to condenser units. The phantom power source is usually built into the mixing console.

Phantom powering makes use of the principle that opposite polarities are self-canceling just as in the case of induced noise signals. The phantom power supply is applied to the balanced line by means of a voltage divider network. At the microphone a similar voltage divider passes DC voltage to the element and/or the amplifier in the microphone. For the signal line, the DC voltages are self-canceling; for the DC input and output, the audio signal is self-canceling. Figure 6-9 shows a schematic diagram of a phantom power supply system.

FIG. 6-9: PHANTOM POWER FOR MICROPHONES

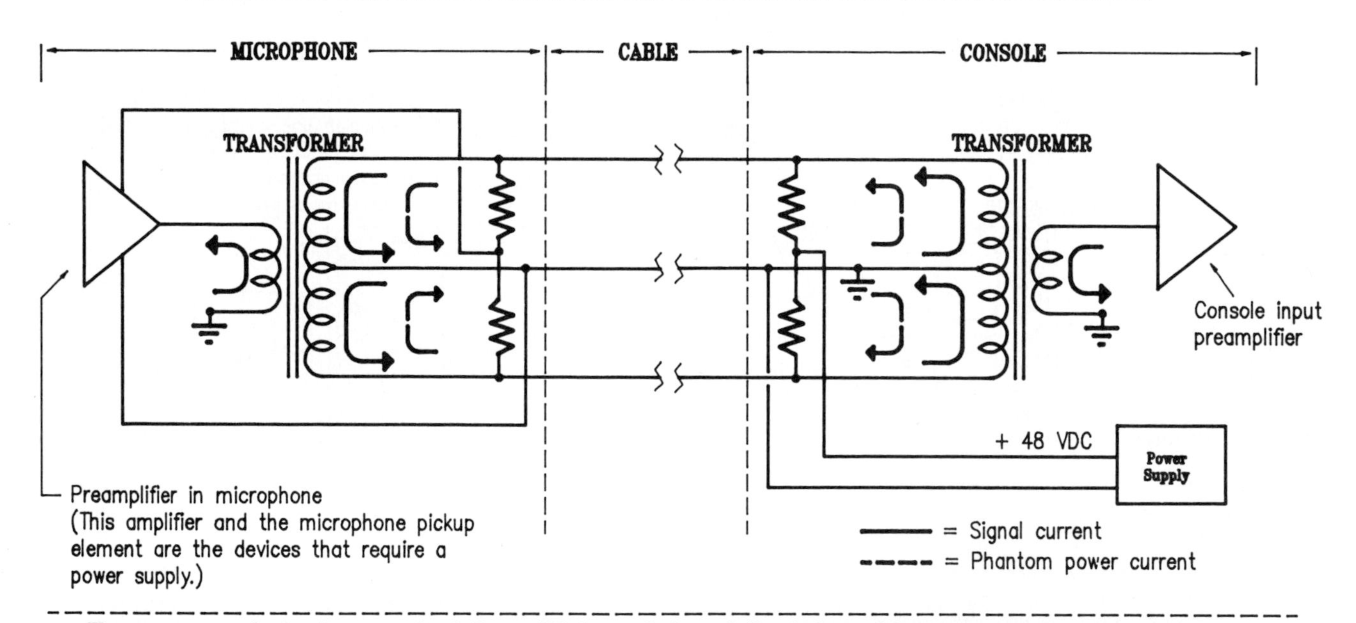

The power supply in the console delivers DC to each leg of the balanced transmission line through resistors. In the example shown, each leg would stand at a potential of approximately +45 volts (supply voltage minus drop across resistors). Microphone preamplifier receives approximately +42 volts through resistors at other end of line.

The resistors isolate the power supply from the line and also prevent damage to the power supply in the event of a short circuit at the microphone connector. As with induced noise, power supply current in each leg of the transmission line is oppositely polarized. Consequently, the power supply voltage is canceled out in the transformers at either end of the line.

[1]Electret pickup elements, though they do not require polarizing voltage supplies, still require power supplies for internal amplifiers.

Microphone Preamplifiers

The amplifiers that receive signals from microphone transmission lines are specialized amplifiers that are designed to sense an extremely low-power signal and amplify it to a satisfactory working level. Because a microphone signal is so small, the danger is that it may not be significantly greater in magnitude than the level of residual random noise in the input circuit and in the active elements of the preamplifier. The preamplifier itself, therefore, must be carefully designed to have the lowest possible noise factor, and it must amplify the input signal quickly to a level that is well above the noise factor of subsequent stages of amplification.

Because the sensitivity of a mic preamplifier input must be so great, microphones with especially high output, such as condenser microphones, can sometimes produce overload of the preamp input and clipping (excessive amplitude) at the preamplifier output. Therefore, preamplifiers often contain **pad networks** in the input circuit, a **gain control** as part of the preamplifier's feedback loop, or both. A pad is a resistive network that introduces a certain amount of voltage drop, thereby reducing signal level at the preamp input, whereas a gain control reduces or increases the amount of signal amplification from input to output of the preamp.[2] Because input overload and excessive output are somewhat different matters, the best microphone preamps have both kinds of protective devices.

[2]A gain control is different from an attenuator or fader. An attenuator is a passive control that merely reduces signal level because of its nature as a variable voltage divider. The attenuator cannot increase signal level above the maximum that the module amplifiers produce. A gain control, by contrast, regulates amplifier gain. It affects the magnitude of signal amplitude increase.

Chapter 7
Amplifiers

Audio depends on amplifiers—the essential elements in all active audio equipment, from the preamplifiers that pick up the tiny signals generated by microphone and phono transducer elements to the power amplifiers that drive loudspeaker systems. Usually the purpose of an amplifier is to increase the gain of a signal; but amplifiers can also relay signal without changing level or can even decrease signal level. The kind of amplifier used depends on its function and place in the audio chain.

Since the development of digital audio, the grounds for a discussion of amplification have changed. Most of this chapter concerns analog devices. Principles of amplification in a digital system are different from those applicable to analog audio. Analog considerations illustrate the basic principles of amplification, so we shall consider analog amplifiers first. At the end of the chapter we will briefly examine the modification of amplification principles to accommodate digital signal processing.

Basic Principles of Analog Amplification

Amplification is the use of a small voltage and/or current to control a larger voltage and/or current. A change in the small voltage or current (called the input) is reflected as a proportional change in the main voltage or current (called the output). The changing voltage and/or current is called the **signal**. The amount of increase in output signal over input signal is known as **gain** (measured in dB).

Originally, most audio amplifiers were made of individual vacuum tubes connected in *stages*. Each stage accomplished some aspect of signal handling, usually increase in signal level. Later, vacuum tubes were replaced by transistors. Transistors were much smaller and did not generate as much heat as vacuum tubes. Also, transistors were much more reliable than tubes. Transistor amplifiers, like vacuum tube amplifiers, were constructed in stages.

Transistors were subsequently replaced by integrated circuits (also called *chips*). The first integrated circuits were simple, single-function devices. Although integrated circuits contained several stages internally, the chip itself tended to look much like a single transistor. Amplifiers were constructed as stages of integrated circuit chips.

As chip technology advanced, larger numbers of transistors and, consequently, more functions could be incorporated into one device. The most recent chips are made by a process called *large-scale integration* (*LSI*) that permits literally thousands of transistors to be contained in a single integrated circuit package. Thus, an entire amplifier that would have needed several cubic feet of space during the vacuum tube era can now be condensed into a single chip. Such miniaturization has made possible the extensive control and signal processing devices currently available, as well as the small, portable radios and tape recorders we now enjoy.

Most audio equipment users need only a minimal understanding of how amplifiers actually work. The important characteristics of amplifiers for the purposes of most users are gain and **configuration**. An understanding of amplifier gain helps the user to maintain good signal quality; knowledge of configuration is essential to proper interconnection of equipment. Gain has already been discussed in Chapter 3, so let's concentrate on configuration.

Amplifier Configuration

Amplifier configuration depends on polar relationships within the amplifying device and the power supply system. As discussed in Chapter 2, difference in electrical potential is fundamentally a condition of unbalance—a region of negativity (electrons) opposed to a region of positivity. The difference in potential is impressed across the external circuit. A circuit that uses only a single power supply and passes signal only along the hot leg of the circuit is known as an *unbalanced circuit*. A circuit can be balanced by grounding some point midway between the most negative and most positive poles of the power supply.

Thus, one half of the circuit assumes a positive potential while the other half assumes a negative potential. (Refer to Fig. 2-7 to review the basic principles of balancing a circuit.)

As noted in Chapter 6, a balanced transmission line (one that has two hot legs and a centered, common neutral) has the advantage of rejecting induced noise. The traditional method of balancing an input or output in an audio device is the use of transformers, as shown in Fig. 6-8; however, a balanced line can be connected to a differential amplifier without the use of a transformer. Manufacturers' literature often refers to such inputs as transformerless.

A **differential amplifier** is a relatively recent development in audio technology.[1] A differential amplifier has two inputs and one output, and the device responds only when a difference exists between the two inputs. Like the balance shown in the mechanical analog in Fig. 7-1, as long as the two weights connected to the balance arm are equal, the weight representing the output remains still. Only if a difference develops between the balance weights will the output weight move. If input 2 becomes heavier, its weight will drop and so will the output weight. If input 1 becomes heavier, its weight will drop but the output weight will rise. Input 1, therefore, is an *inverting* input, whereas input 2 is a *noninverting* input. Similarly, the electrical output of a differential amplifier will follow a waveform applied to its noninverting input, but will invert a waveform applied to its inverting input.

Although differential amplifiers are commonly used for balanced microphone inputs, they may also be used for line-level inputs. A differential amplifier can receive a signal from either a balanced or an unbalanced line. Applying an unbalanced signal to a differential amplifier is like adding or removing weight on only one side of the mechanical balance arm in Fig. 7-1. A balanced signal applied to a differential amplifier is like simultaneously removing weight from one side of the balance and adding weight to the other. Connecting a balanced line to a differential input is simple: Each hot leg connects to one of the two inputs while the center neutral connects to ground. An unbalanced connection is not quite as obvious. Recall that the differential amplifier works only when there is a difference between the two inputs, not when there is a difference between either input and ground. Accordingly, an unbalanced line must be connected between the two inputs with *no* connection to the input amplifier's ground.[2] Under no circumstances should input ground and the neutral leg be tied together. Figure 7-2 shows how to connect both balanced and unbalanced lines to a differential input.

Transformerless outputs (sometimes called differential outputs) are also common in modern equipment. Such outputs

FIG. 7-1: DIFFERENTIAL AMPLIFIER

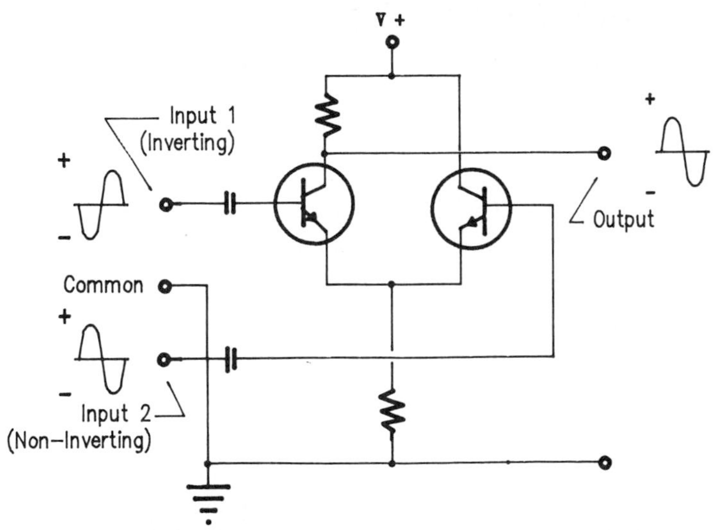

As long as the two inputs are equal (identical signal applied to each), the output will be zero. Only when there is a difference between the inputs will a non-zero output result.

MECHANICAL ANALOG

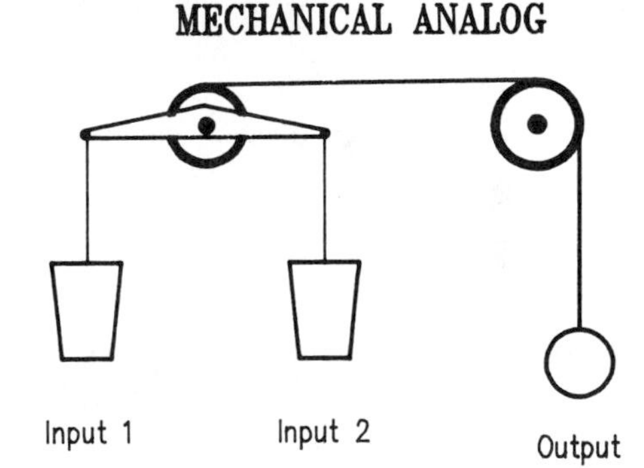

As long as the two weights representing the inputs remain equal, the weight representing the output remains still. When a difference exists between the two input weights, the output weight will move. Direction of output movement will be identical to Input 2 but opposed to Input 1. Input 1, then, is an inverting input, whereas Input 2 is a non-inverting input.

---

[1] Differential amplifiers were possible before the development of the transistor; however, they required complex arrangements of expensive components in order to keep vacuum tube elements operating at nearly identical no-signal conditions. Modern differential amplifiers are small, inexpensive integrated circuit chips.

[2] Shields for transmission lines are connected to the amplifier ground, but at one end only. Refer to the table in Chapter 14 for information on which end of a shield is connected in various line configurations.

FIG. 7-2: CONNECTING TO A DIFFERENTIAL INPUT

UNBALANCED TO DIFFERENTIAL

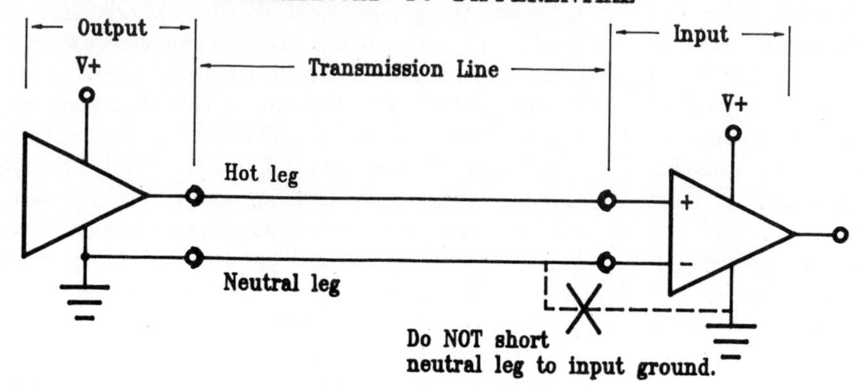

Normally, the hot leg of the output is connected to one of the differential inputs and the neutral leg is connected to the other. No connection is made to the input ground. Observe that the two devices MUST have separate power supplies; otherwise you get the forbidden short to ground.

BALANCED TO DIFFERENTIAL

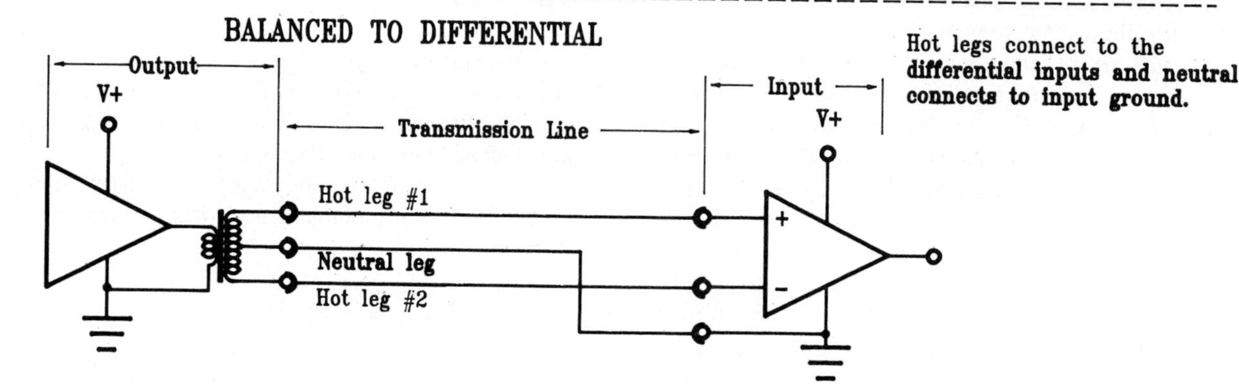

Hot legs connect to the differential inputs and neutral connects to input ground.

may be connected to either unbalanced, balanced, or differential inputs. Refer to the table in Chapter 14 for connections.

Line Amplifiers

Line amplifiers, like most amplifiers used in audio equipment, are devices designed to increase voltage level. Signal handling in almost all audio equipment is a process of passing change in voltage from one device to the next. Current is kept at as low a level as is reasonably possible, as current passing through the various nonlinear (reactive) elements of an audio circuit can cause distortion in the signal waveshape.

The standard signal level used by most audio manufacturers is based on a maximum average signal voltage of approximately 1.4 volts across an impedance of 600 ohms, yielding a power level of 3.3 milliwatts, which represents a level of +4 dBm (refer to Chapter 5 for a review of dBm). Most line amplifiers used in audio equipment operate somewhere near this level. Line amplifiers isolate inputs and outputs and restore gain after various signal processing operations such as mixing and equalization. Amplifiers that isolate sensitive portions of devices are called *buffers*; those that restore signal level are called *boosters.*

In present-day audio equipment line amplifiers are usually integrated circuit devices called *operational amplifiers.* As explained in Chapter 5, an operational amplifier is a device where gain is a function of the amount of feedback from output to input. The value of the operational amplifier is that it can be configured to serve many purposes; thus, line amplifiers now routinely combine the functions of summing node (mixer) and booster.

Unity-Gain Amplifiers

A special kind of line amplifier, called a **unity-gain amplifier**, forms the basis of many kinds of signal processing equipment. The purpose of a unity-gain amplifier is to relay the signal without either increasing or decreasing its signal level. Unity-gain amplifiers are used as buffers and as the basic signal handling element in automatic gain control devices.

Preamplifiers

A **preamplifier** is used to raise extremely low-power signals to line amplifier levels. Most transducer devices (phono pickups, microphones, tape heads) generate extremely low signal power, on the order of millionths of a watt. Preamplifiers are especially designed to receive very low power inputs. Preamplifiers for microphones are normally incorporated in mixing consoles; phono pickups and tape heads usually require preamplifiers mounted close to the pickup head; therefore, preamplifiers are built into tape machines and are usually mounted into the base of phono turntables. Present-day preamplifiers, especially those used as microphone preamps, are usually IC op-amps using a differential input. The differential input allows a balanced input to be connected directly to the amplifier without the use of a transformer.

Preamplifiers are usually built into consoles, tape recorders, and other large pieces of audio equipment. As built-in equipment, they rarely include any kind of control specific to themselves. An exception is the variable-gain microphone preamplifier used in good quality mixing consoles. A gain pot is provided to allow the operator to set the amount of signal level increase that the preamplifier will provide.

Power Amplifiers

Power amplifiers are designed to increase current instead of voltage, in contrast to preamplifiers and line amplifiers, which avoid increase in current level. Power requires current amplification, as power increases as the square of the current (refer to Chapter 2 for review). The typical use for a power amplifier in an audio system is to drive loudspeakers. Power amplifiers must take signal input at operating (line) level (a power level of about 3.3 milliwatts) and boost it to an output of about 1 or 2 watts to as much as 100 to 200 watts. Because voltage output from a power amp is rarely more than 60 to 80 volts, an immense gain in current is required to produce a significant increase in power.

Increase in current is usually accomplished in stages leading up to the final output stage. The stages preceding the output are called *drivers*. In solid-state power amplifiers the current amplification stages are usually bipolar, complementary-symmetry arrangements of linear power transistors. Even under the most ideal circumstances, power transistors dissipate a considerable amount of heat; therefore, large, finned heat sinks are usually evident in a power amplifier. Often, the heat sinks are a part of the outside casing.

Because a power amplifier is very nearly a pure current source it is also highly susceptible to short circuits. Miswiring a connector or fusing elements in a driver coil can cause a severe **overcurrent** condition (meaning that the maximum rated output current for the amplifier is exceeded). To avoid overcurrent damage, automatic protection circuitry is usually part of the design of a power amplifier. (Recall, in this context, that parallel connection of loudspeakers *lowers* impedance, and that, because loudspeaker impedance is not large, rarely above 16 ohms, paralleling quickly drops speaker impedance to dangerously low levels.)

In combination, a power amplifier and loudspeaker become a system that takes on its own characteristics. One of these characteristics is the ballistic nature of the speaker itself. Once in motion, the diaphragm or cone has an inertial tendency to keep moving, which means that the power amp must expend increased energy to force the cone to change directions. But it also means that any inertial movement that does not result from current output from the amplifier *acts as a generator to produce counter emf in the circuit between the amplifier and the loudspeaker.* In other words, the loudspeaker produces spurious source signals which it reflects back into the power amp. High quality power amplifiers have circuitry built in to sense any countercurrents generated by the loudspeaker and to oppose a very high impedance to them. The effect of the high impedance is to **damp** the spurious oscillations in the loudspeaker. The degree to which the power amplifier corrects for spurious oscillation in the speaker voice coil is known as the amplifier's **damping factor**. The higher the damping factor the more effectively spurious signals are suppressed.

Power amplifiers may or may not come equipped with input level controls. Input level adjustments are convenient because they permit compensation for acoustic factors at the power amplifier itself, leaving the console operator free to set up output level controls in a way that reflects the intended function of the sound rather than the actual power levels needed to accomplish the function. Large power amplifiers often include output VU meters. Smaller power amps may not provide output VU meters but may include LED indicators to indicate operational status. Green LEDs signify normal levels, amber indicates levels near peak, and red denotes peak. Operating at peak, the amplifier is likely to **clip** the signal. Clipping occurs when the output voltage swing attempts to exceed the power supply voltage. Because the amplifier *cannot* produce a voltage output greater than that of the power supply, the waveform is clipped (flattened at the top or bottom of the swing). Clipping is a form of distortion that is particularly obnoxious and is stringently avoided.

Digital Amplification

Digital signal processing (DSP), as noted several times in this text, is a process of manipulating audio information as numeric data. At the input to a digital system, analog signals are converted into data words by analog-to-digital converters. The quality of the signal, initially, is determined by the sampling rate and the resolution of the system, as discussed in Chapter 4. The signal is handled entirely as manipulation of data *for as long as the signal remains in a digital domain.* At present, most digital devices provide both digital and analog inputs and outputs; some provide only analog connections.

DSP includes aspects of signal handling: amplification, equalization and filtering, compression and expansion, and delay. Of these, all except the last become matters of mathematical operations on one or more aspects of signal intensity. Delay requires that we examine the relationship of both analog and digital signals to time.

The aspects of DSP that deal only with alteration in signal intensity are amplification, equalization, filtering, compression, and expansion. Basic gain and amplification in analog audio involves strategies to make a large current or voltage change at the output of a device follow small current or voltage changes at the input. In a digital signal, each attribute of a sample is represented as a number. Because the number mainly describes the intensity of the signal, intensity can be varied simply by scaling the number. Increasing the value of the number equals positive signal gain (increase in level). Decreasing the value equals negative signal gain (decrease in level).

Compression and expansion constitute processing problems similar to basic amplification. Compression reduces the dynamic range of a signal, whereas expansion increases the dynamic range of a signal. Compression and expansion are concerned with the maximum and minimum limits of signal intensity throughout the entire duration of signal variation. In analog systems, compression and expansion are accomplished by modifying signal strength above or below a *threshold*, reducing or increasing amplifier gain whenever signal strength crosses the threshold value. The same principle can be used in digital processing, except that the operational principle is mathematical scaling rather than variation in amplifier gain.

Equalization and filtering also deal with intensity. However, with timbral modification, we are concerned not with intensity in general but with variation in intensity over a specific period of time. In order to focus processing on a particular frequency, the system must be instructed to look for cyclical intensity change that has a precise periodicity. For example, in order to modify a signal of 1000 Hz, the processor must be instructed to screen the signal for variations in intensity that complete the basic cyclical process in 0.001 second. If the signal meets that preliminary qualification, then subsequent processing to scale (boost or cut) that portion of the signal can proceed. Clearly, the success of digital signal processing that relies on ability to detect small changes inside an overall waveform pattern is directly dependent on adequacy of sampling rate and resolution.

Delay requires repositioning the signal in time. An analog signal is a variation through time in the value of some particular parameter. Analog delay implies lengthening the signal path through space in order to accomplish an offset in time. Digital delay implies storing data describing attributes of a signal for the period of the time offset, then recalling them and passing them on through to the output of the system. The maximum delay value that can be achieved is a function of the amount of memory available for storage. Here, again, sampling rate and resolution become primary factors. A sampling rate adequate to describe a complex waveform having a base frequency of 20 kHz is at least 80 to 100 kHz. Resolution should be at least 16 bits. One hundred thousand words of sixteen bits each works out to 1,600,000 storage cells for each second of sound. Computer memory chips are relatively cheap, but the costs of long delays are considerable. The possible strategies for reducing costs include storing every other sample (halving the number of storage cells required), or downscaling resolution in the delay device. Downscaling means converting a 16-bit word to a smaller value — say, a 12- or an 8-bit word. At the delay output, the signal would be converted back to a sixteen-bit format, but some signal detail would inescapably have been lost.

The interrelationship between the signal as space and the signal as time is one that becomes very important when we talk about audio storage systems in Chapter 10. So far, including the present discussion, all forms of audio signal have been essentially locked to a time base. Even in the process of delay, elaborate methods must be employed to offset sound in time. And delay is no more than an offset. The sequence of events that makes up the audio waveform is not truly altered. In the discussion of storage, we shall see that the newest methods of sound recording do offer the possibility of uncoupling sound from time.

Digital Electronics

The earlier part of this chapter spent a great deal of time in discussion of analog audio hardware. For analog processing a certain amount of understanding of the electronic devices involved is necessary. Digital processing removes some necessity to understand what goes on "under the hood." The paragraphs that follow are intended to cover the most elementary technology of digital signal processing.

As we know, conversion of an analog signal to digital format requires a device called an analog-to-digital converter, often abbreviated as A/D converter. Usually an A/D converter contains all of the electronics to sample a signal and convert the sample to a binary word. The input to the converter is an analog signal; the output is digitized data. At the output the material difference is whether the data is fed out as a parallel stream or as a serial stream. Parallel data is fast, but if the word width is 16 bits, then the output path must provide 16 parallel conductors. Serial data is slower, but requires only two conductors. Like so many aspects of digital audio, the choice is made on the basis of the ultimate cost of the product.

At some point in a system (the power amplifier if not before) digital signal processing must end and the analog signal must be reconstituted. Passing a signal from the digital domain to the analog domain is accomplished by a digital-to-analog converter, or D/A converter. The input to a D/A converter may be either serial or parallel, depending on the design of the system architecture. In either case, the input is digital data, and the output is an analog audio waveform.

Between the input and output extremes of the digital system, a number of processes occur based on a variety of forms of digital electronics. First, the system requires storage registers, which form a part of the signal path for the converted data. A register is a bank of memory cells. Data is inserted into the cells based on availability of cell space. A memory controller is responsible for organizing the passage of data through the register. Data words move out of the register and into processing space, usually into immediate data registers in a central processing unit (CPU). In the CPU, the processor consults a list of program instructions to find out what action to perform on the data. The kind of processing involved depends on the nature of the particular piece of audio equipment or the particular kind of processing specified if the device is capable of multiple operations. For example, a device such as a digital equalizer would incorporate instructions to enable the processor to find particular frequency components of a signal. Once found, the processor would look at the instruction set again, to find out what action to perform and how to perform it. Once the action is performed, the word is moved out to be sent along to the next processing stage or to the output.

The output of a digital system can be either digital data or analog audio. Currently, most manufacturers provide both formats. Because several different digital data protocols exist, some equipment provides more than one digital output. Lower cost equipment generally provides only analog audio.

Chapter 8

Basic Signal Processing

Rarely can audio signals for use in theatre sound be prepared without making some kind of selective alteration in the quality and character of the signal. Alteration of audio signal characteristics is accomplished by means of a variety of devices called collectively, **signal processors**. Processing occurs as a modification of the controllable properties of an audio signal. Perceptually, signal processors modify intensity, frequency, timbre, and duration (time). Processors that control intensity are compressors, limiters, expanders, and noise gates. Those that alter complexity (timbre) are equalizers, filters, and phasers. Frequency is modified by harmonizers, phasers, and delay devices. Equipment made to control the time parameters of a signal include delay devices and echo/reverberation generators.

Despite the perception that processing alters the full range of signal characteristics, audio signal processors really manipulate only two properties: intensity and time. Compressors, limiters, expanders, and noise gates are designed to regulate the intensity of the *total* audio signal. Filters and equalizers also operate on intensity, but confine their action to specific bands of frequencies. Controllers that act to alter the time relationships of a signal, called *time-domain processors*, include harmonizers, pitch shifters, phaser/flangers, delay lines, and echo/reverberation generators.

Our study of these important devices begins in this chapter, with filters, equalizers, and automatic intensity controllers—the group of processors that act mainly on the intensity of a signal. As a group, intensity controlling devices present concepts which should be understood before we engage in an examination of consoles and mixing systems. Because they are much more complex, and because some other considerations should be investigated first, time-domain processors are considered in Chapter 11.

Filters and Equalizers

Filters and **equalizers** are about the simplest of the signal processors, at least operationally. Filters and equalizers are alike in that both alter the intensity of some part of the audio spectrum, but the effect of equalization is very different from the effect of filtration. Filters and equalizers act on groups of frequencies, called **bands**. If the device strengthens or transmits a group of frequencies, the band is known as a **passband**. If the device weakens or removes a group of frequencies, the band is known as a **reject band**. A very narrow passband is called a **peak**; a very narrow reject band is called a **notch**.

Difference between Equalizers and Filters

A filter absolutely excludes frequencies that fall outside a passband and absolutely removes the frequencies lying inside a reject band. An equalizer, on the other hand, simply attenuates the frequencies within a reject band but does not eliminate them. Equalization boosts the level of frequencies within a passband without affecting the other frequencies of the spectrum. In principle, an equalizer is intended to correct an imbalance in the audio spectrum—excessive or insufficient level in some part of the frequency range. The purpose of a filter is to exclude some portion of the audio spectrum. Filters rarely boost signal; true filters only cut parts of the audio spectrum.

Examples of equalizer and filter usage may serve to clarify their differences. Proximity effect, the bass boost characteristic of dynamic microphones when less than 2 feet or so from the sound source, is corrected by equalization. An equalizer attenuates low frequencies to the same degree that the microphone boosts them, leaving a flat response. Sixty-cycle, AC hum is handled by means of a filter. We don't want the noise at all, so simple attenuation is not enough. A narrow-band notch filter tuned exactly to a center frequency of 60 Hz is used to remove the noise from the signal.

Filters

True filters are relatively hard to find in catalogs of studio processing hardware; most recording and reinforcement equipment manufacturers make equalizers, not filters, because most audio problems can be solved with EQ. Thus, the greater market is for equalizers. Musical instrument amplifiers and analog synthesizers use true filters, so that sometimes for theatrical purposes one has to adapt devices designed expressly for musicians. Although equalizers are useful and necessary, they don't really substitute for filters, and filters are very necessary items for theatre sound.

Filters come in four varieties: **bandpass, band reject, low-pass,** and **high-pass.** (See Fig. 8-1.) Band pass and band reject should be obvious concepts at this point. The former passes the included band of frequencies and cuts out the rest of the spectrum; the latter passes everything except for the band. A high-pass filter is just what it says: It passes everything above some specified frequency and excludes all frequencies below that point. The low-pass filter does exactly the reverse.

Consider the high-pass filter shown in Fig. 8-2. At about 140 Hz, the response of the system to lower frequencies begins to decrease. At 100 Hz, the response is down by 3 dB from "flat" response. This is called the **cutoff point** of the filter. From 100 Hz down, the output of the filter steadily decreases. The **slope** of the filter (i.e., the sharpness of cutoff below the 3-dB down point) is measured as the number of decibels the response drops over an interval of an octave. Although it wouldn't be a very sharp cutoff, if the response of the filter were down 5 dB at 50 Hz, the slope would be 2 dB/octave (5 dB represents a drop of 2 dB from the cutoff point). Most filters cut off at a rate of about 12 dB/octave, so that at 50 Hz (one octave below 100 Hz), the output of the filter in Fig. 8-2 would be -15 dB, and at 25 Hz (two octaves below cutoff), it would be -27 dB. A low-pass filter does the reverse of the high-pass unit, cutting off frequencies above the cutoff point and passing those below.

Observe that peak and notch devices may be configured from high-pass and low-pass filters whose bands either overlap or are slightly separated. For instance, if we *parallel* high-pass and low-pass filters, each having similar slopes, with cutoff points separated by a few hertz as shown in Fig. 8-3, we get a notch. Each of the two filters passes its respective band, but the frequencies lying between the cutoff points are out of the range of either filter. Placing high-pass and low-pass filters with overlapping cutoff points in *series* yields a peak. The low-pass unit excludes everything above its cutoff and the high-pass knocks out all frequencies below its 3-dB down point. The only frequencies that get through are those lying between the two cutoffs (i.e., the frequencies that

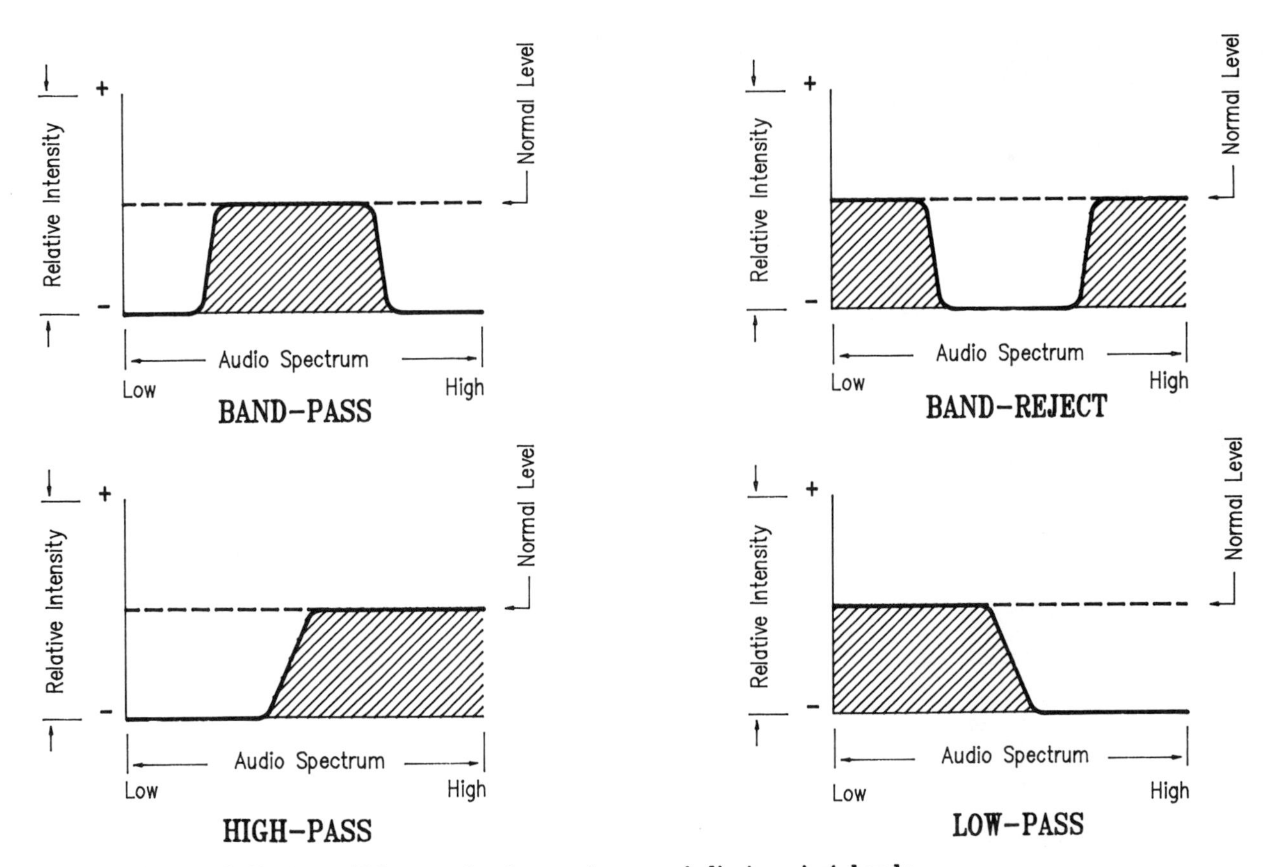

Shaded areas indicate audible pass bands; empty areas indicate reject bands.

FIG. 8-2: HIGH-PASS FILTER

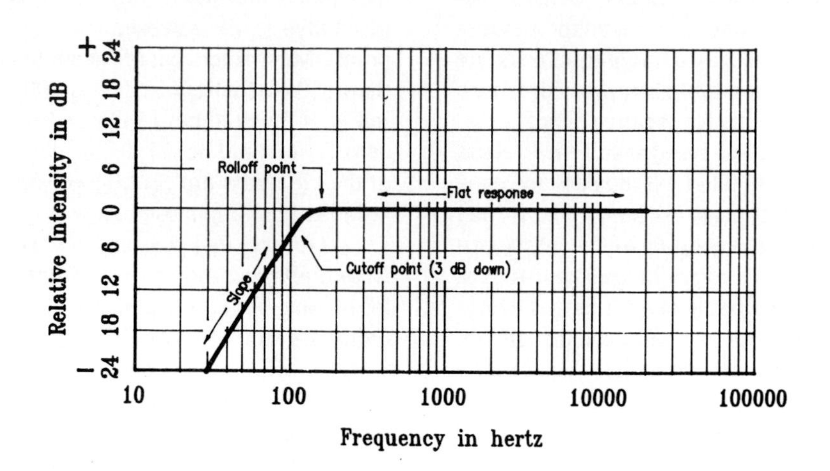

FIG. 8-3: COMBINING FILTERS

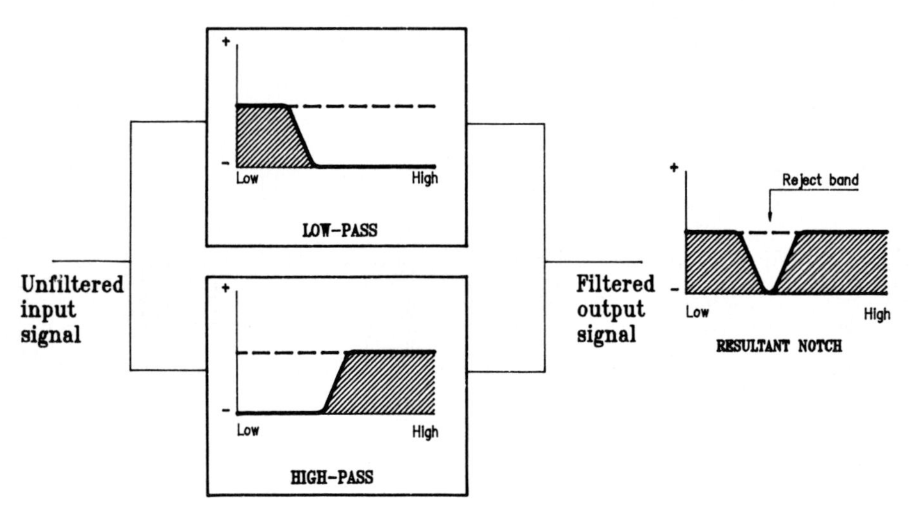

LOW-PASS AND HIGH-PASS FILTERS IN PARALLEL

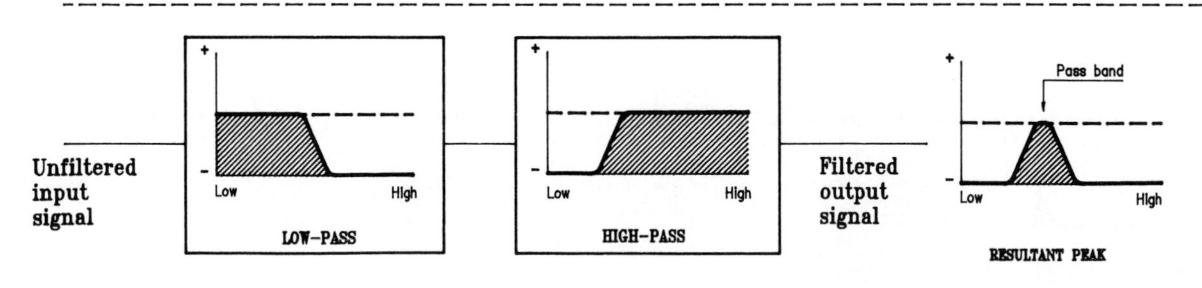

LOW-PASS AND HIGH-PASS FILTERS IN SERIES

the two filters pass in common). Both filters and equalizers are classified according to the ways in which they act on the spectrum and the fineness or coarseness of the band that can be tuned by the device.

Equalizers and filters may be built as fixed devices for specific applications or as variable units for more general use. Older broadcast systems used fixed, passive equalizers to provide RIAA compensation for phono pickup cartridges; filters in permanently installed reinforcement systems are often fixed. EQ controls built into reinforcement and recording consoles and filters used in temporary reinforcement systems are variable.

Equalizers

Two common versions of equalizers are **graphic equalizers** and **parametric equalizers**. Graphic equalizers are usually made in octave, two-third octave, and one-third octave versions. Controls for a graphic EQ are usually sliders. Each slider regulates the response of a bandpass/band-reject filter section. (See Fig. 8-4.) Each band overlaps the adjacent bands to some extent. The normal position for the control is at the center of its range. In the center position the control does not affect the signal at all. Raising a slider will boost frequencies within the passband that it controls; lowering the slider will attenuate the group of frequencies. Most graphic equalizers can provide at least 12 dB boost or cut at the center frequency of each band. With all controls set, the visual configuration of the knobs forms a graphic representation of the equalization curve imposed on the audio spectrum. (See Fig. 8-5.)

Octave graphic equalizers provide very coarse equalization. Two-third and third-octave devices provide a much finer degree of control and are generally more useful. The International Standards Organization (ISO) center frequencies for graphic equalizers are shown in Appendix I-B. The octave graphic provides 10 bands of control, the two-thirds octave provides 15, and the one-third octave 30 bands.

The parametric equalizer differs from a graphic equalizer in that it provides much coarser control of spectral response, and controls its bands in a much different way than does a graphic EQ. Parametric EQs come in two forms: **full-parametric** and **quasi-parametric**. Full-parametric EQ provides a selector for the center frequency of each band, controls depth of boost or cut, and allows the width of the band (called Q by electronic engineers) to be varied from a wide

FIG. 8-4: GRAPHIC EQUALIZER

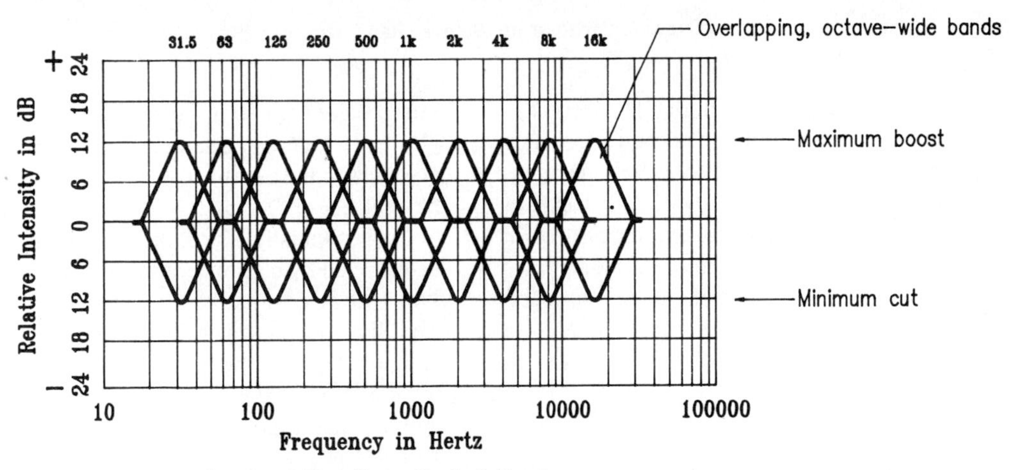

Graph of Equalizer Control Bands

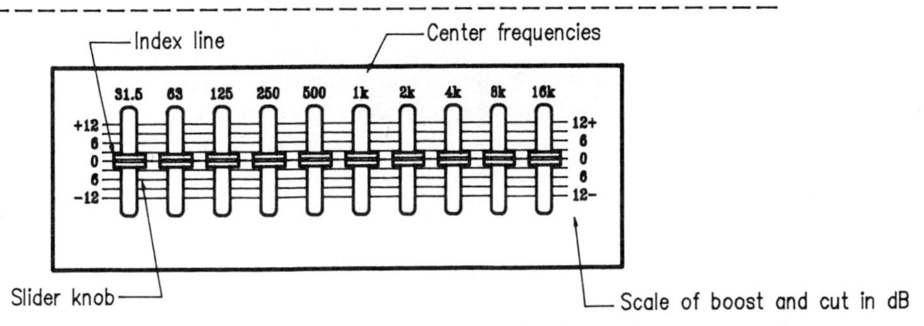

Octave-Band Graphic Equalizer Faceplate and Controls

FIG. 8–5: SETTING RESPONSE WITH A GRAPHIC EQUALIZER

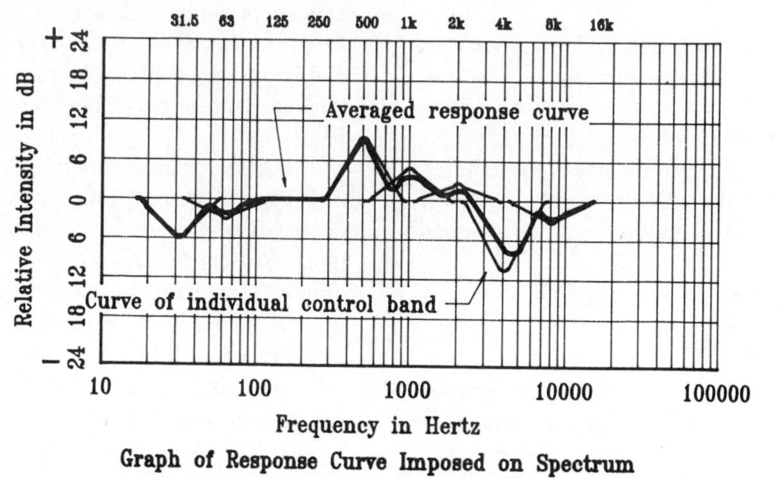

Graph of Response Curve Imposed on Spectrum

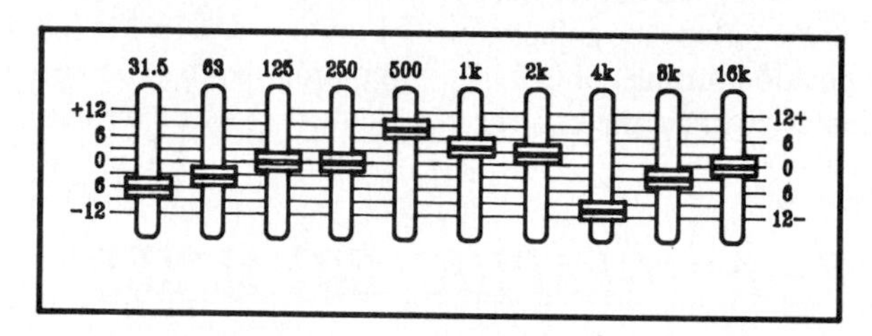

Control Settings on Octave–Band Graphic EQ

band (low Q) to a narrow notch or peak (high Q). Quasi-parametric equalizers provide only control of center frequency and depth; bandwidth is not controllable on a quasi-parametric EQ. Parametric equalizers usually provide four bands of control and can shape the overall, generalized response of an audio channel. By its nature, a parametric can sometimes accomplish results that are not easily done or that take too much time using a graphic equalizer. Figure 8-6 shows a typical four-section parametric equalizer and the ways in which its control bands can be varied.

Automatic Intensity Control Devices

Variation in the loudness or softness of sound is called **dynamics**. The maximum and minimum intensity limits of a sound are known as its **dynamic range**. The term also applies to the maximum and minimum intensity limits of an audio system or piece of equipment. Technically, system dynamic limits are important, because exceeding those limits damages signal quality. Artistically, both the limits of signal dynamic range and the lesser interpretive dynamics are important. Audio signals vary constantly in dynamic level—the result of interpretive crescendo and decrescendo in music, for example. Normally, dynamic change is good, but situations occur, both artistically and technically, in which dynamic variation must be controlled.

Technically, the limits of an analog audio system are determined by components within the system. Digital system limits are determined by the resolution of the encoding system. The dynamic range of any system as a whole is set by the component with the smallest dynamic range. Analog tape recorders are frequent culprits. Most amplifiers, including those inside a tape recorder, can handle a dynamic range of 120 dB (as great as the dynamic range of the human ear), but magnetic recording tape (when used for analog recording) typically handles no more than 60 to 70 dB.

Artistically, some discrepancy usually occurs between the dynamics of a sound as recorded and the dynamics of the sound a theatrical production needs for a particular dramatic situation. For example, level variation in underscore music and sounds (underscore refers to sounds played *during* the course of a scene while actors are speaking lines) can mask dialogue, affecting intelligibility for the audience. In order to keep the sound from interfering with dialogue, the dynamic variations must be minimized or removed. That means fading down when the sound gets loud and up when sound gets too soft.

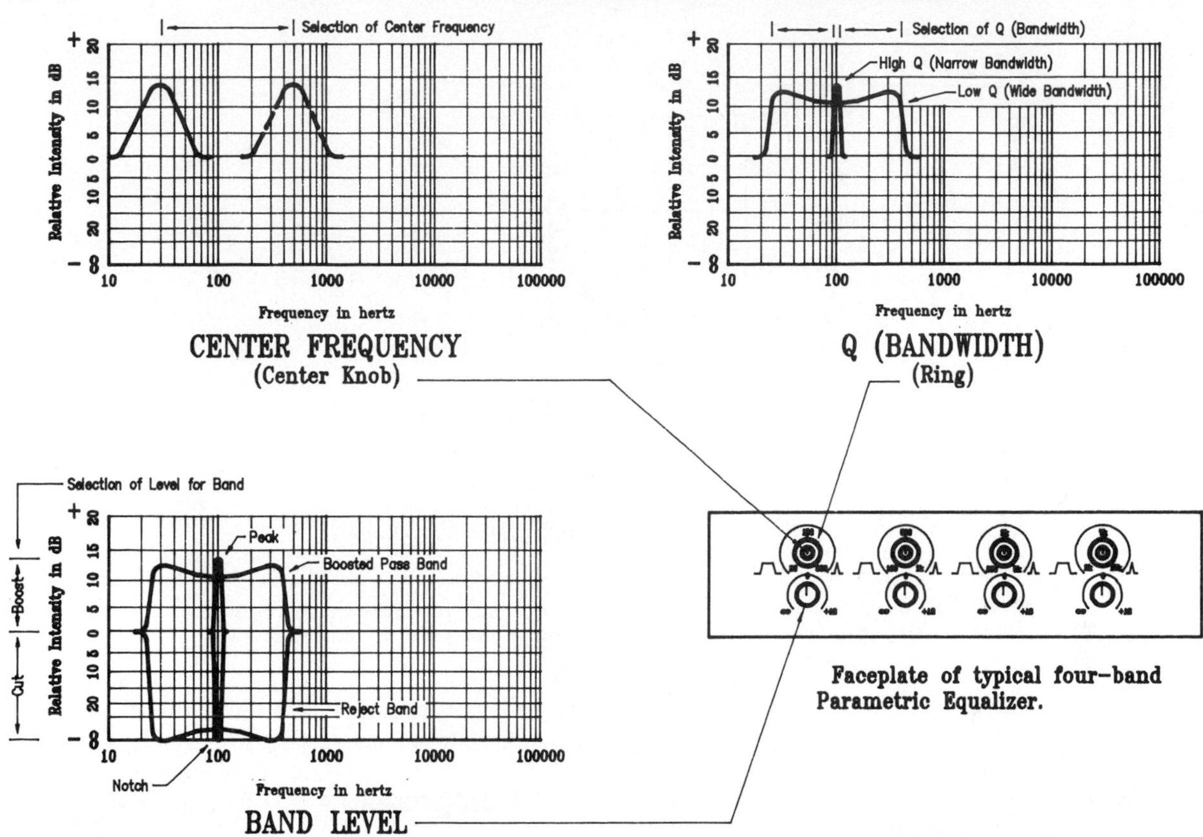

Whether keeping within the dynamic limits of a tape recorder or controlling small interpretive dynamics in underscore music, intensity correction must never be obvious. Subtle adjustments are necessary, and humans don't make subtle level corrections very well. In the first place, we usually don't detect the need for the adjustment until far too late; and when we do detect the need, our reflexes are not quick enough, or else we overreact. Electronic equipment is far better at subtle control. Equipment designed to regulate intensity falls into the category of automatic intensity control. Whether for technical or artistic purposes, automatic intensity control devices are useful.

Operational Principles

Automatic intensity control devices fall into three categories: compressors, limiters, and expanders. Within each category, variations in implementation exist, but the principle of operation is essentially the same: Intensity control devices modify the dynamic range of the signal. Compressors and limiters *reduce* dynamic range; expanders *increase* dynamic range. All automatic intensity control devices operate by modifying amplifier gain. The point at which modification commences is known as the **threshold**. A threshold is a designated condition (a particular level of intensity, in the case of automatic level control) that triggers activation of a particular function.

Negative Feedback

All automatic level control devices involve **unity-gain amplifiers**. A unity-gain amplifier, as its name suggests, has a gain ratio of 1. It does not increase or decrease the level of the signal. (See Fig. 8-7.) If the input level increases by 1 dB, so does the output level. In analog automatic level control equipment, a **negative feedback loop** is added to the amplifier to *reduce amplifier gain*. Gain reduction is triggered whenever input signal intensity crosses threshold; otherwise, the amplifier functions as a normal, unity-gain device. When the loop is active, output intensity is less than the intensity of the input. For a 1-dB increase in input, the output would increase by some fraction of a decibel.

Attack and Release

Time parameters are important in control of gain. If the feedback circuits act too quickly or too slowly, the action of the automatic intensity control device becomes obvious. **Attack time** is the speed with which the gain reduction circuits sense and act

FIG. 8-7: UNITY GAIN

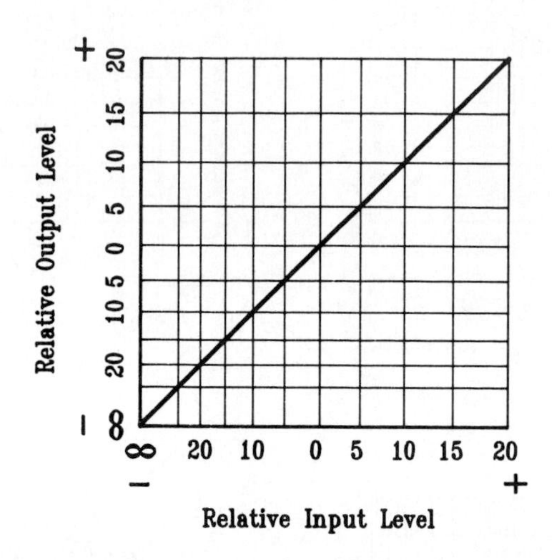

Note that output level is same as input level. Any increase or decrease in input level is matched by an equal increase or decrease in output level.

on a signal level rising above threshold. **Release time** is the rate at which gain returns to normal after signal level drops below threshold. Attack and release times are critical to smooth control in a gain reduction device. Most automatic intensity control devices provide a range of attack and release times.

Attack time can be very fast or relatively slow. Release time is always much longer than attack time, but can be made relatively short or very long. The difference between short and long for attack and release is considerable. Short attack time typically means 2 or 3 milliseconds; short release time usually means at least 50 milliseconds. Long for attack time is rarely more than 50 milliseconds, whereas long release time may be as much as 5 seconds.

Very short attack time produces little noticeable effect when a signal gradually creeps above threshold, but, given a short, sudden burst of intensity, the reaction from the compressor is violent, making the output sound as though someone had slammed the fader almost to the bottom of the slide, then brought it back up again. A longer attack time will let a sudden, loud signal get somewhat above threshold without significant compression for a few milliseconds, but the subsequent recovery is virtually unnoticeable. A short release time makes an obvious change in level—like raising a fader. A long release time sounds as if signal dropped suddenly, then faded back up to normal. When attack and release times are matched to average rates of signal variation, regulation is not discernible.

Operating Controls

Most compressors and expanders provide controls for input and output level, for setting compression or expansion ratio, for switching gain reduction circuitry on and off, and for setting attack and release times. Some compressors and expanders also provide a control for varying threshold level.

Compressors

A **compressor** is a device that reduces amplifier gain gradually as input signal intensity increases *above* threshold. Below threshold, input and output levels are the same; above threshold, the output level is always *less* than input level. The ratio of input level to output level is known as the **compression ratio**, also called gain-reduction ratio. Most compressors provide controls to vary the compression ratio. Standard ratios of compression are 2:1, 4:1, 8:1, 12:1, and 20:1. Figure 8-8 graphs a compression ratio of 2:1. Given a 2-dB increase in the input signal level above threshold, the output increases by only 1 dB. Gain-reduction ratio for compressors is rarely less than 2:1 or more than 8:1. Compression ratios of 12:1 or greater constitute limiting rather than compression. Many manufacturers make devices that may be used either as compressors or as limiters.

Limiters

Functionally, there is very little difference between a compressor and a **limiter**. Both are unity-gain amplifiers with automatic gain reduction circuitry that goes into action above a set threshold level. The difference is that a limiter effectively **clamps** the output. (See Fig. 8-9.) The input signal may increase by many times, but the output level will effectively remain at threshold level. A limiter is fully capable of eliminating almost the entire dynamic range of an audio signal, flattening it to very nearly one constant intensity level.

FIG. 8-8: 2:1 COMPRESSION RATIO

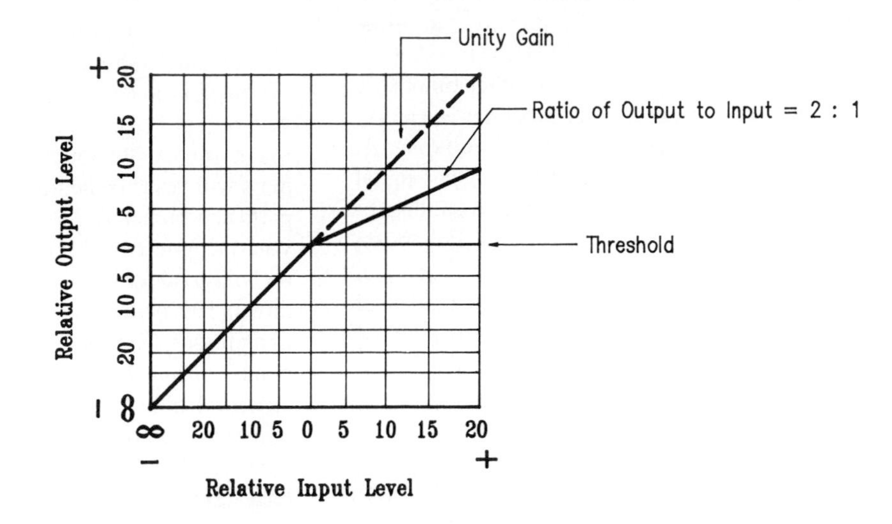

Below threshold, input and output levels are equal and change in equal amounts. 1 dB increase in input equals 1 dB increase in output. Above threshold, output level ceases to follow input level. An increase of 1 dB in input level produces only a 1/2 dB change in output level.

FIG. 8-9: LIMITING

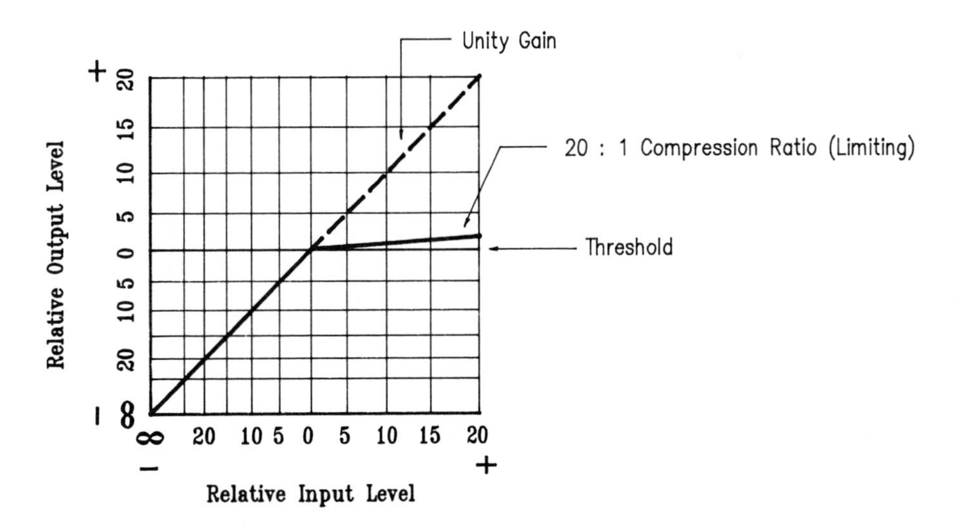

Above threshold, output level effectively does not increase. A 20 dB increase in input level will cause no more than a 1 dB increase in output level.

Devices intended exclusively as limiters often have a fixed compression ratio. Broadcast operations almost always use fixed limiters to prevent audio signal peaks from over-modulating the transmitter. Some kinds of reinforcement installations require fixed limiters to prevent the system from exceeding a determined sound pressure level. Most theatre sound applications are better served by variable devices that can function either in compression or limiting modes.

Expanders

An **expander** (Fig. 8-10) is the reverse of a compressor. A compressor reduces gain *above* threshold in order to *shrink* dynamic range. An expander reduces gain *below* threshold in order to *expand* dynamic range. Like the other automatic intensity control devices, the heart of an expander is a unity-gain amplifier with a negative feedback loop. In an expander, however, the gain control element is inactive when signal level is *above* threshold. When the signal drops *below* threshold, the feedback circuit begins to reduce gain. Expansion ratios determine the amount of gain reduction. For example, given an expansion ratio of 2:1, a 1-dB decrease in input level would result in a 2 dB drop in output level. The further below threshold the input level drops the greater the reduction in output signal level.

A **noise gate** (Fig. 8-11) is a special purpose form of expander. Its threshold is set just above the defined minimum program level for the system and gain reduction is set at maximum. Signal levels below threshold are essentially switched off. When signal appears at the channel input, the gate turns on if the signal is above threshold. Once on, the gate functions as a unity-gain amplifier. Noise gates are used to suppress noise by automatically disabling momentarily inactive channels.

A Practical Example

The use of a compressor may not be immediately obvious, so let's use an example to clarify. Say that you need to use a piece of music in a stage production, but the piece contains some very soft passages. You don't need a very loud level overall, but if you play the loudest parts at a suitable intensity, the soft parts will never be heard. If you play the music so that the soft parts can be heard, the loud parts will be too loud. What do you do? You could ride gain manually, but inevitably some part of a loud passage will slip by before the fader can be pulled down. The better choice is to use a compressor. Using the compressor, first set gain reduction *off*; then adjust input control so that the softest passages come through at a satisfactory level. Now turn gain reduction *on* and set the compression ratio until the loudest passage stays within manageable limits (usually staying at or below 0 VU). Make any adjustments in attack and release times to get the smoothest sound, and then set the output level for desired playback intensity. Once the output level is set, the dynamic range of the signal will stay within boundaries. You will probably have compressed the dynamic range from something like 90 or 100 dB to within 40 dB or less.

FIG. 8-10: EXPANDER

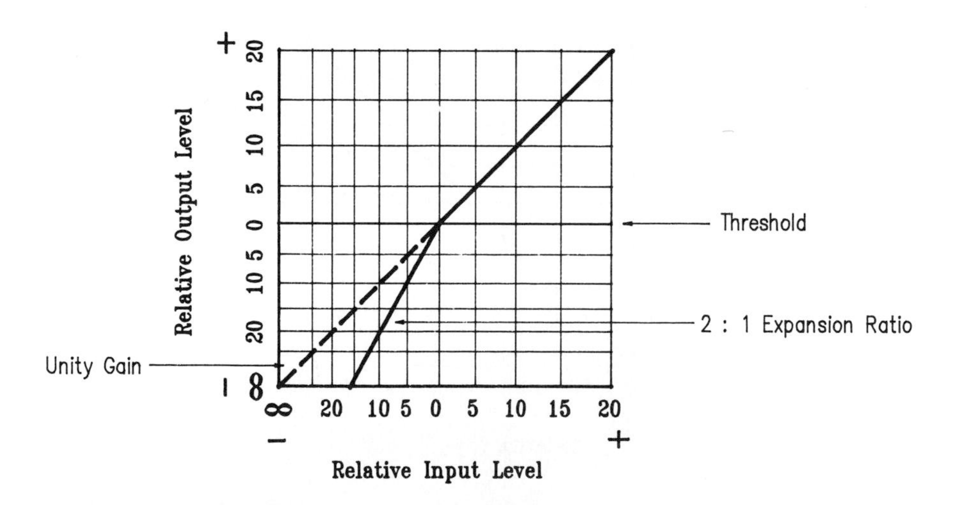

Above threshold, output and input follow unity gain relationship. 1 dB change in input level produces 1 dB change in output level. Below threshold, output level change is twice input level change. 1 dB change in input level produces 2 dB change in output level.

FIG. 8-11: EXPANDER USED AS NOISE GATE

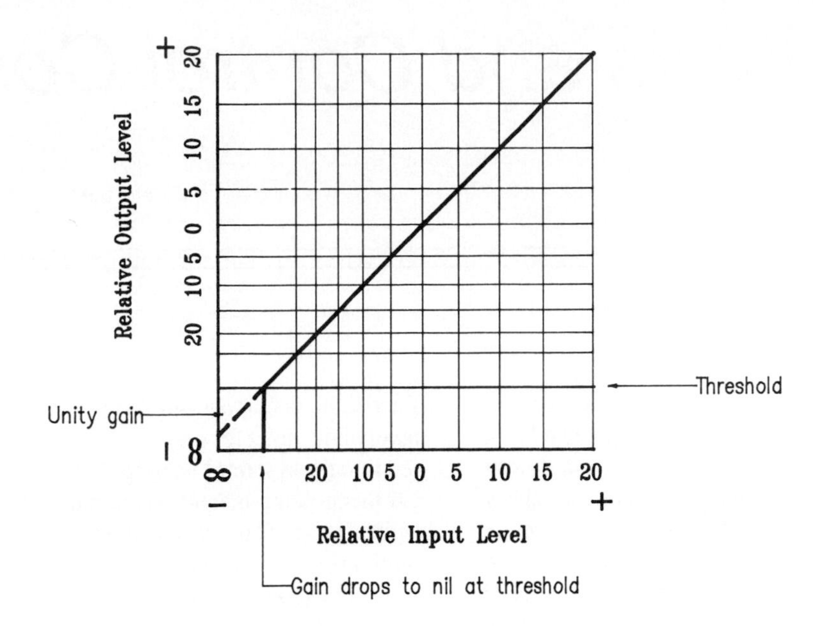

Above threshold, device behaves as simple unity gain amplifier. Below threshold, gain drops to nothing and amplifier turns off. In effect, amplifier is off for low-level noise but turns on for signal.

Overall, even though signal processors are only one part of an total audio system, there is no more useful group of devices for the theatre sound artist. Without them, the best part of what can now be done to provide environmental enhancement for dramatic production would be, at best, extremely difficult if not impossible.

Chapter 9
Mixers and Control Consoles

The heart of an audio system is the special configuration of amplifiers and controls that make up the **mixing console**. A mixing console is exactly what its name implies—a device that can mix a number of input signals together to make a single, composite output. Most consoles, especially those for reinforcement, actually provide a number of outputs; but the principle is the same as if only a single channel were involved: Several input signals are mixed together to create the composite output for any given channel.

Of all audio components available, consoles are probably the most specialized for particular purposes. Broadcast operations, dramatic and concert reinforcement, monitor mixing, studio recording, and public address—all have their own particular arrangement of controls and components. Some areas require features and functions that others do not. Consoles for theatrical effects playback are among the most specialized because they require a rather different configuration than do most other consoles.

Consoles for the majority of audio applications are, as we shall see, devices that combine a large number of inputs to make a single output. For the theatre and for background music and paging systems, the console must distribute a single input to a number of outputs. Combining signals into a single composite output is known as *mixing* or *mix-down*; we shall refer to distributing one signal to a number of outputs as *fan-out*.

Basic Characteristics of Mixing Consoles

All mixing consoles have at least one common attribute: the ability to receive two or more input signals and mix them together to form a composite output signal. Figure 9-1 is a diagram of a rudimentary mixer. The system shown is highly simplified and *passive*—meaning that it has no amplifiers. Each input is fed to a variable potentiometer that regulates the level of the input for the mix independently of any other signal fed into the system. The output of the system is taken from yet another potentiometer that regulates the level of the composite (mixed) signal output. The junction of all of the inputs and the output is called the *mixing bus*.

The passive system of Fig. 9-1 is not a practical system.[1] Schematically, however, the figure shows the basic operation of a mix-down console. If we replace the resistor network of Fig. 9-1 with the amplified system of Fig. 9-2A, we get something that begins to look like a typical mixing system. Each input branch has an input amplifier, a level control, and a booster or line amplifier. All of the booster amps connect to the mixing bus, which, in turn, feeds the first stage amplifier of the output circuit. After the first stage amp is the master level control, followed by another booster amp.

To complete the console as a reasonably finished device, we could add microphone preamplifiers as alternate input elements and a VU meter at the output to provide a check on the level of the composite signal (Fig. 9-2B). Completed in this way, the console of Fig. 9-2A becomes a basic public address mixer, just about the simplest mixing system one normally encounters. Most audio systems require somewhat more control capability.

Traditional Console Design

To understand the operation of mixers, let's examine the characteristics and construction of a traditional console designed for the most basic kind of signal handling in the audio industry—broadcast audio. Traditional broadcast consoles serve the need to select one or more signals quickly and easily from among a small, fixed group of inputs and to feed the

[1]The system shown in Fig. 9-1 wouldn't work well enough to be worth trying, not even as an experiment. Passive mixers are possible, of course, but not without a lot of fancy engineering. Passive mixers can be more difficult to build than active mixers, especially now that IC op amps are available. Most mixing devices in present-day consoles are constructed as some form of summing node using operational amplifiers.

FIG. 9-1: PASSIVE MIXER

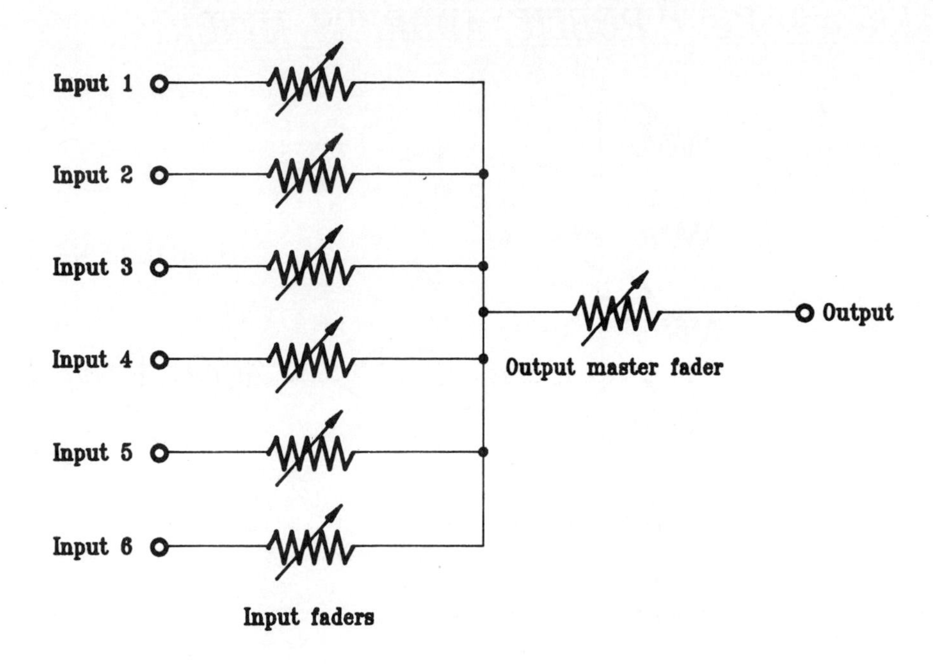

FIG. 9-2A: ACTIVE MIXER

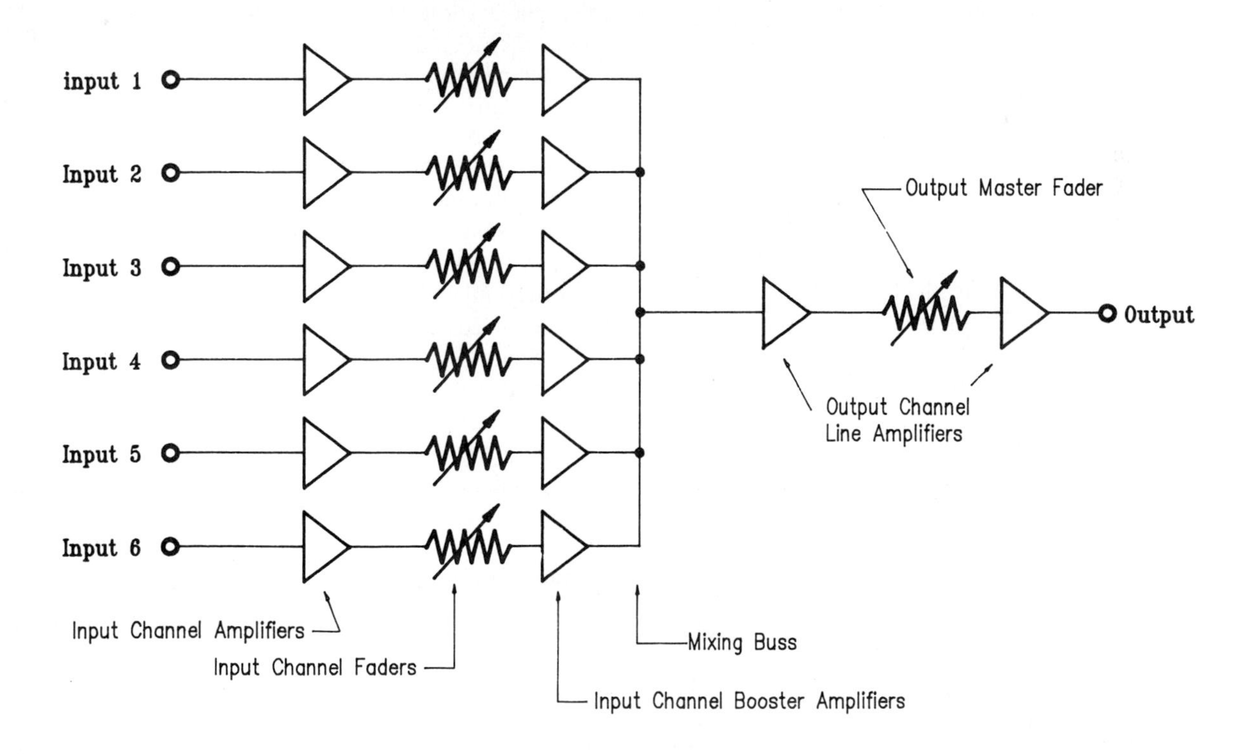

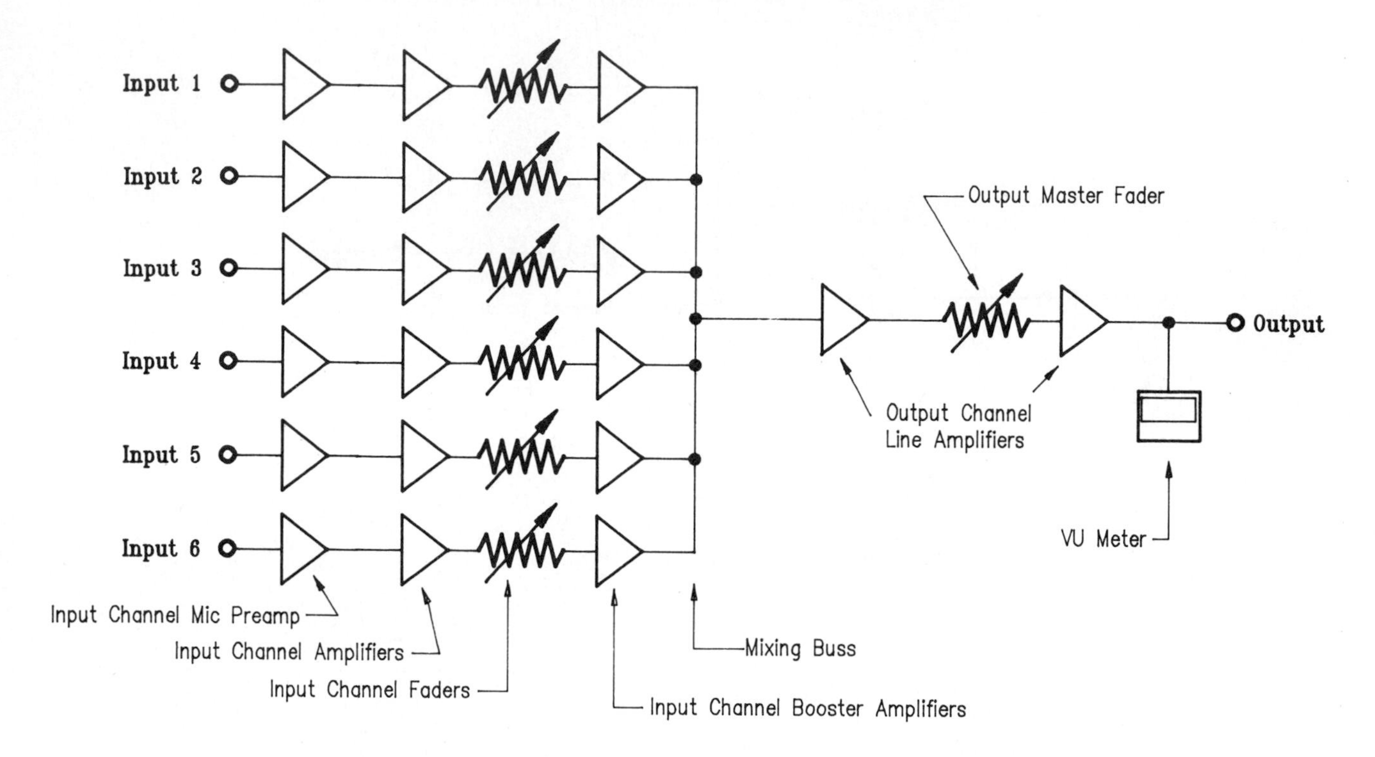

selected signal to a specific destination, usually to a transmitter or a network line.[2] The broadcast console shown in Figs. 9-3A and 9-3B is typical of consoles dating from about the middle of the twentieth century. Although such consoles are certainly obsolete, they serve to illustrate basic principles of mixing systems in general.

Recall the simple mixer in Fig. 9-1. You can readily see its basic pattern in the block diagram of a typical broadcast console. (See Fig. 9-3A.) The mixer has several inputs that may be added together on a common output. In the case of the broadcast console diagrammed, there are two mixing busses and two outputs. Switches are used to connect each input to one or the other of the two mixing busses.

One of the main characteristics of a traditional broadcast console is the use of *dedicated inputs*—inputs permanently assigned to microphones, turntables, tape recorders, and network or telephone lines. Dedicated controls are ideal for broadcast use because the operator becomes accustomed to a set location for each control function. Observe the control panel of this console in Fig. 9-3B, and notice that the controls are placed so that the most frequently used are closest to the operator's hand. Under the pressure of live broadcast schedules, any convenience that makes operation habitual is an advantage.

Note in Fig. 9-3A that inputs specifically intended for microphones or turntables have the appropriate preamplifiers and equalization devices built into them. The source devices (microphones or turntables) may even be hard-wired to the inputs. (*Hardwiring* implies a permanent connection that cannot be plugged or unplugged at will.) Present-day consoles do not use dedicated inputs, because free inputs are much more flexible. However, if a source requires equalization, impedance matching, or preamplification, those needs must be provided by resources outside of the console itself. For example, a phonograph pickup requires outboard preamplification and RIAA equalization.

One output channel of a broadcast mixer is always used to drive the transmitter. Most broadcast operations need at least one additional channel, however. Consequently, most broadcast mixers incorporate two mixing busses and output channels. The second channel is usually called the *audition line* or the *preview line*. The names derive from the practice of verifying program material prior to transmitting it. The audition line can also serve as a backup for the main channel in an emergency. Traditional consoles use a lever switch called a *key*, or *keyswitch*, to connect an input to either of the two mixing busses; there-

[2]Traditional broadcast consoles are designed specifically to facilitate real-time live broadcast operations. Studio recording consoles, however, are much more flexible than are traditional broadcast audio mixers, especially for the process of program assembling and editing. Consequently, offline production work is better handled using recording consoles. Large stations and network operations use recording consoles as offline mixers, keeping traditional broadcast consoles as online operational units. Small or low-budget stations frequently use a studio recording console in place of a standard radio/TV audio mixer simply because the one unit can handle both jobs, thereby reducing the cost of hardware installation.

FIG. 9-3A: TRADITIONAL BROADCAST CONSOLE

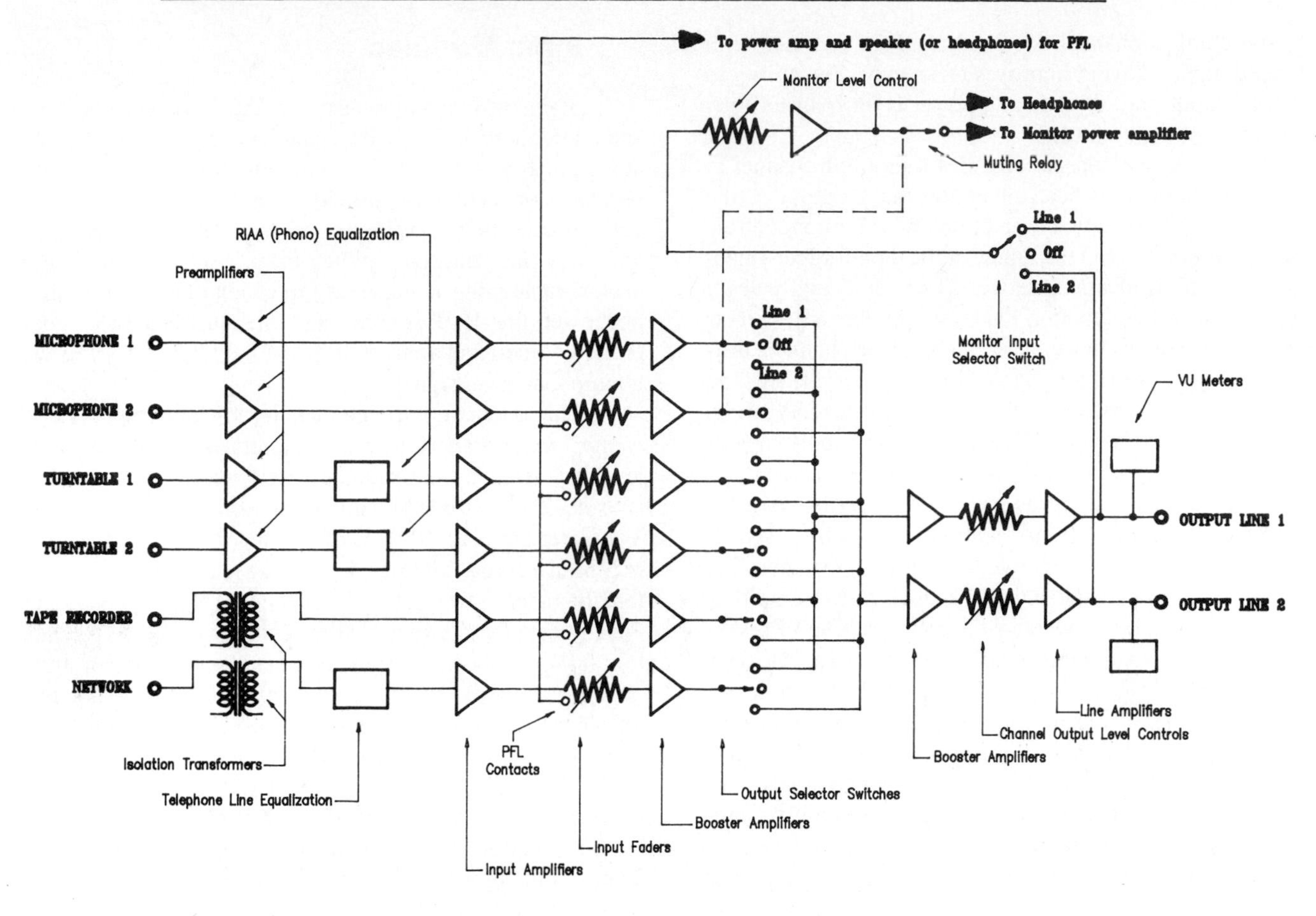

FIG. 9-3B: TRADITIONAL BROADCAST CONSOLE

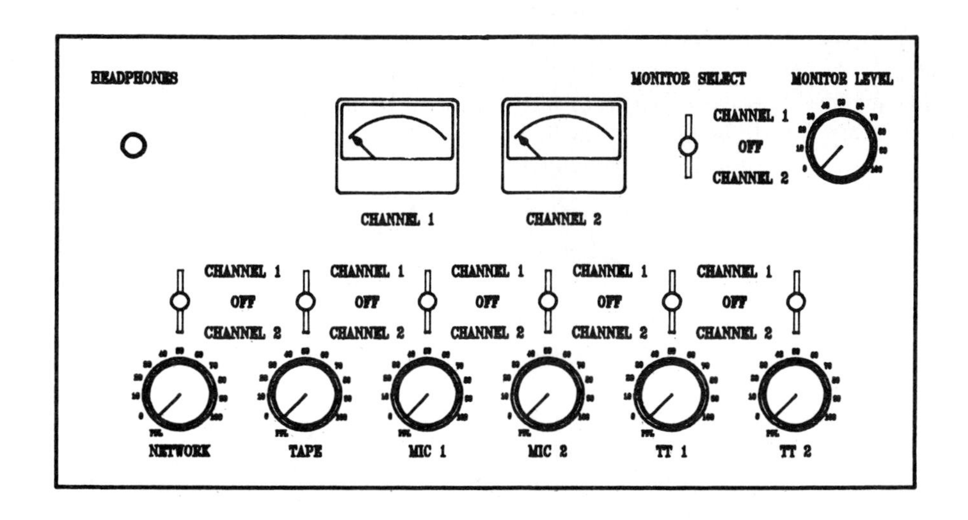

fore, switching a channel is often called *keying*. (As with dedicated inputs, keying improves efficiency. The ability to switch an input onto the bus at a preset level is faster than fading the input up to level.)

The broadcast console provides a monitor channel to permit the operator to hear either the main output to the transmitter or, alternately, the audition line. Monitor controls include a switch to select the source of the monitor feed (main channel, audition channel, network line, etc.) and a level control so that the monitor loudness can be adjusted as needed. Keyswitches that control microphone channels usually also trigger **muting** relays (shown by the dashed line connecting the microphone keyswitches with the relay in Fig. 9-3A) which cut off control room or studio monitor loudspeakers to prevent feedback during the time the microphones are live. Broadcast systems also provide *PFL* (*pre-fade listen*). PFL (also known as *cue monitor*) is usually a switch position at the bottom of the level fader's travel (whether slide or rotation) that routes the audio to a small power amplifier and speaker, or into the operator's earphones. The PFL system is used for positioning records and tapes so that their audio program begins as soon as they are started.

The broadcast console described here is typical of very traditional radio and television operations. Console functions may be integrated into a single chassis; input channels have no EQ or effect-send capability; and no grouping or submastering capability is provided.

Characteristics of Modern Consoles

Most consoles built within the past two decades, including the newer broadcast consoles, tend to be modular. That is, they are designed around a group of removable units. Console modules usually provide an expanded range of controls, allowing each individual element of a mix to be processed according to its own needs. Standard modules include input controllers, output controllers, master faders, slate-talkback-tone generator controllers, and auxiliary send/receive controllers. Inasmuch as modularity is such an important feature of current console technology, we shall examine the most common elements found in modular consoles. Following the discussion of modular elements, we will look at the way the standard units are applied to create consoles for particular aspects of audio production.

Console Mainframe.

The **console mainframe** provides the housing and electrical connections for all of the modules used to assemble the console functions. It usually contains all of the VU meters, all of the input and output connectors, and the *motherboard*—the printed circuit board into which all of the modules are plugged.

Input Modules

Standard input modules (see Fig. 9-4) serve both line and microphone level signals, changing from one to the other at the touch of a button, called the **mic/line switch**. Usually, separate connectors are provided for microphone and line-level sources; however, lower cost modules may be equipped with only one connector. When there is only one input connector, replugging is necessary to change between a microphone and line-level sources, and one must be careful not to plug in a line-level source when the module is set to receive microphone level signals.

The primary control element on any input module is the *channel fader*. Faders are slide controls used to set the signal level delivered by the module to the mixing section of the console. Faders are indexed to correspond to the scale on the VU meter (see Fig. 9-4). With the fader set at 0, a test tone may be used to calibrate the output for a reading of 0 VU. (Note that the fader can be set to produce more than 0 VU output.) Faders act only to *attenuate* the signal level. When the fader is at full (+3), the maximum level that the module can supply is being delivered to the mixing bus. Reducing fader level attenuates the signal strength by introducing resistance between the module amplifier and the mixing bus.

Maximum module signal level is determined by the *gain control*. Gain is the amount of amplification provided by the module electronics. Gain is measured in dB, comparing the level of the signal as received at the module input to the unattenuated signal level at the module output. The gain control regulates the number of dB increase in signal level that the module can provide. Gain control is important to ensure that all microphones and other signal sources behave in similar fashion, thereby simplifying the operator's job in setting levels in the mix. The gain control calibration indicates number of decibels of amplification.

Standard features of input modules generally include equalization, effect and foldback auxiliary sends, inserts, and output switching.

Equalization in console modules follows the pattern of equalizer units discussed in Chapter 8. EQ is provided to balance out any unusual spectral components resulting from microphone characteristics or from the acoustic environment. Equalization may be designed either as shelving or as peaking control and is usually divided into three bands, though some consoles have as many as four bands and some have only two. EQ for most consoles is *quasi-parametric* in that the operator may control some but not all of the equalization parameters. All consoles provide control of depth (number of dB boost or cut), and most provide a sweepable control to select rolloff and/or center frequencies.)

Auxiliary sends are designed to combine signals from individual input modules into alternate, specialized output paths, leading to outboard processing equipment such as an echo or delay line (*effect send*), or to monitor channels for performers to use during recording (*foldback send*). Effect

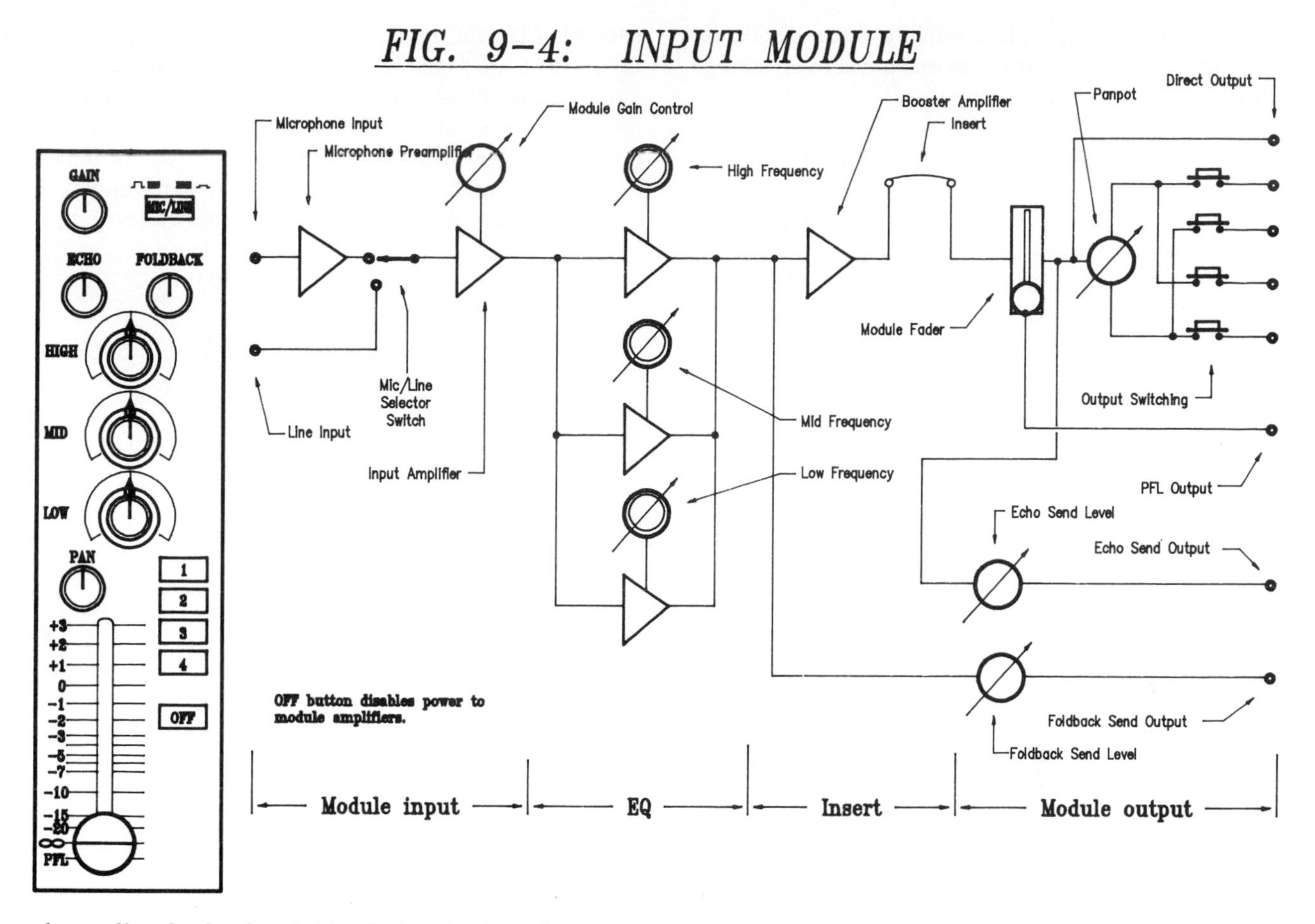

FIG. 9-4: INPUT MODULE

sends usually take the signal either before (pre) or after (post) the module fader. Foldback sends originate pre or post EQ. Some manufacturers provide switches to select origination point; others provide internal *jumpers* that can be changed only if the module is removed from the console. The least expensive systems do not permit change. Auxiliary sends have individual level controls on each module, and the combined auxiliary signal is mastered by a control on a separate auxiliary output module.

The module provides patch points, called **inserts**, which permit an outboard processing unit, such as a delay line or graphic equalization, to be connected into the module's signal path. An insert is made up of two jacks—one to send the outgoing signal and one to receive the returning signal—which are provided with *normaling contacts*. Normaling contacts close against the plug contacts of the jack when the jack is empty. The normaling contacts carry the signal through the jacks and maintain an unbroken signal path. Inserting a plug into one of the jacks breaks the signal flow through the module. (See Chapter 14 for a more extensive discussion of normaling.)

Be sure that you recognize the difference between an insert and an auxiliary send. The insert provides a path to outboard equipment for the individual input module only; the auxiliary send combines signals from any or all input modules into an alternative signal path, either to outboard processing or to monitor amplifiers.

The final step of the module is its **output switching**, a bank of switches which assigns the signal to any or all of the mixing busses in the console. In present-day consoles output switching usually accesses busses in pairs and is combined with a rotary control called a **panpot** (an abbreviation of the original term "panoramic potentiometer"), which proportionally assigns the signal to either or both of the two busses. Usually turning the pan control counterclockwise shifts the signal to an odd-numbered buss while a clockwise turn shifts the signal toward an even-numbered buss. Because odd-numbered busses are usually left channels and even-numbered busses are usually right channels, panning serves to place the signal in the stereo spectrum. Modules in some systems will also provide an individual line connector called a *direct output*.

Channel faders, mic/line switches, EQ, sends, output switching, and panpots are the main control elements of input modules. Depending mainly on the cost of the system, other features may also be included, such as separate gain controls for mic and line inputs, peak level indicators (an LED that blinks each time the signal momentarily exceeds the defined maximum signal level of the module amplifiers), solo (a switch that excludes from the monitor system or headphones [but not from the mix outputs] all signals except that of the particular input module), mute (a switch that shuts off the output of the module without altering the gain and/or EQ settings), and indicator LEDs that visually signal details of system status, such as solos or mutes that are active, routing of module signals, and the like.

Output Modules

Several varieties of output module can exist for any particular console—auxiliary master, stereo mix output controller, matrix and group controller, among others. The group controller module may be structured as a main output with auxiliaries or as a main output with a matrix. Usually most of the output modules included in a console mainframe will be mixing channel controllers, with one module each for auxiliary, monitor, and stereo mix.

Group Master Module.

Figure 9-5 is a diagram of a **group controller** module. One of these modules is provided for each of the console's main output channels. All of the input modules that are assigned to the group are summed into the controller. The controller module has its own complement of controls, much like those for the input modules. The fader, indexed to match the calibration of the VU meters, sets the level for the output channel. Like the input modules, the group controller has inserts to permit an outboard processor device to be connected into its signal path. Routing in this module is accomplished through the panpot and the group level control rather than through routing switches. Panpot outputs go to the stereo master module. The group output is similar to the direct output on an input module, except that it has its own separate level control.

Matrix Module.

Figure 9-6 is a diagram of a group controller module with a **matrix**. Signal flow for the module is similar to the standard group master module, except that, in place of a direct, group output, this module feeds its composite signal to a matrix of outputs that form, in effect, a small auxiliary mixing system. The composite signal for the group is routed from the module fader to the matrix. The matrix, in this case, has four output busses, so that each module has four level controls regulating the level of the module's signal input to each channel of the matrix. Each module contains one of the matrix's summing busses and the master level control for that channel of the matrix. A matrix output makes possible routing any combination of signals to serve particular needs. For

FIG. 9-5: GROUP MASTER OUTPUT MODULE

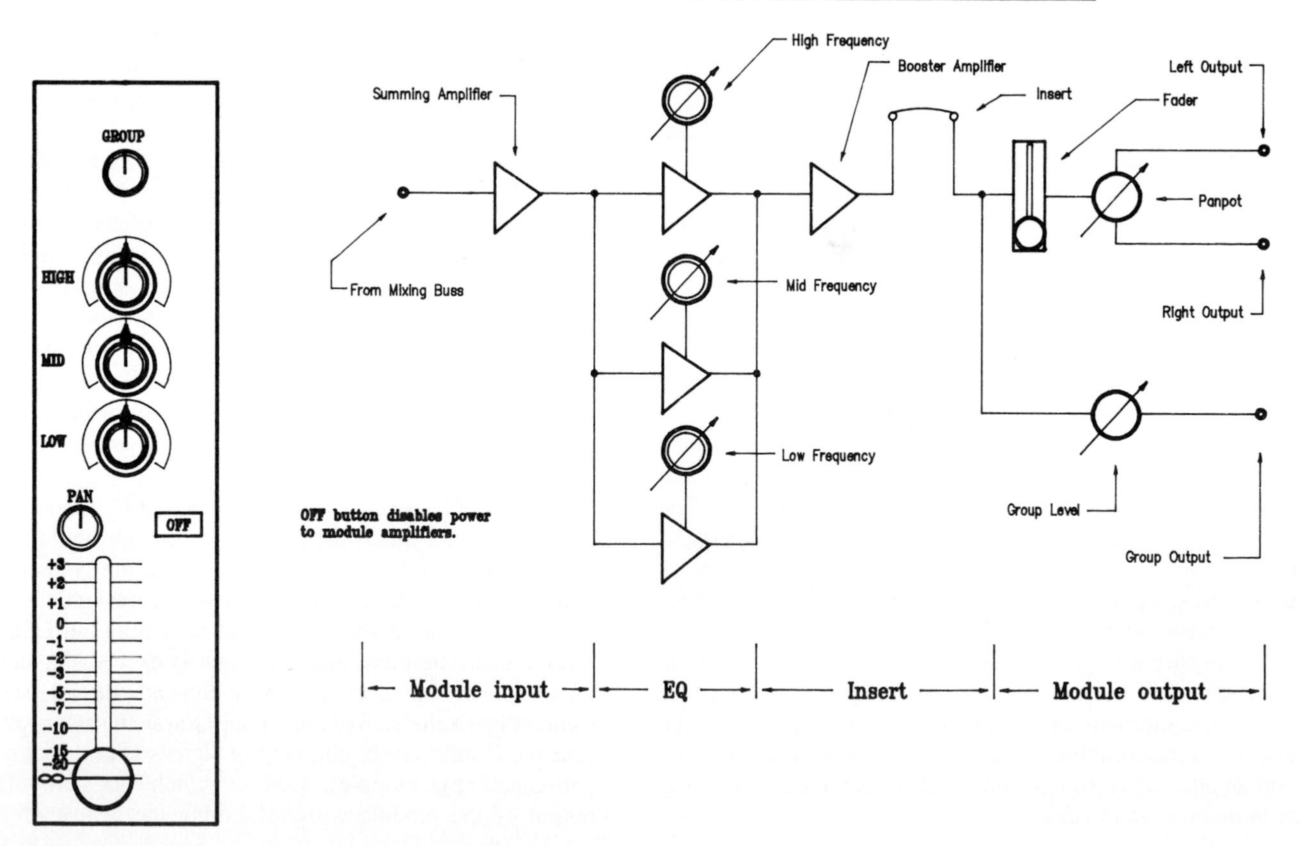

FIG. 9-6: MATRIX OUTPUT MODULE

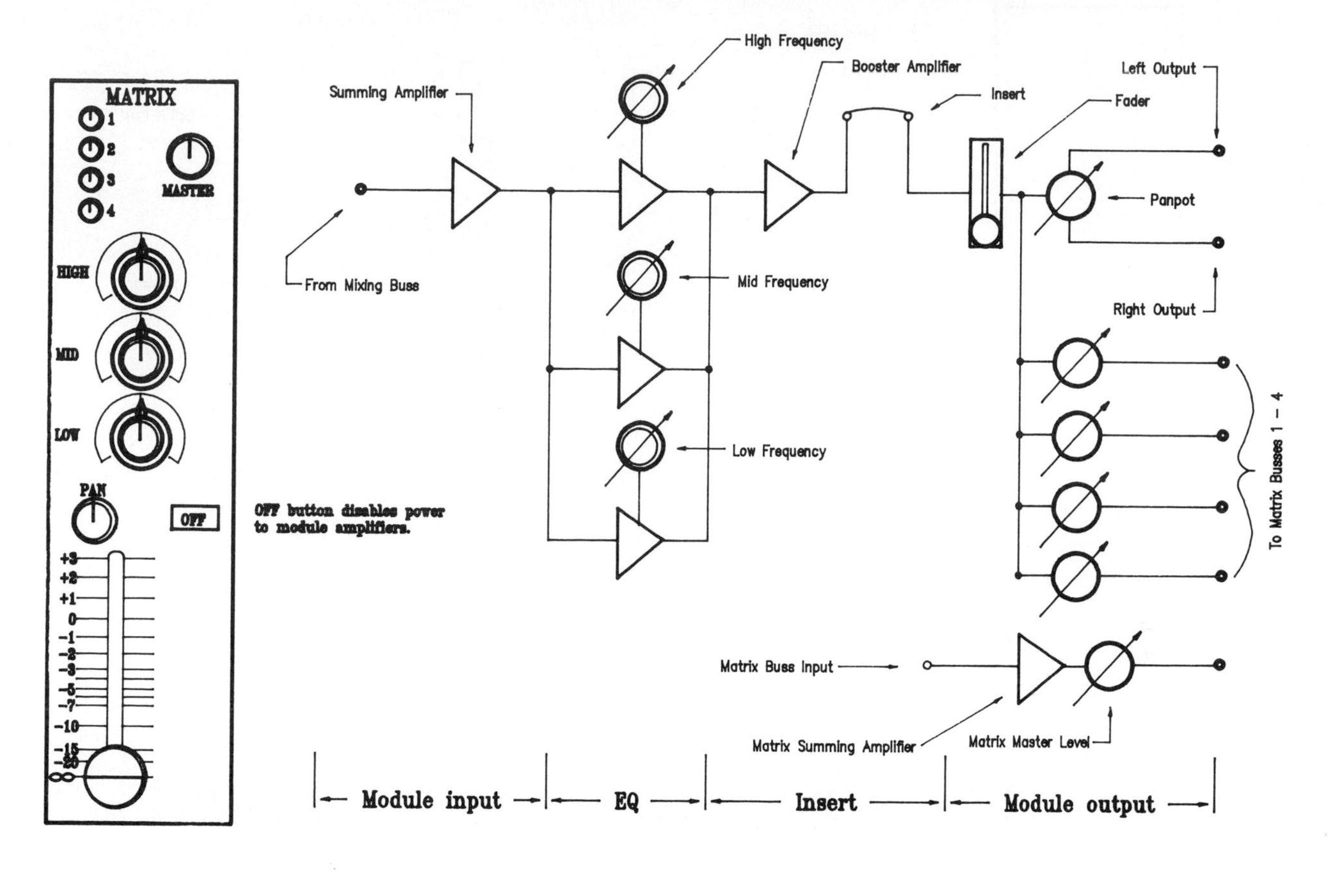

example, if one part of a reinforcement system needs only the feeds from the keyboards, the backup vocalists, and the traps, then those signals are summed into the matrix and fed to the amplifiers for that group of loudspeakers.

Stereo Master Module

The **stereo mix output** module (Fig. 9-7) provides dual faders for left and right stereo channels and a main output for the stereo mix. The module is simple and straightforward in its function: It receives a composite signal from the left and right pan busses and passes that signal on to the stereo left and right outputs and to the console monitor amplifiers. Although not shown in Fig. 9-7, this module sometimes incorporates other functions, such as monitor level controls and mic controls for slate, talkback, or paging facilities.

Auxiliary Output Controller Module.

The **auxiliary output** (Fig. 9-8) is frequently used in studio recording, where the number of secondary outputs is limited. Matrix output modules are more often used in theatrical reinforcement, where a large number of different balances and distributions of signals may be needed. Auxiliary outputs are frequently included in consoles with matrix output configurations, however.

Miscellaneous Functions

Depending on the particular applications of a console, certain kinds of functions may be needed, such as *slate* (a way to send an identification signal simultaneously to all channels of a recording), *talkback* (a means for voice communication from console to a remote location such as an enclosed recording studio), *headphone level control* and *headphone jack*, and a *setup tone generator* (a steady-state signal to allow calibration of VU meters). When functions such as slate and talkback are required, the module either must provide a built-in microphone or must allow an external microphone to be connected to the module. Figure 9-9 shows a module that incorporates all of the functions mentioned here.

Another element included in almost all consoles is a *meter panel*. The meter panel contains several VU meters, and it may also provide switching to permit the function of one or more meters to be varied according to need. See Fig. 9-10 for a typical meter panel.

FIG. 9-7: STEREO MASTER OUTPUT MODULE

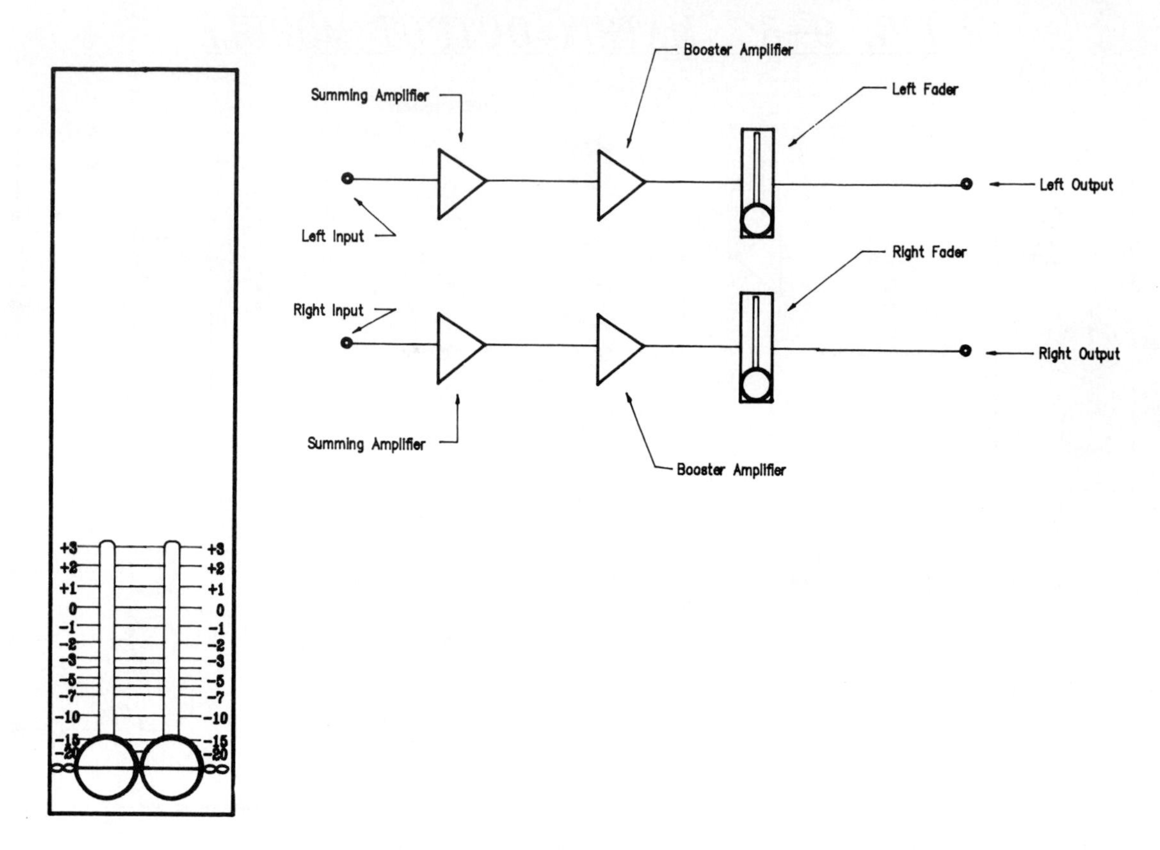

FIG. 9-8: AUXILIARY MASTER MODULE

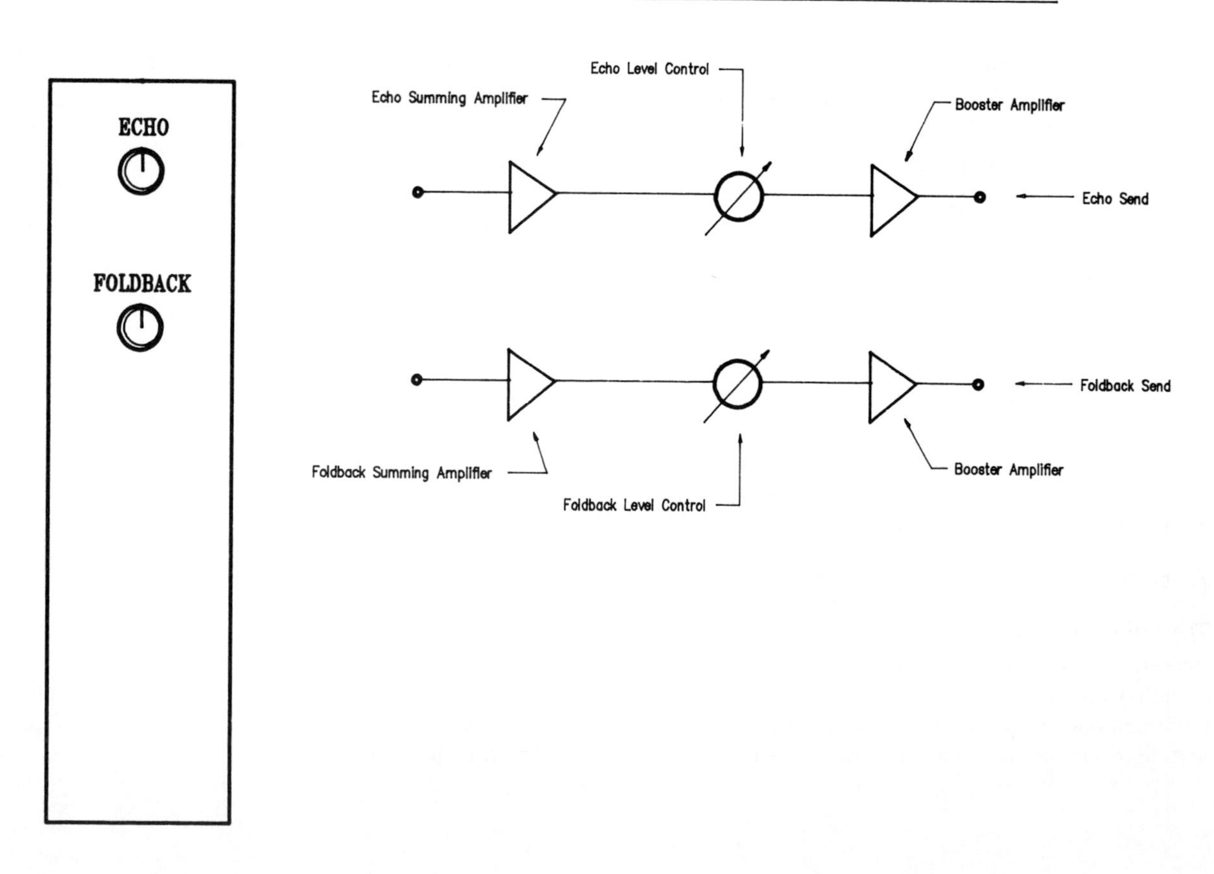

FIG. 9-9: TONE/TALKBACK MODULE

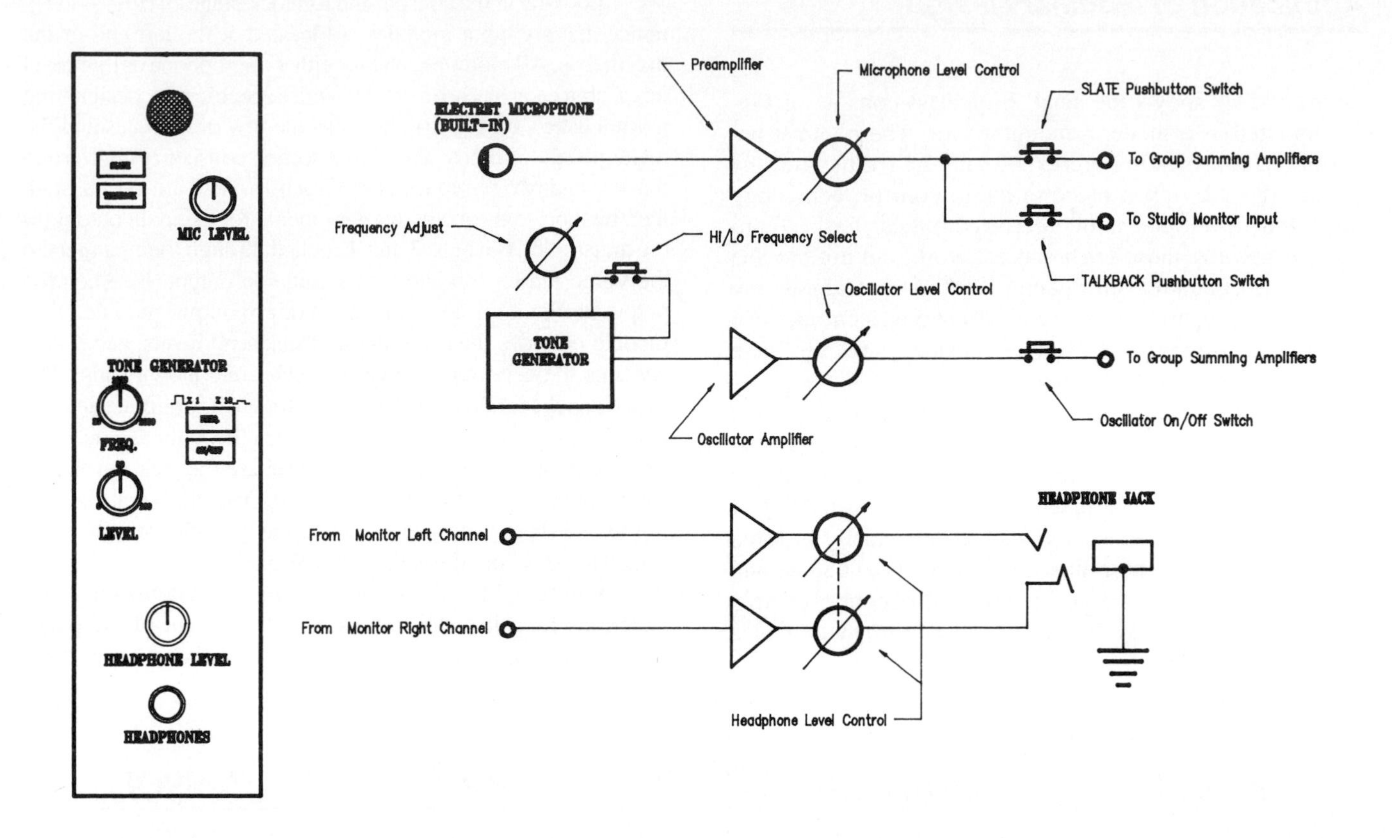

FIG. 9-10: VU METER PANEL

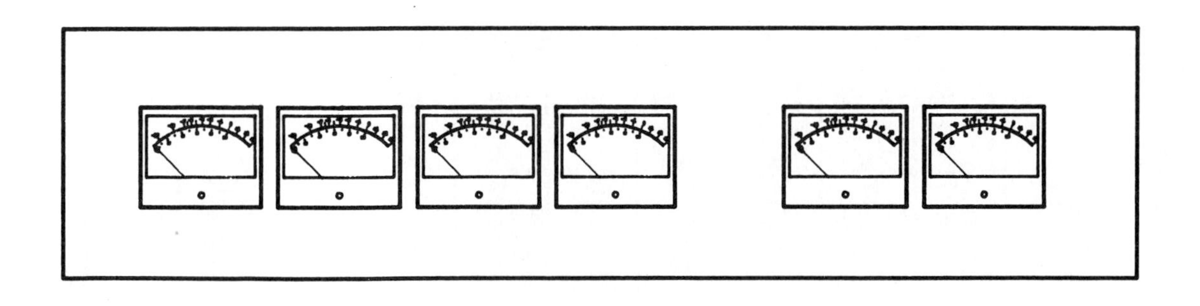

Application of Modular Design

Figure 9-11A shows the small broadcast console of Fig. 9-3 updated as a modern, modular unit. The contrast between the consoles in Figures 9-3 and 9-11 illustrates the change that has taken place in audio control technology since the development of microelectronics. Earlier consoles, especially those for broadcast work, did not provide the variety of controls now commonly found on almost any console, due in part to the size and cost of the electronics that would have been required. Earlier consoles, however, were much simpler and less cluttered.

Although present-day consoles are only slightly more difficult to learn than older consoles, the beginner sees what appears to be a complex system with an intimidating number of buttons and knobs. The essential concept to grasp is that of modularity. Each module of a given kind has exactly the same controls. Therefore, rather than a large number of buttons and knobs to be mastered, one has a small group of control devices repeated many times. When one module is learned, the rest is easy.

Looking at the console and its block diagram (Fig. 9-11B), notice that six input modules are located at the left end of the mainframe. All modules can take either microphone or line-level inputs; but each has been labeled with a specific title, designating normal usage, on the strip between the row of modules and the meter panel. Similarly, the output section consists of four group masters and two stereo masters. Groups 1 and 2 are tied through their panpots to the stereo master module for the main output (to the transmitter.) Groups 3 and 4 are tied through their panpots to the stereo master module for the audition output. Between the output and input modules is an auxiliary output module. This module provides the echo and foldback send levels, and it also contains the echo return level and echo return switching. The meter panel contains one VU meter for each group master and two meters that can be switched between the main and audition stereo outputs. Also on the meter panel are the jack, level, and source switching for the monitor system. Trace the block diagram in Fig. 9-11B to follow how the various modules are interconnected to constitute the entire console system.

Figure 9-11C shows the input/output connector panel of the modular broadcast console. At the right of the panel (behind the input modules) are six microphone connectors

FIG. 9-11A: MODULAR VERSION OF BROADCAST CONSOLE

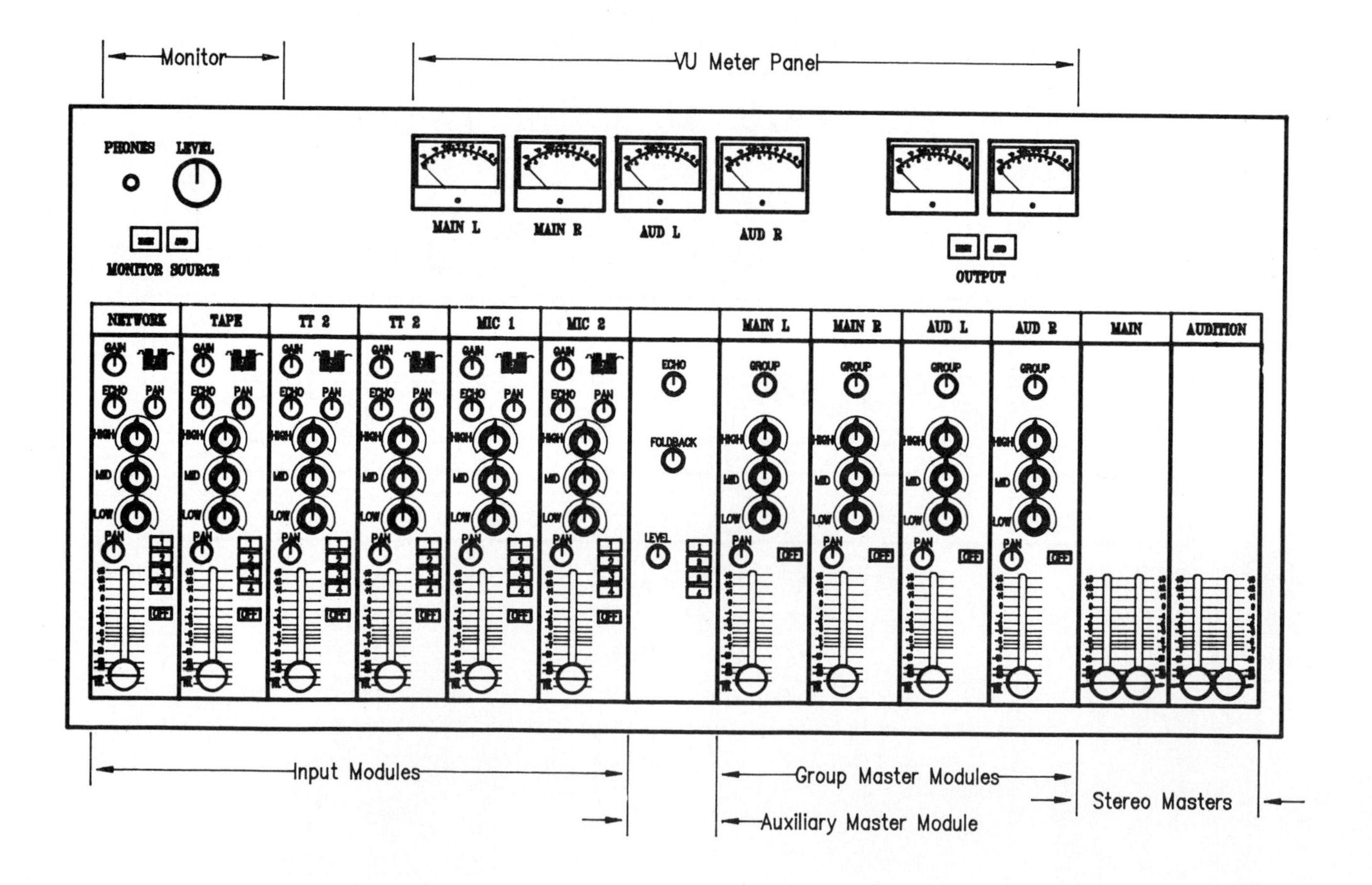

FIG. 9-11B: BLOCK DIAGRAM OF BROADCAST CONSOLE

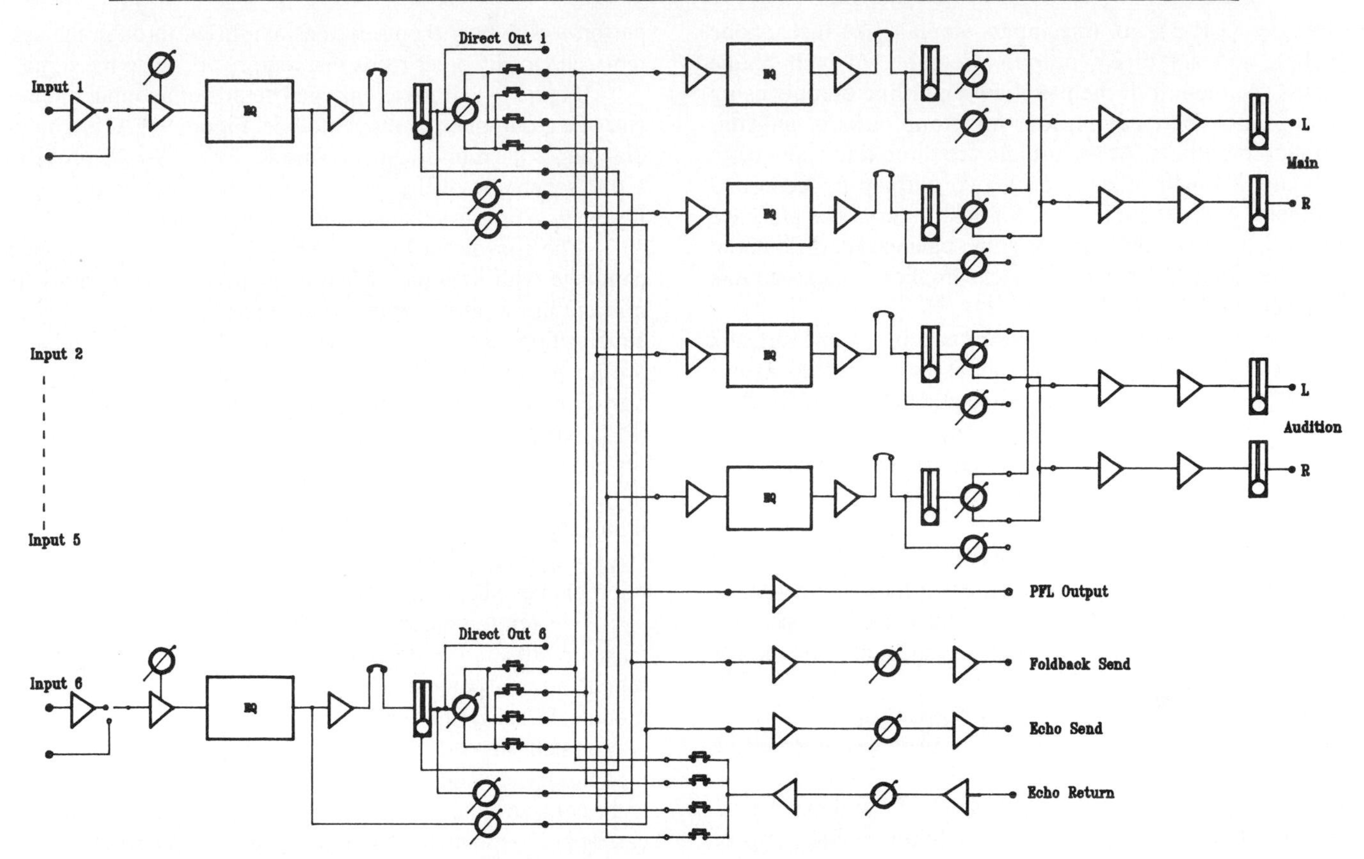

FIG. 9-11C: BROADCAST CONSOLE CONNECTOR PANEL

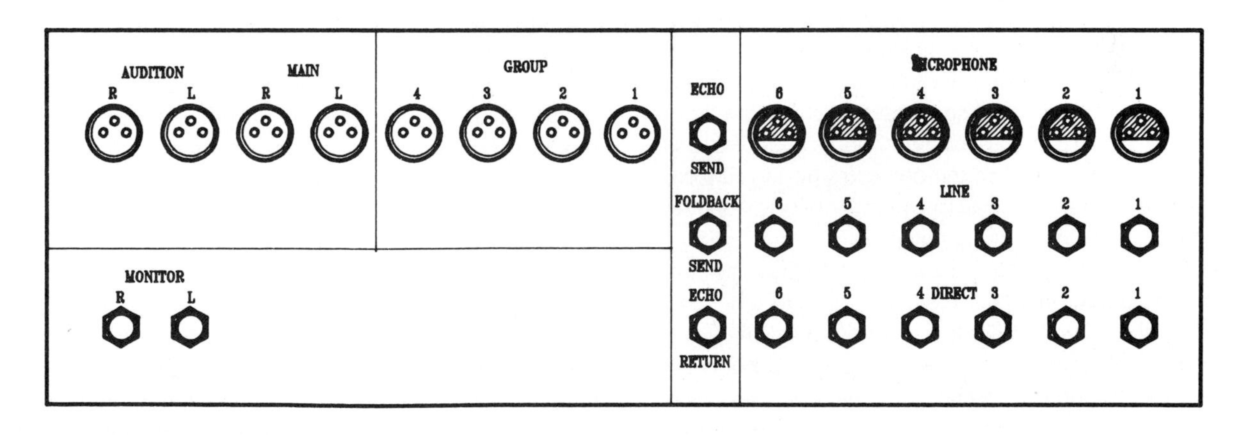

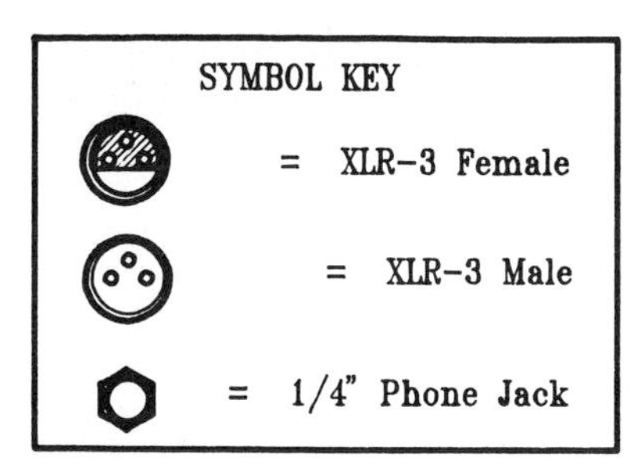

(female XLR-3), six line inputs (female 1/4-inch phone jacks), and six direct outputs (female 1/4-inch phone jacks). At the left of the panel are eight line outputs using male XLR-3 connectors, four for group outputs and four for stereo outputs. At the middle are three line connectors (female 1/4-inch phone jacks), two outputs for echo and foldback sends and one input for echo return. The presence of XLR-3 connectors suggests either balanced or differential configuration; 1/4-inch phone jacks implies unbalanced line connections.[3]

All consoles require connector panels of some sort, and connector panels tell a great deal about the nature and capabilities of the console. As we examine other consoles, we shall look at the connector panels to determine what kinds of connectors and line configurations are needed in order to use it with other pieces of equipment.

Let's now proceed to examine the ways in which standard modules are combined to create consoles for two other major aspects of audio—acoustic reinforcement and studio recording, both of which are very closely related to audio needs in theatre.

Acoustic Reinforcement Consoles

Figure 9-12A shows the control panel of a full-sized reinforcement console. Reinforcement requires a larger array of modules than does a broadcast system, because reinforcement demands a large number of microphones, a much more detailed output section, and a more complex monitoring system.

Reinforcement depends on microphones, but it also requires the ability to accept line-level feeds from devices such as instrument amplifiers and possibly from other consoles. Typical practice in reinforcing musical performances involves the use of wireless microphones for major roles, cabled microphones used as general cover pickups or placed to catch voices from secondary roles or chorus members, and cabled microphones placed in the orchestra pit to reinforce instrumental musicians. Such a large number of microphones requires many microphone input channels.

For the most complex productions, more than one console may be required, in which case, one of them will serve as a master with others slaved into it on one or more line-level input channels. For example, one board might handle only the orchestra pit microphones, another the cabled (i.e., nonwireless) cover and special purpose microphones, while a third takes control of all the wireless units. Each console would have its own operator, and one of the three would serve as the master. The master console would handle the distribution of the signal to the primary loudspeakers, to the stage monitor system (which in this case means a foldback that permits the performers to hear themselves as amplified through the system), and to any other feeds that require access to the signal.

Figure 9-12B gives enlarged details of an input module and of a group\matrix output module. Figure 9-12C is a block diagram of the reinforcement console. Figure 9-12D provides a line schematic of the modules and their interconnections. Figure 9-12F shows the console connector panel.

The console in Fig. 9-12 has 24 input channels, each complete with line/mic switching, separate gain control for mic and line inputs, output switching and panning, four-band equalization, and two sends with one (for echo) switchable pre- or post fader and one (for foldback) switchable pre- or post-EQ. Each input module also has an off/on switch, a solo (PFL) switch, a peak program indicator, a phantom power off/on switch (see Chapter 6 for a review of phantom powering), a switch to insert a 20-dB loss pad into the microphone input, and a bass-cut switch (to reduce frequencies below 100 Hz, compensating for proximity effect in dynamic cardioid microphones.) Each input module provides a direct output.

The reinforcement console provides eight group/matrix master modules. The group control can act as an independent output or as a panned submaster to a stereo master output module. Equalization for the group/matrix modules is a three-way system instead of the four-way EQ found on the input modules. Each module can be soloed AFL (after-fade listen) to the console monitor output. The group section of the module can be fed to the matrix system pre or post fader. The matrix consists of an eight-way mixer. (See diagram, Fig. 9-12E.) Each horizontal row of pots constitutes one mix channel. Each group module provides a level control knob for each of the eight rows. These level pots admit the group signal to the matrix mix busses. Masters for each matrix row are placed on a separate module at the left end of the group\matrix modules.

On the meter panel for the console a transfer switch allows the eight VU meters to be switched between group outputs and matrix outputs. Two other meters on the panel monitor the output of the stereo master module. The stereo master module provides the left and right channel faders, and also contains the monitor master level control.

Notice particularly the way in which the solo (PFL, AFL) system functions. Each solo switch is a double-pole switch that connects the module's audio output signal to the solo buss and also applies console DC supply voltage to the solo relay in the monitor system. The relay transfers the input to the monitor line amplifier from the stereo master output to the feed from the solo buss. The result is that the operator may elect to hear any module or modules accessing the solo buss through the monitor, but the main feed from the system proceeds to normal destinations without interruption.

Each input and output module contains a separate pair of insert connectors to provide an alternate output from the

[3]One-quarter-inch phone jacks usually imply unbalanced lines, but not necessarily. Balanced lines can also use 1/4-inch phone connectors, so always check the manufacturer's specifications to know exactly what configuration does lie behind a phone jack. A 1/4-inch phone plug is clearly a two-conductor or three-conductor device.

FIG. 9-12A: REINFORCEMENT CONSOLE CONTROL PANEL

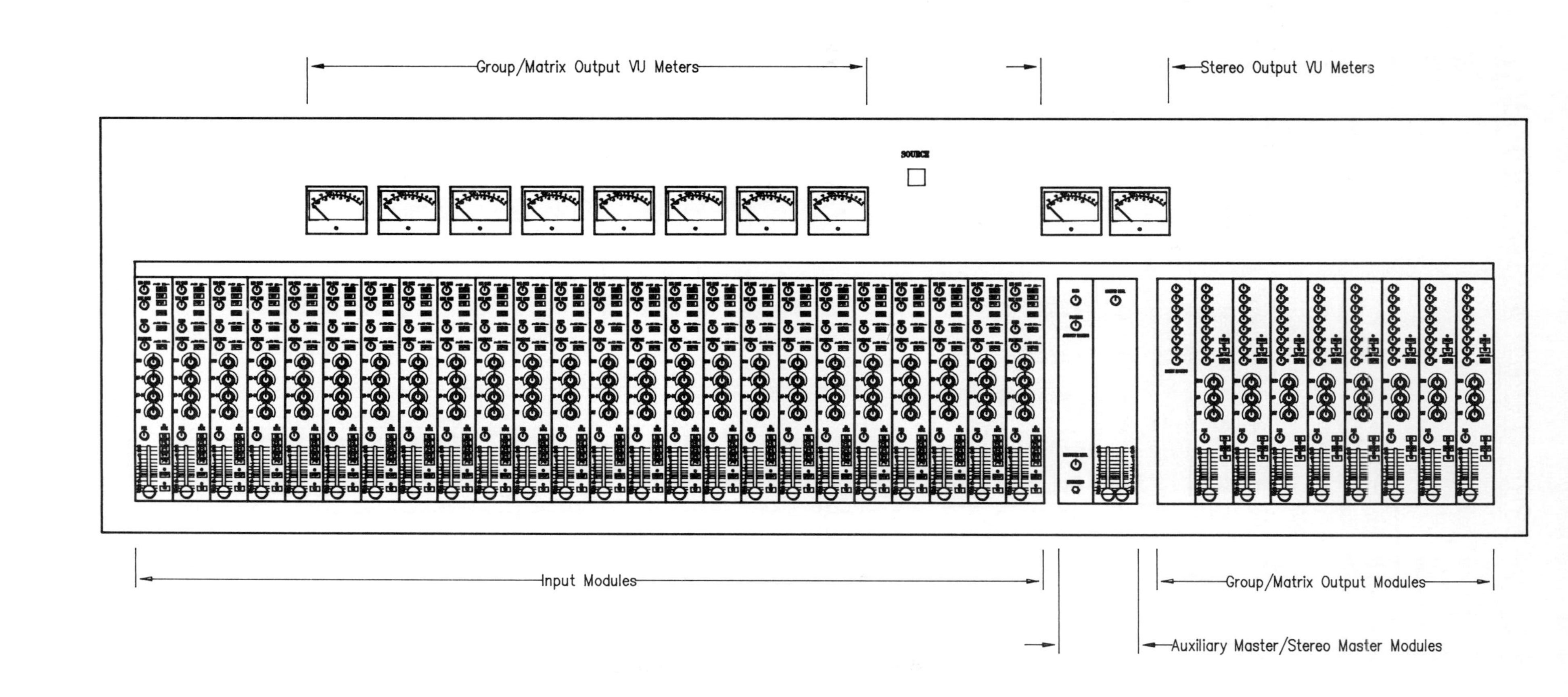

FIG. 9-12B: INPUT AND GROUP/MATRIX MODULES

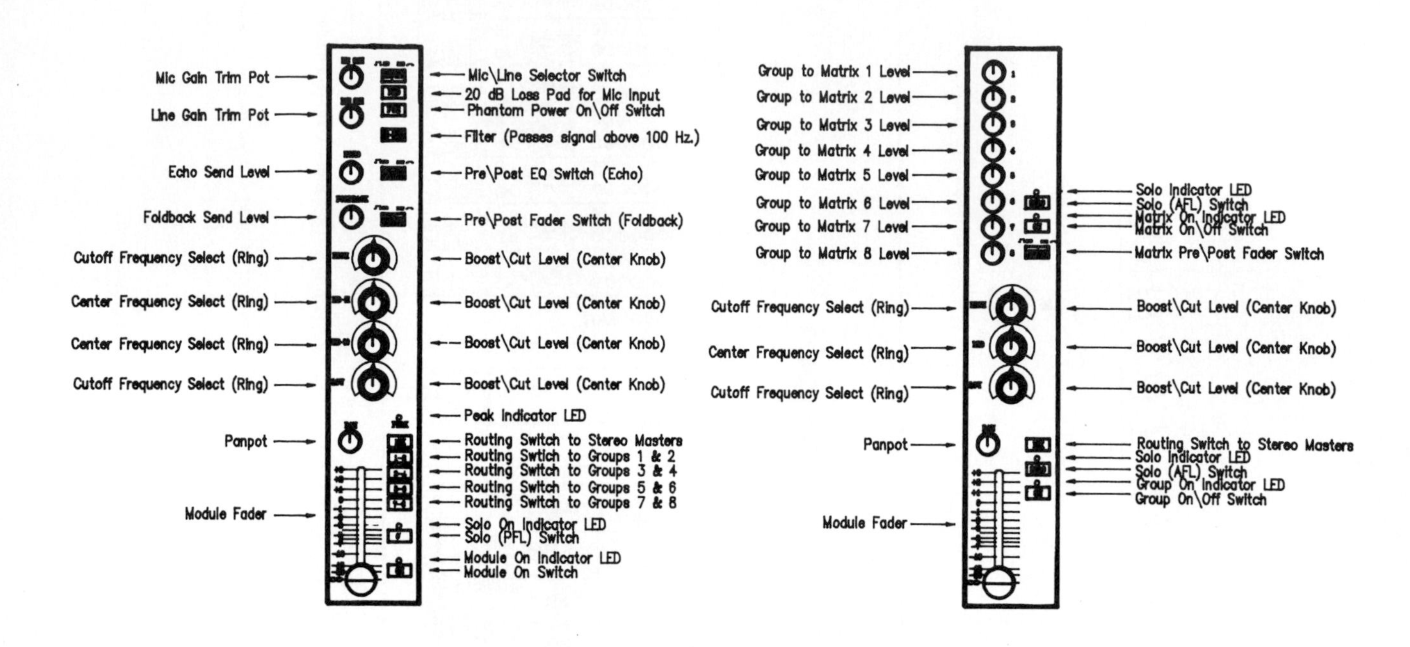

FIG. 9-12C: REINFORCEMENT CONSOLE BLOCK DIAGRAM

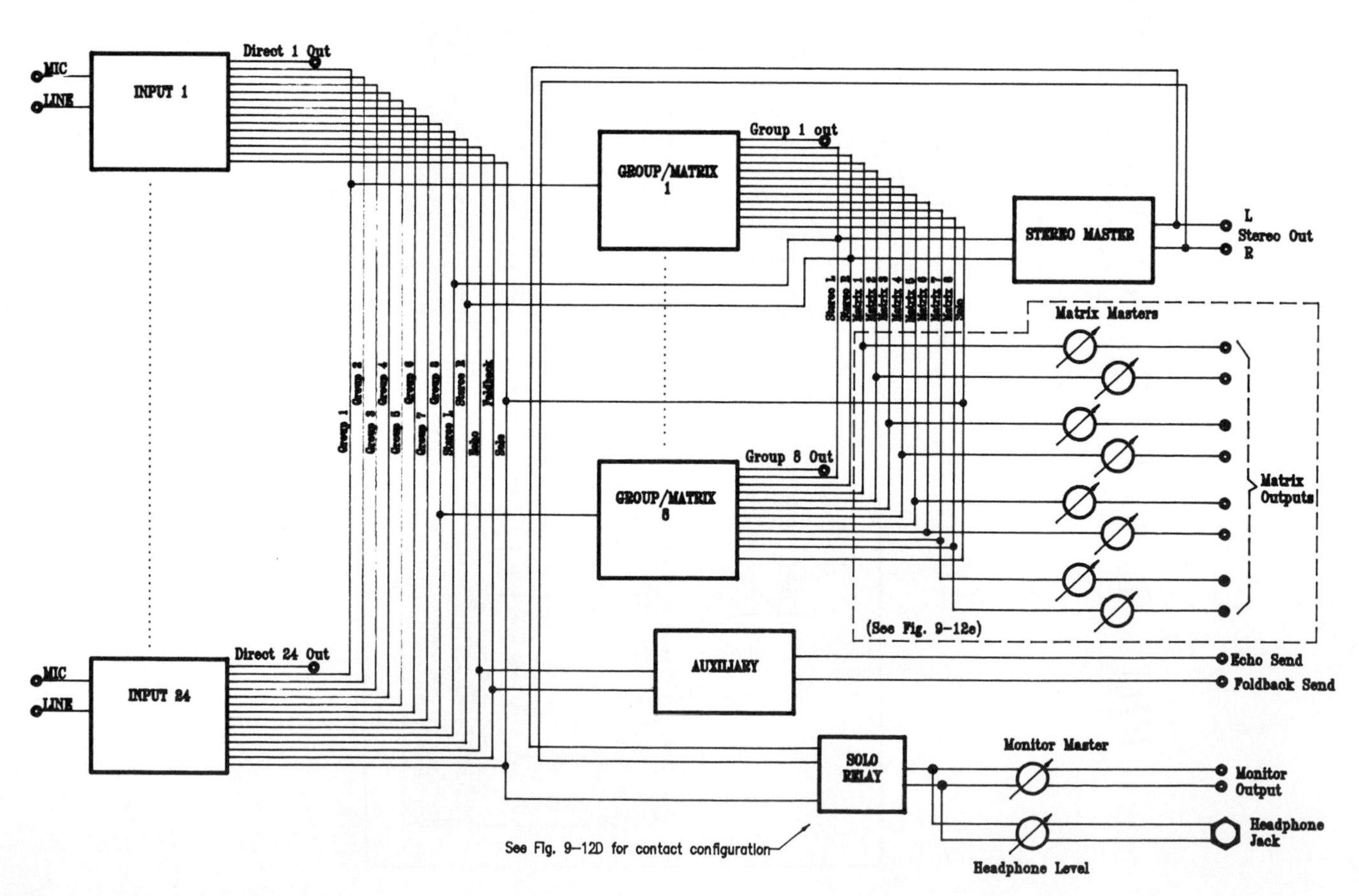

FIG. 9-12D: REINFORCEMENT CONSOLE -- BUSSING DIAGRAM

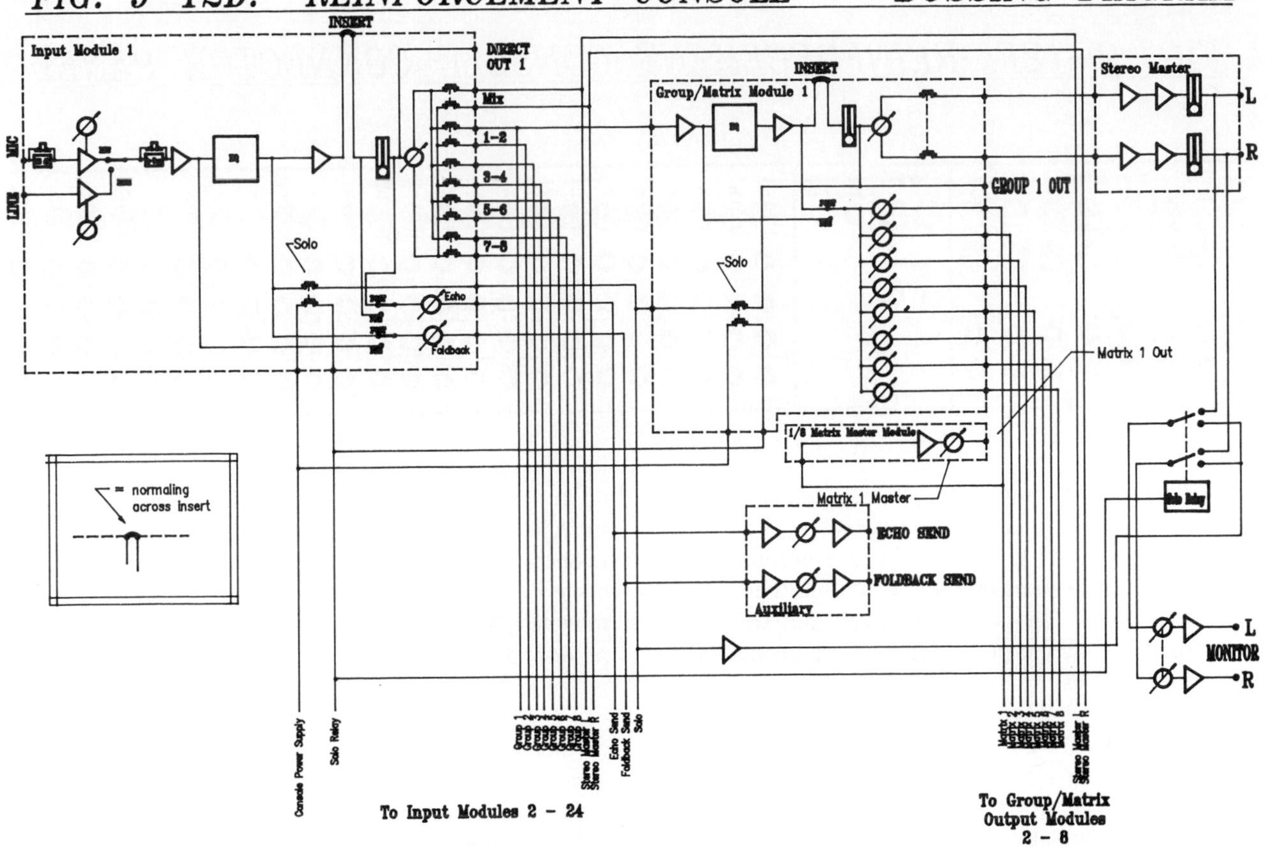

FIG. 9-12E: REINFORCEMENT CONSOLE MATRIX SYSTEM

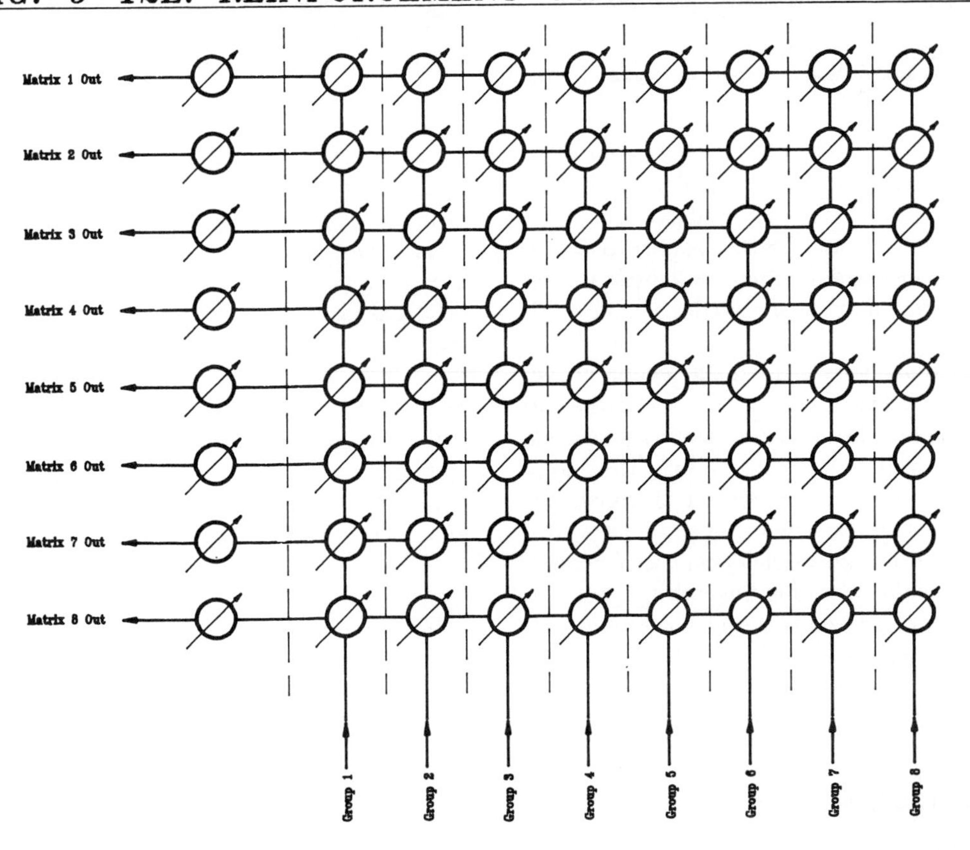

FIG. 9-12F: REINFORCEMENT CONSOLE CONNECTOR PANEL

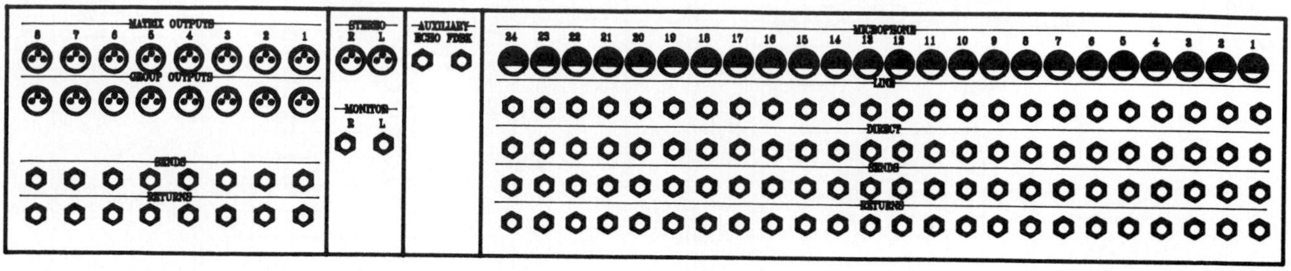

module and to permit the insertion of an outboard processing device, such as graphic equalization or delay.

The auxiliary master module provides level controls for the echo and foldback send outputs. Note that there is no specific return for the echo send in this console. The usual method of return is to connect the delay device to one of the standard input modules.

The connector panel for the console (Fig. 9-12F) provides 24 female, XLR-3 connectors for microphone inputs and twenty 24 quarter-inch *balanced* phone jacks for line inputs. Direct outputs from the 24 input modules, insert connections for all modules, and send and monitor outputs also use quarter-inch balanced phone jacks. Stereo, group, and matrix outputs use male XLR-3 connectors.

Studio Recording Systems

Recording consoles are not radically different from reinforcement consoles except that output requirements are less complex. Studio recording does not demand as much flexibility in sending signals to a variety of destinations as does a reinforcement console. Recording consoles almost always feed the inputs of a multitrack tape recorder. Figure 9-13 illustrates a typical recording console.

Except for auxiliary mixes, the input section of the console is virtually identical to that of the reinforcement mixer. Note, however, that the number of auxiliary sends is much greater than for the reinforcement console. The studio recording arrangement provides four echo send busses and four foldback busses. Input to an echo bus can be taken pre- or post-fader, and input to a foldback bus is switched for either pre- or post-EQ. The multiple send arrangement permits up to four different groups of echo and delay characteristics and up to four separate foldback mixes.

Eight group master modules, an auxiliary master module, and a stereo mix module provide the primary outputs for the console. The group master modules can provide independent output or can act as panned submasters for the stereo master module. Equalization is quasi-parametric and can be switched in or out of the signal path. Group modules also contain a solo switch and a module off/on switch. The auxiliary master modules provide eight level controls, four designated for echo sends and four for foldback sends. The stereo master module provides dual faders for main left and right channel outputs.

Other significant differences between the recording console and the reinforcement system are the presence in the recording setup of more detailed monitor elements, left and right returns for a stereo tape recorder, and an elaborate talkback system. The separate stereo outputs and level controls are provided for control room and studio monitors, permitting the operator to regulate level or mute monitor loudspeakers in either area separately. The monitor source switch allows the monitor feed to come either from the stereo master module or from the two-track (master stereo recorder) return inputs. Return for the master tape recorder permits the operator to hear the progress of a stereo mix as the recording proceeds without having to touch any of the input modules. The talkback module includes a small, electret condenser microphone, with routing switches labelled slate, talkback, and comm (communications). *Slate* sends the microphone signal through all of the group master modules for the purpose of identifying each take on all channels of a multitrack recording. *Talkback* directs the microphone signal into the studio monitor amplifier. *Comm* sends the mic signal to a jack on the connector panel of the mixer. This connector allows the signal to be sent to an external destination such as an observation booth.

The recording console also includes a test tone generator to permit internal and external[4] amplifier gains to be set to a uniform 0 VU level.

Console Automation and Computerization

Consoles used in large recording facilities and in postproduction studios for processing film and television sound generally

[4] External amplifiers are usually those in tape recorders and signal processing equipment.

FIG. 9–13A: RECORDING CONSOLE CONTROL PANEL

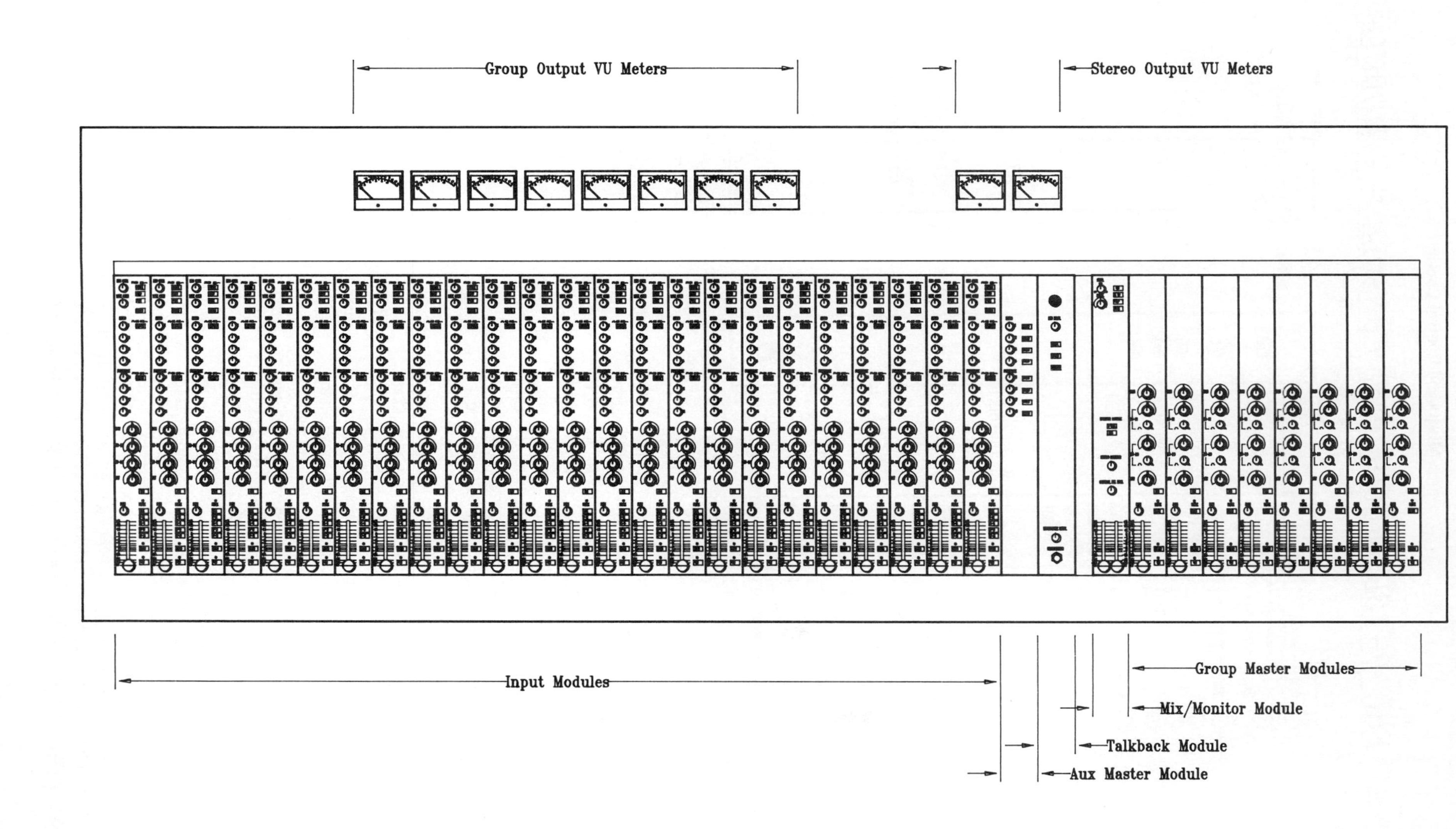

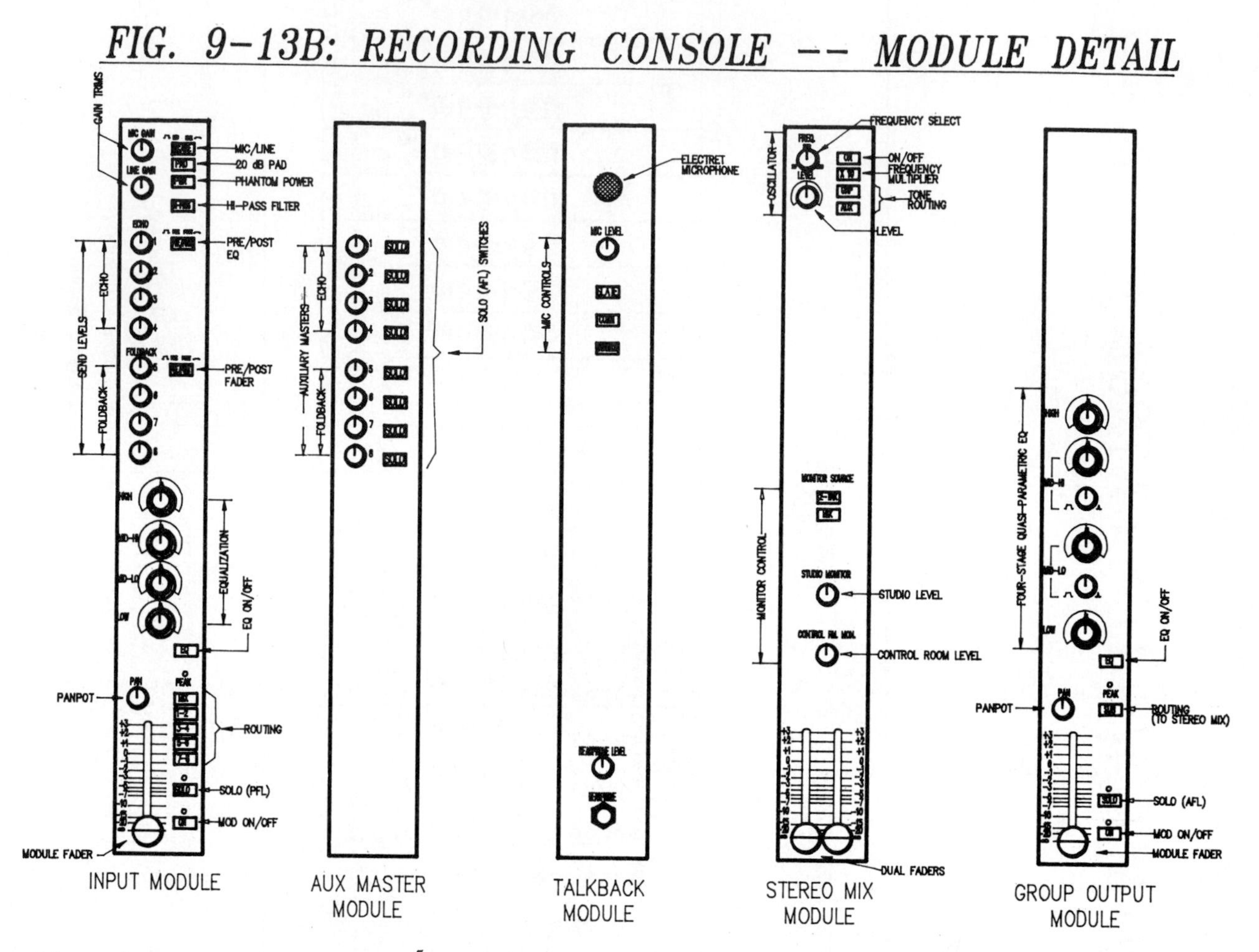

are much larger than any unit described so far.[5] Such consoles may require more than 100 input channels and 24 to 48 output channels. Activities involved in record mastering and film postproduction require precise levels and minute timing, both in fade initiation and in fade duration.

Consoles for mastering and postproduction are frequently equipped with memory elements to record control settings. In their simplest form, memory elements provide storage of "cues"—control settings associated with a particular portion of a song or a film sound track. Usually, memory elements in such consoles are able to vary control settings through voltage controlled amplifiers, motorized ("flying") faders, or both. If consoles are fitted only with cue memory, responsibility for activating each cue, and therefore for event timing, may remain in the hands of a human operator. More advanced consoles include a capability to lock cue timing to SMPTE Time Code recorded on the master tape (discussed in Chapter 10.) With the use of tape-based time code, a given "cue" will occur very precisely at a specific point in the program recorded on the tape.

The most advanced consoles are fully computerized and automated. Such systems provide extended control over tape motion and time code, permit labeling and modification of cues, and facilitate the operation of complex fade processes and timing. Although fully automated consoles can be useful in theatre, they are seldom used because of their cost. State-of-the-art recording and postproduction consoles usually sell at prices well over $100,000.

Theatre Sound Scoring Consoles

The majority of consoles available on the market today are mix-down systems. Consoles for replaying sound scores for theatre require a **fan-out** configuration. Fan-out systems essentially distribute a few inputs to a large number of outputs. The basic arrangement of a fan-out console is shown in Fig. 9-14. Compare the diagram with Fig. 9-1, and you will see that the two are exact opposites in the function that each performs. The mixing system reduces multiple inputs to a single output; the fan-out system distributes one input to multiple outputs.

[5] Postproduction refers to the variety of processes used to complete the final audio component of a film or videotape. Postproduction includes adding "looped" dialogue (dialogue recorded by the actor in a soundproof booth, sychronizing words with the lip movements recorded on the film or videotape), environmental effects, Foley effects (sounds matching events in the film, such as falling objects, noises or cutlery and glasses, etc.), and musical underscore.

FIG 9-13C: RECORDING CONSOLE -- BUSSING DIAGRAM.

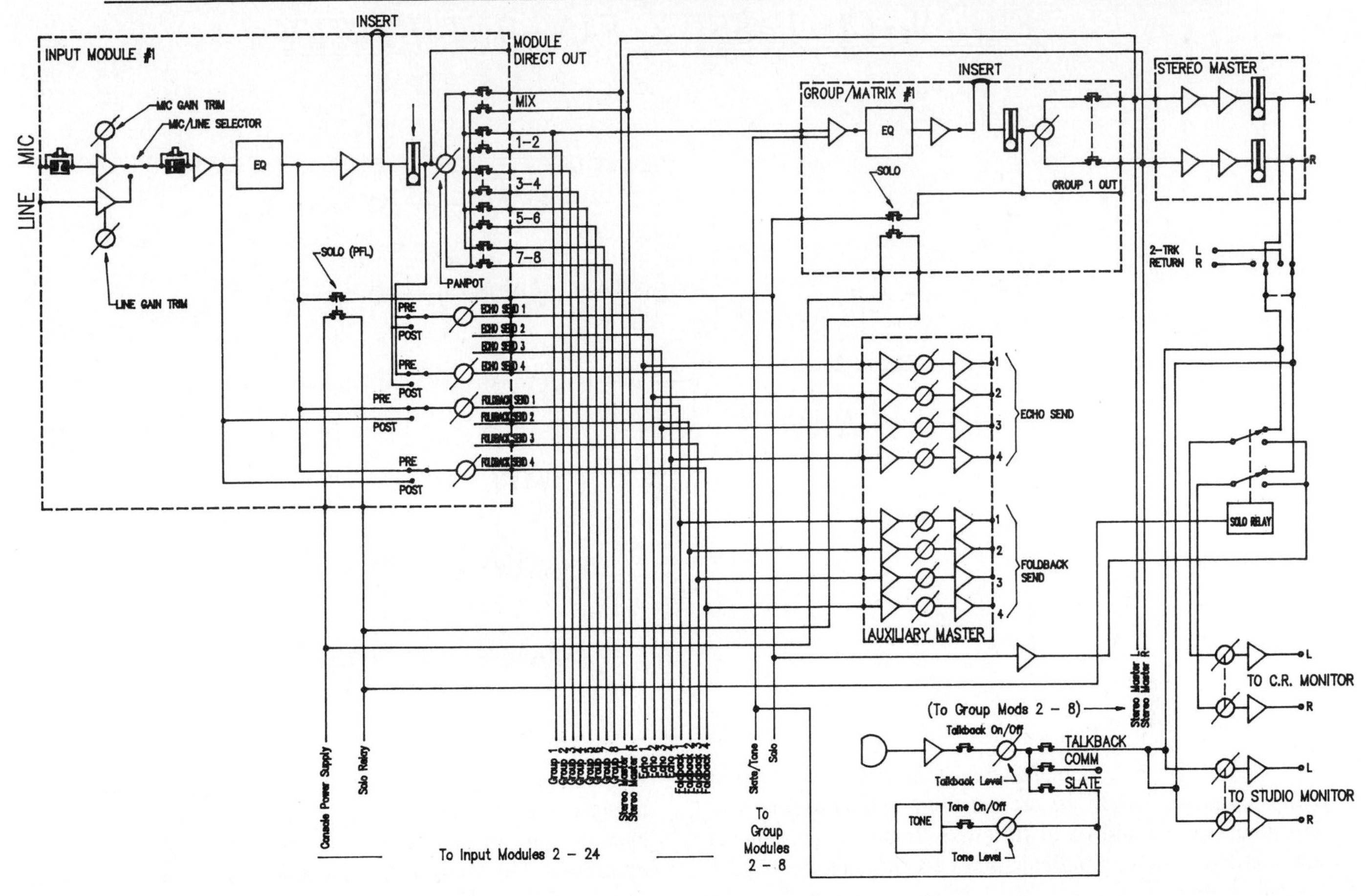

FIG. 9-13D: STUDIO RECORDING CONSOLE CONNECTOR PANEL

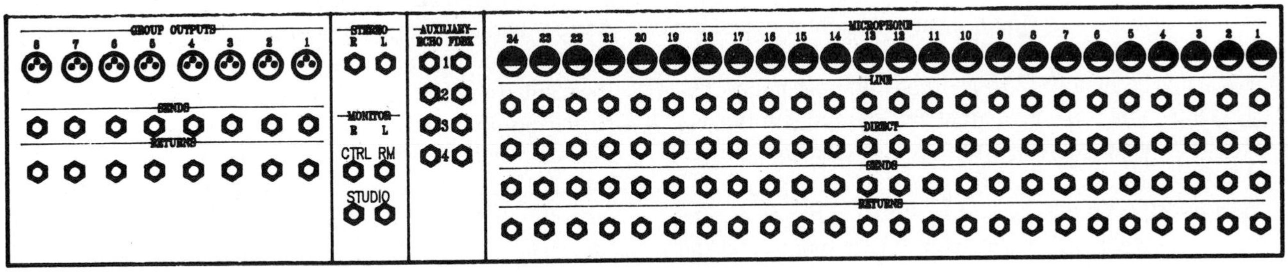

GROUP OUTPUTS
8 7 6 5 4 3 2 1
SENDS
RETURNS
STEREO
R L
MONITOR
R L
CTRL RM
STUDIO
AUXILIARY
ECHO FDBK
1 2 3 4
MICROPHONE
24 23 22 21 20 19 18 17 16 15 14 13 12 11 10 9 8 7 6 5 4 3 2 1
LINE
DIRECT
SENDS
RETURNS

FIG. 9-14: PASSIVE FAN-OUT SYSTEM

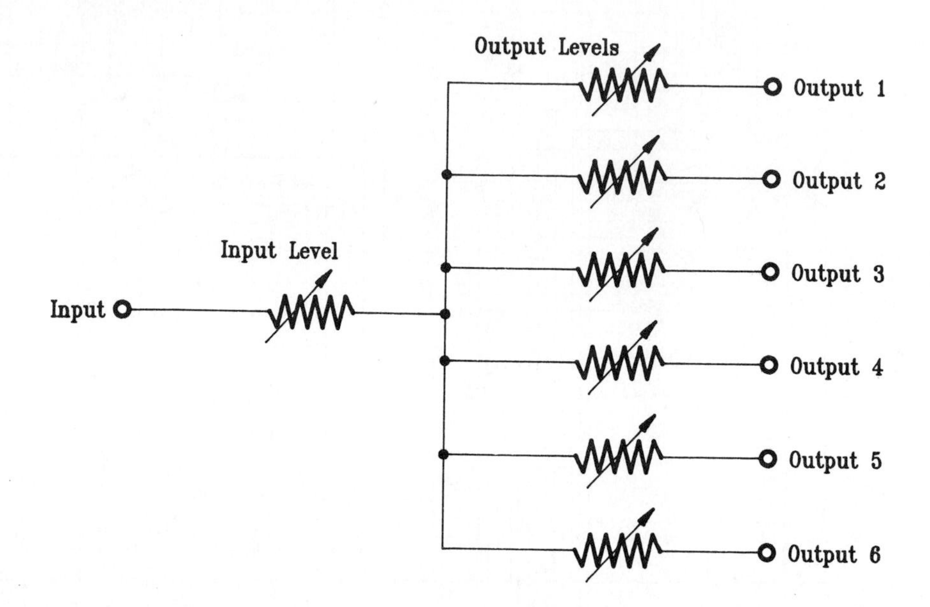

As in Fig. 9-1, the passive system of Fig. 9-14 is not really practical. Amplifiers are required, and a console with only one input would be of little use. Figure 9-15 shows a schematic of a more practical sound scoring console. In the system of Fig. 9-15 any or all of the four inputs may be connected through the switch matrix to any or all of the eight outputs. The arrangement allows for very flexible patterns of directional source sounds, auditory surrounds, and apparent motion of sound sources; but the operator must be well practiced and highly skilled in rapid fader motions.

More practical still is the fan-out system shown in Fig. 9-16 in which a fader matrix is substituted for the switch matrix, and instead of a single group of eight outputs, the system provides four presets of eight output controllers each. Each input can be switched to one of the four preset lines. The preset arrangement provides a great deal more flexibility, but still requires considerable operator skill to execute complex sound motion.

An improved preset system is shown in Fig. 9-17. The improvements include a master fader for each preset and an expanded input system using two control faders and a panpot for each input. The panpots and dual input faders are the features that make this console more flexible and easier to use than those shown in Figs. 9-15 and 9-16. Let's consider an example to illustrate how this configuration can be used.

We intend to use a loudspeaker configuration that surrounds the audience. All incidental music and some environmental effects are to be played through the surround speakers. For the opening of the first act we want the intro music to fade down out of the house speakers and to localize to a stereo system that appears onstage as part of the furniture in an apartment. The loudspeakers surrounding the audience are on outputs 1 through 6; the speakers in the prop stereo are on outputs 7 and 8. The music begins with input channels 1A and 2A up nearly full and routed to presets 1 and 2 (1A to preset 1; 2A to preset 2). Preset 1 has channels 1, 3, 5, and 7 at full; preset 2 has channels 2, 4, 6, and 8 at full. Preset 1 forms the left channel of the stereo surround, and preset 2 is the right channel. At rise, the operator slowly fades panpots 1 and 2 from A to B. Input channels 1B and 2B are set at about 60 percent and are routed to presets 3 and 4. Because presets 3 and 4 have outputs open only to the speakers in the prop stereo, the result is that the sound fades out of the house and condenses to the property unit onstage with a single turn of the two panpots. On either of the previous sound scoring consoles, the operator would have had to move a lot of faders quickly and smoothly to accomplish the same effect. Fig. 9-18 shows the faceplate of the console, giving the positioning of the faders and indicating the panpots to be used for the fade.

One problem with the console in Fig. 9-18 is that the operator must use two hands to accomplish the fade. Each hand must turn a panpot. Although not a major problem, modern technology does afford better alternatives. Figure 9-19 shows a console that alters the arrangement of both the control faceplate and the circuitry, providing electronic fades from one preset to another. Assume the same situation—the

FIG. 9-15: SOUND-SCORING CONSOLE SCHEMATIC

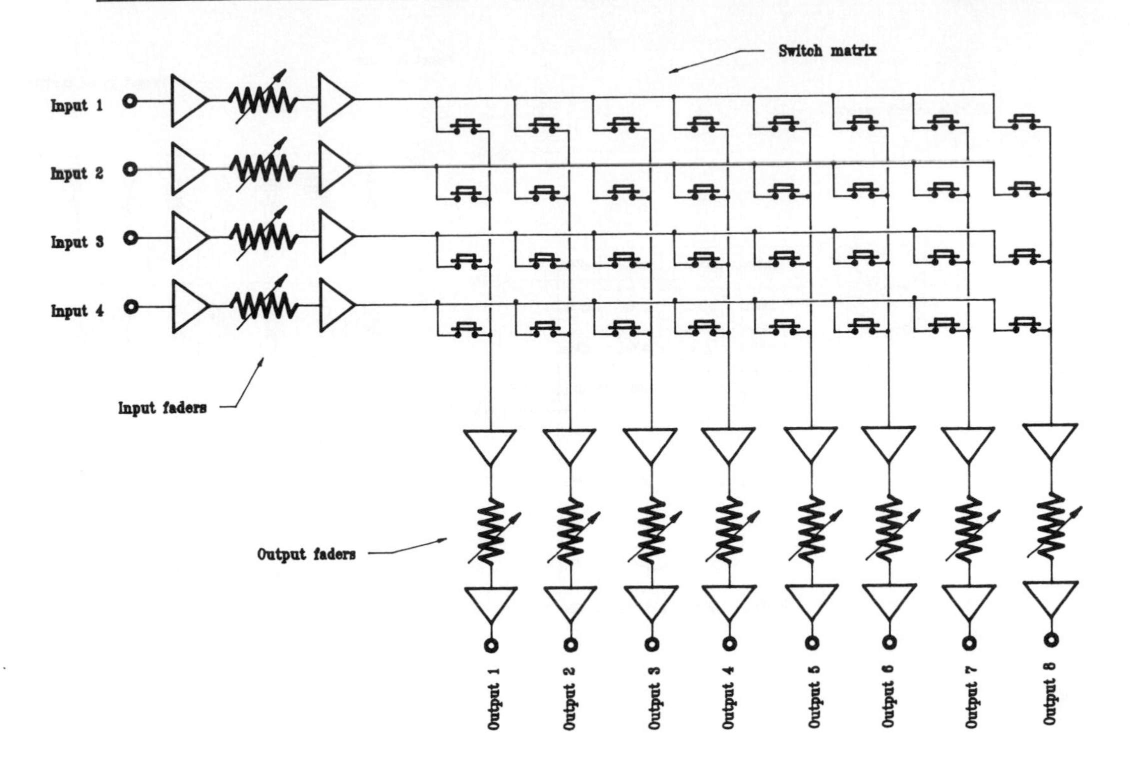

FIG. 9-16: PRESET CONSOLE

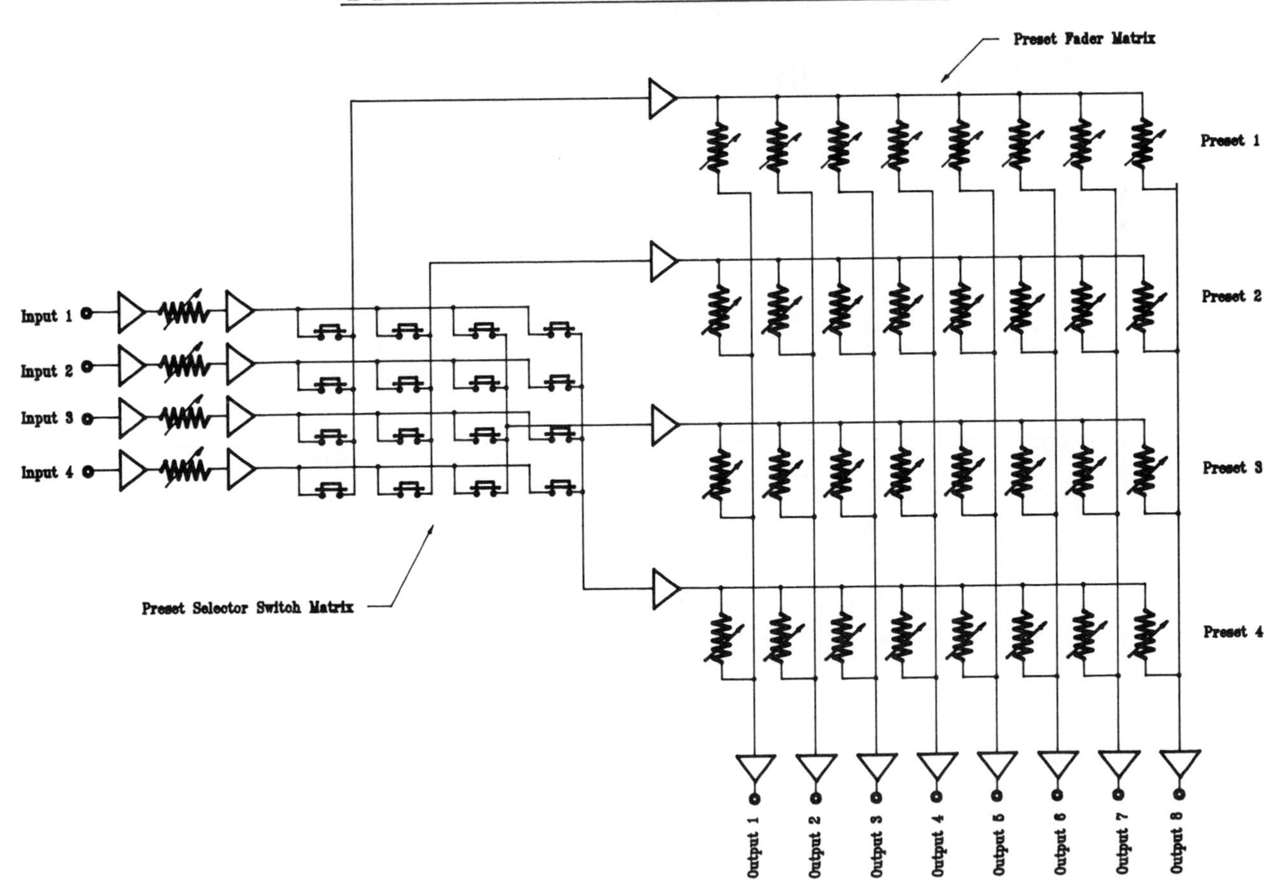

FIG. 9-17: IMPROVED PRESET CONSOLE

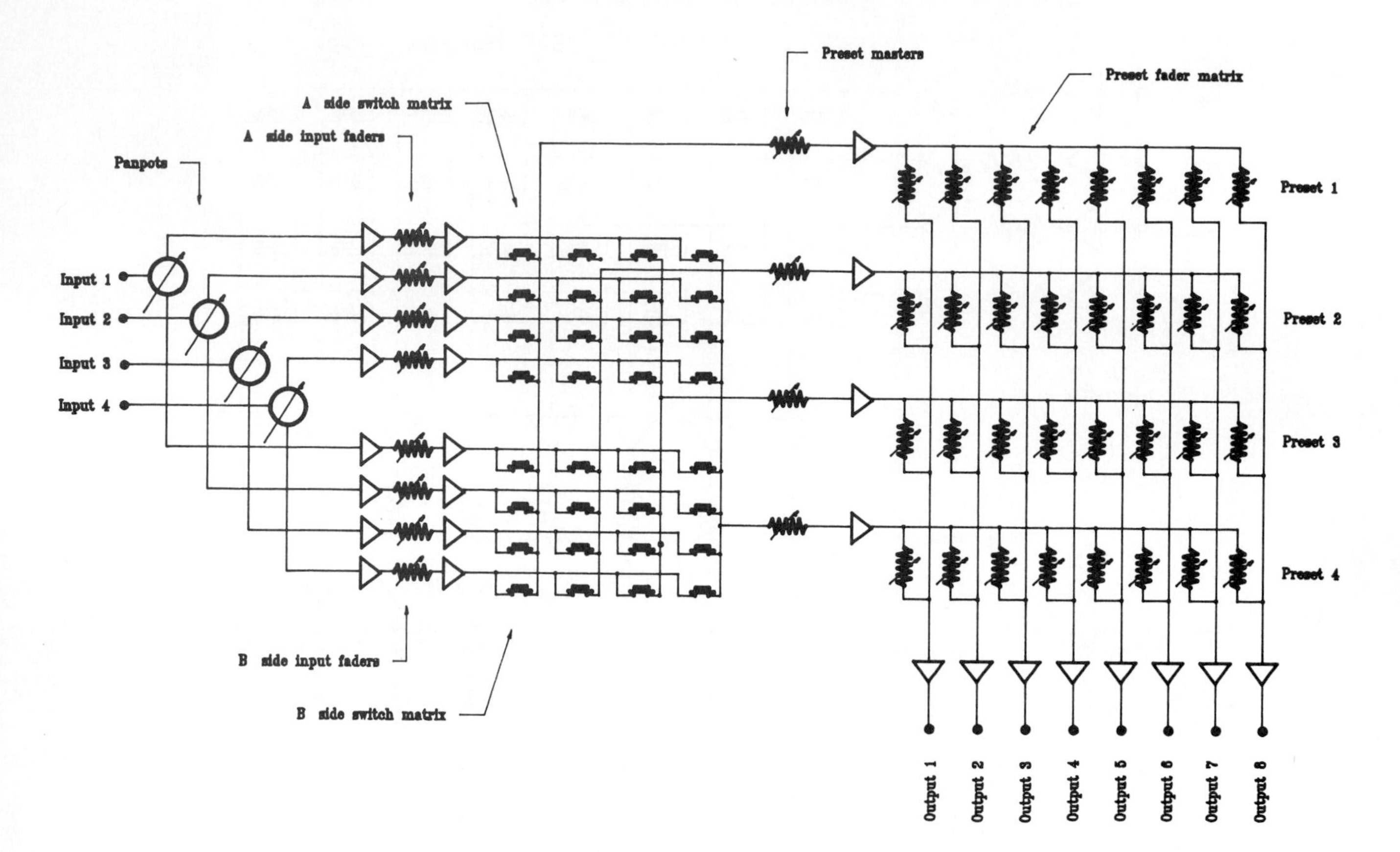

FIG. 9-18: FADE USING IMPROVED CONSOLE

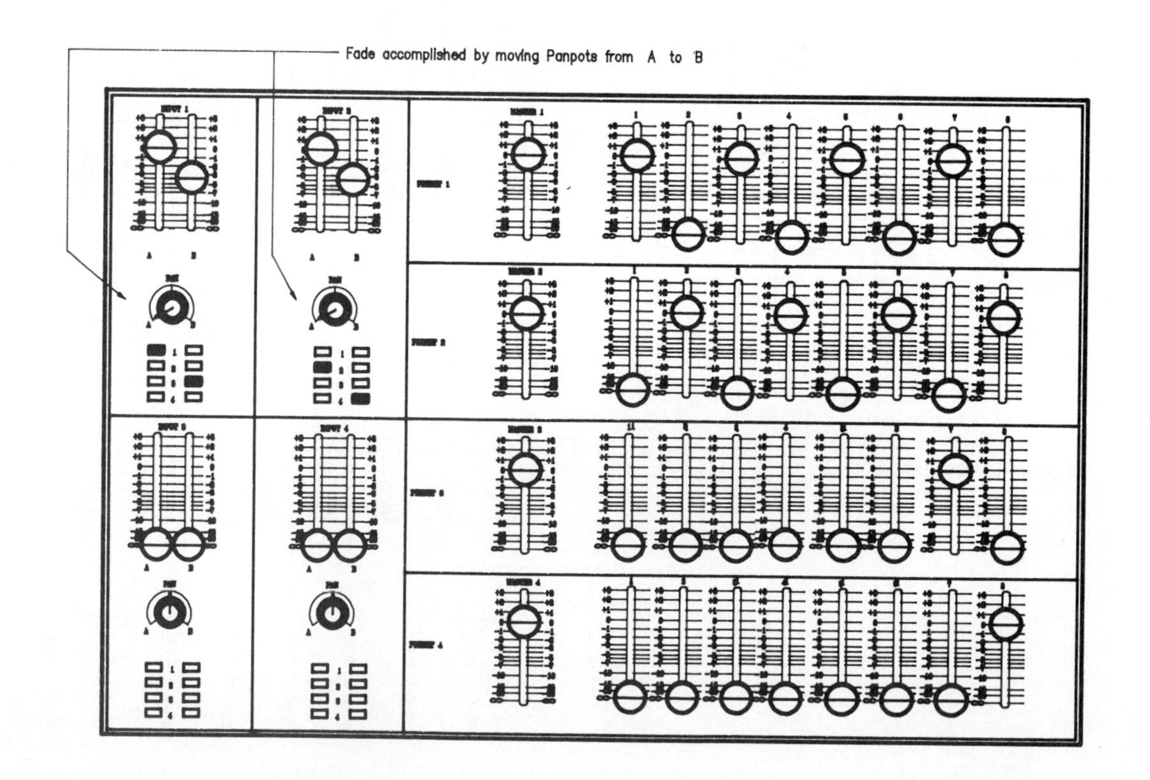

FIG. 9–19A: SOUND SCORING CONSOLE WITH IMPROVED PANPOT

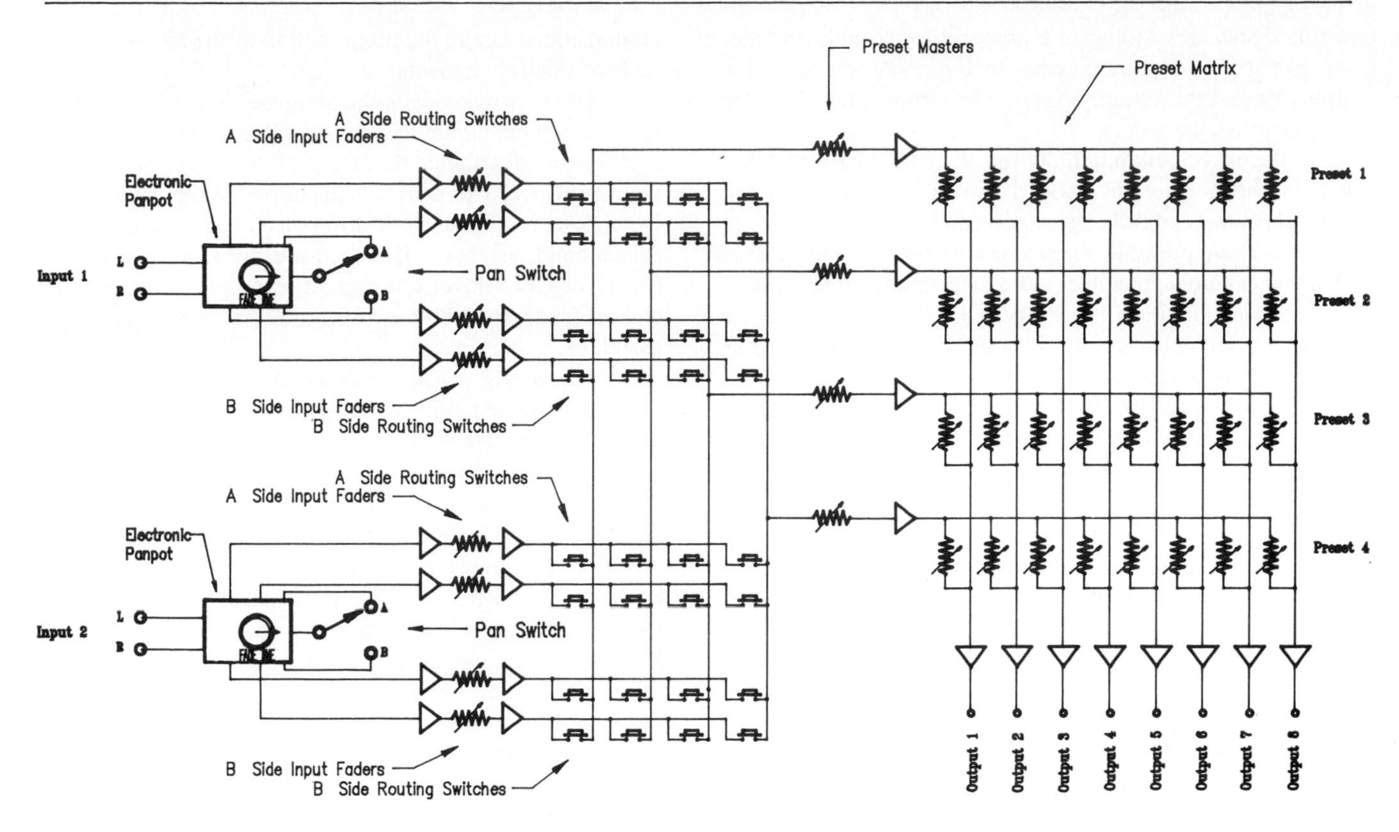

FIG. 9–19B: SOUND SCORING CONSOLE WITH IMPROVED PANPOT CONTROL FACEPLATE SHOWING FADE

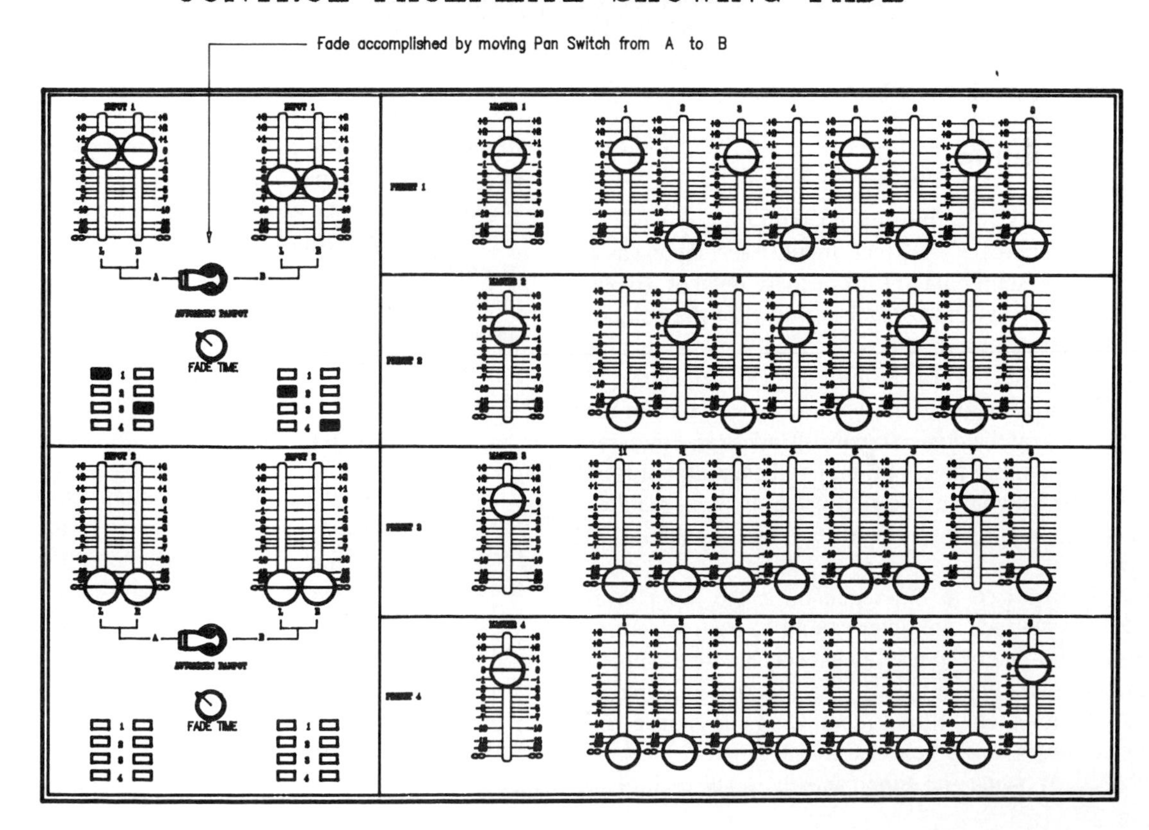

surround speakers fading to a property stereo unit onstage. Observe that, using the system in Fig. 9-19, the operator simply throws the automatic pan switch from A to B in order to accomplish the fade.

Electronic panning, however, is by no means the ultimate in improvement of a sound scoring console. Memory-assisted control and full computer mediation of console functions has made possible a generation of highly flexible systems that can do a great deal more than simple crossfades.

Computerization began to make its way into sound scoring equipment in the 1980s with the development of at least two consoles designed primarily for theatrical purposes. Only one survived into the 1990s, and that system typically uses a format that can be configured in any manner that suits the user's purpose. The most usual configuration is quite like the console described in Fig. 9-19, although much more complex.

A typical configuration of a computerized sound scoring console is shown in Fig. 9-20A. The audio electronics for this system are contained in a rack-mounted chassis that presents only a blank faceplate to the outside world. The control elements of the system are contained in a software-generated, graphic interface displayed on the video screen of the system computer.

For those who are primarily familiar with mix-down consoles, the arrangement of components in this system requires some reorientation of ideas about audio system structure. For example, the output control elements are arrayed in redundant banks like presets in a lighting console, and the entire control system is built on microprocessor-operated, digitally-controlled attenuators (DCA's). To understand the operation of a sound scoring console, it's best to divide the console architecture into functional sections.

Input Section

A selector device permits the operator to choose one of 16 lines as the source for a given input. Each input has a selector, so the system contains 16 of these devices. (Note, however, that each selector branches to *two* of the components on the next control module.) Next, each input contains a fade rate timer. The fade rate timer is a DCA designed to fade from full attenuation to unity gain or vice versa in the interval selected by the device control. Understand that this device does *not* control level but, rather, the amount of time required to fade a signal in or out. Since each input requires two fade timers, the input section of the system contains 32 such devices.

The third device in the sequence is the input level control. Just as the fade rate timer cannot be used to control level, this device cannot be used to execute a fade. It simply determines the maximum level of signal that will be admitted to the rest of the system for that input. On a cue change, the level device *jumps* to the next setting. A fade can only be accomplished by using the timer and the level control in combination. Again, the input section of the system contains 32 level control elements.

The fourth device is the assignor switch matrix, which routes the input signal to a preset. (Recall the way in which signals were distributed through presets in the system of Figs. 9-17 and 9-18.) This device is the mirror of the input selector. Note, however, that the one difference in connection between selector and assignor is that each selector feeds two fade rate timers and two level controls. The output from one level control feeds one assignor; therefore the system contains 32 assignors.

The wiring of the input section may seem unusual to those accustomed only to standard mixing consoles. Why should an input selector drive a *pair* of fade, level, and assignor elements? Think back to the console shown in Figs. 9-17 and 9-18 and to the cue that was illustrated on that console. On that console we used input 1 as the left channel for the input signal and input 2 as the right channel. Each input had two faders, however, and the panpot was used to fade the input signal from fader A to fader B. The purpose of that arrangement was (1) to allow a lower signal level to be set on the B faders and (2) to permit each fader to be routed to a different preset. (The A faders went to presets 1 and 2; the B faders to presets 3 and 4.) Similarly, in the computerized system, we gain the possibilities of (1) two different levels, (2) routing to two different presets, and (3) different rates for upfading and downfading signals.

The assignors route signals to eight presets in this system; therefore only half of the outputs from each assignor are used.

Output Section

Each preset begins with a master fade rate timer and a master level control, just as in the input sections. Here, however, there is no pairing of channels. Each buss from the assignors drives one and only one preset input.

The output from each level control device feeds one row of a matrix of 8 rows by 16 columns of faders. Each row of the matrix forms one preset of output settings; each column represents the mix of signals fed to a single output device, usually a power amplifier driving a loudspeaker.

Control Interface

The control interface of the sound scoring system contains a graphic icon for each component of the system: selectors, fade timers, level control, assignors, preset master fade timers, preset master level controls, and a matrix of individual level controls for the output presets. All controls are manipulated by means of a mouse. The operator uses the mouse cursor to point to a control. Clicking on a control icon brings up a detailed screen showing all of the control details of that element. The operator points to and clicks on pushbutton switches and clicks and drags fader and timer controls. Once

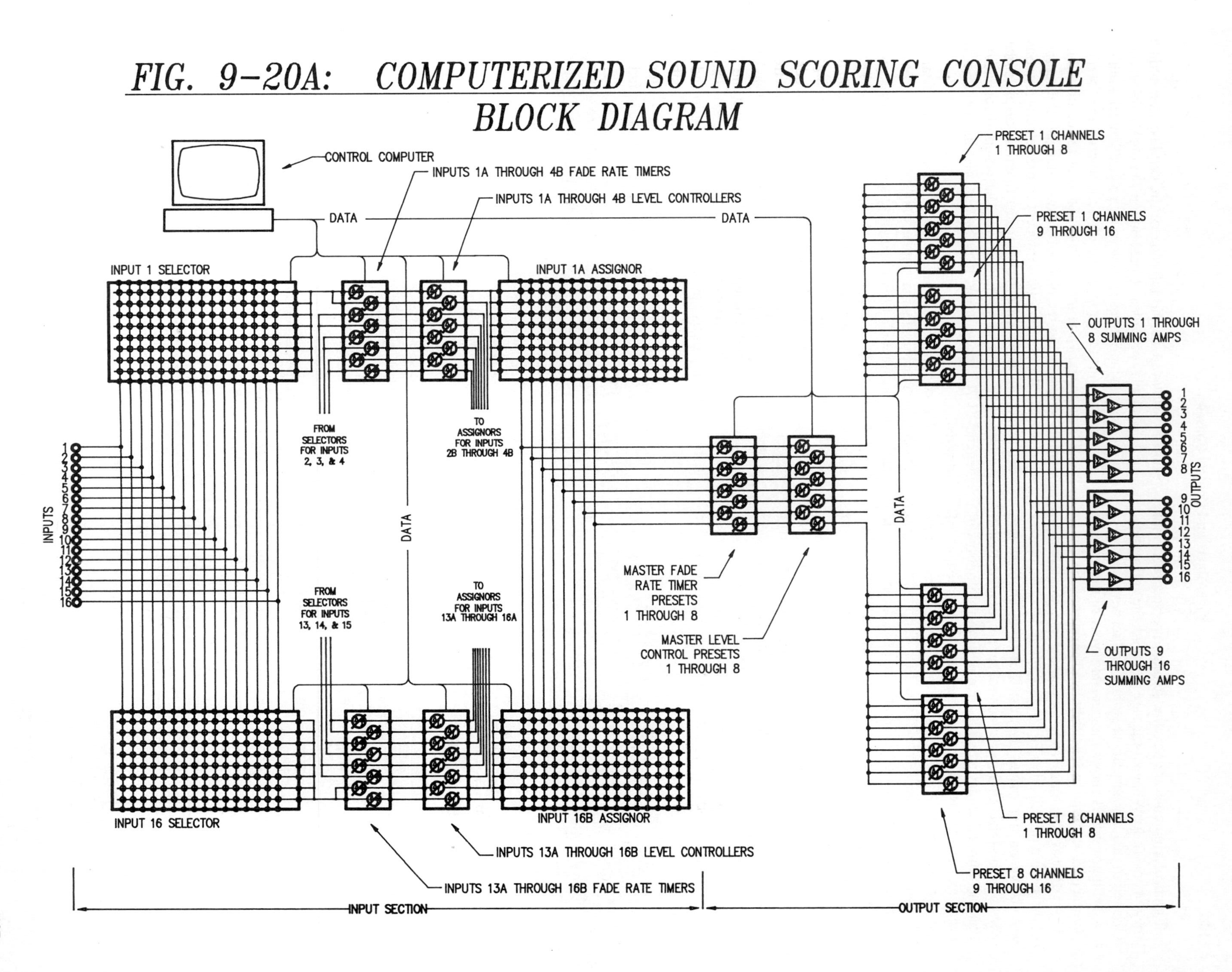

FIG. 9–20A: COMPUTERIZED SOUND SCORING CONSOLE BLOCK DIAGRAM

all control elements are set, the settings are recorded into memory as a cue. When all cues are recorded, the entire show is saved to disk for permanent storage and retrieval.

Figure 9-20B shows the general layout of the graphic interface for the system. The right hand portion of the screen shows one page of controls. For the particular implementation of the system described in Fig. 9-20A, all of the eight presets are shown on this page. The first four input channels are shown, although eight rows of controls are displayed. Recall that each input channel requires a parallel pair of controls in order to make crossfades possible. Avoid the mistake of associating input 1-A with preset 1, even though they occupy the same line of the display. The four columns of input controls should be considered as absolutely separate from the eighteen columns of output controls. Any input can be assigned to any of the eight presets. As pages are shifted to display input controls 5-A through 16-B, the screen may continue to display a matrix of controls in the position of the presets, but these controls have no effect. Any attempt to activate one of these controls should bring up a "not used" label somewhere on the screen.

Note that the screen shows that the selector for input 1-A has been selected, and a detailed control panel for that unit is displayed on the left side of the display screen. The detailed display provides 16 pushbutton switches to select one of the sixteen input lines. A button is "pushed" by clicking on its symbol with the mouse, then by clicking on the "on" or "off" button shown at the lower left side of the control. The outline around either the on or off button indicates the setting recorded in the current cue. Adjacent to the selector buttons are two label strips. One, showing the function, displays the legend "line 1." The other, showing the group, displays the legend "1-A." Both labels may be changed by the user, if desired.

The control screen for the assignors looks identical to that for the selectors. In this system, only the first eight buttons are usable, as we have only eight presets.

Other controls are shown in Fig. 9-20C. The control diagram for the fade timer shows a toggle button for fade direction (near the bottom of the module, showing a reversible arrow), a slider to set time interval, pushbutton switches to select max or min (maximum or minimum time intervals), and a label strip to identify the group position of the control. The second label strip automatically displays the rate setting for the unit. The setting changes as the control is dragged from the bottom to the top of its range. The small representation of the

FIG. 9-20B: COMPUTERIZED SOUND SCORING CONSOLE GRAPHIC INTERFACE

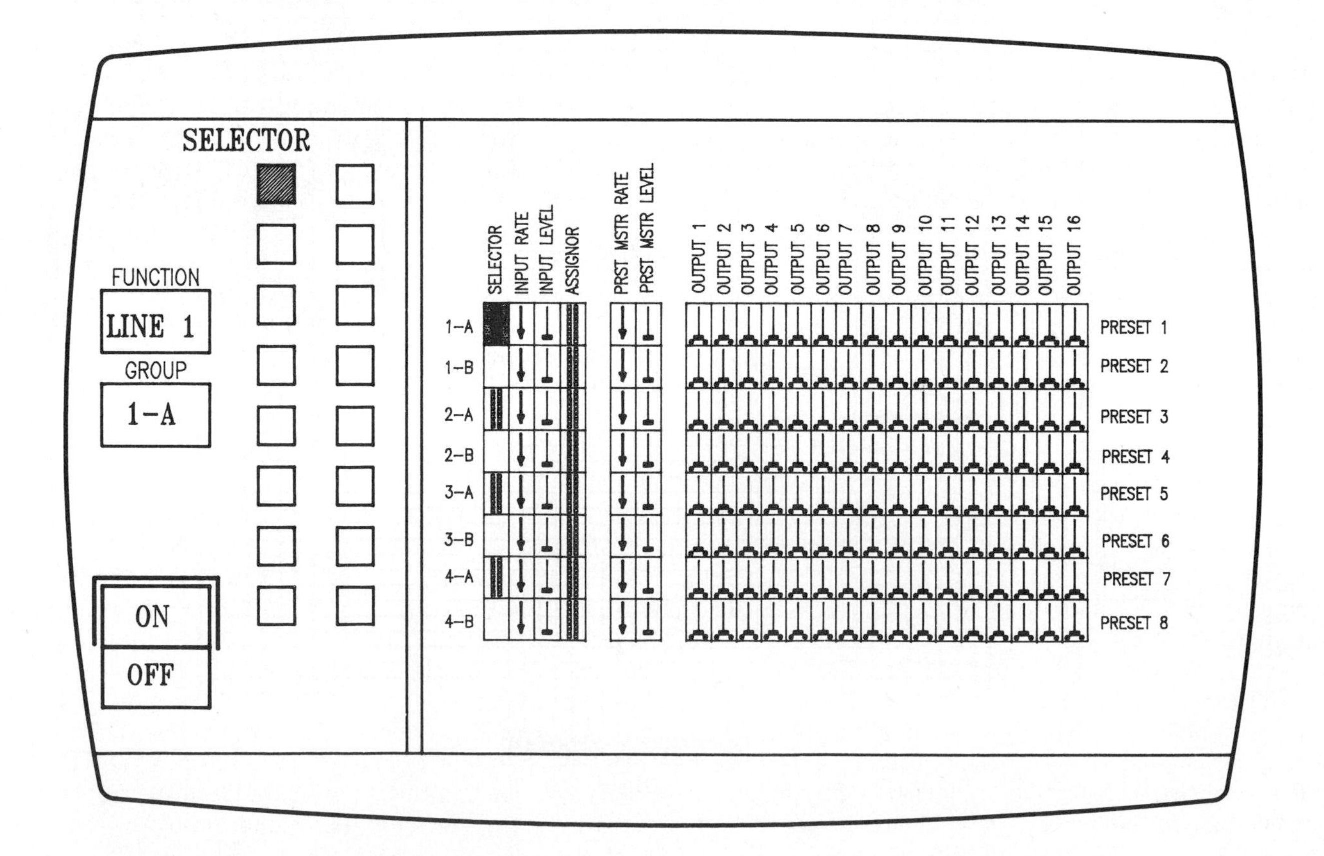

FIG. 9-20C: COMPUTERIZED SOUND SCORING CONSOLE FADE TIMER AND LEVEL CONTROL DETAILS

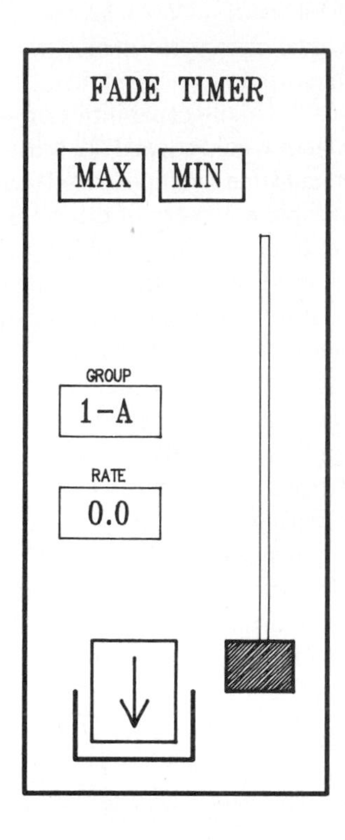

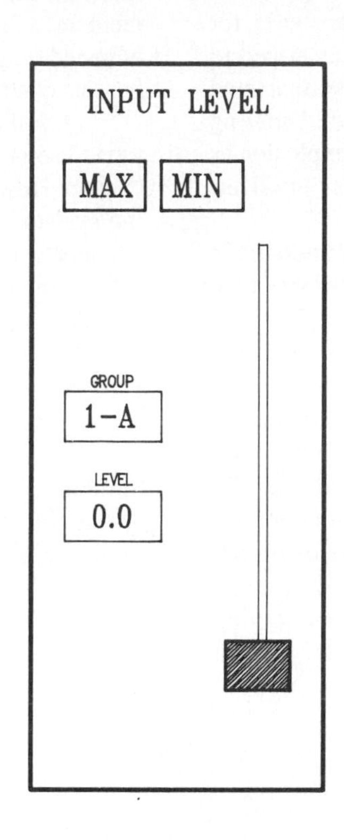

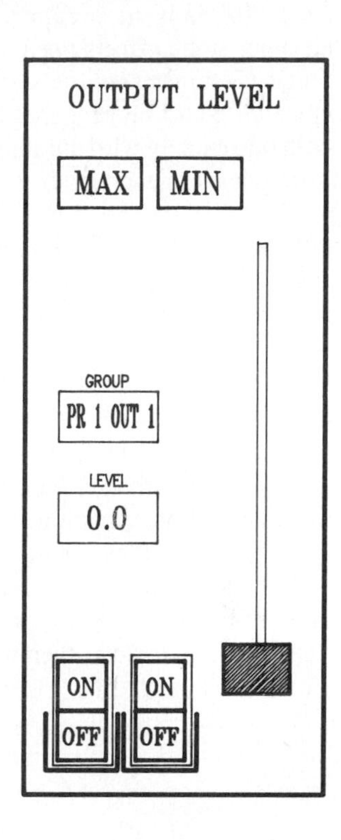

fade timer in the matrix at the right side of the screen displays only the arrow indicating the fade direction.

The level control screen for the input section and the preset master shows a fader to set level and max and min pushbuttons. The label strips provide group position identification and a level strip that automatically shows the setting as the fader is changed. The level control screen for the output faders adds two sets of on/off push-buttons to control digital switches. These switches can be used to control peripheral equipment. Each output control implemented in the preset matrix provides two switches. In the system diagrammed in this chapter, the preset matrix contains 64 faders and therefore provides 128 digital switches. The pushbuttons on the control panel may be either alternate or momentary action switches. When outlined with a solid border, the switch is momentary (i.e., pushing a button will close the switch only so long as the button is "held down" by a mouse button). Clicking on the border toggles it on and off. When the border is off, the switch is alternate action (i.e., will remain in the state corresponding to the button last "pushed").

The top and bottom lines of the display screen are used for information, including system name, show name, cue number, and the like. When the mouse cursor is placed on the top line, however, it changes to a menu area. Clicking on a menu title pulls down the menu. These menus permit loading, copying, deleting, and saving cues, as well as a number of other functions. Menus also change the screen to permit blind editing of cues and to bring up an operating screen and a tracksheet. The operating screen displays a list of cues and settings for whatever control the operator decides to monitor. The tracksheet displays the setting of a given control over a sequence of cues. Settings can be modified in the tracksheet—frequently a time saver when the value of a particular control needs to be changed consistently over several cues.

In performance, each cue is activated by pressing a key on the computer's keyboard. Cues can also be programmed to go at a particular setting of the system clock. For example, the system clock can be started with the activation of, say, cue 1. Cue 2, programmed to start automatically 45 seconds after the start of cue 1, is activated when the clock counts off 45 seconds.

A useful feature of the computerized sound scoring system is the digital switches that can be used to control equipment associated with the sound system. For example, starting a cue can also cause one of the system switches to activate the tape recorder from which the audio for the cue is to be replayed. The system switches are not limited to control of sound equipment, however. They can also extend control to a variety of other elements. For example, a switch could be used to trigger a cue on the lighting console or to start a motorized effect on stage. A switch event can also be set to happen at a given timing on the system clock, just as any other cue can be made to run automatically from the clock. Thus, the sound scoring system can be used to synchronize a number of events, especially in circumstances where cues for lighting and other stage effects need to be closely paced to sound or music. For instance, the visual effects associated with an explosion could be triggered by the sound scoring system to coincide exactly with the sound of the explosion as it is played from tape or generated by a sampling synthesizer. (See Chapter 12.)

As with any new technology, certain accommodations are necessary. Only a small segment of the control interface is visible on screen at any given time, and locating a given control with the mouse can take several seconds. Once the control is selected, the system displays the detailed control screen. If the selector button of the mouse is held, the mouse pointer is automatically placed on the primary control element of the module; otherwise, the mouse cursor must be moved to the push-button, direction switch, or other control to be changed, and the mouse button clicked to begin the change. Considerable time must be reserved to read in cue settings. In rehearsal, the process of changing a setting using the mouse is often far too slow; therefore, a flexible system of MIDI-controllable faders is often added to the sound scoring system. (MIDI—Musical Instrument Digital Interface is discussed in Chapter 12.) Special fader elements in the graphic interface can be assigned both to a MIDI fader and to any fader element within the control matrix. When levels need to be adjusted quickly in rehearsal or in performance, the operator has that control through the MIDI fader assignments.

A full discussion of the many capabilities of a computerized sound scoring system would exceed the limits for this chapter. However, understand that a computerized sound scoring system can make possible a number of effects not easily accomplished with standard equipment—intricate sound movement, looped cues (cue sequences that are connected tail to head, so that they repeat for some fixed number of times), and closely timed sequences of multiple sounds. The latest version of sound scoring software has relatively complete MIDI implementation and can sequence synthesizers, samplers, drum machines, and other MIDI-controllable devices. Similarly, MIDI output from other devices can control cue operations in the sound scoring system. At present, consoles for theatre sound have begun to achieve the levels of theatrical lighting consoles—flexible devices that enable designers and operators to concentrate on artistic elements, and leave the manipulation of technical elements to the computer.

Chapter 10

Audio Signal Storage

Audio equipment can amplify and return a magnified signal to an immediate environment with relatively little difficulty, but the real impact of audio lies in its ability to store and retrieve sounds so that they can be used at any time, not just at the time the sound source creates them. A number of ways exist to freeze and hold an audio signal, the principal methods being mechanical, magnetic, and optical. Mechanical storage is now obsolete, but versions of magnetic and optical storage systems appear in both analog and digital audio processing formats. Analog versions are now in decline as digital signal handling becomes the dominant technology in the audio industry. In order to understand how signal storage is accomplished, we begin with a closer examination of the audio signal.

Analysis of Time-Varying Energy Patterns

An acoustic stimulus is a pattern of variation in atmospheric pressure. The pressure differential (the difference between compression or rarefaction in acoustic signal pressure and the average steady-state pressure of the surrounding air) constitutes a source of energy sufficient to induce mechanical movement in a compliant object such as a human eardrum or the foil diaphragm of a microphone. Terms such as *movement* and *variation* suggest a sequence of change. *Sequence* implies time, and acoustic energy is a time-varying signal.

A *time-varying signal* is one that exists only when energy fluctuates periodically above and below an average norm. The progression of values that constitute the instantaneous states of signal energy are the *attributes* of the signal. At the most elemental level of an acoustic signal, only two attributes are significant: relative magnitude and rate of change. All perceptual properties of sound result from these two quantities. The term *relative magnitude* encompasses both amount (intensity) and direction (polarity) of change in energy level; *rate* indicates speed of energy variation.

Usually, when we refer to intensity, we mean the overall loudness level of sound. Our conscious experience deals with *macroacoustic* events—the sound of an engine, the chirp of a bird, the melody and harmony of a piece of music. A time-varying signal, however, is made up of *microacoustic* events—the extent and number of changes in magnitude and polarity of signal energy that make up the vibrations that we perceive as sound.

A simple tone can be described as a given number of excursions between positive and negative extremes of pressure in a given period of time. Successful storage of sound implies that we must be able to freeze and hold the smallest perceptible change in magnitude and the shortest perceptible segment of time. Human hearing cannot detect pressure change that exerts a force smaller than 0.0003 dyne per square centimeter on the eardrum; nor does it respond to pressure variations that occur in a time interval shorter than 50 microseconds. A storage system, therefore, must be able to capture at least those minima. At the opposite extreme, human hearing can tolerate pressure magnitudes up to one trillion times greater than the minimum required to induce significant eardrum motion. Similarly, audition fails to discern meaningful information when one cycle of pressure change is longer than 62 milliseconds.

Storing high intensity and low frequency usually is less difficult than storing low intensity and high frequency; however, small changes are critical to audio fidelity. Most sounds are complex, not simple. Timbre depends on complexity. A complex sound waveform, instead of varying smoothly between pressure maxima and minima, undergoes small internal variations in direction and rate of pressure change. These small internal variations may be analyzed as the combined effect of a number of sinusoidal waveforms acting on the atmosphere simultaneously. Thus, within a maximum period of 62 milliseconds (the period of the lowest frequency we can recognize as tone) internal variations as short as 50 microseconds (the period of the highest frequency human hearing can detect) may exist. Internal variations may also consist of very small changes in relative magnitude, approaching the minimum to which the ear can respond. Therefore, for a storage system to capture all aspects of a signal, it must be responsive to very small variations in intensity and to very rapid rates of change.

A time-varying signal can be stored in either of two ways: as an analog of the original variable or as an encoded description of the physical attributes. An analog recording stores fluctuating state levels in a physical variable that can substitute for acoustic pressure. An encoded recording writes a table of values that can be reconverted into variation in acoustic pressure. *Either method substitutes extent in space for extent in time.*

Analog Storage

Media that transmit sound must be *elastic*. An elastic medium can be distorted momentarily, but tends to snap back to its basic shape whenever the distorting force is removed. (See Fig. 10-1A.) In an analog storage system the recording medium must vary in proportion to magnitude and polarity of the original signal, but it must hold each position into which signal energy drives it. A medium that can hold a record of spatial position is the opposite of an acoustic medium in one important sense: It is *plastic* rather than elastic. The storage medium must provide a number of plastic spatial domains equal to the number of temporary positional changes induced into the elastic medium. The states of signal energy in the elastic medium extend through time; the states of signal energy in the storage medium extend through space. (See Fig. 10-1B.) The storage system forms an array of spatial domains.

In order to recapture the signal, we must move the spatial domains of the storage array past a reading device. Motion is used to reconstitute the element of time within the spatial storage medium. The reading device converts the stored states back into energy, which, again, reconstructs temporary positional variation in the elastic medium.

Encoded Storage

Encoded storage converts signal states into a description in symbols, usually numeric. The symbols can be stored in a variety of ways, but the most common methods utilize a magnetic medium, such as a computer disk, or an optical medium, such as a compact disc. Encoded storage is, in effect, a two-dimensional array of logical cells rather than a linear array of spatial domains. The rows of the array represent signal attributes; columns represent position in time. (See Fig. 10-1C.) Each cell presents a "snapshot" of one attribute of signal state at a particular moment relative to the beginning of the recording. Retrieval of the signal requires that a master controller consult a list of cell addresses (sometimes called a table of contents), sample the value contained in the next listed cell, and deliver the value converted into one state of an audio

FIG. 10-1: ELASTIC MEDIUM VS. ANALOG STORAGE

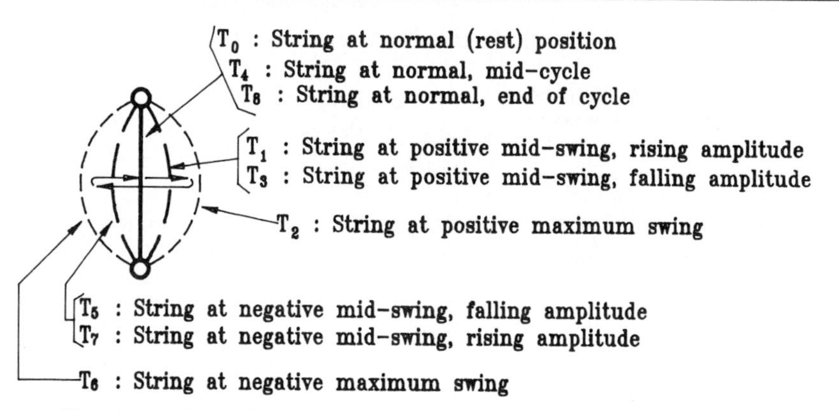

A. String vibrates changing position through time, but does not hold any position except rest position.

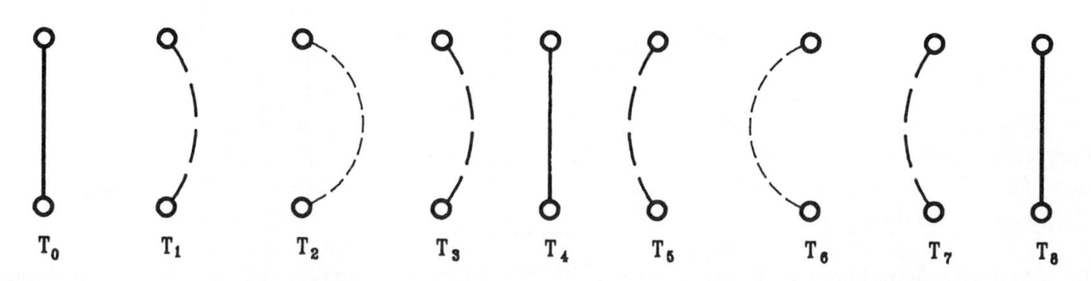

B. Analog of string positions stored in adjacent domains of linear space.

FIG. 10—1C: ENCODED ATTRIBUTES

Attribute/Time	T_0	T_1	T_2	T_3	T_4	T_5	T_6	T_7	T_8
Tension (psi)	100	102	104	102	100	102	104	102	100
Deflection (mm)	0	+5	+10	+5	0	–5	–10	–5	0

signal. The time element is resynthesized, not by motion but by a logical pattern of cell sampling operating under the command of an internal clock pulse.

Note that logical sampling opens the possibility of altering sequence (taking cells out of original order) and time relationship (altering response to internal clock) in ways that are impossible for analog storage, with attendant creative opportunities that are almost unlimited. Encoded storage is the basis of modern digital recording technology.

Overview of the Development of Signal Storage

Mechanical Storage

The earliest forms of audio recording used mechanical storage devices. Mechanical storage usually means applying energy transduced from the audio signal to reshape some element of the storage medium. The primary system of mechanical storage is the phonograph record. To create a phongraph recording, audio signals are inscribed into the surface coating of a master disc by a cutting stylus. The stylus is driven by a power amplifier that raises signal current and power to a level sufficient to force the stylus to carve a trace into the disc surface. Small variances in the form of side-to-side swings from the center line of the groove represent the magnitude and polarity of the original signal. The cutter head moves from the outside of the disc to the inside, tracing a spiral groove as the disc rotates beneath it. The master is used to make a metal mold to cast multiple copies of the recording, usually in vinyl plastic.

Playback is the reverse of the cutting process. A stylus carried by a *tonearm* reads the shifts in magnitude and direction of groove swing. The stylus motion induces current in small coils of wire located in the pickup cartridge, and that current forms an analog of the original signal. The tiny current produced by the stylus must be boosted by a preamplifier to standard line level.

The speed at which the groove is turned beneath the stationary playback stylus reconstructs the time component of the original signal. The spacing between adjacent turns in the spiral groove effectively determines the maximum amplitude of groove deflection from center, which limits the dynamic range that the disc is capable of storing successfully.

The earliest discs had very coarse grooves and used a rather high rate of rotation (78 revolutions per minute), because the available surface materials could not follow rapid changes in direction (necessary to represent high frequencies) within a physically small space. High speed meant that a relatively heavy pressure was needed to keep the playback stylus in the groove; but significant stylus pressure also meant erosion of groove, which would eventually wear away the trace of the audio signal. The 78 rpm discs were made with a very fine abrasive substance embedded into the surface medium, which meant that the groove would wear down the stylus faster than the stylus could abrade the groove. Styli had to be replaced frequently, but the logical choice was to have the stylus deteriorate rather than the recording.[1]

A high rate of rotation limited the amount of time available on a single disc. Recording a long symphony or an opera, for example, could require an album of 10 to 20 12-inch discs (20 to 40 sides) and a break in performance (to flip or change discs) every five minutes or so.

The first steps toward a more satisfactory and economical phono disc format came with the introduction of 45 and 33.3 rpm discs. Aside from extended playing time, lower playing speed has a natural advantage in that it reduces the frequency of noise produced by the friction of the stylus against the groove wall. Developments in electronics, metallurgy, and plastics technology combined to make possible lighter stylus pressures. Lighter tracking pressure enabled thinner grooves with closer spacing, making both extended dynamic range and higher frequency response possible. Most of the 45 rpm recordings were 5-inch discs that played up to about 10 minutes on a side. 33.3 rpm discs were 12 inches in diameter and could store 30 minutes or more per side.

The 33.3 rpm long-playing record was the standard medium for commercial distribution of recordings between 1955 and 1980. During that period, magnetic tape developed into the primary recording medium for the studio. The open-reel format for tape never became commercially successful, however. By the late 1970s, tape in a cassette format had become viable, and tape cassettes almost became the standard that replaced the 33.3 rpm disc recording.

[1]The fact that 78 rpm discs contain coarse grooves and a built-in abrasive is the reason that such discs should not be played with equipment expressly designed for 33.3 rpm long-playing discs. Styli are available that permit 78 rpm recordings to be played on modern equipment, and these, not the styli for 33.3 rpm LPs should be used when reproduction of an old disc is necessary.

Magnetic Storage

Phonograph discs carry a mechanical analog of an acoustic signal; tape recording creates a **magnetic** analog of the audio signal. In order to store energy in a magnetic format, one needs some kind of magnetically permeable material that retains the image of the last magnetic polarity impressed upon it. Like the phono disc, the time component of a magnetic recording must be synthesized by motion. The ability of the medium to capture and hold a wide range of frequencies is determined by the magnetic characteristics of the medium and the speed of motion.

Soft iron is a substance that can hold a record of magnetic polarity, but soft iron alone is not suitable as a medium because there is no simple form into which it can be shaped that will allow it to serve as a practical storage device. During the 1930s and early 1940s, experimenters did try iron wire as a recording medium, and it almost worked. Iron wire could be wound back and forth between two spools, but it was hard to handle, subject to rust, abrasive to the magnetic write/read heads, and if it were ever snapped or kinked it became unusable. Some wire recorders were built and sold, but they were just too difficult to be practical.

Plastics made the development of tape recording possible. An obvious solution to creating a magnetic recording medium had long been recognized. If iron could be ground to a fine powder and bonded to a flexible ribbon, a satisfactory recording medium could be achieved. Until the development of cellulose acetate, however, no substances suitable for an economical flexible ribbon existed. Cellulose acetate provided a backing upon which fine-ground soft iron could be deposited and that was sufficiently flexible and strong to stand repeated spooling back and forth from one reel to another. Cellulose acetate did break fairly easily, but, unlike iron wire, it could be spliced together using adhesive-backed tape.

Magnetic tape dominated the audio industry during the last half of this century, and it continues to do so, although the format in which tape is used has changed considerably. Tape now stores data rather than analog audio. The open-reel format is almost completely limited to studio use as of 1990. Eighth-inch cassettes, in lengths ranging from 45 minutes to 120 minutes total playing time, have become the consumer recording standard. Recently, a new digital cassette format, called DAT, has become available. DAT cassettes have potential applications in both home and studio markets.

Optical Storage

Although tape became the dominant recording medium for the second half of the twentieth century, one form of sound storage was much older, having been used for many years by the motion-picture industry. That form is optical and relies on the fact that either the reflective or transmissive properties of certain materials can be used to hold a record of signal variation.

In the film industry, variable transmissivity of motion-picture film stock is used to record an analog of sound. A band of varying width and transparency, called the *sound track*, placed near one edge of the film controls the amount of light that gets through to the optical reader. The optical reader is a light-dependent resistance that varies current flowing in the audio circuitry in proportion to the amount of light falling on its surface. As with all of the storage devices we have examined, motion of the medium synthesizes the time dimension.

A more recent application of optics to sound storage is the compact disc, currently emerging as the dominant format for mass marketing of audio recordings. Compact discs (CDs) use reflectivity rather than transmissivity, and, unlike film optical sound tracks, the recording is not an analog of the original signal. CDs are strictly digital. The reading (as well as the writing) device is a laser beam controlled by a dedicated computer.

Tape Recorders

Since the 1960s magnetic tape has reigned as the primary storage medium for the audio industry. Accordingly, a significant portion of this chapter is given over to a discussion of tape recording devices and to matters concerned with the storage of sound on tape. The technical principles of recording on a magnetic medium are essentially the same in both analog and digital equipment, although significant differences exist between digital and analog in matters such as longevity of storage and ease of copying and editing. Most tape machines in use at the beginning of the 1990s are probably still analog; however digital tape recorders, which began to assume a significant position in the industry during the 1980s, are now becoming the dominant means of recording for all major studios.

Theory of Magnetic Recording

Magnetic recording is a process of locking a pattern of variation in the polarity of a magnetic field into a **permeable substance** (a substance that can be magnetized.) The most usual substance used for the recording medium is finely powdered soft iron bonded to a ribbon of flexible plastic. Chromium dioxide has recently achieved a significant place as the medium of choice for high quality audio cassettes.

Molecules of permeable metal respond in a very particular way to a magnetic field: They align themselves with the polarity of the magnetic field. If the direction of the field shifts, the molecules realign to follow it. If the field is suddenly removed, *the molecules retain the last alignment imposed by the field.* Ability to hold the most recent field alignment is what makes magnetic tape recording possible.

The device used to store and retrieve the acoustic image is a specially designed electromagnet called a **head**. When used to write an audio signal onto the tape, the device is known as a **record** head. Used to retrieve the signal, it is called a **reproduce** head or a **playback** head. Tape recorders also include a third head, called the **erase** head, which is used to destroy previously recorded data before new signal information is written to the tape. The erase head produces a very strong, high-frequency signal that effectively randomizes molecular alignment.

One major problem in writing an audio signal onto magnetic tape arises from the fact that, when the molecules of a permeable substance track changes in an electromagnetic field, they do not follow those changes instantaneously. A lag always exists between the electrical signal and the state of the field. The lag creates a pattern like that shown in Fig. 10-2. The pattern is called a **hysteresis loop.**

As signal current increases, generating a magnetic field in a given polar direction, there is an inertial lag in the oxide molecules of the tape. The molecules follow the polarizing field toward one extreme of alignment. As the direction of the signal current reverses, the molecules continue to preserve the initial alignment for a short time; then they follow the polarity reversal until, at the other extreme there is a similar inertial event. If the audio signal is a sine wave, the signal field written to the tape would be distorted as shown in Fig. 10-3.

Naturally, a signal recorded with hysteresis distortion would be worthless. The solution to the problem is to introduce into the recorded signal a high frequency signal called **bias**— the same signal used to erase the tape. Bias is a magnetic field oscillating at a frequency equal to at least twice the highest frequency to be recorded. (In premium quality tape recorders the bias frequency is as much as five to seven times the highest frequency to be recorded.) The magnitude of the bias current is determined both by the design of the head and the permeability characteristics of the tape. Audio signal is **modulated** onto the bias waveform causing the DC level[2] of the bias signal to shift in proportion to the audio waveform. The peaks of the bias signal are made large enough to stay within the *linear portion* of the transfer curve. Because DC level of the peaks carries the audio signal, the audio signal is undistorted. (See Fig. 10-4.)

FIG. 10-2: HYSTERESIS LOOP

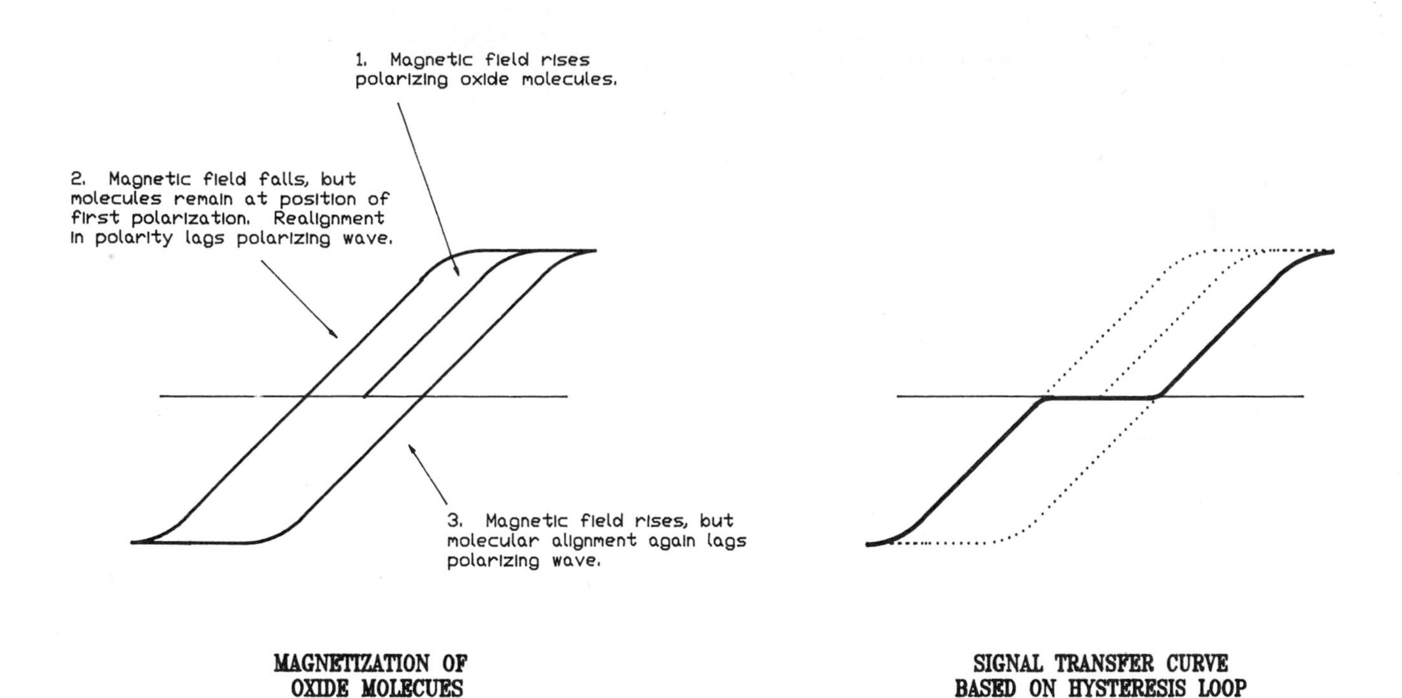

[2]DC level refers to the average voltage level of an alternating signal. Given a steady-state alternation, the average voltage level is zero. (Positive voltage equals negative voltage and number of positive peaks equals number of negative peaks in any given unit of time; therefore, average voltage equals zero.) The audio waveform, which will always be much longer than the bias signal waveform, adds a positive or a negative offset to the bias signal, creating periods of net positive and net negative DC level.

FIG. 10-3: HYSTERESIS DISTORTION

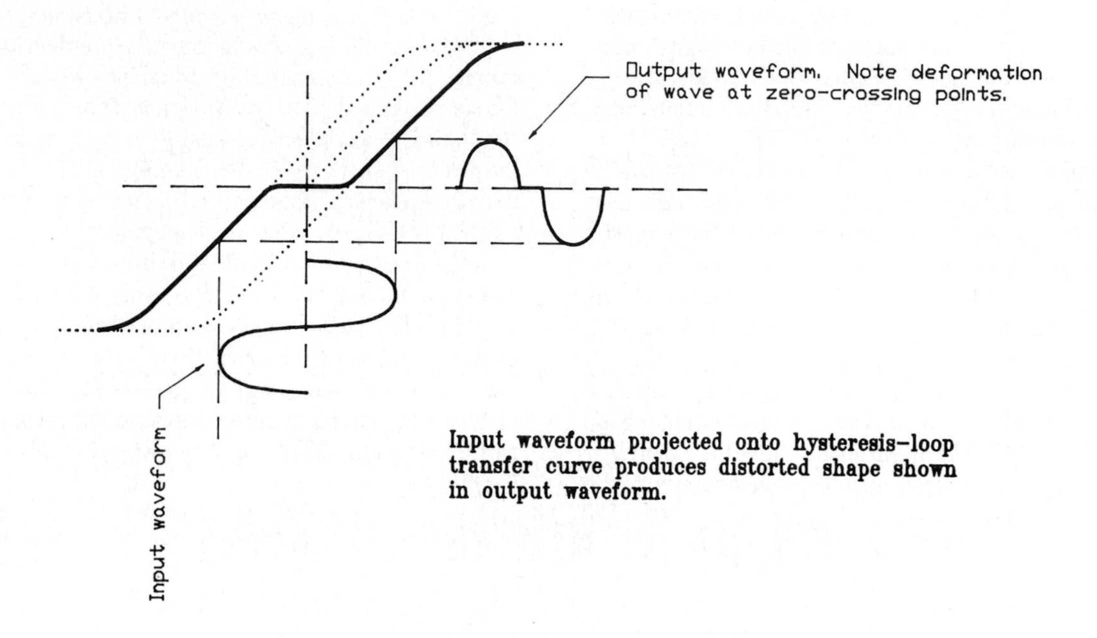

Input waveform projected onto hysteresis-loop transfer curve produces distorted shape shown in output waveform.

FIG. 10-4: USING BIAS TO CORRECT HYSTERESIS DISTORTION

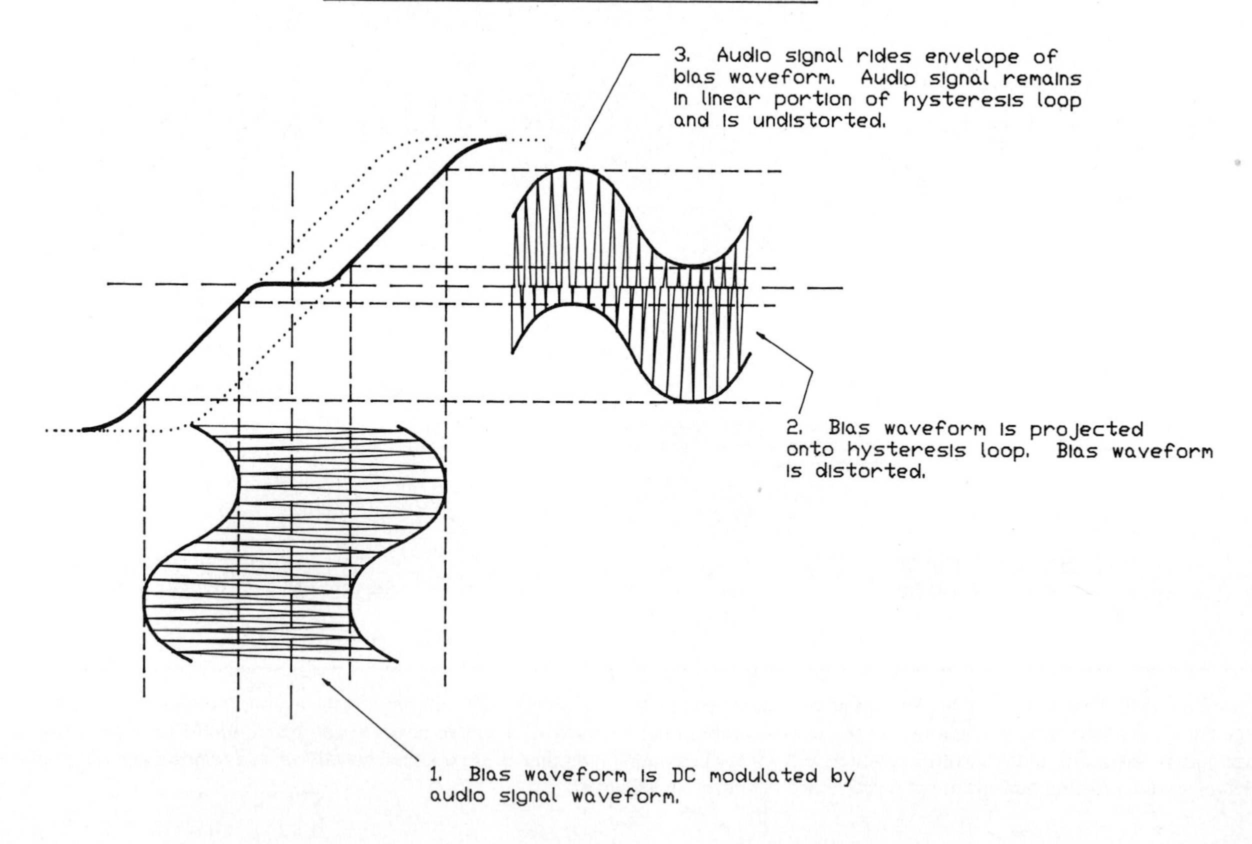

The tape itself is a ribbon of plastic with a finely ground formulation of iron oxide (or for the most recent high quality cassette tapes, chromium dioxide) bonded to it. The molecular structures within the oxide are called **domains**. As a domain passes over the tape head, the domain must align with the polarity of the magnetic field at that instant and then hold that polarity permanently. In order not to be affected by the next instantaneous state of the magnetic field, however, the focus of the tape head must be very narrow, and the tape must be separated from the head as quickly as possible. Head surfaces are rounded to reduce both the space and time for contact between head and tape. Relatively high tape speed generally produces the best quality signal storage. Audio recording tape is made in a variety of grades and types. The quality of the tape and its performance characteristics depend primarily on the oxide formulation and the degree of adhesive bonding of oxide to backing.

Essentially, the resolution (highest recordable frequency) of a tape is dependent on the size of the oxide molecule used as the storage medium. The smaller the molecular structure, the higher the upper frequency limit. Molecules of iron oxide tend to be relatively coarse; therefore, research is always underway to find permeable compounds with smaller molecules. The chromium dioxide molecule is somewhat smaller and permits higher resolution. Each particular formulation of recording medium requires its own particular level of bias current, however, so that for optimum signal capture, bias needs to be adjusted for each kind of tape used.

The magnetic medium needs to be bonded very securely to the tape backing, because friction between oxide and heads is relatively severe. Oxide, which scrapes off the tape, contaminates the heads and degrades both recording and playback performance; loss of oxide ultimately creates regions of tape that have no storage capability. Such regions are called **dropouts**. A tape with too many dropouts is a candidate for the trash barrel.

Basic Parts of a Tape Recorder

All tape recorders are essentially made up of three basic parts: the tape transport, the head assembly, and the electronics assembly.

The Tape Transport

The **tape transport** is the mechanical assembly responsible for moving the tape from reel to reel. The normal configuration of a tape transport consists of two turntables that spin the tape reels, a drive system that is responsible for pulling the tape through the head assembly at constant speed during recording and playback, and a tape tensioning system designed to keep the tape in firm contact with the heads and eliminate jitter, vibration, or momentary aberration in tape motion. Two styles of tape transport are shown in Fig. 10-5. Differences between the two are explained shortly.

The most characteristic element of a tape recorder is its reels. The reels are mounted on two turntables, one for the feed

FIG. 10-5: TAPE RECORDER TRANSPORT ASSEMBLY

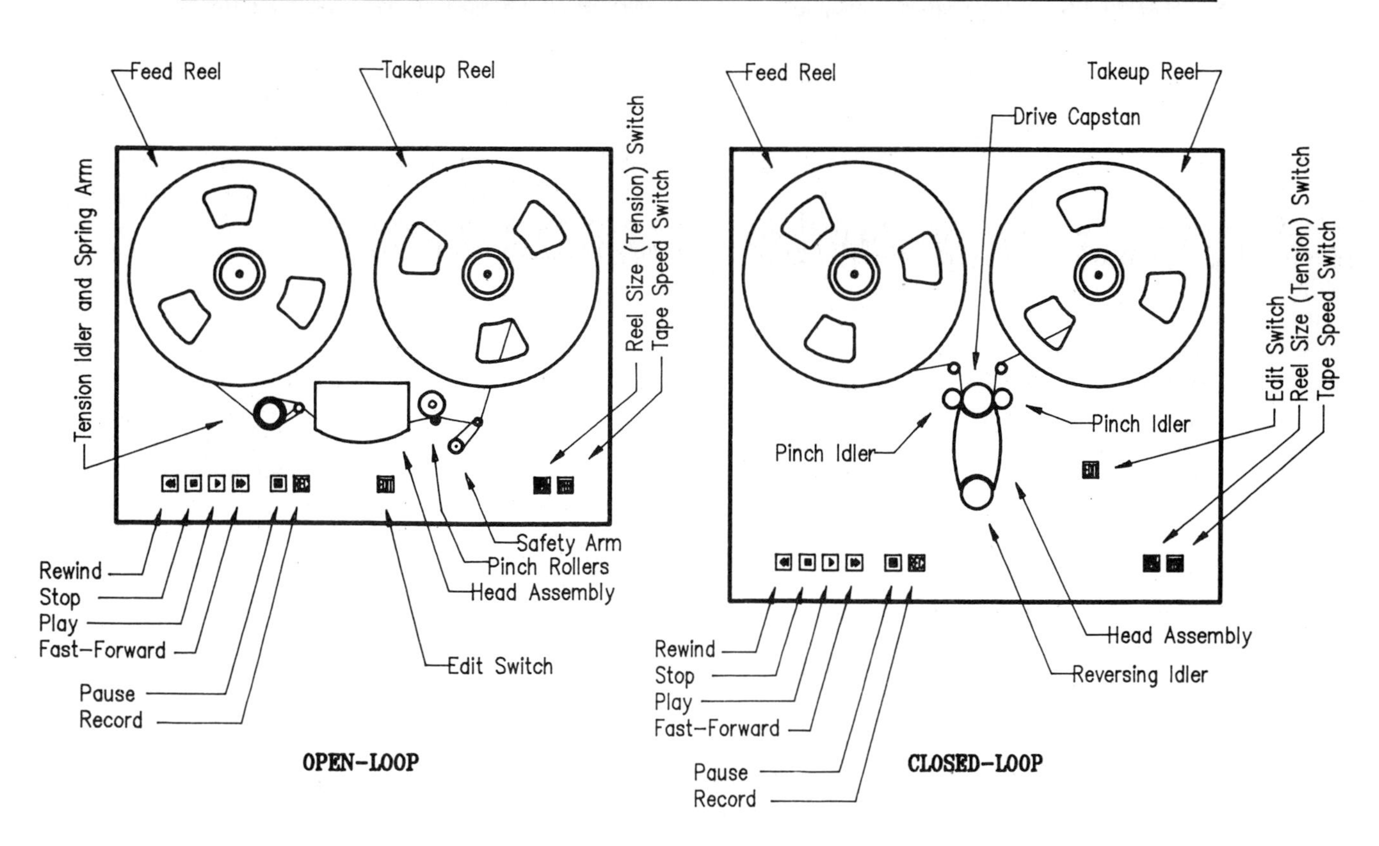

reel, one for takeup. In professional quality tape recorders each turntable is driven by a separate motor. The motors always pull in opposite directions, the rewind motor always pulling backward and the takeup motor pulling forward.

Tape speed during play and record is controlled by the drive system, a finely machined capstan combined with a rubber pinch roller. (Machines that use the closed-loop system shown in Fig. 10-5 have two pinch rollers.) The capstan is powered by the main drive motor; the pinch roller is an idler (a roller that moves with the tape but is not powered.) During play or record, the pinch idler is closed tightly against the capstan and tape is pulled by friction between the two rollers. Capstan diameter combined with motor speed determines the rate at which the tape is pulled past the heads. Most professional tape recorders change speed by increasing or decreasing motor speed. Older machines and many consumer-grade tape recorders use some form of mechanical linkage to change speed.

The tensioning scheme of most tape transport systems consists of the drive opposition from the turntable motors combined with a series of spring- or flywheel-loaded devices, as shown in the open-loop version shown of Fig. 10-5. The purpose of the tensioning system is to keep the tape surface tight against the contact area of the heads and to correct for any sudden decrease or increase in tape tension caused by uneven spooling.

Spooling tape back and forth between reels sometimes results in an uneven wrap. As the tape is pulled through the heads, regions of loose wrap can cause tape tension to drop, sometimes quite suddenly. The spring-loaded arms and tension idler usually counteract sudden changes. The idler tends to keep turning at a constant speed because it is loaded by a massive flywheel. The two tension arms spring apart during a drop in tension or pull in to absorb an increase in tension. Together, these components act to keep the tape at constant tension as it passes through the head assembly. One of the tension arms controls a safety switch to stop tape motion in the event of a complete loss of tension.

In the closed-loop system, two drive rollers are used instead of spring- and flywheel-loaded devices. One, at the entrance to the head assembly, runs at standard speed; the second, at the exit from the heads, runs just slightly faster. The speed mismatch pulls tape tight as it travels through the head loop. The closed-loop system also provides a small tensioning system (located in the tape support rollers between reels and drive assembly). As in the open-loop system, one of the tension arms also acts as a safety switch.

The transport provides standard motion controls: rewind, stop, play, fast forward, and pause switches. The record switch is located next to the pause button. Tape speed and tension (reel size) controls are also located on the transport deck. Most professional machines provide an edit switch to disable the takeup reel motor and tape lifters. (The functions of the edit switch and tape lifters are explained later.)

The Head Assembly

The **head assembly** contains the electromagnetic write/read devices that place the signal on tape and retrieve the signal for playback. The heads are usually enclosed in a nonpermeable metal housing that shields them from stray magnetic fields. The customary order of heads, proceeding in the direction of tape travel (usually from left to right), is erase, record, and playback. (See Fig. 10-6.) Professional quality tape recorders include all three heads; low-cost machines usually do not. In a two-headed machine, the order is erase and record/playback. The second head is switched to serve two functions, an economy measure that limits the usefulness of such a machine for most applications in the theatre.

In order to transfer the magnetic field from the head to the tape, the head must present a high magnetic impedance to the field just at the point where the tape contacts the head. A high magnetic impedance is created by placing an **air gap** (Fig. 10-6, inset) in the head core. The tape, therefore, presents a lower impedance to the field, and the flux jumps from the head into the oxide around the gap. On replay, flux recorded on the tape induces signal currents into the playback head as the recorded domains move past the head gap.

Tape heads require careful cleaning and maintenance, which should be performed daily, if possible, and certainly before any extended recording session. With each pass of tape over the heads, oxide tends to scrape off the tape and coat the head surfaces. In older machines, head gaps are open and will become clogged. (Modern heads do not have this problem because head surfaces are physically solid.) Tape heads are cleaned with cotton swabs and denatured alcohol.

The other important procedure is *degaussing*. Tape heads accumulate residual permanent magnetism. As permanent (DC) magnetism builds up, the ability of the head to respond to rapid field changes deteriorates so that high frequencies are lost (i.e., either not recorded, if the record head is affected, or not reproduced, if the playback head is magnetized). DC magnetism also permits a higher level of residual tape noise (pseudo-signal caused by random polarizations of small domains of oxide molecules), and it can also cause partial erasure of high-frequency program information as the tape passes over the record or reproduce head.

Degaussing is done with a special tool, essentially an AC electromagnet. The tape machine should be turned off, and there should be no tape on or near the machine while the process is carried out. The degaussing process starts with the degausser some distance away from the head. Moving in a small arc from side to side, the operator moves the degausser toward the head. When magnetic pull can be felt between the degausser and the head, the arc is narrowed until the tip of the tool sweeps just above the head surface. Then the tool is retracted slowly until the pull between tool and head is no longer noticeable. The head is now demagnetized. The process is repeated for each head.

FIG. 10-6: TAPE HEAD ASSEMBLY

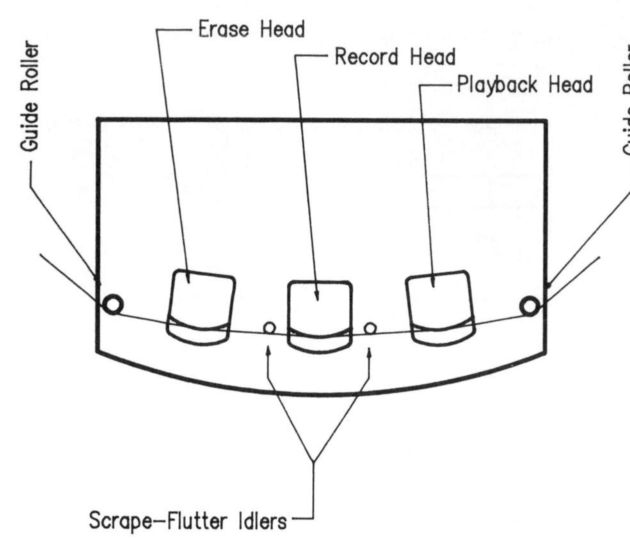

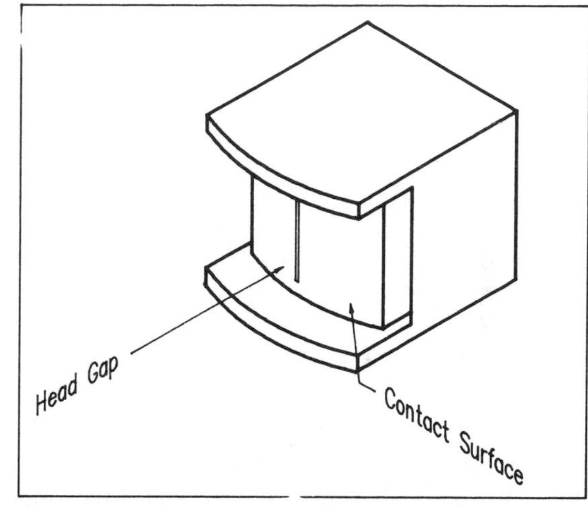

The Electronics Unit

The **electronics unit** contains input and output amplifiers, preamplifiers, VU meters, the bias oscillator, and the driver amplifiers for writing signal to the tape. All controls not associated with tape motion are also located on the electronics assembly. The controls include all input and output level knobs, the meter/output transfer switch, record safety switches, and switches to adjust bias and or record EQ.

The bias oscillator is the source of the high-frequency alternating current that drives the erase head and supplies the bias frequency to the record amplifier. Bias is delivered full strength to the erase head. The record safety switch, located on the face panel of the electronics unit, disables the bias oscillator and the mixer/driver, preventing accidental erasure or overwriting of a tape. The mixer/record driver receives the bias signal, attenuates it to a level matched to the oxide formulation for the tape[3] to be used, and mixes the bias and the input signal together (i.e., modulates the DC level of the bias signal, as described earlier). The mixer/driver then equalizes the modulated signal to match record head transfer characteristics (the circuitry that accomplishes this is called a *preemphasis network*), and boosts signal power to a level suitable to drive the record head.

The microphone preamplifier and line amplifier receive input signals for the recorder. Not all machines provide both line-level and microphone-level inputs, but most do. Those that do not provide line input rather than microphone input. Electronics units usually include separate input level controls for both microphone and line inputs. The playback preamplifier processes the signal picked up by the playback head. The preamplifier boosts the playback signal to near operating level. The preamplifier also contains a bias trap, to remove the bias frequency from the signal, and a **deemphasis** network, which removes the equalization imposed by the record amp.

The final stage of the tape recorder's amplification system is the line amplifier, which raises the signal to standard operating level and sends it to the output of the tape recorder.

[3]Bias current level must be set by a competent technician for each different kind of tape to be used. Normally, one kind of tape is used almost exclusively, so that bias can be calibrated for that particular oxide formulation. Occasional use of other brands of tape will not create serious difficulties. If several different brands or different kinds of oxide are used with some degree of frequency, a machine with adjustable bias compensation should be used. Several manufacturers provide either switch-selectable settings or a bias level control knob. The level control knob should be calibrated for normal usage, then varied for best results on each different kind of tape.

FIG. 10-7: TAPE RECORDER ELECTRONICS ASSEMBLY BLOCK DIAGRAM

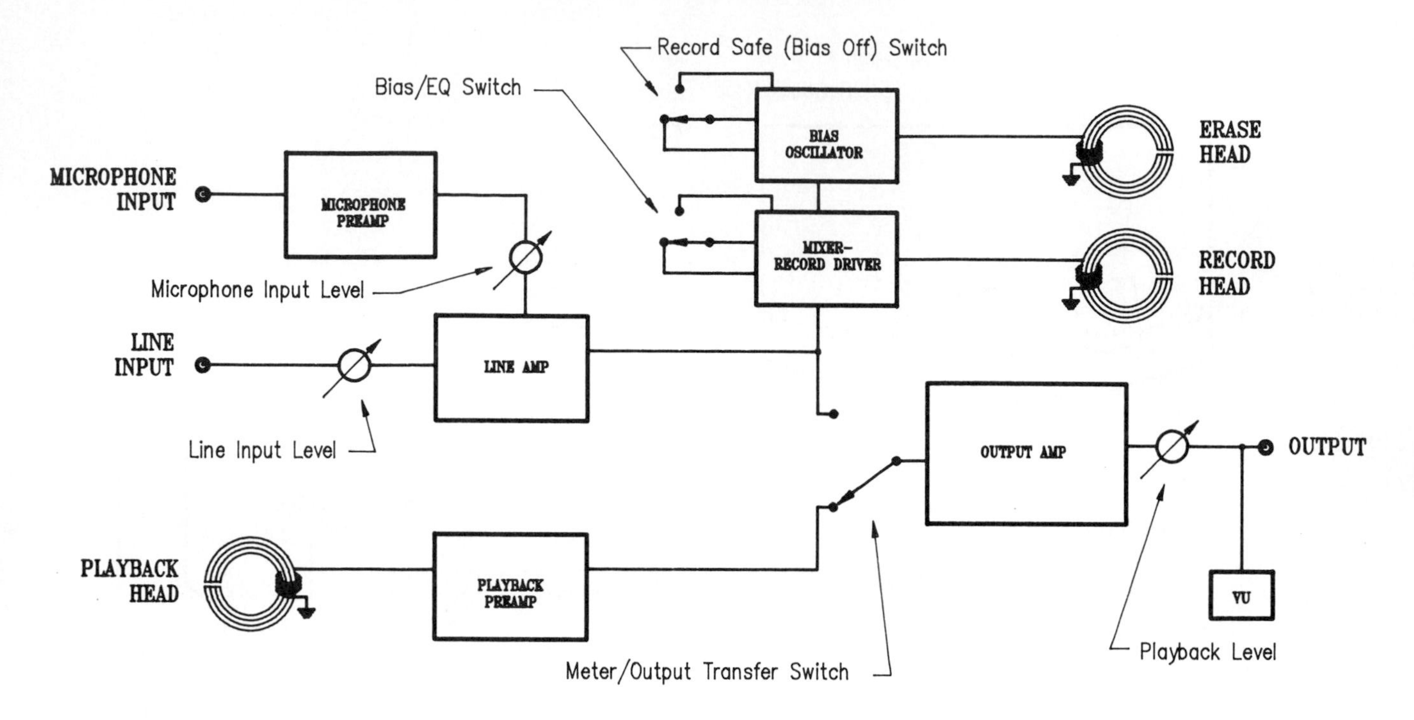

The line amp also provides the circuitry to drive the recorder's VU meters. At the input to the line amp is a switch that permits signal to be taken either from the record amp or the reproduce amp. This switch is called the meter/output transfer switch, because it permits the operator to hear either the signal in the record amp (before the signal is written to the tape) or the signal from the reproduce amp.

Figure 10-7 shows a block diagram of typical tape recorder electronics. The system in the figure shows only a single audio channel. Except for the bias oscillator, multichannel tape recorders must have one of each element for each channel used. Electronic assemblies for 16- and 24-channel recorders, therefore, are quite large.

Observe in Fig. 10-6 that switching the meter/output transfer switch back and forth between input and tape has no effect on the recording itself. A common misconception among novices is that to record the machine must be set to input and that turning the selector to tape will interrupt the recording. The switch is merely to allow input and output levels to be compared and to verify recording quality during the record process.

Tape and Tracking Formats

Recording tape is made in widths ranging from one eighth of an inch (for cassette recorders) to 2 inches (for large, 24 track studio recorders). The tape, physically, is very nearly the same, no matter the size; the number of tracks recorded on tape are determined by the record head.

Originally, all recorders were **monaural**, which means literally "one eared." Everything was all mixed down to a single track. (See Fig. 10-8A.) The normal tape width was one quarter of an inch. For consumer machines, recording time could be doubled by making only half the record and playback heads active, storing signal on only half the tape at each pass. When the top half of the tape was recorded, the tape could be flipped over and recorded in the reverse direction. (See Fig. 10-8B.)

The first step into multidimensional audio was the development of stereophonic sound using two tracks. (See Fig. 10-8C.) Two tracks occupied nearly the full width of the tape. Only a small guard band was left between left and right channel tracks. Many consumers, however, still wanted the double play capability provided by half-track monaural. For home use only (professional machines never employed this practice), manufacturers developed a head configuration that recorded two tracks with each track occupying slightly less than one sixteenth of an inch. (See Fig. 10-8D.) The heads were again divided in half, each track writing to approximately one quarter of the width of the tape. The left channel occupied the first quarter; the right channel the third quarter. On turning the tape over, the second and fourth quarters of the tape covered the active head zones, permitting recording in the reverse direction of the tape.

Reduction of head size to quarters of tape width, however, suggested the possibility that more than two tracks could be recorded on tape. First, a kind of superstereo, called quadraphonic sound, was attempted. In quad recordings, the first

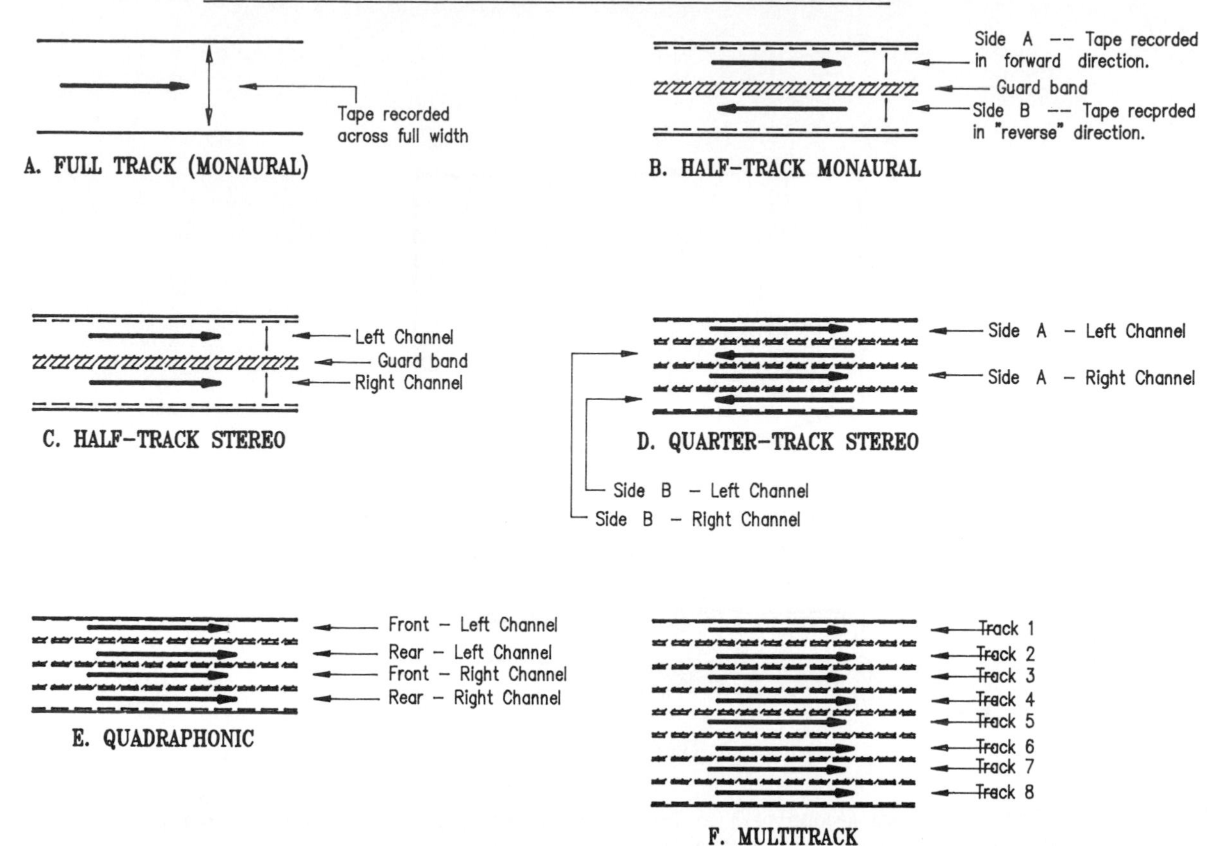

and third quarters of the tape carried standard stereo information. The second and fourth quarters carried ambience information picked up from the recording environment. On playback, the standard stereo loudspeaker system was supplemented by two rear speakers to reproduce the ambience tracks.

Engineers soon discovered that all four tracks could be used for individual elements of a recording, and both they and musicians were quick to see the possibilities of multiple simultaneous tracks. (See Fig. 10-8E.) Using multiple tracks, some of the instruments or voices in a musical ensemble could be isolated and recorded separately; afterward, these isolated tracks could be processed to achieve some kind of enhancement or modification of sound. (See Fig. 10-8F.) Multitrack started at four tracks and grew to 24 or even 32 tracks. (Currently, 24 is the maximum number of tracks provided by most professional, multi-track recorders.)

Multitrack recording introduces a number of technical problems, the most basic involving tracking synchrony. The standard three-head configuration is useful because it permits the recording engineer to monitor both source and tape alternately during the recording process; but because recording and reproduce heads are set a few centimeters apart, tape playback will always lag slightly behind the input signal. (See Fig. 10-9.) One can even feed the output of the playback amps back into the record amps and generate a kind of reverberant echo called **slapback**.

Potentially, one advantage of multitrack tape was that a musician could record one part, say, the melody to a song, then play along with the original track to record a harmony part on a second track. Clearly, however, the lag between input and tape playback would make it very difficult to synchronize the two tracks.

An obvious solution to the synchrony problem was to let some tracks of the record head serve as a pickup for tracks that were already recorded, while other tracks recorded new sounds onto the tape. Multitrack electronics units usually permit each element of the record head stack to be switched between the record amp and the reproduce amp. A switch associated with the electronics for each track determines whether the playback amp gets its signal from the standard playback head or from the record head. (See Fig. 10-10.)

Varieties of Tape Recorders

Tape recorders are made in a variety of formats, some useful for theatre and some not. Some of the least expensive (and also some of the most expensive) are manufactured for the home audiophile—the person who likes to listen to high quality sound in his or her home. Consumer-grade, home

FIG. 10-9: LAG BETWEEN RECORD AND PLAYBACK HEADS

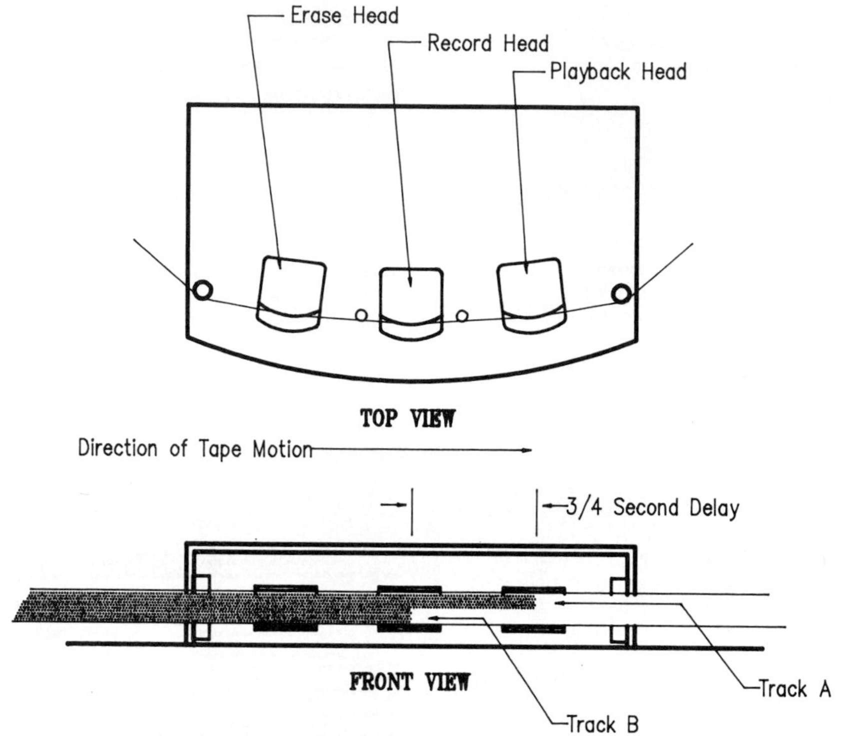

Result of trying to record second part on Track B while listening to previously recorded Track A. Second part lags first part by almost one second. Tracks are out of synchrony.

entertainment recorders usually combine all elements—transport, head assembly, and electronics—into a single unit. The least expensive home-grade machines have only two heads, switching one head to serve both recording and playback functions.

Recently, a number of products aimed at the musician and the audio amateur have been placed on the market, and multitrack tape recorders (four to eight tracks) are included among these products. Such tape recorders are designated as semiprofessional machines, in that they have many features of the large, studio tape recorders but usually incorporate all elements in a single chassis. Their specifications are not as good, nor are they as rugged as full-featured, professional machines; but they provide an excellent trade-off between price and quality. Most such machines are usually quite satisfactory for theatre use.

Professional studio tape recorders are large, ruggedly built, and are engineered for wide frequency range and dynamic response. Almost all have the capability for complete remote control and usually incorporate search and locate functions under an industry-standard time encoding system. Transport and electronics are generally separated. Full sized, professional recorders are too expensive for most theatres, but they are the most flexible and durable.

Other forms of tape recorder include cassette and cartridge machines and sprocket-driven magnetic film recorders. Magstock (magnetic film) recorders are primarily the domain of the film recording industry and play very little part in theatre sound, but cassette and cartridge players are useful in theatre.

Cartridge players were developed originally for the broadcast industry. The cartridge is a continuous loop of tape, arranged so that the tape spools up around the hub of the single reel and rewinds onto the outside of the spool. The obvious advantage is that the tape never needs rewinding. Coupled with a way to stop the tape automatically, the device is self-cuing. Cartridge recorders use two heads: one for record and the other for playback. Cartridge machines are available in monaural or stereo versions. The monaural version uses a two-track head. One track records and reproduces the audio signal; the other records and senses a cue tone. The stereo model has a three-track head. Two tracks are for signal and one is for cue tone.

In the record mode, when the start button is pressed, an oscillator writes a 1-second burst of tone onto the cue track of the tape. Pressing the start button also disables the cue read head for about 5 seconds so that the just-recorded cue tone will not stop the tape prematurely. After the 5-second lag, the read head is enabled, and when the tape has completely spiraled around to the starting point, the cue tone reaches the read head

FIG. 10-10: SYNCHRONOUS RECORDING

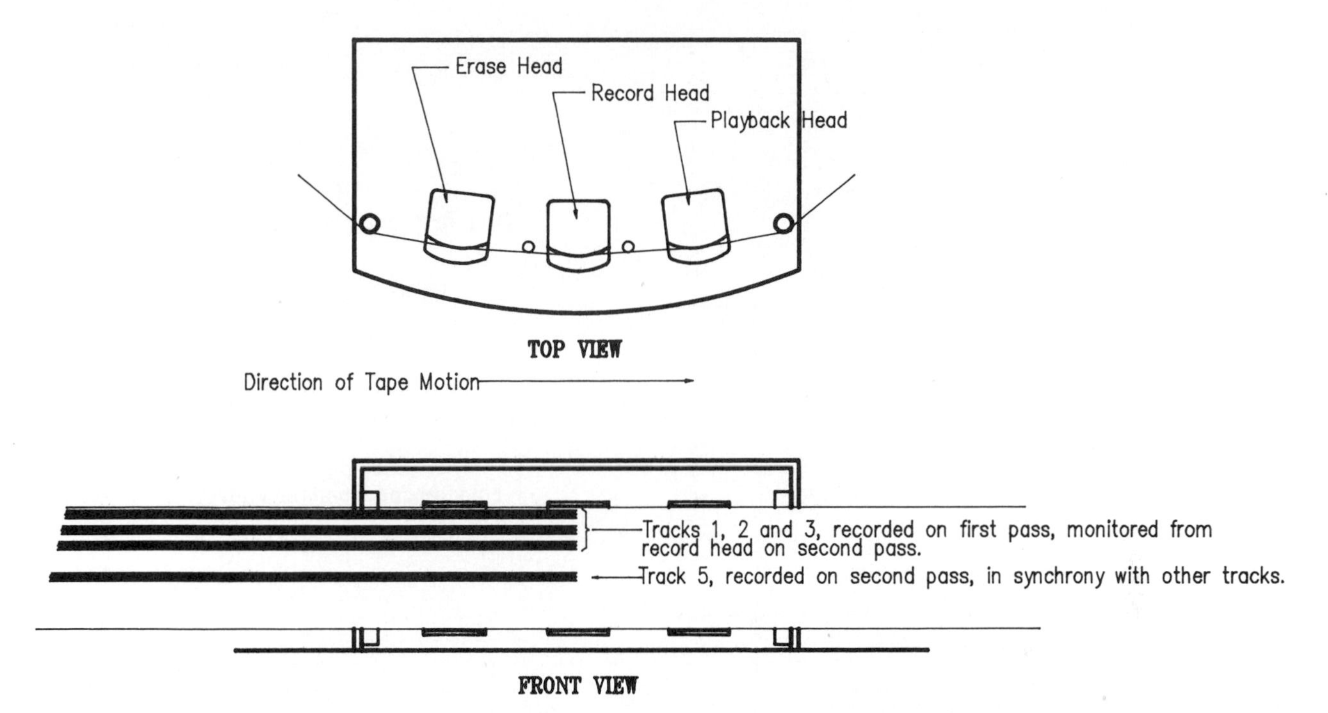

Multitrack recorder using synchronous recording. Tracks 1, 2, and 3 of the record head have been switched to act as read heads. All other tracks remain as write heads. Program recorded onto track 5 during second pass is played as performers listed to base tracks previously written onto top three tracks. All tracks are synchronized with no lag.

and trips the stop circuits. If the cue tone is disabled, however, the cartridge can be used as an endless loop—a feature that can be very useful in theatre. For example, if a general background such as low crowd noise were placed on a cartridge used as a loop, the crowd sound could be kept going indefinitely, for as long as the scene requires.

Cassette recorders have become excellent sources of compact, high quality sound in recent years, but their greatest problem for the theatre is that tight cuing is difficult. They can be useful, however, for situations in which long cuts of sound are needed with no particularly tight cuing involved. They are also good for recording archival tapes of performances. A few experimental cassette recorders using a special form of time coding have been marketed, but with little success. Time-coded cassettes can be cued reliably, but the encoding circuitry makes the cost of the machine too high.

Editing Stored Audio Signals

One of the real advantages of analog tape recording over earlier storage systems is the fact that tape can be cut and spliced. By marking the precise location on tape where a segment of sound begins, then marking the end of the segment, a portion of a sound can be completely removed from the recording. Parts of sounds can be moved and rearranged to form totally new sounds.

The traditional procedure for analog tape editing (Fig. 10-11) is as follows: The starting point for the edit is found and marked. Finding the point to mark is a matter of moving the tape back and forth over the playback head until the end of the lead segment is just beyond the playback head. In order to hear the effect of rocking the tape back and forth, the tape must be in contact with the playback head. Most machines have tape lifters that extend to push the tape away from the heads while the machine is in fast motion. Most machines provide a tape lifter defeat that allows the tape to remain in contact with the heads. A marking pen is used to place a line on the tape backing at the edit point. Usually the tape is marked as it sits in place on the reproduce head. Alternately, one can mark some specified distance away from the head, using a measuring "jig" to transfer the mark to the exact splice point. (Marking away from the head probably does help keep the head clean but is less accurate for locating edit points.)

Next, the tape is placed in an editing block, marked directly over one of the guide grooves, and cut. Ordinarily, the cut is diagonal because a diagonal cut makes a smoother,

FIG. 10-11: TRADITIONAL TAPE EDITING

Sentence to be edited: "We had--umm,lots, umm--lots of fun."
Sentence to be heard: "We had lots of fun."

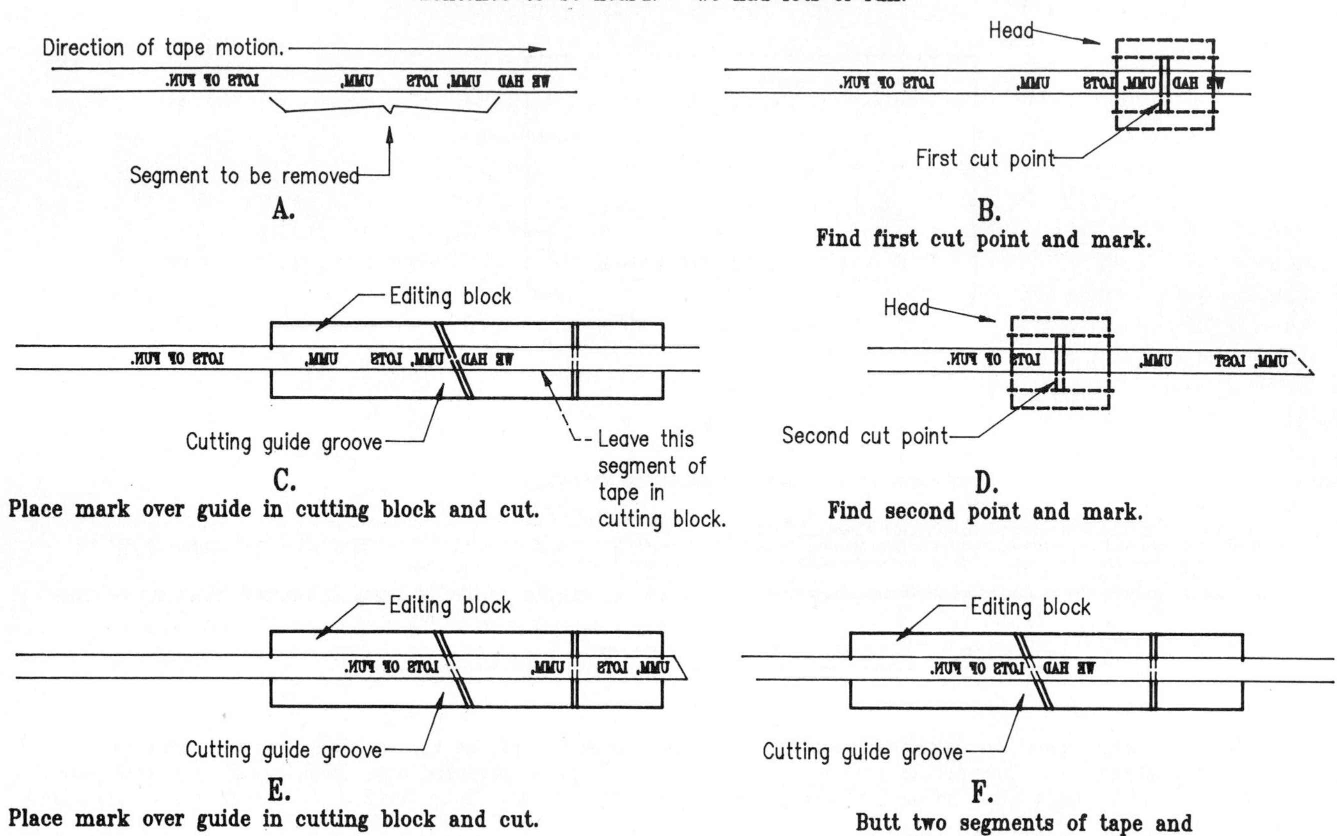

stronger, and (usually) less noticeable transition across the splice. Most splicing blocks also include a straight cutting slot. At times the splice is simply too fine to permit a diagonal cut—a splice between two eighth notes of a piece of music, for example. The diagonal cut would blur the splice, because one would hear part of the note following and the note preceeding the mark points for the splice.

When the first cut is made, the takeup end of the tape is left in the splicing block. Now the segment of tape to be removed must be pulled off the feed reel. Usually, the drive system is engaged and the tape is played through to the next edit point. The loose tape is allowed to spill into a wastebasket. The problem is how to handle the tape in the splicing block that is still attached to the takeup reel. Also, since as tape tension between the heads and the takeup reel is nil, the safety switch will have dropped, and the machine will not run at all until tension is restored. Most machines provide an edit switch. The purpose of the edit switch is to disable the takeup motor and the safety switch. Turning the edit function on is sometimes referred to as going into "wastebasket mode."

When the end of the discard segment is found, the edit point is marked and the tape is placed in the editing block. The mark is centered over the guide groove, as before. The tape is cut, the discard segment is dropped into the wastebasket, and the two edit points are spliced together in the editing block. Finally, the splice is checked by replaying it to be sure that it is accurate and contains no extraneous noises.

Tape editing is one of the primary processes of preparing sound to enhance the dramatic environment of a theatrical production, so we shall have occasion to talk much more about splicing and editing in later portions of this book.

Cuing Tape for Theatrical Purposes

Virtually all tape machines have either footage or turns counters associated with the transport. A turns counter (the most commonly found) counts the number of revolutions of one of the two reels (usually the feed reel). An elapsed-time counter tracks the length of tape (length divided by rate of motion equals elapsed time) that has spooled past the heads. Counters are useful for the grossest kind of location, that is, hitting somewhere within two or three minutes to either side of a starting point in the tape. Counters are *not* satisfactory cue markers for use in theatrical sound operations.

Leader Tape and Labelling

Cues, in theatrical sound usage, are segments of tape that contain an effect designed for a specific aspect of the dramatic auditory environment. To be able to run a cue correctly, a sound operator needs to be able (1) to identify clearly the cue name, (2) to identify clearly the beginning and end of the cue, and (3) to be able to set the tape so that the sound starts precisely with no "scooping" (sound produced by hearing a tape start and get up to speed) and no extraneous noises (such as tape hiss or auditory background noise) before the actual cue sound begins. The most satisfactory method for separating and identifying tape cues is through the use of **leader tape**. (See Fig. 10-12.)

Leader tape is *nonmagnetic*, usually white but also available in colors, and can be spliced to standard audio tape. Leader tape is made in both paper and plastic versions. Plastic leader is certainly much stronger and less liable to tearing and breakage, but it also tends to carry static charges, which can set up random noise in the playback head. Paper leader can and does tear, but it does not hold static charges sufficient to create noise while the leader is passing over the playback head. Where skilled, competent operators are running the tapes, paper leader is desirable because of its immunity to noise.

Time-Code Locator Systems

Since the explosion of small computers and computer related electronics, a new form of tape control has become available through the use of a time code placed on one track of a tape. Together with a microprocessor-controlled device to read and act on the signals from the time track, time code controllers are capable of providing precise control of tape motion, including very precise positioning. (See Fig. 10-13.) Time code may be used to control multiple machines operating together or in staggered sequences, or, using an automated console, the controller may regulate functions such as level and equalization. The time-code system that has effectively become the standard throughout the world is one designed jointly by the Society of Motion Picture and Television Engineers (SMPTE) and the European Broadcast Union (EBU) and is known as the SMPTE (or EBU) Time Code. A number of manufacturers make code controllers to SMPTE/EBU standards.

Digital Storage of Audio Signals

The term *digital* refers to discrete, countable entities, like fingers (comparison chosen deliberately, as the word *digital* comes from the Latin for "finger"). In order to "digitize" an

FIG. 10-12: LABELING LEADER TAPE

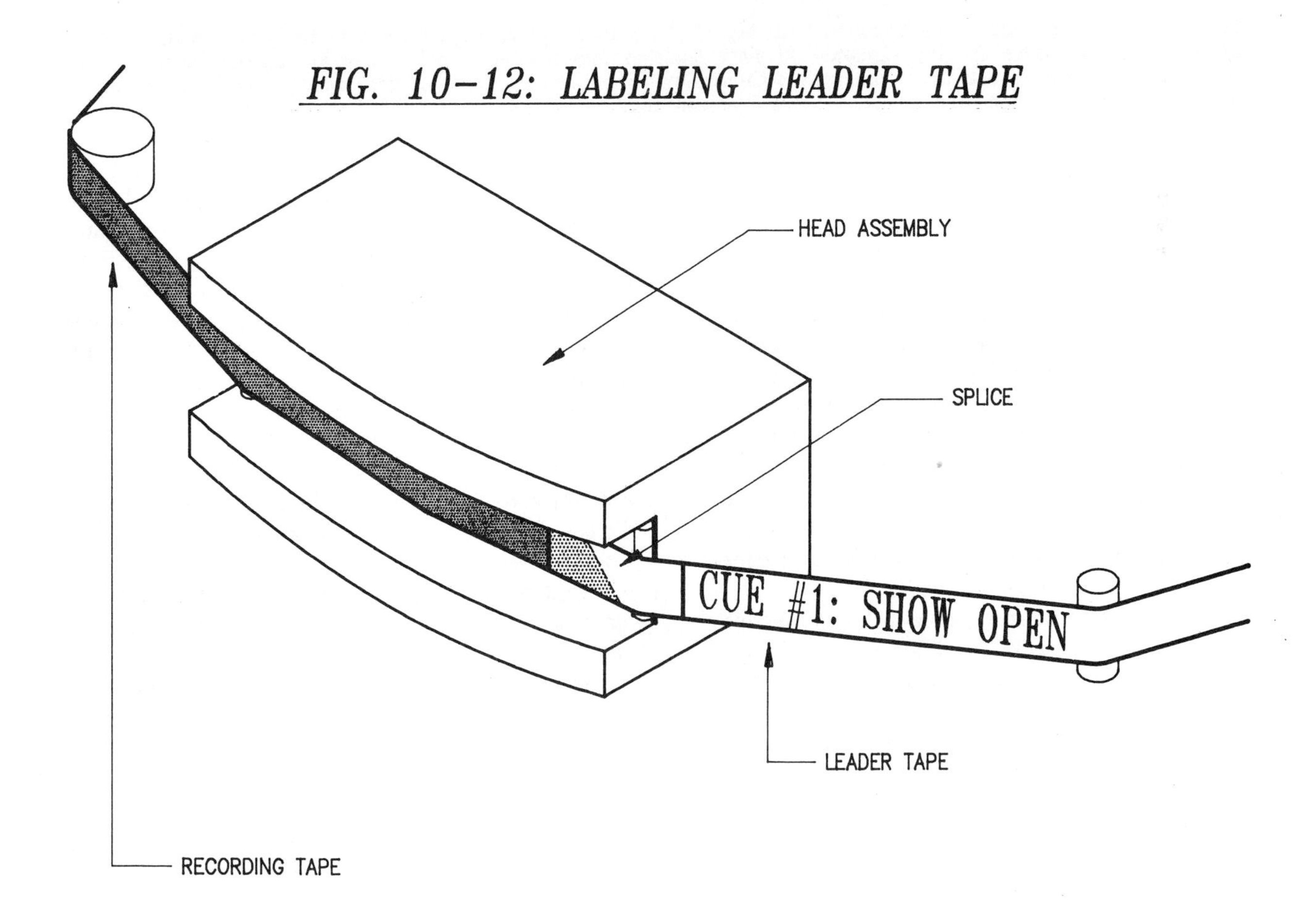

FIG. 10-13: SMPTE TIME CODE

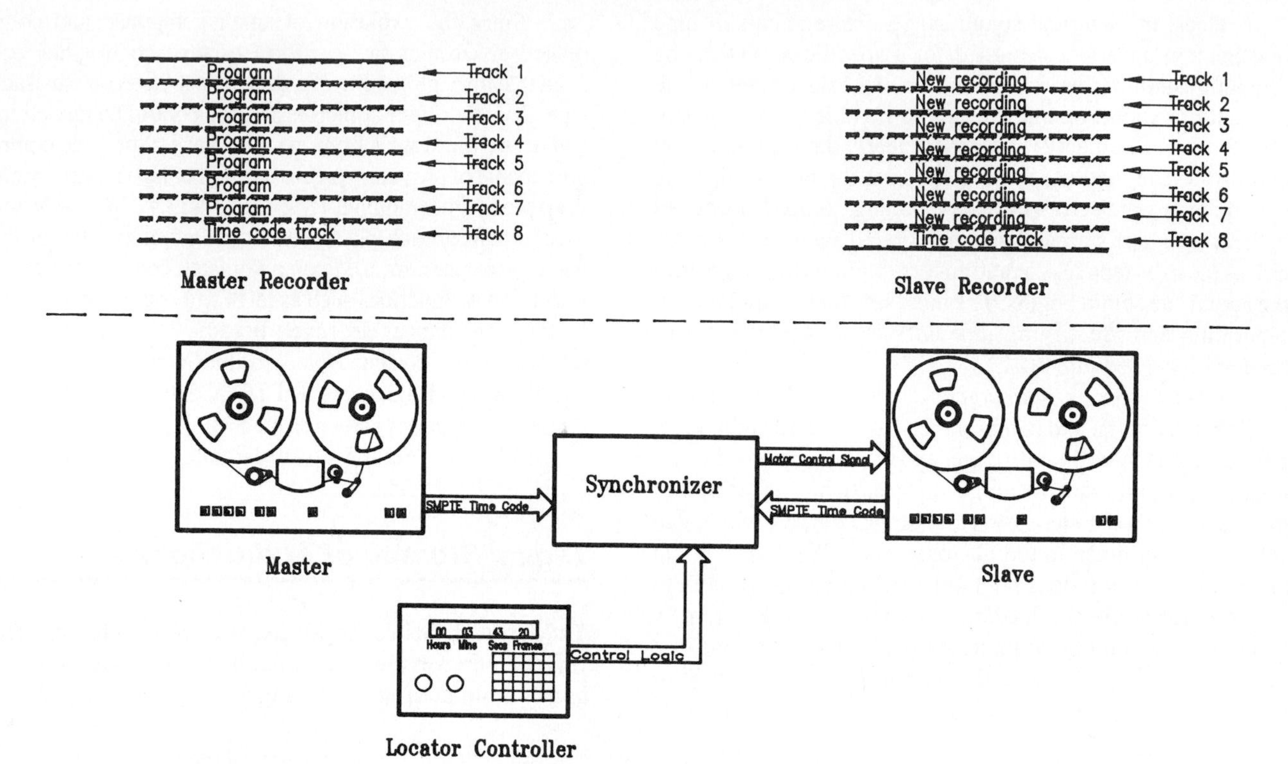

Time code is striped onto one track of both tapes. Both master and slave feed code to Synchronizer. Synchronizer compares the two streams. If slave leads or lags master, Synchronizer accelerates or slows motor to bring slave back into synchrony with master. Locator controller provides readout of location in hours, minutes, seconds, and frames. Control permit tape to be indexed to any time in tape.

analog signal such as an acoustic waveform, we have to pretend that we can break up a constantly varying energy pattern into a number of discrete parts and still keep a reasonably exact replica of the original sound. In fact, slicing the analog signal into samples, however small, destroys the continuity of the signal. Discontinuity produces *granularity.* Granularity means that the signal is not a smoothly varying waveform but a series of incremental steps. The steps create additional high-frequency components that alter the timbre of the digitized signal slightly. Thus, resynthesis of the signal as an analog waveform never yields an exact replica of the original input. Fortunately, the approximation can be made to come so near to the original that most of us do not notice the difference.

A neat analogy for digitizing sound is motion-picture film. Motion pictures are really just a series of still shots snapped one after the other in such close succession that, when displayed by a projector, they capture a reasonable approximation of continuous movement. The human eye responds relatively slowly, so we don't see the momentary blackout between frames, and the visual processing system smooths out slight frame-to-frame positional discrepancies, giving an illusion of smooth, unbroken action.

Although the analogy is useful, digitization of an audio signal is not as simple as taking a series of still pictures. An audio recording system must sample energy over a spectrum that is 11 octaves wide. The highest frequency in the range is more than 10,000 times greater than the lowest frequency. Each half-cycle of signal frequency must be sampled at least once or else the digitizing system generates a spurious signal. That implies that the sampling rate should be at least twice the highest frequency of the audible spectrum. A sampling rate equal to twice the highest signal frequency yields an immense number of samples per cycle at the lowest frequencies of the audio spectrum. The digitizing system takes fewer and fewer samples per cycle as frequency increases, and fails to capture any measure of sufficient detail concerning intensity and complexity at the upper limit of the audio spectrum. (See Fig. 10-14.)

Superficially, the answer to the problem of adequate sampling seems to be quite simple: Increase the sampling rate. Practically, every sample captured must be encoded as a digital "word," and each word requires its own group of storage locations—one for each bit in the word. Because each word, in effect, encodes a floating-point number representing the physical state of the signal during a given sample, the

FIG. 10–14: NYQUIST FREQUENCY

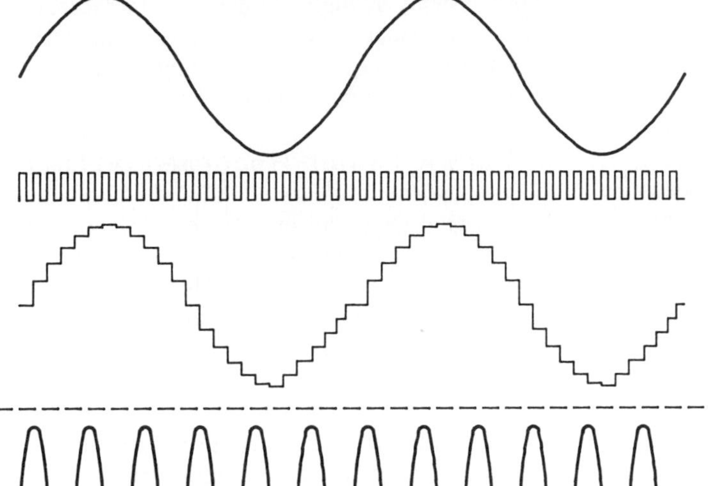

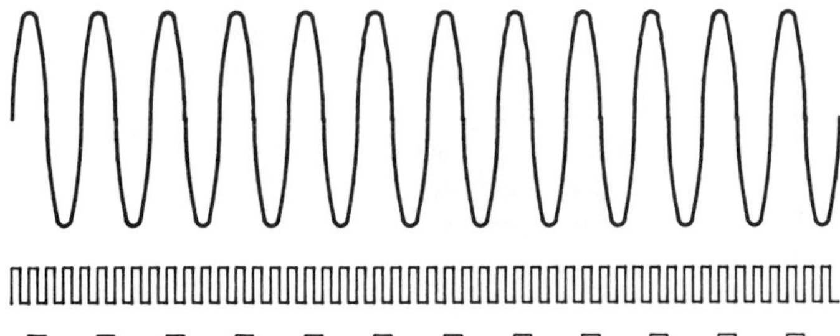

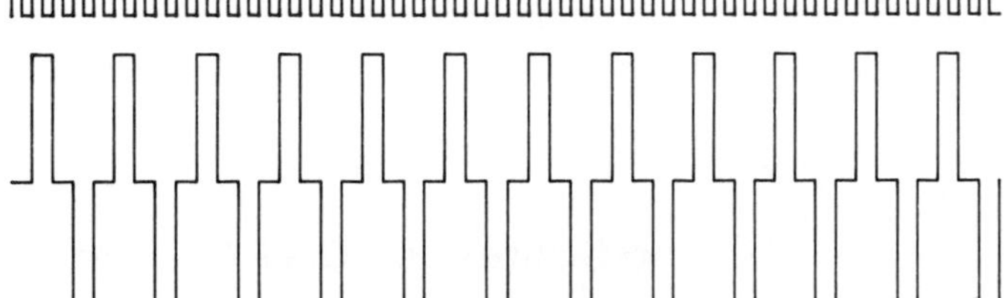

number of bits must be fairly large if the system is to record fine detail. Perfect fidelity all the way up to 20,000 Hz requires at least a 16-bit word and a sampling rate of at least 200,000 Hz. That, unfortunately, requires 200,000 individual 16-bit word-storage locations for every second of sound. Recording a mere 10 minutes of acoustic signal would demand 120 million word locations.

In most digital systems, the effects of inadequate sampling and of granularity are smoothed out by filtering. The filter, placed in the output of the system, is a low-pass device with a cutoff frequency set somewhat below the highest reproducible frequency. The low-pass system smoothes the output waveform and minimizes distortion.

Each digital word generated in an encoded representation is simply a string of 0s and 1s. Once stored, the bits become merely a series of recorded domains. Without some kind of synchronizing scheme, playback of the recording would be difficult. In order to reconstruct the audio signal, the start and end of each word must be clearly marked, and a constant time reference (a clock) is needed to keep the system oriented with respect to the beginning of the recording.

In spite of the problems involved in digitizing the signal, the output of a good digital audio system is almost as free of noise as is the signal delivered to the system from the microphone. No matter what happens along the signal path, so long as the encoded data maintains its integrity, the output contains no noise components that were not present at the original input. The noise immunity of digital audio is due to the fact that the audio signal passes through the system in completely encoded form; thus, any noise picked up by the electronics, though it may attach to the digital pulses that represent the audio, cannot affect the encoded data carried by the pulses.

Noise immunity facilitates copying, and for theatre sound, the ability to copy with no significant degradation is an immeasurable benefit. We often need to generate several copies of a sound in order to recombine it, with itself or with other sounds, into a specialized effect for some particular aspect of the dramatic auditory environment. Analog audio storage systems tend to generate noise, so that only an original recording can be as completely noise-free as the system will allow. Copies contain noise generated by replaying the original recording. Copies of copies become progressively worse. A digital recording, on the other hand, can be copied as often as desired, and a copy of a copy is essentially as good as the original.

Digital Tape Recording

Currently, two main digital audio storage formats exist: magnetic tape and compact disc. Of these two, only magnetic tape can serve as a medium for general-purpose recording. Digital tape recording is, superficially, very much like analog tape recording. A transport is required to move the tape through a head assembly, and an electronics unit contains the amplifiers, power supplies, and control elements. The heads write and read magnetic signals impressed onto the permeable oxide coating on the tape. Beyond surface appearance, however, the differences are considerable.

Analog recording aligns the domains of the magnetic medium in a form that mimics the continuous variation of the acoustic signal. Because the signal is continuous from beginning to end of the recording, an analog tape recording inherently provides its own time reference. Digital recording uses the polarity of magnetic domains only to store a pattern of 0s and 1s forming the encoded attributes of the audio signal. As noted, the digital signal needs to be organized into word groupings, each bracketed with a starting and ending marker, and the entire recording must maintain a carefully controlled timing signal.

The first machines for digital tape recording looked very much like analog machines. (Most were analog transports fitted with digital electronics.) Gradual modifications developed with sophistication in handling digital data, but most large, digital, studio recorders still look very much like their analog counterparts. Most recording studios have one or more digital multitrack recorders, and at least one digital mastering (two-track) machine.

The major concern in research and development for digital recording has been to find efficient ways to store high-resolution data in the smallest possible amount of storage space. Various data compression and encoding schemes have been tried. The industry now has several competing signal handling systems. All produce creditable results, and we are now into a period of waiting for usage to determine which will survive.

DAT Recorders

The newest entry into the digital tape market is the DAT, or R-DAT, digital cassette recorder. Just now becoming available in the United States and Canada, the DAT recorder offers the possibility of extremely flexible and precise sound creation. The DAT system uses sealed cassettes in sizes that can store up to 120 minutes of sound. Sound quality is high, and DATs have the capability to move very quickly from one point on tape to another. DAT recorders use a rotary head system, somewhat like video cassette recorders, that writes a strip of signal information diagonally across the moving tape. The encoding scheme is similar to that used for compact discs, and information can be transferred from CD to DAT directly in digital format.

Currently, the greatest obstacle to the use of DAT machines for theatre is cost. Compared to analog machines, prices for DAT systems are high. The cost is further increased by the fact that a specialized editing system is required. Prices continue to fall, however, so DAT technology should soon be fully competetive with analog systems.

Editing Sound Recorded on DAT

A digital recording in open-reel format can be edited much in the same manner as standard analog audio tape—with a razor blade and cutting block. The process is not simple because of the synchronization data that is recorded along with signal data. A good edit in a digital tape really should fall precisely at a clock mark or between word markers, but an edit can work reasonably well without this degree of precision. DATs, however, are sealed cassettes. One cannot reach the tape to mark or cut; therefore editing requires a somewhat different process.

The usual process of editing DAT recordings is by means of a *waveform editing station.* The general procedure is to load the segment of program to be edited into the memory of the waveform editing system. The standard tools of the editor (discussed in Chapter 12) are used to find, remove or splice, modify, or join the parts of the program that need rearrangement. When editing is complete, the program is recorded back to the DAT cassette.

Cuing Sounds on DAT Cassettes

With any cassette-based tape system, the problem in cuing sound segments for theatre is that normal visual means of identifying segments and for positioning the top of a cue are not available. Splicing leader into a cassette is difficult; and once placed in the deck, the tape is not readily visible, so that use of leader is of relatively little practical value.

DAT cassettes are sealed, which absolutely prevents cutting and splicing. DATs can have program or segment identification marks recorded at any point in the tape, either at the time of the original recording or afterwards, during processing. Identification marks make location of specific points on the tape easy, and the control system is capable of relocating from one identification mark to another in a remarkably short time. Logically, the capacity to find and move quickly to an identification mark should make DAT recorders excellent playback machines for theatre sound. However, some DAT machines require several seconds after the start button is pushed to position tape around the head, find the cue point, and begin play.

A nice feature of the better DAT systems is the *search wheel*, sometimes called the *jog wheel.* The wheel permits the operator to scan backward or forward through the tape at variable search rates. Turned as far as it will go to one side or the other, the wheel scans the tape at high speed. Turned only slightly from its center rest position, the tape runs forward or backward at a very slight increase above normal speed. Use of the jog wheel helps locate cue points and edit points in the tape, like rocking tape back and forth on an open-reel machine.

When DAT technology is commonly available, one great advantage will be the possibility of transferring audio in digital format directly from CD to DAT. DATs will then offer much more flexibility in joining, processing, and editing sounds than can the CD, or even the CD combined with analog tape. In conjunction with a waveform editing system (discussed in Chapter 12), DAT can be expected to constitute a significant addition to theatre sound technology.

Compact Discs

Compact discs are small, plastic wafers that contain a sequence of microscopic depressions in their surfaces, carrying the sequence of digital words used to encode the audio signal. The reading device is a laser beam that can detect the difference between normal surface reflectance and the reduced reflectance of the depressions. The audio signal, of course, is coded into the surface of the disc, but so, too, is a great deal of file control information (called a table of contents) that tells the system the location of each section of audio program. Compact discs are capable of storing an immense amount of data. In fact, the rest of the computer world has looked enviously at the storage density and speed of CDs and has adapted a version of the compact disc (CD-ROM) for storing banks of permanent data.

At present, compact discs can be recorded once, and standard, commercial CD recordings can be produced only with very specialized and expensive duplicating equipment. User-writable CDs are available, but both recording machines and media are expensive. An erasable compact disc for audio purposes is probably not far in the future but is not currently available. One of the great advantages of the nonerasable compact disc, however, is that the data, once stored, appears to be almost indestructible. Even after damage directly to the disc surface, the reader usually can retrieve a clear, noise-free reproduction of the stored program.

Much to the advantage of the theatre, CD players can be programmed to replay any portion of a compact disc, in any sequence, and with any number of repetitions. A huge amount of music and an increasingly large library of sounds are available on compact disc. Using a CD player, a sound designer can easily program repetitions and other effects that once required intricate and tricky tape editing. With two or more CD players, rather elaborate joins and splices can be accomplished with a minimum of effort.

Most professional quality CD players incorporate some or all of the following features: random access programmability, indexing, a search wheel, pause, instant-start cuing, and looping. *Random access* allows the operator to select any band of a compact disc at will. *Programmability* allows automatic replay of a group of bands in any sequence. *Indexing* permits the operator to step through to subordinate segments of a band. (Indexing is used on some CD sound effects libraries to permit one part of a group of sounds to be accessed separately.)

The *search wheel* allows the operator to scan a particular portion of a recording, much like fast-winding tape over the heads to locate a specific portion of taped program. The wheel can be quite precise. Using the search wheel, an operator can find an exact point on a band. The *instant-start* cue feature (which may appear under a variety of names) usually works in conjunction with the pause control and the search wheel to set the reading device to the exact head of a band, an index mark, or a point located by the search wheel. When instant start is active, the reader is set exactly to the desired cue point. The program starts immediately when the start button is pressed. *Looping* forces the reader device in the CD to jump from a predetermined point on a disc directly back to the beginning of a band or to an index point and repeat that segment of the band.

Although the user cannot record his or her own program material onto compact disc at present, CD players are rapidly becoming the best way to play back prerecorded music for such purposes as incidental music for theatrical performance. In conjunction with tape recorders, CD players make possible easy creation and assembly of relatively complex sounds. The combination of CD and DAT seems especially promising.

Hard Disk Storage

The rapid growth of computer technology is changing the audio industry in a number of ways. One development in particular—direct -to-disk recording—will have an enormous impact on signal storage and processing. The possibilities for creation of sound for dramatic environments are particularly exciting.

The magnetic data disk has been the standard method of data storage in the computer industry for some years, but even the largest hard drives lacked sufficient speed and capacity to handle the requirements of audio storage. During the 1980s, all disk formats, including the hard disk, achieved major gains in both speed and in storage capacity. The late 1980s saw the earliest announcements of direct-to-disk systems, but most of them were rather expensive and limited. At the beginning of the 1990s, several manufacturers brought out practical, well-developed direct-to-disk recorders that offered capabilities equal to or better than the best tape recording systems. Price per unit of storage also began to decline to levels that could make direct-to-disk recording practical for a large segment of the audio industry, including theatres.

Direct-to-disk recorders are made in a variety of formats. Some present the user with a control deck superficially like the transport of a tape recorder (minus reels of tape, of course). Other formats utilize controls embedded in a computer graphic interface. Controls can be manipulated with a device such as a mouse or with assignable faders and switches that serve multiple purposes.

The actual process of recording using a direct-to-disk system is very much like standard tape recording. Levels are calibrated, the record button is pressed, and the machine converts electrical signals from the mixing system into domains of polarity on the surface of a magnetic medium. The difference is that the signal is stored in a format that depends completely on *electronic logic*. The domains are not locked into any form of spatial order. Instead, each element of a signal is assigned an address in the storage array, and that element can be retrieved at any time, independently of any other element of the original signal. Normally, no reason exists to alter the order in which samples of a signal were originally stored, but the possibility of processing individual components of a signal separately implies enormous power to control all elements of an audio program.

For example, under some circumstances processing technicians would like to be able to "slip" one or more tracks on a tape. *Slipping* means moving the signal on a track ahead of or behind the signal recorded onto other tracks. On standard multitrack tape, tracks are locked in a constant time relationship, because the recorded information is written to the tape in parallel tracks. Another technique that both musicians and engineers like to use on occasion is *reversing* a track—that is playing one track backward while others play in the normal forward direction. Varying the time relationships or the direction of one individual track on a standard tape is almost (though not completely) impossible. Direct-to-disk recording, however, permits tracks to be slipped or reversed easily. Indeed, the individual samples of signal on a particular track can be recalled in any order whatsoever, without altering the time positioning of other parallel tracks.

Another interesting benefit offered by direct-to-disk storage is that all processing can be done within the storage system itself, under software control. Discrete, hardware-based devices (EQ, delay, compression and expansion) can be eliminated. For instance, delay and echo become extremely easy; we merely retrieve the same information a second time at a specified time lag. Equalization and filtering are simply matters of altering the exact order of 1s and 0s in a sequence of digital words. All of these possiblities—time manipulation, direct signal processing within the recording workstation—offer especially exciting opportunities for sound designers.

Currently, direct-to-disk recording systems exist in two dominant forms: multitrack recording systems and mastering systems. The multitrack systems are hard disk versions of digital multitrack tape recorders, and they do essentially the same thing as do the multitrack machines—store individual, parallel tracks of information for later processing and mastering. Mastering systems are essentially two-channel recording systems designed to accommodate all of the processes commonly used in the final stages of production of a commercial recording—echo, reverb, equalization, compression, limiting, and editing. A number of manufacturers provide mainly time manipulation features in direct-to-disk multitrack systems and waveform editing capabilities in direct-to-disk mastering systems.

A number of lower cost disk storage systems take a somewhat different approach to storage of sounds, but one that is, in some ways, more useful for theatre sound design than substitution of the hard disk for a reel of tape. These systems store MIDI information, rather than audio, using outboarded samplers and synthesizers to produce the actual sounds involved in a program. Many such systems provide extensive waveform editing capability to permit the operator or performer to shape the samples to be used by the outboarded devices. Such MIDI-based systems are primarily super-sequencers. (See Chapter 12 for a discussion of synthesis, including MIDI.) A technique explored by some manufacturers is to incorporate direct-to-disk, MIDI, and waveform editing all in the same package. Such combined systems can offer a great deal of flexibility at reasonable cost, but one is wise to check out the capabilities of integrated systems to be sure that the necessary features in each area of production are adequately covered.

Backup for any disk storage system is essential. Most disk systems can store only a limited amount of program, and hard disk systems can crash and lose data. Most manufacturers either provide or suggest DAT recorders as the logical backup device. Alternatively (as digital transfer from one system to another exacts no penalty in signal quality), one could record directly onto DAT or to an open-reel digital machine, then transfer to the disk system for processing. The processed signal can be returned to tape for permanent storage.

Choosing a direct-to-disk recorder is a matter of defining the purpose of the system. Most disk systems provide some amount of waveform editing capability (see Chapter 12) in addition to storage. If most operations require no more than audio recording, editing, and playback, a system that provides only these capabilities should be satisfactory. However, an operation that requires use of MIDI would be better served if the disk recording system also provided MIDI capability. Having MIDI in the recording system facilitates control and eliminates outboarded processing equipment and transmission lines.

Chapter 11

Time-Domain Processors

Time domain is an area of signal processing that, for all practical purposes, came into existence in the mid-1970s and grew to maturity during the 1980s. The arrival of digital technology—integrated circuits, chip-based memory, and dedicated microcomputers—made time-domain signal processing possible. Accordingly, most time-domain equipment is digital.

Principles Underlying Time-Domain Processing

Compared to time-domain processing, the forms of signal processing—equalization, filtration, compression, limiting, and expansion—that we studied in Chapter 8 are all relatively simple and easy to understand. They are all based on some form of intensity control. Time-domain processors, even in analog formats, require a bit more background to understand.

Time-related aspects of sound were always challenging to sound designers, partly because the possible effects were interesting and partly because, with pre-digital technology, they were so difficult to achieve and to control. The physical aspects of sound concerned with duration and frequency are explained in detail in Chapter 1, but here we need to review a few basic concepts as we begin our examination of time-domain processors.

Echo and Reverberation

Echo and reverberation are acoustic phenomena. Acoustically, when a sound source, as in Fig. 11-1A, exists in an environment with a significant boundary surface (such as the wall shown in the figure), an observer will hear two images of the signal: the first, a direct sound wave from the source, and the second, the sound wave reflected from the wall. Varying the spatial relationships will vary the time relationship between direct and reflected sound. The echo is always later than the original sound wave, of course, but a reflection path that is only slightly longer than the direct path creates a quick echo that almost appears to be a doubling of the sound. Lengthening the reflection path delays the echo and separates it from the direct sound.

Reverberation is a condition of multiple echoes. The echoes may be reflections from a number of boundary surfaces, or else may be multiple reflections passed back and forth between two or three surfaces (like the walls of a long hallway or a tunnel). Each echo has its own path length and, therefore, its own time delay. (See Fig. 11-1B.) Taken all together, the multiple echoes appear as the persistence of sound through time that we call reverberation.

Audio simulation of echo and reverberation requires division of the audio signal into two separate, parallel paths. (See Fig. 11-2.) One path simulates the direct signal from source to listener; the other, the reflection from source to boundary to listener. With older equipment, the division in signal path had to be accomplished by the console. Recent time-domain devices accomplish the signal path division internally.

Looping delayed output back to the head of the delay path—a process called **feedback**—simulates multiple echoes (reverberation). Most presently available delay devices incorporate feedback. Feedback lacks the spatial character of true reverberance. True reverberance is three dimensional—the result of multiple reflection paths from many different locations within an environment. Feedback is one dimensional; it simply loops one signal through the same delay path many times.

Pitch Modification

Good musicians can **transpose** a piece of music from one key to another, that is, they can raise or lower the overall pitch of the music. When musicians transpose, the **tempo** (the speed of performance) does not necessarily change with the pitch. Musicians can also vary tempo without necessarily transposing. Good musicians can play **in tune** with other musicians, which means that the fundamental frequency and major harmonics of each note that a musician plays are so

FIG. 11-1A: SINGLE ECHO

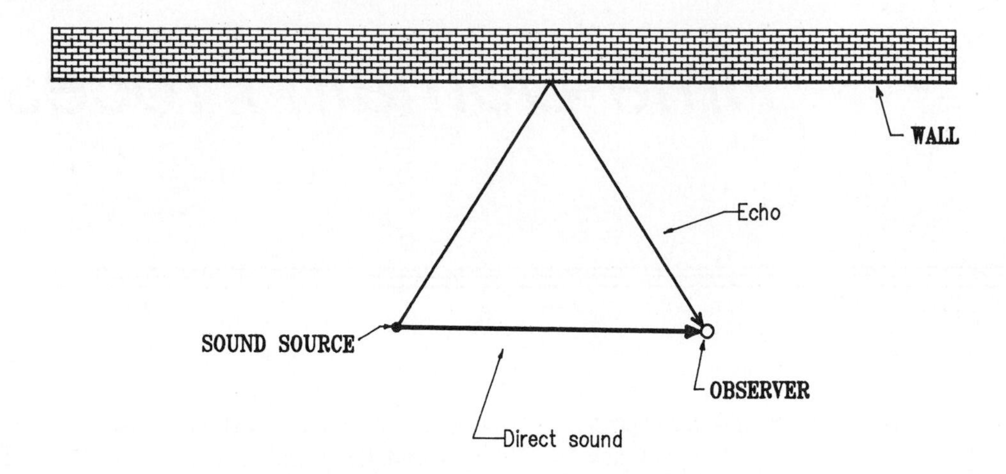

The wall reflects a single echo to the observer. The echo will be later and softer than the direct sound. The clarity of the echo as a separate sound will depend on the distance the reflected sound travels. The longer the distance, the later the echo arrives. The later the echo, the more it is heard as a separate, distinct echo.

FIG. 11-1B: REVERBERANCE.

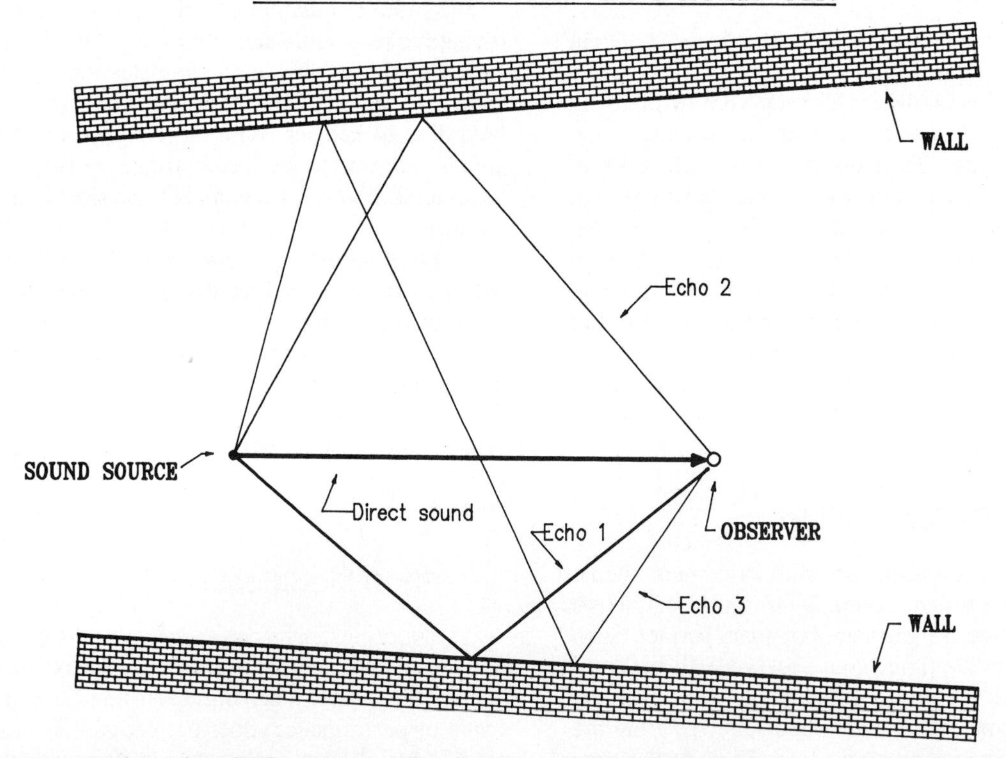

Reverberance consists of multiple echoes. The observer would probably hear more than three echoes, but the three echoes shown are typical of the kinds of reflections involved. Echoes 2 and 3 are later and softer than echo 1. All echoes are later and softer than the direct sound. Reverberance is perceived as longer and louder sound rather than as a single sound with multiple echoes.

FIG. 11-2: AUDIO SIMULATION OF ECHO AND REVERBERATION

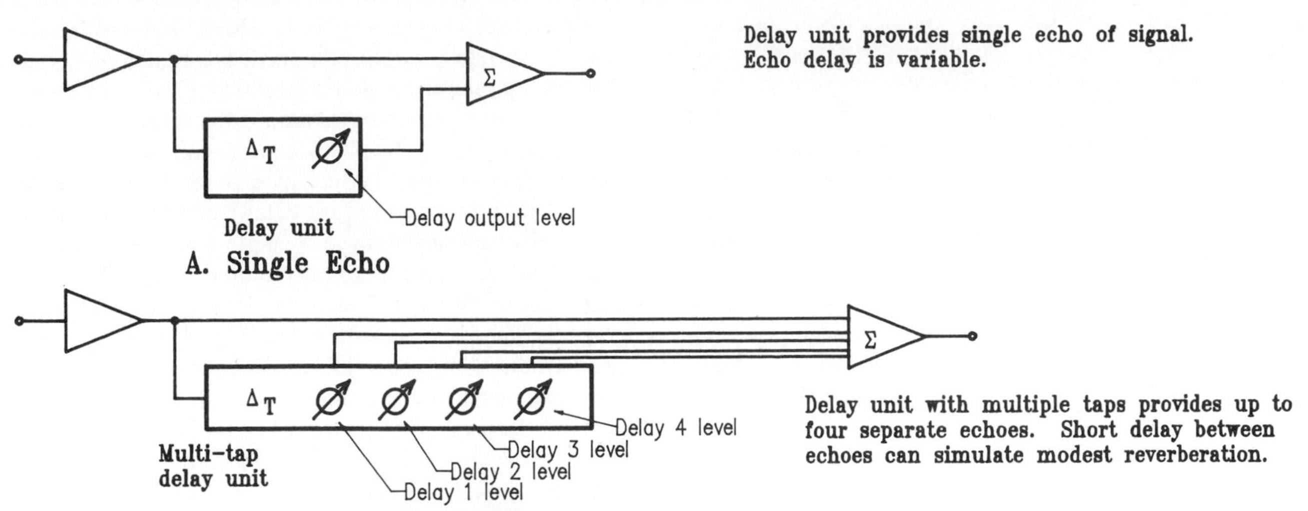

A. Single Echo

B. Multiple echoes using delay element with multiple taps.

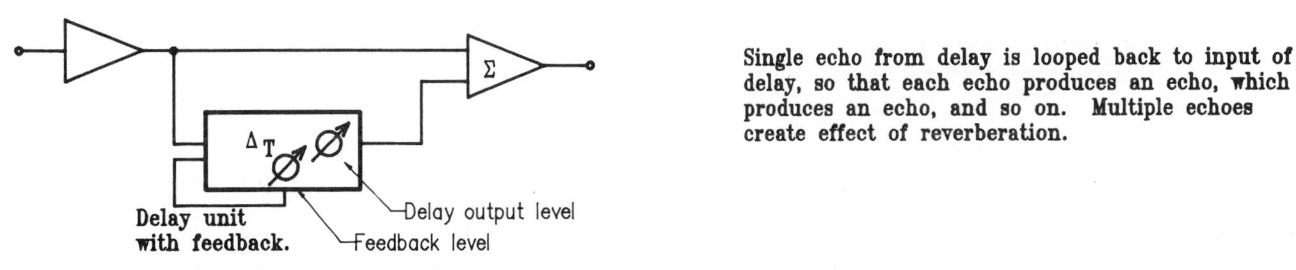

C. Reverberation simulated by single echo and feedback.

closely compatible with notes played by other musicians that all the sounds go together with no discord. Good musicians generally have no trouble **harmonizing**—playing or singing harmony to another musician's melody. Harmony is a complex of several compatible tones that produces a musical structure known as a **chord**.

Over the years, engineers searched for electronic methods to vary pitch and tempo independently. Before the age of digital signal handling, however, audio lacked the necessary technology. Basic amplifiers merely act on signal gain. Combined with frequency-sensitive components (capacitors and inductors), amplifiers can modify timbre, but they cannot alter pitch. Motion-based storage devices can change the pitch of a recording, but they also change its tempo. Traditional recording media lock pitch and tempo into fixed space/time relationships.

As early as the 1960s some experimental methods had actually managed to produce pitch shift without speed change, or speed change without pitch shift. Most of the devices involved complex arrangements of moving mechanical elements added to traditional storage devices. Almost all created undesirable signal modifications and distortion. The most successful attempts were so costly to produce that they were economically impractical. One device, however, was extremely simple and surprisingly successful, although for a very limited application. The device was the Leslie vibrato used on electronic organs during the 1950s and early 1960s.

The Leslie vibrato employed a very simple principle—Doppler shift—to vary perceived frequency very slightly. In the Leslie technique, a loudspeaker was mounted on a rotating platform. The speed of rotation could be varied to control the rate of vibrato. (Review Chapter 1 for the principles of Doppler shift.) As the platform rotates toward the listener, the apparent frequency shifts slightly above true pitch. When the speaker faces directly toward or away from the listener, the perceived pitch matches true frequency. As the speaker rotates away from the listener, the perceived pitch drops slightly.

Digital Signal Handling and Time-Domain Processing

Digital technology represents the first successful method of uncoupling stored sound from its original time base. Digital audio samples instantaneous states of a sound, then encodes the attributes of each sample as a word of data. Each word is stored in a unique memory location. Access to that location is under the control of a central processor directed by a control program. Like any computerized operation, software control makes possible alternative actions. For example, the processor can retrieve the value from each memory unit in normal sequence, or choose some other sequence of retrieval—repeat the value of each memory unit twice, skip every other unit, go back to a given location and retrieve so many units forward, and so on.

Computer controlled retrieval of encoded sample attributes breaks the fixed relationship between signal energy states and time. When sample sequence is a matter of logical selection of storage cells and not a matter of a fixed space/time continuum, a stored signal can be transposed without changing tempo or played at a different tempo without transposition. Tuning and harmonization are possible because the processor can convert one pitch into another or generate several pitches from one.

Development of Echo and Reverberation Devices

Mechanical Methods

The earliest attempts to simulate echo and reverberation for audio used very crude and generally nonelectronic methods. Echo, in fact, was perhaps the more difficult of the two kinds of reflection to simulate. The so-called echo chambers used in early radio sound effects produced mainly reverberation rather than echo. Echo chambers were large, hard-surfaced rooms with a microphone at one end and a loudspeaker at the other. Playing sound through the speaker created acoustic reflections that the microphone then captured and returned as reverberance. Changing the relative positions of loudspeaker and microphone could alter somewhat the quality and duration of the reverberance.

One of the early efforts to produce echo without an acoustic chamber achieved good results but was limited to sounds recorded on a phonograph disc. A phono player was fitted with two pickup arms, one on each side of the turntable. Each pickup had its own separate channel and level control. By varying the intensity of the second channel, the effect of a sound and its echo could be created. Of course, the styli had to be carefully positioned so that both were riding the same groove on the record, but, having done this, one could achieve a real echo. Specially designed effect turntables enabled the operator to change the position of the second tonearm to shorten or lengthen the delay of the echo.

Tape recorders can generate both echo and reverberation. When tape recorders with separate record and reproduce heads became available, technicians quickly learned that feeding the output of the reproduce head back into the input to the record amplifier would generate a combination of echo and reverberation. The technique is called **slapback**. Keeping the feedback signal low gives a modest echo with a fringe of reverberance; a higher level of feedback produces equal echo and reverberance. Letting the feedback signal get too large, however, produces an almost frightening runaway condition in which the reverberation swamps the signal and becomes loud, pulsing noise. Slapback echo sounds very mechanical but can be a very useful effect, even in its runaway condition.

A tape recorder used as a delay path paralleling the main signal channel can produce a single echo. Again, separate record and reproduce heads are required. The signal is written to the tape by the record head. Transit time between record and reproduce heads creates the delay. Signal from the playback head produces the echo. The tape output is mixed back into the main signal path at a suitable intensity level. Figure 11-3 shows the procedure. Tape speed is set for the most acceptable delay length. Fifteen inches per second (the highest speed for most home and semiprofessional tape recorders) produces something like doubling of the signal rather than echo; 7.5 inches per second gives a quick echo, and 3.75 inches per second will create a moderately long echo delay. Given the technical means to vary the speed of the tape drive, greater flexibility in control of echo delay is possible.

Tape recorders can be modified to provide a wide range of echo effects. For example, the Schober Reverbatape (marketed by the Schober Organ Company in the 1950s and 1960s) produced reasonably good quality echo and reverberance. It consisted of a loop of tape that passed over one record head and multiple playback heads. Some of the playback heads could be adjusted to permit variation in delay time. All of the playback heads had individual level controls. Using one reproduce head gave a single echo; using several heads created reverberation.

The spring reverberation generator was one of the first successful electromechanical delay devices. The device contains a group of steel springs that are driven by a vibrating transducer. The element responds to signal frequency and intensity. On the other end of the spring, a pickup transducer turns the motion of the springs into an audio signal. Some forms of spring reverberation use variable damping pads to shorten reverberation time. The springs are capable of no more than a very crude image of the actuating signal, and they also impose their own basic resonant frequency and vibratory characteristics. A spring reverberator tends to produce a very metallic kind of reverberance, but it can be useful for particular kinds of effects.

A more sophisticated form of electromechanical reverberation device became the standard of the professional audio industry for many years. The device contains a large steel plate against which an actuator (like a loudspeaker magnet and voice coil) is positioned. The actuator induces vibrations at signal frequency into the plate. The actuator creates waves that, when they reach the edge, are returned to the middle of the plate, much like conditions in an enclosed, acoustic space.

FIG. 11-3: SLAPBACK ECHO

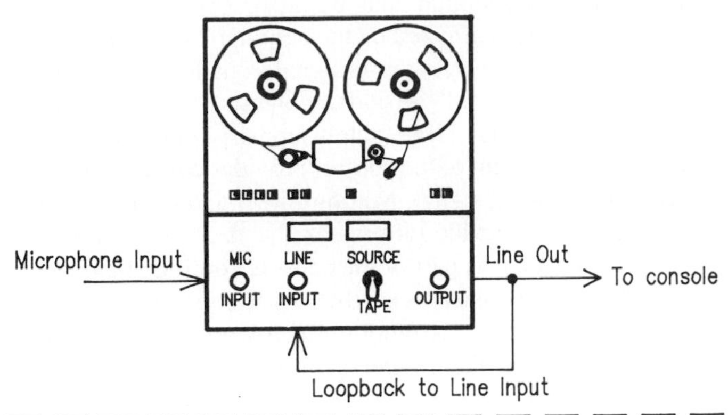

Slapback echo is produced by patching the output of the tape recorder to the input. The machine should be set to replay the recorded tape (meaning that a two-headed machine will not work for this application.) The line input control is used to regulate the amount of feedback. Too much feedback can cause a runaway condition that is potentially hazardous to the amplifiers. The speed of the tape combined with the distance between the Record and Playback heads determine the amount of delay. The slower the tape motion, the longer the delay and vice versa.

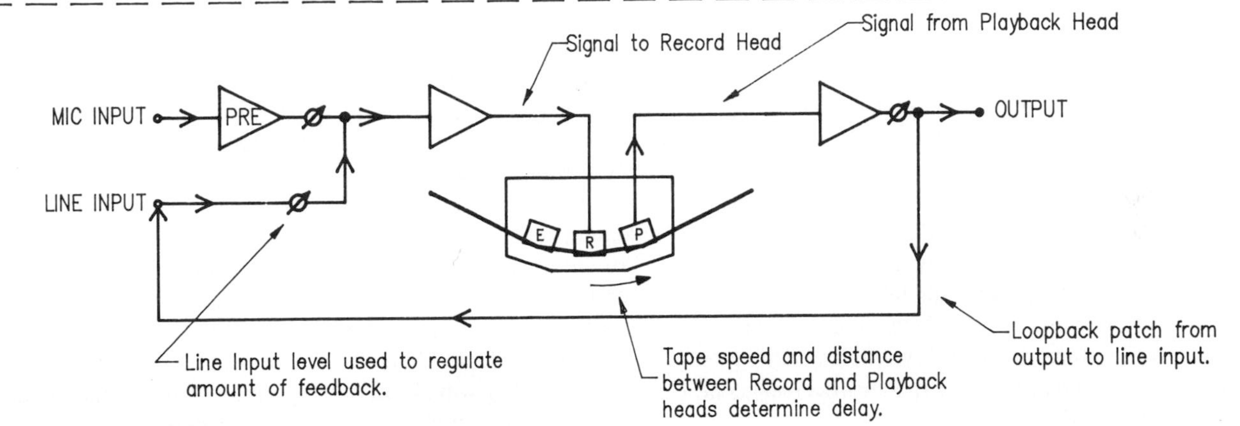

The reflections are captured by pickup transducers located at various points on the plate surface. The output signal is very clean and clear. The plate introduces very little resonant character of its own. Movable damping panels physically alter the area of plate surface, thereby lengthening or shortening the delay time and number of reflections.

Electronic Delay Devices

Electronic delay systems began with the development of an analog integrated circuit chip consisting of an array of *charge-coupled devices (CCDs)*. CCDs are used to make **analog delay lines**. Like the digital delay systems, analog delay lines *sample* the signal. (Unlike digital systems, the sample remains a voltage or current analog of the signal. The sample is not converted into a digital word.) A clock pulse, called the **strobe**, regulates the action of the CCDs. At each pulse, the CCD accepts a new sample of the input signal and passes its old sample to the next device in the string. Because of its method of passing the signal—from one device to the next to the next—a CCD delay device is also called a **bucket-brigade delay**.

Like any sampled signal handling system, bucket-brigade delay produces a stepped waveform rather than a smooth replica of the original signal. Between samples, some part of the original signal slips by without being recorded. High-frequency signal is most likely to suffer. A clock frequency equal to twice the upper limit of signal frequency is the minimum necessary to ensure that not too much of the waveform is lost between samples. A fast strobe, however, means that there won't be much delay in the output signal unless the number of storage cells runs into the hundreds of thousands, and even then the delay will only be a few milliseconds. A slower strobe increases delay length but reduces frequency response. Heavy filtering is required to eliminate the high-frequency distortion caused by sampling error. Still, analog delay lines were useful, and they were the first fully electronic delay devices to produce high quality echo and reverberation.

A **digital delay line**, like any digital device, encodes magnitude and polarity attributes of each signal sample into data words. Like the bucket-brigade delay, digital delay devices use a strobe pulse to clock the progress of data through the delay line. The signal is sampled and words are stored in memory at each pulse of the device system clock. Unlike CCD devices, digital delays do not have to use the bucket-brigade method. A digital delay can use memory just as a computer does: Each new datum is stored in the next available memory location. Data words are written to memory until the storage space is full, at which point new data must overwrite old or else be lost. If an overwrite occurs before old data has been retrieved and delivered to output, then the old data is lost. The amount of storage required is a function of sampling rate and length of delay. Fast sampling and long delay add up to megabytes of storage! Decreasing clock rate minimizes the demand on memory and increases length of delay, but slowing the clock also requires high-frequency filtering to counteract the effects of sampling error.

Standard Controls for Electronic Echo/Reverberation Devices

Standard electronic delay devices usually provide controls for input level, delay length, clock rate, feedback level, infinite repeat, output mixing, modulation waveshape, modulation rate, and modulation depth. The purpose of each of the most common control functions is explained in the following paragraphs.

The **input level** trims the intensity of the audio signal as it enters the delay unit. Because delay devices are commonly connected to effect sends, the maximum input level for most delay units is less than standard line level. -4 to - 6 dBm is typical. Many manufacturers provide LED indicators to help the operator keep input level within acceptable bounds.

Several controls affect **delay** in a digital echo device. First, a step-up/step-down switching system selects basic delay length. A variable control fine-tunes that delay period, lengthening or shortening as needed. Some devices incorporate a delay doubler switch that multiplies the delay time by two.

Feedback amount adjusts the level of reverberance. With no feedback, the delay line produces a single echo of the input; as feedback increases, the delayed signal is returned to the input of the delay line and passed through again. The higher the delay, the longer the "persistence" of any given portion of the signal.

The **infinite repeat** switch temporarily switches the input of the delay away from the external audio feed and connects it to the output of the delay line. The result is a continuously recirculating segment of sound. The captured sound segment will repeat for as long as the infinite repeat switch is on. Turning off the infinite repeat will restore normal audio.

The infinite repeat switch works in conjunction with clock rate. The clock rate control determines the number of clock pulses per second as a fraction of total delay time. The length of the repeat loop is a function of the number of clock pulses that will occur in a given delay period. A long clock period coupled with a short delay time yields a short loop; a long clock period and a long delay time give a long loop. A short clock period combined with a short delay time yields a very short loop; a short clock period with a long delay time gives a medium length loop. Rate of repeat is also a function of clock period and delay length. Length and rate of repeat can be fine-tuned with the variable delay control.

Output mixing is a variable control that determines the ratio of direct signal to delayed signal. Most delay units provide at least two outputs: direct and delay. The direct output is a simple feedthrough that bypasses the delay circuitry altogether and is unaffected by the output mixing control. The delay output, however, takes its signal from the output mixing control. Set to "direct," the output consists only of undelayed signal; set to "delay," the output consists only of delayed signal. The midpoint of the control provides equal amounts of direct and delayed signal. Mixing direct and delayed signal inside the echo device eliminates the need for a discrete, parallel delay path as part of the console configuration.

Modulation is a process that has no equivalent in acoustic echo and delay; it is purely an audio effect. Modulation momentarily alters the frequency of the signal, shifting it to a higher or lower value than normal. Modulation takes advantage of the fact that changing the delay length as a signal is passing through the device momentarily shifts the pitch of the signal. In modulating a signal, a low-frequency control waveform is applied to the circuits that generate the system clock, the same circuits that the variable delay control affects. The modulating signal serves as an automatic rate control, speeding up or slowing down the clock. The waveshape, wavelength, and amplitude of the control signal determine the pattern of modulation. Three control elements are required: modulation waveshape, modulation rate (control of wavelength), and modulation depth (control of amplitude).

When the modulation rate control is at lowest setting, the modulating waveform runs very slowly (at less than 1 Hz). Maximum setting yields control frequencies of 20 to 30 Hz. The depth control regulates the amplitude of the control voltage. At the minimum setting, the modulation voltage is effectively off. Increasing the control's setting increases the peak level of the control voltage.

A small amount of modulation (modulation depth control set near minimum) produces a tremulant effect, like a solo musical instrument's vibrato. Delay devices can be used to add vibrato to recorded music or to the output of a synthesizer or other electronic instrument. Increasing modulation depth yields a more violent effect, called pitch bending. The speed of vibrato or bending is set by the modulation rate control. A low rate (control near minimum) produces a very slowly varying vibrato or bend. A high rate of modulation produces a warble. Extreme settings of rate and depth controls can make a recorded signal virtually unrecognizable.

Modulation waveshape controls the contour of the vibrato or pitch-bend effect. Most manufacturers provide three modulation waveforms: ramp (sawtooth), triangle, and sine. Usually the ramp is a positive-going wave. That is, it rises gradually from minimum level to its peak then drops back to minimum and rises again. The triangle wave rises gradually to peak and falls gradually to minimum. Both ramp and triangle are linear waveshapes, varying smoothly from minimum to maximum; the sine wave, as its name suggests, increases slowly from minimum, rises sharply toward peak, then rounds off as it approaches maximum. Each waveshape produces a vibrato or bend that follows the character of the modulating waveform. Many units permit continuous variation from ramp to triangle to sine, allowing a variable symmetry to the vibrato or bend.

The most recent (and most expensive) of the digital reverberation devices go well beyond simple echo, reverberation, and modulation. They can certainly provide any reverberant quality, but they can also be programmed to simulate the reverberant character of particular concert halls and auditoriums. These very sophisticated devices are especially useful in large, multipurpose halls that need a variety of different reverberant characters, and wish to choose the best characteristics for any given occasion.

Electronic Pitch Control

Phasing, at least according to popular audio history, was more or less an accidental discovery by studio technicians trying to create sound doubling (almost simultaneous echo). They reasoned that they should be able to produce the effect by playing two copies of a tape with one running just slightly behind the other. Starting the two recorders one after the other could not provide a sufficiently precise degree of control, however. So they resorted to starting both machines simultaneously, then stalling one machine for a fraction of a second by holding a thumb against the flange of the feed reel. Instead of doubling, what the technicians got was a hollow-sounding notch that swept up and down the audible spectrum cyclically, varying the timbre of the recorded sound. Because of the method of stalling the delayed machine (putting pressure on the flange of the reel), they called the effect **flanging**. Later the term phasing was adopted as a more accurate name for the effect. (The story may well be spurious, but flanging is still frequently used for the effect.)

The phasing effect is produced by a phenomenon known as **comb-filtering** (see Fig. 11-4). In flanging, the effect of the comb filter is produced by the fact that offsetting one tape a few milliseconds creates a condition in which some wavelengths between the two signals are 180 degrees out of phase, some are in phase, and the remaining wavelengths range between 1 and 179 degrees out of phase. Varying the relative timing offset between the two signals alters the precise group of frequencies affected, producing the comb-filter effect. Given that a delay line, by definition, retards a signal, any delay device can create the time offset that produces the comb-filter effect. Phasing using a delay unit is quite simple to produce, by mixing equal parts of direct signal and signal delayed by a few milliseconds.

FIG. 11-4: COMB FILTER

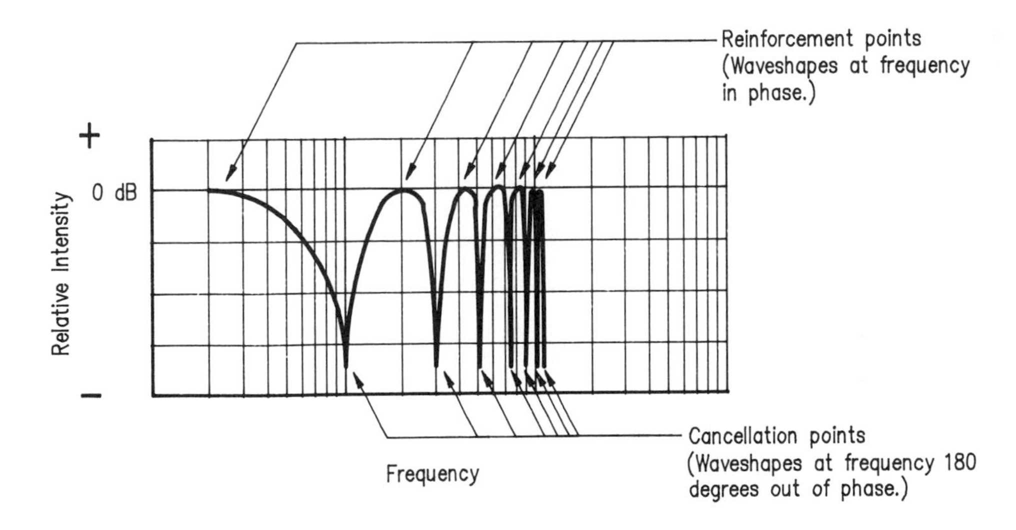

Comb filter is produced by introducing slight phase shift in signal (i.e., signal split between a normal path and a delayed path.) At some frequencies, waveshapes will be in phase and will reinforce. At others, waveshapes will be 180 degrees out of phase and will cancel out. Frequencies between 0 and 180 degrees phase will produce varying amounts of signal attenuation.

Altering delay will shift reinforcement and cancellation points, producing a sweeping effect moving up and down through the spectrum. Using modulation on a delay device can produce a constantly–cycling sweep.

The modulation controls of delay devices produce temporary pitch shift. A device designed to change and maintain the rate at which the signal waveform is played out of the device, however, can produce an entirely different pitch than that of the original signal. The effect is like playing a tape recorder at half or double speed, except that the tempo of the original signal doesn't change. A machine that can alter pitch without changing tempo makes a number of useful effects possible. One effect is to bring an out-of-tune instrument into accord with other components of a musical group. Some pitch control devices include a electronic music keyboard, so that a performer can sing through the harmonizer and, by playing the keyboard, produce simultaneous chords. Recent products condense pitch-shift logic into small packages that create and store harmonies under performer control or under MIDI control from a synthesizer or sequencer.

Harmonizers can be extremely useful devices for theatre sound processing. For example, supernatural persons can be made to speak in "chorus" voices, or with inhumanly low pitch. Sounds can be made to bend in pitch. Some harmonizers provide outputs to drive the capstan motor of a tape recorder. The frequency of the output controls the speed of the tape. The signal from the tape is then processed by the harmonizer to correct the pitch. Newer pitch-shift devices can adjust frequency and tempo without varying tape speed; however, the quality of digitally adjusted sound deteriorates as pitch shifts further from the original key. Servo-controlled tape speed combined with digital frequency correction provides better quality sound when a large shift in pitch is needed.

Time-domain processors are potentially among the most valuable and useful devices for theatre sound designers, offering a range of effects obtainable in almost no other way.

Chapter 12

Synthesizers and Samplers

Most people in the Western world, certainly those under the age 25, know what a synthesizer is. We have all seen the banks of electronic keyboard instruments that are so universally used in most contemporary popular music groups. We have all heard their sounds in any number of film and television sound tracks and certainly in scores of commercial jingles.

Synthesis implies creation from assembled components. For musical synthesis, the components are sinusoids or other waveforms combined to produce various timbral qualities. Acoustic physicists recognized a very long time ago that almost any timbre could be created by a mixture of sinusoidal components. One had only to generate the requisite number of partials, then regulate their respective intensities and phase relationships correctly. Combining sinusoids to produce a complex sound is known as *additive synthesis*. An alternate way to derive a complex sound is to generate a waveform that implicitly contains all linear partials of a given fundamental, then filter out those partials not needed for a particular timbre. This latter procedure is known as *subtractive synthesis*.

The Development of Practical Electronic Synthesis

When audio experimenters were just beginning to build electronic musical instruments, adding sinusoids together was very expensive and not very practical.[1] Because of the difficulty and expense of additive tone production, most electronic instruments developed before 1960, primarily electronic organs, used subtractive synthesis.

Additive or subtractive, elemental synthesis produces steady, unvarying tones. Engineers build electronic oscillators to be as stable as possible, as instability in most applications would make the device unusable; but a sound that drones on with no timbral variety quickly becomes boring. Traditional musical instruments produce timbral modulation naturally. In fact, good string, woodwind, brass, or vocal quality is interesting because of subtle timbral change that occurs between initial attack and final decay of the tone. Electronic organs of the 1940s and 1950s were acceptable because traditional organs tended to produce a fairly unvarying sound and because religious services, for which electronic organs were often used, encouraged a solemn tone quality. Even so, all electronic organs provided means of creating both *vibrato* (a slight but steady variation in frequency) and *tremolo* (a rapid fluctuation in intensity level).[2]

Aside from organs, electronic instruments were novelties forty years ago. Most people did not care for the artificial sounds of these instruments, and not many serious musicians were interested in learning to play them because of their limited appeal and, in some cases, because of the odd and difficult techniques involved. Beginning in the 1930s, however, a small group of composers began experimenting with sounds that could be produced by electric and electronic devices: tuned resonances, regenerative oscillators, and random energy produced by arc discharge. Later, when tape recorders became available, techniques such as slapback oscillation, tape loops, and sound splicing developed. Finally, as vacuum tube technology advanced toward its peak, a few experimental music studios in large universities assembled huge banks of oscillators and filters that could be driven from punched-card or paper-tape readers to produce musical output.

The various experiments of the early years produced a new genre, called *electronic music*, which deliberately sought

[1] In the early years of electronics, the only way to generate a tone was by a technique called *regenerative feedback*. A regenerative amplifier couples its output back to its input through a tuned, resonant circuit, producing a sustained oscillation. Regenerative oscillators are inherently unstable. Any small change in tuning will completely disrupt the circuit's function. Additive synthesis potentially involves a large number of oscillators. Mixing a number of regenerative oscillators together creates frustrating interactions, with consequent unpredictable behavior from the oscillators. A change in one oscillator can easily upset all the rest.

[2] Vibrato and tremolo became staple characteristics of electronic organ music used for radio soap operas and early television shows.

to explore strange and unusual sounds. They also developed fundamental knowledge about electronic synthesis, but most of the early techniques were too cumbersome, too expensive, and too limited in appeal to be widely useful. Above all, the resulting music could be reproduced only from a recording. Live performance was virtually impossible.

In the 1960s the transistor revolutionized electronics. Transistors were small, required very low operating voltages, did not need heat to generate a flow of electrons, and could be configured into circuits that were not possible with vacuum tubes. More important, however, these devices did not necessarily depend on regenerative feedback to produce oscillation. Transistors could function readily as *relaxation oscillators*—devices based on the capability of a capacitor to charge and discharge at different rates, dependent only on the immediate resistance present in the circuit during the various phases of conduction. Because transistors are also electronic switches, they could be easily used in *multivibrator circuits*—oscillator circuits that simply switch a given level of current or voltage on or off or alternate between positive and negative polarities.[3]

Relaxation oscillators produce not sinusoidal waveforms, but ramps—sawtooth waveforms. The sawtooth waveform implicitly contains all partials of its fundamental; therefore, it is the most useful for subtractive synthesis. Multivibrators produce rectangular waveforms. Rectangular waveforms contain all of the *odd* harmonics of the fundamental and are readily adaptable to synthesize sounds produced by conic radiators in which one end is closed (such as the clarinet). Altering the duty cycle of the rectangular wave (i.e., the amount of time on or in positive polarity, as opposed to the time off or in negative polarity) creates considerable timbral modulation, even without filtering. Also, rectangular waveforms can be turned into sawtooth waveforms by *integration* or into triangular waveforms by *differentiation*. Triangular waveforms can be shaped by various kinds of networks into pseudo-sinusoids. Thus, the multivibrator ultimately became the primary form of tone generation for electronic music prior to the development of digital synthesis.

Using transistor technology, Robert A. Moog developed a collection of circuits for electronic sound synthesis, including oscillators, filters, controllers, and amplifiers. Moog's system was introduced in the late 1960s. Moog was not the only person to develop such circuits for purposes of synthesis, but he was the first to put them in a form capable of producing an electronic musical instrument with a real chance for popular appeal.

Moog's synthesizer was composed of individual modules that could be flexibly interconnected by patch cords (the reason that a selection of parameters constituting a particular timbre is called a "patch" in today's digital synthesizers.) Each synthesizer consisted of a keyboard and a flexible configuration of modules. The basic complement of modules contained two full-range, voltage-controlled oscillators (VCOs), one low-frequency oscillator (LFO), one voltage-controlled filter (VCF), one bank of fixed filters, and two voltage-controlled amplifiers (VCAs) with an envelope generator for each VCA. Envelope generators are "one-shot" multivibrators that produce the ADSR (attack-decay-sustain-release) characteristic of a sound. (Refer to Chapter 1 for an explanation of ADSR.) Low frequency oscillators, usually operating in the sub-audio range, are used to modulate other controllable elements in the synthesizer. Low-frequency modulation will produce effects such as vibrato, tremolo, warble, and repetitive filter sweeps.[4]

Additional support modules necessary for basic operation included a four-way, active mixer, a white and pink noise source, a high-pass filter, a low-pass filter, an inverting/noninverting amplifier (to control signal amplitude and phase), and several commoned patch points that served to combine and divide signals and control voltages.

Control voltage in Moog's system conformed to a rigorous but simple standard: A change of 1 volt equaled a pitch difference of one octave. All audio voltages produced by the oscillators were fully compatible with the system control voltages and could be used as control voltages. Except for system trigger pulses (needed to activate the envelope generators), any voltage obtainable anywhere in the synthesizer could be patched to any part of any module for use either as signal or as control.

Because the Moog synthesizer was modular in design, a system could be customized to fit individual needs. The more modules the system contained, the more complex and interesting were the timbres and timbral progressions that could be generated. Of course, patching such large synthesizers is time consuming. Initially, finding a timbre required a lot of experimentation. Every last detail of the hookup had to be recorded; and the next time the synthesist wanted to use that sound, the hookup sheet had to be retrieved from the files and the modules repatched to create the timbre.

Moog's oscillators (both full range and LFO) could produce four basic waveforms—triangle, sawtooth, rectangular, and a pseudo-sinusoid. Timbral production was usually subtractive,[5] using the filters to subtract unwanted partials from the complex waveforms. Signal from the filters was fed

[3]Vacuum tubes could be used in relaxation oscillators and multivibrators, of course, but not so neatly or so readily as transistors. Similarly, transistors could be used as regenerative amplifiers, but relaxation oscillators and multivibrators are much more stable and trouble-free.

[4]One interesting advantage of Moog's LFO was that it was not just a sub-audio oscillator. It could produce the full audio range; but it offered a finer degree of control in the sub-audio range than did the standard, full-range oscillators.

[5]Additive timbres were possible, but degree of complexity was more or less limited, according to the number of oscillators provided and the number of mixing channels available to combine the individual tones.

to one or more voltage-controlled amplifiers, which gated the sounds and shaped their amplitude (ADSR) envelopes following control signals from the envelope generators.

Figure 12-1 shows several typical patch patterns for a basic Moog synthesizer, which we examine in detail because they show how synthesized sounds are built. The process of constructing synthesized sounds is essentially constant, no matter what the method of generating the sound. The latest generation of analog synthesizers hides the patching process but does not change the architecture of a sound in any way that differs greatly from a Moog patch. The sounds produced by digital synthesizers result from similar architectures, except that the entities that create the sounds are *virtual devices.* Virtual devices are logical equivalents of the oscillators, filters, envelope generators, and so on, used in analog synthesizers. Virtual devices, like actual modules, must be patched in order to generate sounds.

The first pattern simply directs the sawtooth waveform output of a full-range oscillator to a VCA. The VCA, controlled by its envelope generator, produces an amplitude envelope with a moderately steep attack, a short decay to the sustain level, and a medium length release. The keyboard interface module provides trigger pulses to the envelope generator each time a key is struck and sends a pitch control voltage to the VCO. The patch illustrated is a very basic setup that will simply allow the performer to play notes from the keyboard. Each note will hold for as long as a key is depressed. The synthesizer will produce the full ADSR envelope only if keys are released completely between notes. Legato playing (not releasing one key before the next is depressed) will produce a smooth change from note to note at the sustain level.

Figure 12-1B illustrates a somewhat more complex arrangement. It simply plays a sequence of notes from the keyboard but produces a more interesting sound. The LFO has been added to the patch of Fig. 12-1A. The LFO sends a low-amplitude, 10-Hz sinusoid to one of the control inputs to the VCO. The LFO acts in parallel with the keyboard controller to vary the frequency of the VCO slightly. Now, instead of steady tones, the synthesizer will produce tones with vibrato. Increasing the amplitude of the LFO signal will change the sound from a single tone with vibrato to a warble (a rapid alternation of two tones).

In Fig. 12-1C, filters are added. Output from the VCO is patched to the fixed filter bank. The fixed filter bank is set to emphasize upper mid-range partials of the tone, producing a nasal quality something like an oboe.

The patch of Fig. 12-1D uses the pink noise source, the fixed filter bank, and the VCF to produce a percussive effect something like a bass drum. VCA and envelope generator are included to create the necessary ADSR characteristic.

Creating music on one of the early synthesizers was rather laborious. Because each synthesizer could produce only one "voice" at a time, creating anything more complicated

FIG. 12-1A: BASIC MOOG PATCHES

BASIC PATCH FOR ENVELOPE AND KEYBOARD CONTROL

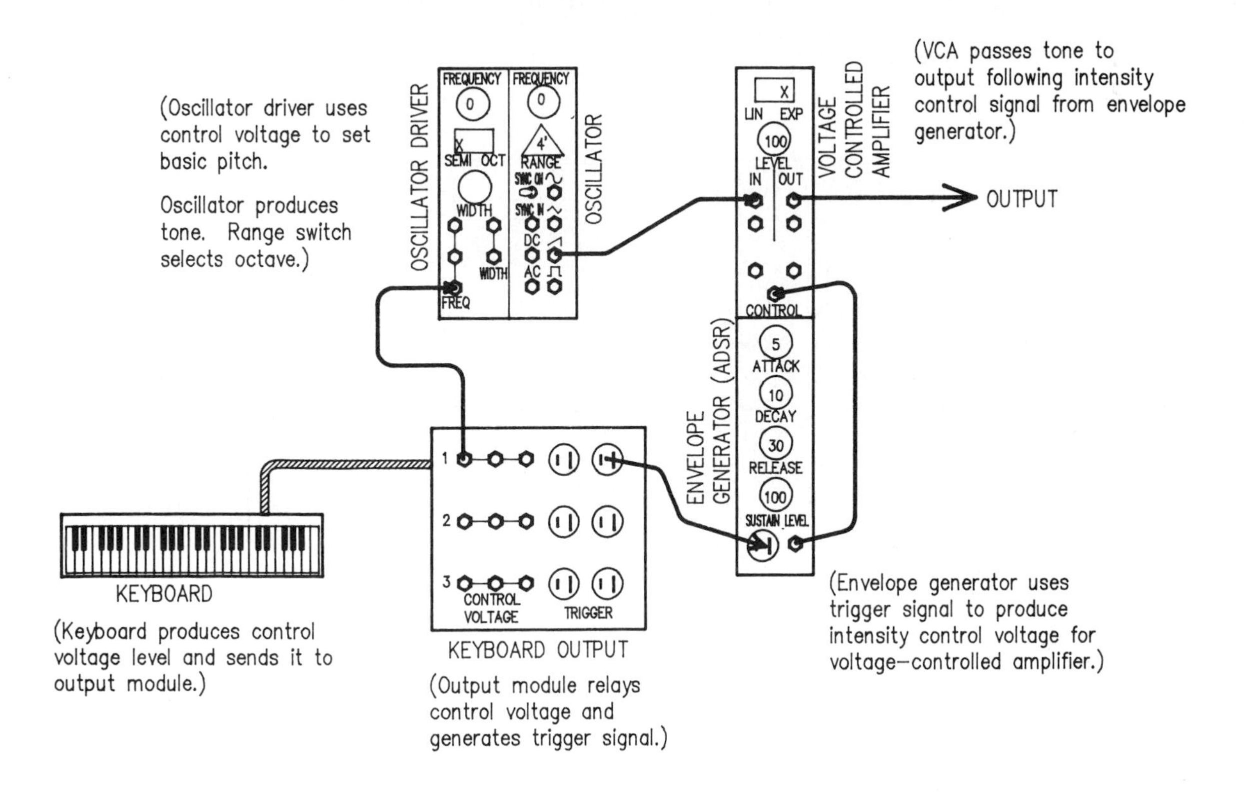

FIG. 12–1B: BASIC MOOG PATCHES

LOW-FREQUENCY OSCILLATOR ADDED TO PRODUCE VIBRATO

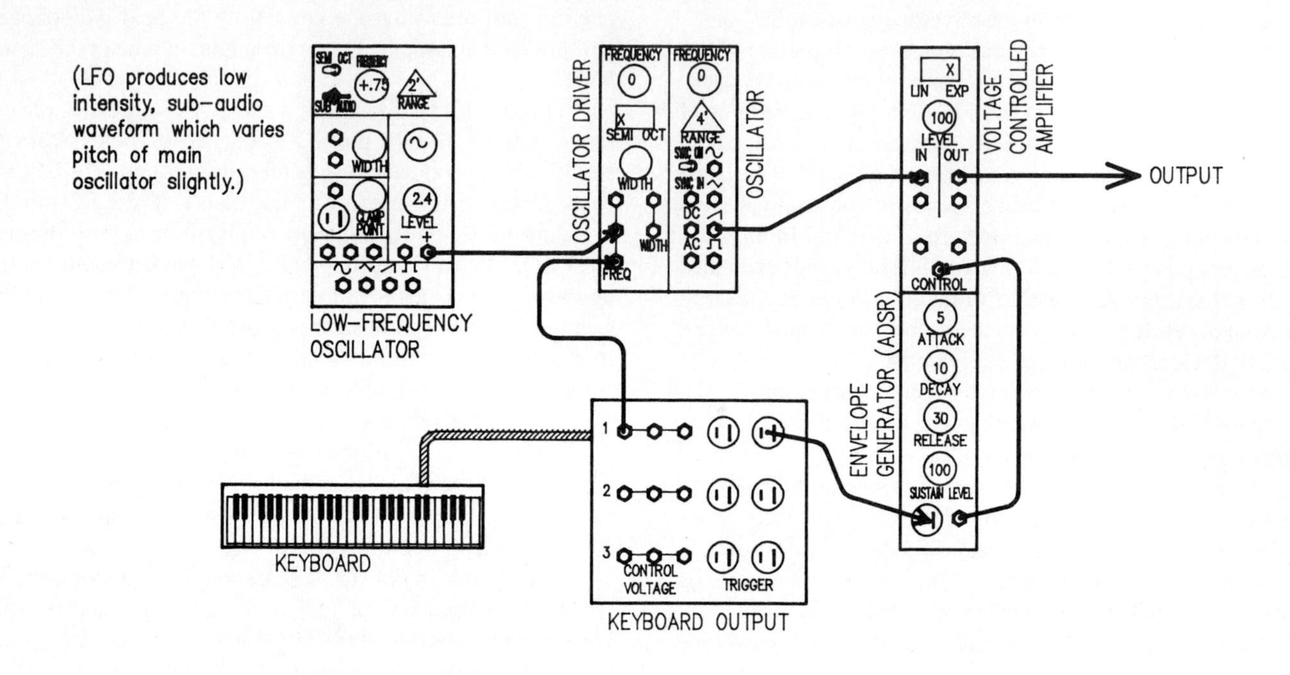

FIG. 12–1C: BASIC MOOG PATCHES

FIXED FILTER BANK ADDED TO PRODUCE OBOE-LIKE TIMBRE

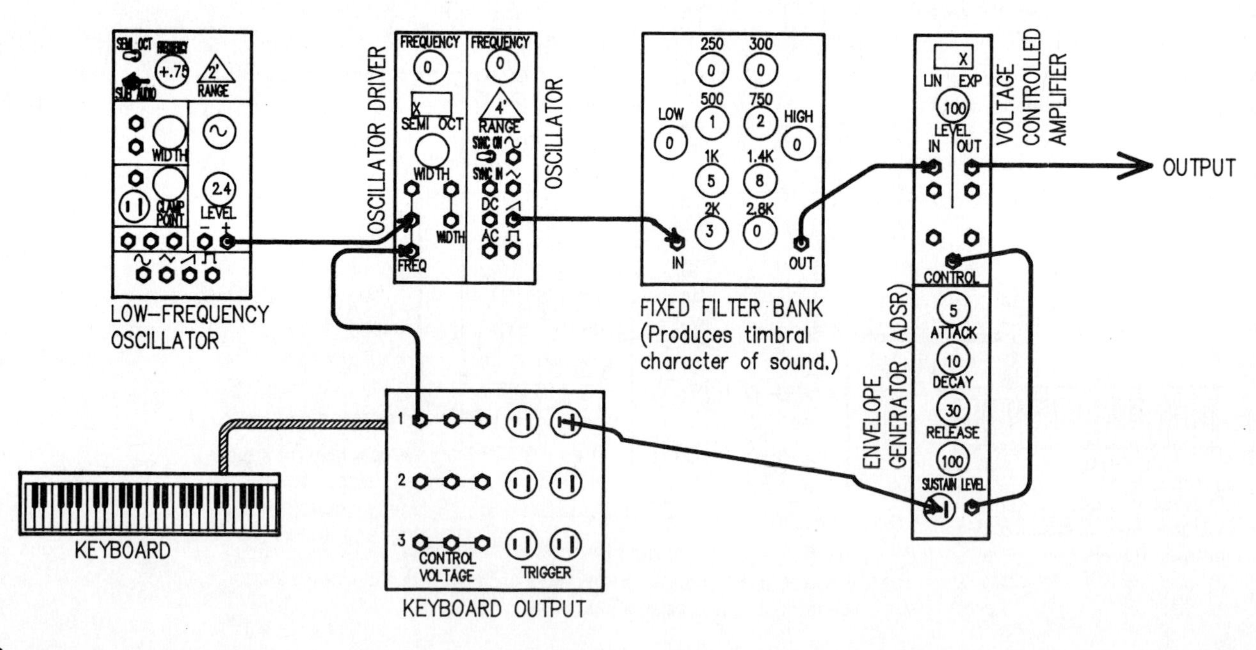

FIG. 12-1D: BASIC MOOG PATCHES

NOISE USED AS BASIS FOR DRUM SOUND OR GUNSHOT

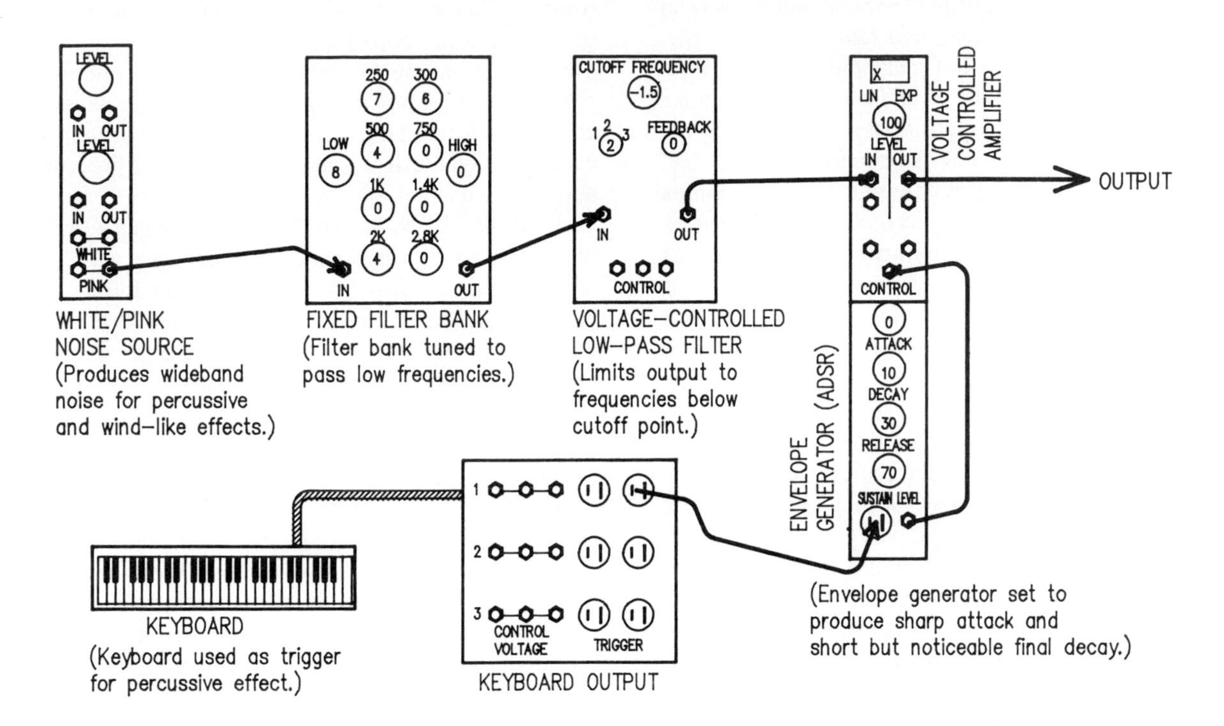

than a single melody line meant hours of synchronous multitrack recording. Because of the one-voice limitation, these synthesizers were known as *monophonic* synthesizers.

Synthesis and Performance

Monophonic performance synthesizers began to appear in the early 1970s. Intended for live music, they usually featured a variety of preset timbres, plus controls for rapid selection of waveform, filters, modulation, and envelope. Most performance devices offered a rudimentary kind of *sample and hold* circuitry designed to make short, repetitive effects possible.[6] They incorporated a control element (either a touch-sensitive pressure strip, a knurled wheel, or a joystick) that could be used to detune the oscillators momentarily for pitch-bending effects. All controls were usually placed to the left side of the keyboard, so that the performer played the keyboard with the right hand and manipulated controls with the left.

Timbral control was much faster on performance synthesizers than on the early patch-controlled devices, but unless they were very simple, installing new patches was hardly an instantaneous procedure. The need to have multiple timbres simultaneously available in live performance led to use of racks containing a number of separate synthesizers.

In spite of the fact that no other instrument could produce so wide a range of sounds, early synthesizers had limiting characteristics that performers generally disliked—among them, lack of sensitivity to touch, inability to build up and use dense voicing and timbral colors, and lack of means to record and recall settings for performer-developed timbres automatically.

Touch is important to keyboard performers. A piano can create subtle differences in the sound by the force and rapidity with which the performer strikes the keys. In the 1960s and the early 1970s, most synthesizers made the same sound regardless of touch. Performers generally felt that synthesizers would be much more useful if qualities of ADSR, timbral modulation, and even pitch could be made to respond to touch or to some other transient input.

Dense layers of sound using the original monophonic synthesizers meant hours of overdubbing, very expensive and elaborate arrays of oscillators, filters, and VCAs, or multiple keyboards and synthesizers. Any one of these processes was costly, time-consuming, and not very practical; yet the very character of electronic sound suggests that some of the most interesting and useful sound qualities would result from "layering" of sound on sound.

[6]Sample and hold on these early live performance synthesizers meant sampling a short segment of control and trigger signals from the keyboard interface or from the low-frequency oscillator used to generate such functions as vibrato and warble. Using sampling, a particular effect could be repeated indefinitely while the performer continued to produce other sounds using the keyboard.

Still another valuable aspect of synthesis is the ability to tailor sounds to individual needs; but, during most of the 1970s, there was not much way, apart from written patch sheets, to record performer-created timbres, and no way to recall them other than repatching and retuning. Most performers wanted a way to record and hold settings in the instrument and recall them quickly with the touch of a button.

By the late 1970s, a variety of developments in the world of synthesis had occurred. First, integrated circuits had replaced transistors, making possible much more complex circuitry in ever smaller packages. Second, several manufacturers had entered the field, accelerating the competitive race to produce synthesizers that were easier to use, practical as performance instruments, and economical to purchase. Activity focused on two main goals: ability to change timbres quickly and easily and the development of a true *polyphonic* synthesizer—one that would produce as many voices as the player cared to or was able to play at one time. Another, though somewhat less important, goal was the achievement of a *polytimbral* synthesizer—one that could not only change timbres quickly but could produce multiple timbres simultaneously.

A few practical, but only marginally affordable, polyphonic devices began to appear near the end the 1970s. These devices could produce multiple voices, usually all using the same timbre. A few very large and very expensive instruments were both polyphonic and polytimbral. Some manufacturers began to incorporate features specifically geared toward live performance. A particularly important development was key pressure sensitivity (sometimes called "aftertouch") to enable the performer's touch to control parameters such as timbre, intensity, and envelope.

The Development of Computer Control

By the early 1970s, electronic music laboratories at major universities were experimenting with the use of minicomputers to control synthesis. The effects that could be achieved through the combination of synthesizers and computers were useful and potentially were capable of overcoming problems in live performance. The available synthesis equipment was analog, however, and the computer was digital. In order for a computer to control a synthesizer, digital-to-analog converters were needed; in order for the computer to receive input from the synthesizer keyboard, analog-to-digital conversion was required. The control interface was inordinately cumbersome. Worse, each new experiment required a new program. The singular achievement of experiments in the early 1970s, however, was that computers actually controlled synthesizers, could produce rapid timbral shifts, and could operate many different groups of modules simultaneously.

The development of the very small, very fast microprocessors that made personal computers a reality also brought about a new phase in synthesis. The microprocessor made computer control affordable, often as a small, dedicated processor that could integrate directly into the synthesizer's electronics. Under microprocessor control, modules can be grouped and regrouped quickly into various configurations. Each group can be updated rapidly by the microprocessor, passing on the latest instruction from the keyboard or other controller. Although microprocessors used digital signals to control the synthesizer, sound generation and processing were still mainly analog.

In the 1970s, the audio industry began to experiment with ways to encode sound as digital data which could be handled, processed, and stored entirely as numbers. As digital audio developed, so did digital synthesis. With the emergence of fully digital sound production, more extensive and subtle control became possible. By the early 1980s, fully digital synthesizers had come on the market, but they were horrendously expensive. Gradually, as both digital audio and microcomputer technology advanced, prices fell. Fully digital musical instruments now exist, varying in form from small, portable units to full, console-sized packages that look like pianos or organs. Even the smallest of the new breed can do far more than any of the first synthesizers could ever accomplish.

The power of synthesis rests on two features: sampling and networking. *Sampling* enables a synthesizer to read a short segment of sound from a real source—a trumpet, for example, or a human voice—and then to use that sound as itself or as a model to generate new sounds. *Networking* permits synthesizers and a host of peripheral devices to talk to and control each other. The communications protocol that has made interconnection of synthesis devices possible is a standard called the Musical Instrument Digital Interface, MIDI for short. Before examining newer synthesizers (including samplers) and their peripherals in detail, we need to become familiar with the characteristics of the MIDI protocol.

MIDI

During the late 1970s, several synthesizer manufacturers began to develop techniques for polyphonic, polytimbral, touch-sensitive control using microprocessors. But each was different, and instruments were rarely compatible. Given that a rig consisting of multiple synthesizers was useful, a performer had to have a complete set of equipment by a single manufacturer, or else had to invest in redundant systems and intricate custom interfaces to obtain all of the performance qualities he or she wanted. In 1981, at a conference of the Audio Engineering Society, a proposal for a musical instrument intercommunication and control protocol was put forward. Within a year a standard, now known as the Musical Instrument Digital Interface, was developed and adopted by a committee of individuals representing most major musical instrument manufacturers. Called simply MIDI, the standard is supported and maintained by an industry organization called the MIDI Manufacturers Association (MMA).

MIDI is an intercommunication protocol that permits any synthesizer, sampler, drum machine, sequencer, or other music module to control any other MIDI device, or to be controlled by any other such device, or to be controlled by a computer running a MIDI-compatible program. The MIDI protocol defines a system of "messages" that communicate either musical events or control events, the rate at which messages can be passed through the system, and the hardware required to implement a MIDI network.

MIDI is a serial, digital protocol. In a serial transmission system, data is communicated in the form of electronic words made up of the electrical equivalent of 1s (on states) and 0s (off states), called *bits*. When a bit is on, it is said to be *set*. When a bit is turned off it is said to be *reset*. In the case of the MIDI system, no current = 1 or set; a 5-milliampere current equals 0 or reset. All MIDI words consist of 8 data bits preceded by 1 start bit and followed by 1 stop bit. Start bits and stop bits are always reset (0). The words must be transmitted within a specified time frame. MIDI specifies that each bit must occupy an interval of 32 microseconds, which converts to 31,250 bits per second. Each electronic word consists of 8 bits (called a *byte*) plus the start and stop bits, so that 3,125 words of 10 bits each can be transmitted each second. Each word, therefore, occupies a time frame of 320 microseconds. Note that a transmission rate of 31,250 pulses per second places MIDI data above the range of audio. MIDI data, therefore, cannot be handled or recorded by an audio system.

The hardware used to support MIDI data transmission is extremely simple. A five-pin female DIN connector is used at all MIDI instruments—synthesizers, sequencers, computer interface boards, and so on. A five-pin male DIN plug is used at each end of a MIDI cable. Signals are carried on pins 4 and 5 only. Where used, pin 2 serves as a connection for shield grounding. Each MIDI connector is called a *port*. The specification defines three kinds of ports: MIDI In, MIDI Out, and MIDI Thru. Diagrams of the three MIDI ports are shown in Fig. 12-2. Note carefully that MIDI cables carry MIDI data *only*. Audio is never carried by MIDI interconnections, just as audio lines never carry MIDI.

The protocol defines two general kinds of MIDI messages: **channel** messages and **system** messages. A channel message carries musical information. A system message carries control information used to synchronize and to change

FIG. 12-2: MIDI PORTS

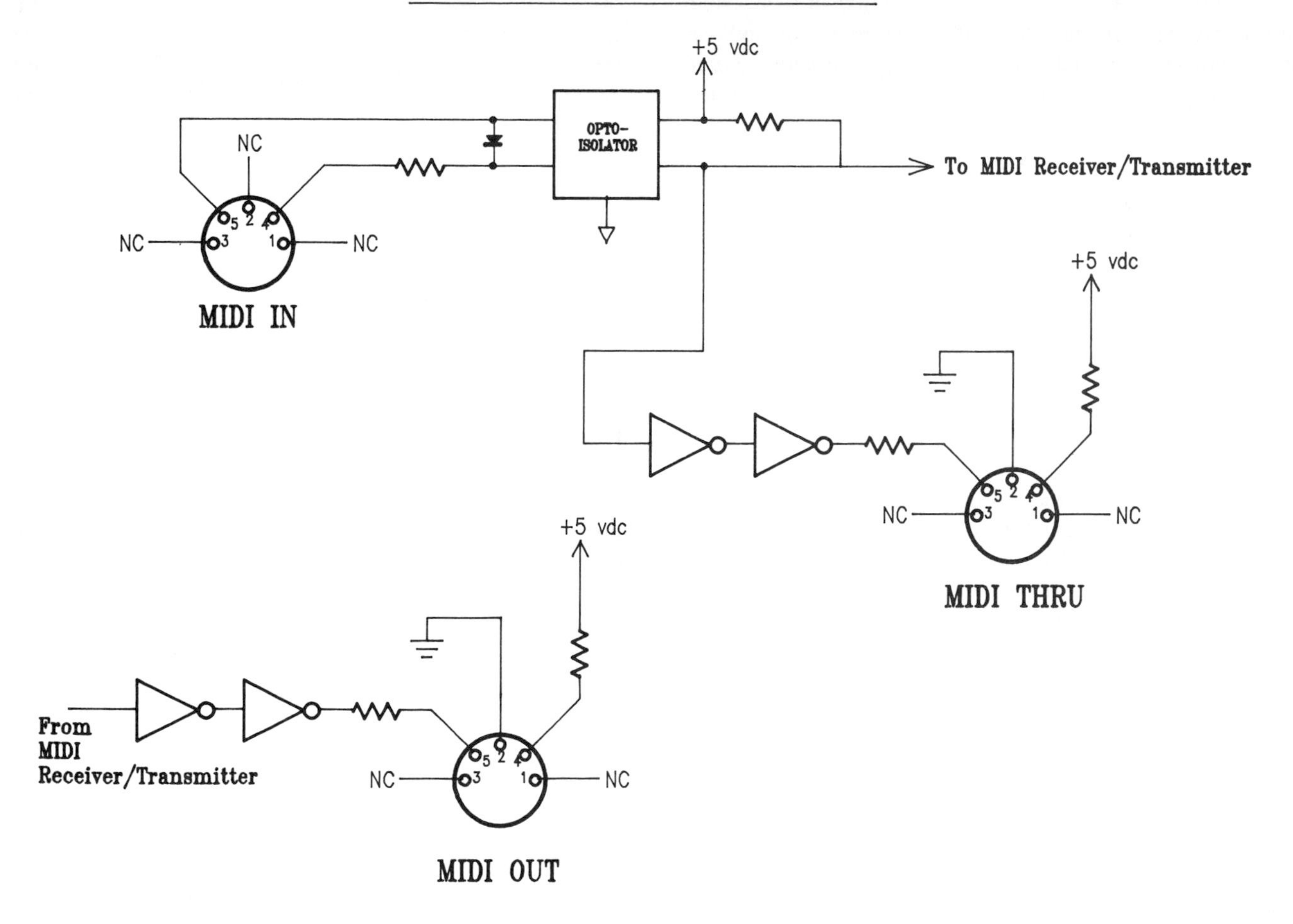

settings for instruments connected to a MIDI network. All messages have a standard form: 1 status byte, followed by 0, 1, or 2 data bytes. Status bytes always have the first bit set (1); data bytes always have the first bit reset (0).

Channel Messages

MIDI messages are carried on channels. MIDI permits 16 channels, and each instrument is assigned to a particular channel. The assigned channel on which the instrument receives its primary information is called the *basic channel*. The basic channel assignment can be set only at the individual instrument. An instrument always responds to its basic channel, and can respond to the other 15 channels, depending on the mode set by the master controller.

Each channel in a MIDI system can be used to control one or more voices. A voice is defined as a particular timbral quality that is created by the control settings of a given instrument. A voice may receive data from all channels or from only one, depending on the operating mode (the way in which a receiving instrument interprets data on channels other than the basic channel).

Channel messages are of two types: **voice** messages and **mode** messages. Voice messages always carry information about a MIDI "event." The MIDI protocol defines an event as one of the following occurrences: note on (a key press), note off (key release), a change in a switch or a variable controller (control change), a variation in pressure on one key (key pressure/aftertouch) or on all keys (channel pressure/aftertouch), or a change in a variable pitch controller (pitch bend). Each channel message consists of a status byte followed by 1 or 2 data bytes. Only one voice message, note off, has only a status byte with no data bytes.

When a key is pressed the status byte signals a note-on event. The first data byte of the channel message specifies the number of the key, which determines the frequency of the note to be sounded. MIDI has a range of 128 notes, from C five octaves below Middle C to G five octaves and a fifth above Middle C.[7] The second data byte specifies the velocity or force with which the key is pressed. A note-on event, which carries a velocity of 0, is functionally equivalent to a note-off event. Logically, therefore, one immediately obvious parameter that velocity may control is intensity.

Mode messages are used to regulate the response of MIDI instruments to voice messages on the 16 MIDI channels. MIDI defines four modes, which are illustrated in Table 12-1. The modes determine whether an instrument responds to all sixteen channels or only one channel and whether all active voices in an instrument or only one voice will respond. Terms used to describe modes are omni, poly, and mono. *Omni* refers to the 16 MIDI channels. If omni is set on, all channels can produce a response from the receiving instrument. If omni is set off, the receiving instrument responds to only one channel. *Poly* (short for polyphonic) indicates that the receiving instrument is expected to respond with multiple voices (though not necessarily different pitches). *Mono* (short for monophonic) indicates that the receiving instrument is expected to respond in only one voice (though not necessarily a single pitch). The modes are responsible for much of the power of MIDI, as the modes enable changes in the voice response of a group of synthesizers and peripherals. Voice changes are equivalent to changes in orchestration of a traditional instrumental ensemble.

TABLE 12-1

Mode	Description
1. Omni on/poly	The usual power-up mode for most instruments. Each voice in an instrument responds to all channels equally.
2. Omni on/mono	One voice in the instrument responds to messages on any of the 16 channels.
3. Omni off/poly	Each voice in the instrument responds to messages only on the basic channel.
4. Omni off/mono	The most complex mode in the system. Each voice in an instrument responds separately to one channel. In effect, the instrument becomes a group of single-voiced synthesizers.

Mode 1 is the simplest configuration, and most instruments default to this mode when first turned on. Any note or notes played on any channel by a controller will sound in all voices on any controlled instrument. Mode 2 sets an instrument to respond to all channels, but only one voice will be active. This mode works in multi-instrument layering, where several individual voices become part of a chorus of sounds. Any note sent by a controller on any channel will produce a response from a receiving instrument, but only in the one active voice. Mode 3 permits an instrument to respond only to one channel, the Basic Channel selected at the receiving instrument. All active voices will respond to a note or notes transmitted, however. Any part assignment used by the particular receiving instrument will determine assignment of notes to active voices.

Mode 4 is the most complex of all and can be executed only by a true polytimbral synthesizer. Omni off in this mode means that each of the 16 MIDI channels can activate only one voice of the receiving instrument. Mono means that each voice of the receiving instrument will listen to only one channel;

[7]Pitch on a synthesizer is not absolute. Both the patch (the way the synthesizer is configured) and the tuning of the instrument will determine the actual frequency that sounds when any particular key is depressed.

however, each active voice in the instrument will respond if the channel to which it is assigned transmits an event. In effect, the receiving instrument is set up as a group of independent, monophonic synthesizers, each playing a different voice and listening to a different channel.

System Messages

System messages are commands that determine the setup and response characteristics of an entire network of instruments and peripherals, and they affect all devices on all channels of a MIDI system. System messages are of three types: common, real time, and exclusive.

System common includes four messages: song select, song-pointer position, tune request, and end-of-exclusive (EOX). The first two are intended for sequencers and for modules that contain internal memory capable of storing event sequences. A sequencer is a device that stores groups of events that make up an individual "song." The *song-select* command sets the instrument to replay that group of events. Up to 128 different "songs" may be stored in a sequencer's memory. The *song pointer* keeps track of which step in the song is currently active. The song-pointer command can direct the song pointer to move to a particular step in a song. *Tune request* applies only to analog synthesizers. Oscillators in analog synthesizers tend to drift, so most analog instruments include a "tune button" that brings the oscillators back into tune with each other. The MIDI tune-request command has the effect of activating the tune button. The *end-of-exclusive* command signals to all instruments on a network that a system exclusive command has ended. Even though this command is directly related to system exclusive commands, EOX is a system common command, because all instruments on a network must know when a system exclusive message terminates.

Real-time messages sychronize the various modules within a system. The activity of all elements is based on the MIDI clock which runs at a rate of *24 pulses per quarternote*. The actual number of pulses per second will vary with the tempo setting, which is set in *quarternotes per minute*. All system real-time messages are based on the clock pulse. Real-time commands are timing clock, start, continue, stop, active sensing, and system reset.

The *timing clock* command sychronizes all instruments on a network with the master clock. Devices such as sequencers and drum machines contain clock circuitry. Either a sequencer or a drum machine can serve the purpose, however, if a network has multiple sequencers and/or drum machines, only one can serve as the system timebase. All such devices should have controls that set them to follow an external clock. The *start* command triggers a sequencer to begin playing through a selected song. The song pointer is advanced with every six ticks of the system clock (although the next event is not played until the clock signal equals the time stored with the event when the sequence was originally recorded.) Stop, naturally, freezes the song pointer and stops replay of the sequence. *Continue* causes the sequencer to resume playing, beginning at the position of the song pointer. (Start would reset the song-pointer to the beginning of the sequence and play everything all over again.)

Active sensing is a real-time command designed to prevent embarrassing accidents. In order to respond accurately to MIDI messages, synthesizers continue the last MIDI command received until a new command comes down the wire. But what happens if the connection between the receiving synthesizer and the master controller is lost? Naturally, the synthesizer simply continues to sound the last note received, indefinitely. Active sense is a dummy status byte that is transmitted at 300-millisecond intervals. A synthesizer will behave quite normally without active sensing pulses; but if active sensing pulses begin, then the synthesizer expects to receive a pulse every 300 milliseconds. If an active sensing pulse fails to arrive, the synthesizer turns off the last MIDI command and waits for a new command. *System reset* causes all instruments on a MIDI network to revert to power-up status.

System exclusive messages carry information that is specific to a particular instrument. In order to use system exclusive, each manufacturer must apply to the MIDI Manufacturers Association for an identification number. That number must appear as part of the system exclusive call. All instruments that recognize that ID number will respond; others will ignore the system exclusive message.

The system exclusive message is different from all other MIDI messages in that the number of data bytes that follow the status byte is undefined. Any number of bytes needed may be transmitted. The end of a system exclusive transmission is signaled by a system real-time call to the EOX (end-of-exclusive) command. The real-time system exclusive is used for theatrical purposes as the basis of a generalized stage control protocol known as MIDI Show Control (MSC), which we examine in more detail later in this chapter.

MIDI and External Timing Methods

The MIDI clock is a tempo-related time base that paces channel events based on the length of a standard musical time interval—the quarter note. The MIDI clock varies with performance speed. The number of quarter notes per unit of clock time (usually a minute) is a metronome like count, almost simulating a conductor's beat.

Clocking used for standard recording systems—audiotape recording, videotape recording, motion-picture film—is fixed, not variable. Recording systems use elapsed time in hours, minutes, and seconds. Film also adds the requirement of tracking *frames*—the number of picture cells displayed per second. The standard time code for film, video, and audio was developed by the Society of Motion Picture and Television Engineers, and is known as SMPTE Time Code. SMPTE has become the standard time base for studio recording, video production, and film post production. SMPTE Time Code is communicated in an 80-bit digital word that tracks frames, seconds, minutes, and hours, and also includes synchronizing data.

The MIDI clock is incompatible with SMPTE; however, MIDI instruments are increasingly used in producing music and sound effects for film and television. In early efforts to synchronize MIDI with SMPTE-based devices, translators called SMPTE-to-MIDI converters were devised. The converters sychronized the MIDI clock stream to the start of the SMPTE code stream, and utilized the song-pointer positioning command to lock particular MIDI events to specific increments of elapsed SMPTE code time. Unfortunately, editing an audio tape or a film occasioned very laborious repositioning of the song pointer to keep channel events synchronized to the desired SMPTE time intervals.

MIDI Time Code (MTC) was developed to overcome the difficulties introduced by MIDI-to-SMPTE converters and to facilitate control of MIDI devices by SMPTE-based systems. Instead of a SMPTE-to-MIDI converter, an MTC generator must be placed between the SMPTE system and the MIDI master controller. The generator receives SMPTE code from a master controller (which may be an audiotape machine, a videotape machine, or a projector running a SMPTE-striped film.) The generator, in turn, sends MTC to a controlling device, usually a computer program that contains an event list, called a *cue sheet.* The cue sheet designates MIDI events or other control events that should occur at particular points in time, based on the SMPTE code stream. The MTC generator may be a stand-alone device but usually is itself a computer program, often the same program that keeps the cue sheet.

MIDI Time Code is based on frame rate, generating four *messages* per frame. Each quarter-frame message constitutes a MIDI system clock pulse and also encodes one field of elapsed time (hours, minutes, seconds, or frames) transmitted by SMPTE Time Code. Eight quarter-frame messages are needed to transmit one SMPTE 80-bit word. As a result, MIDI-to-SMPTE accuracy can be no greater than two frames.

As each scheduled event known to the cue sheet approaches, the cue sheet transmits a setup message. The setup message identifies the type of action to be performed, the precise (SMPTE) time for the action, and any additional information required. Additional information usually specifies either a channel message or a system message required to activate a particular MIDI event (for example, a control change or a "note-on" event). Other events that a setup message can signal include punch-in or punch-out for a tape recorder and triggers for effects devices. Setup messages also include editing commands that insert or take out events and cues in a MIDI Time Code stream.

MIDI Time Code is potentially important for theatre sound as a means for synchronizing MIDI devices with tape recorders or films, although such usage would probably be associated only with very large shows. Staging for concert tours by major popular music groups have used SMPTE Time Code and MTC to synchronize film segments and effects with MIDI devices.

MIDI Show Control

A somewhat different approach to MIDI and stage events is a recently developed addition to the MIDI protocol called MIDI Show Control (MSC). MSC is a system exclusive call using a valid ID number assigned by the MIDI Manufacturers Association. The call identifies specific equipment that should respond. The kinds of equipment currently addressable by MSC include lighting controllers, sound controllers, stage machinery, video devices, film equipment, hydraulic and pneumatic devices, and pyrotechnics. The range of devices is intended to make MSC applicable to all kinds of shows, from traditional staged dramatic productions to theme park displays.

The first part of an MSC call is the system exclusive command, followed by the MSC identification number and an identification number specifying the kind of device to be addressed. Following the device identification, a function is transmitted. The range of functions is similar throughout the range of MSC definitions, but may vary slightly, depending on the kind of equipment addressed. Typical functions include instructions to load a cue number, execute a cue, and alter the setting of a controller. Cues can also be synchronized to MIDI Time Code.

In no case does MSC store or transmit specific cue data for any kind of equipment. MSC handles cue numbers only. For example, an MSC call might instruct a lighting console to execute a particular cue number, or to jump back to a previous cue. The MSC controller has no information regarding the content of the cue. The specific actions to be taken by the lighting console on execution of a particular cue are stored in the console's memory, not with the MSC controller.

Additional data transmitted in a MSC message can include specific elapsed (SMPTE) time at which an event is to be executed, or can carry device-specific information. A recent addition to the MSC specification includes a two-phase commit that provides cue warning, alerting both device and operator that a cue is approaching. The human operator may then decide whether or not to permit the cue to execute automatically. The reason for the two-phase commit concerns safety, primarily in the more dangerous areas such as machinery, hydraulic/pneumatic, and pyrotechnic cue control.

MIDI Networks

Any group of controllers and sound modules connected together form a MIDI network. The simplest network is one keyboard synthesizer connected to another. The MIDI out port of one of the keyboards is connected to the MIDI in port of the other. The former keyboard is called the master; the latter is called a slave. Any MIDI network is set up on a master/slave principle. Only one device can serve as master.

Using the MIDI ports, a network can be set up as a ring. The master feeds the first slave. The MIDI thru of the first slave feeds the MIDI in port of the next slave, and so on, until all slaves are connected. The problem with a ring, or with any configuration that forces several devices to share the same MIDI interconnection, is latency. For example, if a sequencer is connected through a ring to several synthesizers, a string of messages can take a significant amount of time to work its way around the ring simply because MIDI is a moderately slow serial protocol. The last synthesizer on the line may suffer unacceptable delays because of propagation time. Worse, if real-time messages, which demand high priority, are transmitted, channel voice information may be delayed further. The best arrangement is called a "star" network in which MIDI information is sent on separate lines. The master device, of course, must have multiple MIDI out ports to make a star network possible.

Executing MIDI Commands

The MIDI specification defines intercommunication between devices in response to operator actions, not commands that an operator must enter. MIDI is relatively transparent to a system user. For example, when the synthesizer performer strikes a note on the synthesizer's keyboard, the synthesizer sends a channel message (note-on) informing the network that a key has been pushed, the number of the key, and (if applicable) the key pressure. The synthesist does not write a coded instruction. Similarly, using a computer program that cues MTC or MSC events, the operator executes an action within the program—usually cue selection or a choice of cue operations. The most complicated action that an operator may need to perform is the specification of a particular point in the MTC or SMPTE time stream at which a cue is to execute. The program then generates the specific MSC or MTC code that is stored in memory and/or transmitted to the MIDI network.

MIDI System Components

A large group of diverse components may be included in a MIDI network: synthesizers, samplers, drum machines, sequencers, patch bays, and, of course, computers. Moreover, an increasing number of standard audio devices are incorporating MIDI as a means of remote control. Most such nonmusical devices, however, commandeer standard MIDI channel messages to trigger their own actions. Although use of channel messages enables MIDI instruments to effect changes in peripheral audio devices (equalizers, delay lines, etc.), one potential problem is that there is always the danger of a false cue to such a device. Also, if one MIDI channel is restricted to non-MIDI devices, then the performer has one less channel to use for musical events.

Synthesizers

Synthesizers are instruments designed to generate sounds. Usually the sounds are musical tones that can be shaped to various envelopes and timbres, ranging from traditional to new and experimental. Present-day synthesizers may be either digital or analog instruments, and they may be instruments having a standard pianolike keyboard, rack-mounted, electronic modules, or more exotic devices with sounding and pitch selection methods that resemble guitars. In the form of rack-mounted elements, they are usually known as tone modules.

Synthesizers must have a minimum of basic capabilities: They must be able to generate tones throughout (and possibly both above and below) the audible spectrum. Most currently available synthesizers are capable of generating more than one tone at a time (polyphonic). Synthesizers must incorporate the capability to shape tones into any desired timbre, either by waveform addition or subtractive filtering. Some are polytimbral (can produce more than one voice at a time.)

Analog synthesizers, still available, have essentially the same basic elements that were present in the earliest Moog synthesizer. The elements, however, appear as *functions* controllable through faceplate keys, switches, benders, and sliders rather than as modules that must be patched together. The elements are tone generating oscillators (including low frequency oscillators to create effects such as vibrato), filters, voltage-controlled amplifiers, envelope generators, and white and pink noise sources. Digital synthesizers also contain the same basic sound-creating and sound-shaping functions, but these functions are produced not by modules such as oscillators and filters, but by direct generation of mathematical data that can be converted into audio.

Practically, the superficial differences between analog and digital synthesis devices is not great. In effect, modern analog synthesizers will have most of the capabilities of digital synthesizers, except for sampling capability. Some proponents of analog technology argue that the tone of an analog device is warmer and less mechanical than that of a digital synthesizer.

Samplers

Samplers are digital synthesizers which are capable of converting acoustic or audio signals into digital data and storing them for use as a voice. Samplers can also accept predigitized data directly. Recordings of sound or predigitized samples are loaded into the sampler. Each sample can be accessed independently, and a keypress on a master keyboard will determine its pitch. For example, a single tone could be recorded from a trumpet, then used as a trumpet voice under keyboard control. The trumpet sound would behave just as any other sound the synthesizer is capable of producing, but the sound would have the true timbre of a trumpet. It would even contain the timbral modulation that accompanies the onset of a trumpet tone. Even speech, or noises such as city traffic or mechanical sounds, can be sampled and used. Sampling, therefore, becomes a primary means of providing environmental sounds for theatre. As such, sampling is much more flexible than tape, because it can provide instant access to any sound and can generate subtle differences in shading, depending on whether the sample is used at original pitch or at a different pitch.

Most sampling synthesizers come with a library of built-in sounds. The least expensive samplers can use only the built-in sounds; better units are capable of loading new samples, either from floppy disks or from manufacturers' proprietary ROM cards.[8]

The duration of a sample is limited by available memory and sampling rate. Fifteen- to twenty-second sampling limits are fairly long; many samplers provide less. Optional add-on memory units can increase the sampling time. In order to create long sound durations, samples can be "looped." (The name is derived from the tape technique of splicing the tail of a sound to its head to form a loop of tape. The loop is then run through the playback heads to create an effectively endless sound.) Looping is difficult. If the splice is not made at just the right point, differences in timbre, intensity, and pitch can make the splice irritatingly noticeable. Some sampling devices provide "reverse looping" as a technique to overcome joining problems. Reverse looping merely plays the loop backward from tail to head, then from head to tail again, maintaining a continuous, smoothly changing sound. Of course, some part of the sample is periodically played backward, and in some cases the reverse sound can be as irritating as a bad splice. Waveform editing stations, which we examine shortly, can improve the process of splicing loops by making the join process visible and minutely controllable.

Drum Machines

A **drum machine** is a limited-capacity synthesizer/sampler that contains special programming routines to make it capable of complex automatic rhythm production. Very simple drum machines rely on white and pink noise sources, filters, and envelope generators to produce percussive sounds. Most drum machines presently available use samples of real percussion sounds and provide a much wider range of timbres. The best drum instruments, although they contain a built-in selection of stock samples, can read in additional sounds from either ROM cards or disks.

Peripheral Control Equipment for Synthesis

Sequencers

Sequencers record and store MIDI control signals. A performer can play a part (or a group of parts using polyphonic mode) and have the entire sequence of MIDI events recorded by a sequencer. Several parts can be recorded in parallel if the sequencer is capable of storing more than one track of events (and most are). The sequencer can then be triggered to play back one or all of the recorded tracks, either alone or with additional parts played at the keyboard. System common messages are used to synchronize the starting location and clock rate of all sequencers and other memory devices in a system. The song-select and song-pointer commands can be used to control which stored section of performance is to be replayed.

MIDI Patch Bays

A **patch bay** is a device for storing voice information—control settings on a synthesizer that produce a particular timbre and envelope characteristic. As a specialized unit, it is capable of more storage than an individual instrument can provide. Use of a patch bay enables large volumes of rapid program changes under MIDI control. When a performer works out a patch that he or she wants to keep, a command is issued to store the patch. MIDI sends the patch configuration to the patch bay module where it resides in memory until called out again.

MIDI and Computers

Integrating computers into a MIDI system broadens the possibilities of synthesis by an order of magnitude. What a computer can do is determined by its software, and some of the software available to work with MIDI systems is very powerful. For example, using a computer, one can dump the contents of a sampler into computer memory, display the waveform of a given sample, *edit the waveform visually*, store the edited version, and return the result to the sampler. The edited sound can then be used as a voice. Among the possibilities that are immediately evident are smoothing loop splices, picking the best point for a loop splice, and altering timbral

---

[8]ROM stands for read only memory. ROMs are special kinds of computer memory chips that hold information indefinitely. The manufacturer programs sample information onto the ROM(s) and mounts the ROM(s) into circuit boards that are plugged into a receptacle on the synthesizer. So long as the card is plugged in, the device can read and use the additional samples.

qualities very precisely without the need for filtering and equalization.

Computers can also create a sequence of MIDI events, usually for storage and retrieval from a sequencer. Of course, a computer program can also serve as a full-featured sequencer capable of playing many voices at once. A full composition can be built and performed without any use of the keyboard, if desired. Some available software can also turn MIDI control sequences into printed music. Altogether, the application of electronic and computer technology provides a marvelously flexible and resourceful system for creating and managing almost any kind of sound.

Waveform Editors

The waveform editing capability, mentioned in conjunction with sample shaping, was initially limited to very short segments of sound, so that its usefulness extended to material lasting only a few seconds. With the development of very large hard disk storage systems and writeable magneto-optical media, waveform editing has emerged as much more than a sample editor.

The direct-to-disk recording systems mentioned in Chapter 10 routinely incorporate some degree of waveform editing capability. Some have only enough to permit segment editing and track "slipping." Complete mastering systems may offer full editing and signal processing, including EQ, filter, compression, expansion, delay, and pitch processing. "Cut-and-paste" editing is also possible, facilitating copying and moving sounds the way a word processor can copy and move text.

Synthesis and Music

Traditional Multitrack Method

Synthesizers were developed to allow both composer and performer a wider range of new sounds than were available with any kind of traditional instrument, and to permit a single individual to develop and perform multivoiced compositions without the need for a group of performers. The basic process for making music with a monophonic synthesizer is an interaction between the synthesizer and a multitrack tape recorder or between a synthesist and the rest of a group of musicians.

In recording, the first step is to record one track of beats, called a "click track." The click track provides the beat to which all other tracks are recorded. Following the click track, the performer sets the voicing on the synthesizer for one of the other parts (usually the melody) and puts that track on tape. Then any necessary revoicing is worked out and subsequent tracks are recorded until the entire piece is recorded. Note that each additional track after the first one must be recorded in the tape machine's self-synchronizing mode (i.e., using the record head as a playback head for any tracks already recorded). The process of recording synchronized tracks in succession is called *overdubbing*.

Overdubbing, under any circumstances, is a laborious and tedious process. The performer must make sure that the voicing is exactly right; then he or she must play through the entire composition each time a new part is added to the tape. Any small discrepancy in timing or in accuracy of notes requires a punch-in edit at least or a complete retake of the track at worst. Performance using this process involves hours of process time for even a relatively short performance time.

Process time with monophonic synthesizers can be reduced by use of multiple keyboards. Use of multiple keyboards can facilitate simultaneous recording of more than one part and ready access to more than a single timbre.

Polyphonic Method

Clearly, the advantages to being able to record music polyphonically (two or more voices at once) are considerable. Although polyphonic recording requires greater skill as a keyboard performer, the time required to execute a composition is far less than with the monophonic overdub process, and the work usually tends to sound more like a unified piece of music than when each voice is recorded separately. Polyphonic synthesis permits sound density to be built up quickly and easily.

MIDI Method

MIDI synthesizers, samplers, patch bays, sequencers, and disk-based sound recording systems extend possibilities considerably, especially for persons whose musical performance skills are limited. In the first place, the range of sounds accessible to a composer or sound designer is extremely large, in that one may set up customized sounds, record and use samples of external sounds or rely on sample libraries that contain outstanding specimens of instrumental sounds of all kinds. Any of those sounds may be modified to suit needs.

Using a sequencer, any number of tracks can be recorded, step by step if necessary. Control and voice change information can be added, and the composition can be run from the sequencer at performance speed. Using MIDI Time Code, synthesized program can be synchronized with and/or added to recorded audio. The possibilities for composition and especially for adding music to dramatic production are enormous.

Synthesis and Dramatic Effects

The advantages of synthesizing sounds for a dramatic environment are at least two: First, a synthesized sound is "clean"—that is, it has no contaminating environmental background to betray that the sound was recorded in a location other than the environment of the play; second, the designer can tailor the quality of the sound exactly to the mood and character of both the dramatic structure and the implied auditory environment.

Effects Using Older Synthesizers

Although basic synthesizers are demonstrably best at repetitive waveforms and, therefore, best suited to the production of musical sounds, they can be used for the creation of a wide range of nonmusical sounds. Most synthesizers have some form of white and/or pink noise source that may be used as the basis of nonrepetitive, complex waveforms. The purpose of such noise sources in musical synthesis is the production of percussion effects, but these components of a synthesizer can also simulate gunshots, cannons, wind, footsteps, steam locomotives, airplanes, and much more. The newer drum machines can also be adapted to these purposes, of course. Combined with the repetitive waveform generating properties of synthesizers, the noise sources can produce a large number of sounds useful in dramatic environmental ambience.

As long as any of the signal or control points of the synthesizer can be accessed, either by other functions in the synthesizer or by external signals, even the simplest synthesizer can build up quite complex sounds. For example, one oscillator can be used to modulate the pitch or, perhaps, the duty cycle (the percentage of time during one cycle when a rectangular wave is off as opposed to on) of another oscillator. A sound from a tape can drive the control inputs of VCF or a VCA to produce sounds that have the character of the taped sound but the tone quality of a synthesized timbre. Using a multitrack recorder and the overdubbing process, or digital overdubbing using sequencers and MIDI controlled instruments, one can build up multiple sounds in extremely complex layers. With MIDI channels as a means to control a variety of sound sources, sound can be moved in a panoramic arc, bounced back and forth through the dramatic space, or otherwise flexibly distributed.

Coupling a synthesizer with a device such as a vocoder[9], (although stand-alone vocoders are rapidly disappearing with the development of samplers) vocal input can be used to make nonhuman sounds "speak." Vocoders also offer the possibility of giving the articulation characteristics of one sound to a totally different sound. Synthesizers in conjunction with any of the various signal processors (filters, equalizers, compressors, limiters, expanders, and the various time-domain processors) offer a wide range of possibilities for exploration, any one of them potentially useful for modifying and enhancing a dramatic environment.

Effects Using Samplers

With samplers, almost unlimited possibilities are available for the theatre. Because any sound can be recorded and placed in the memory of a synthesizer to be used at will, transposed in pitch, modified in timbre, combined with other sounds, and so on, virtually any sound is potentially available to the sound designer, adaptable to virtually any purpose. Digital synthesis together with samplers and MIDI controllers may very well make the tape recorder almost obsolete in the theatre sound control room, at least in those situations where a play needs complete underscoring or a very full auditory background. Under such demanding conditions, the speed and flexibility of MIDI-controlled synthesis with sampling capability will make tight cuing and flexible movement and distribution of constantly changing sounds a relatively accessible art.

---

[9]A vocoder is a device that will follow the spectral envelope of speech sounds, using them to modify the spectral character of any other sound. Use of a vocoder makes possible imposition of speech on nonvocal sounds.

Chapter 13

Survey of Audio Systems

Now that we've examined individual components of audio systems, let's look at the way various branches of the audio industry have put these components to use. A survey of system configurations can be especially useful because theatres often inherit hardware from other disciplines. The placement of nontheatrical systems in theatres occurs for any number of reasons: because a school or college has an existing system, possibly from a broadcast studio or an old public address installation, because someone responsible for purchasing equipment for the theatre is most familiar with a particular kind of system, or because architects or consultants specify nontheatrical hardware.

In Chapter 4, we listed a number of standard component configurations for particular aspects of professional audio. The applications and their corresponding systems are:

- Public address
- Acoustic reinforcement
- Stage monitor
- Broadcast audio
- Studio recording
- Background music/paging
- Domestic music and entertainment
- Film sound mixing
- Theatrical sound scoring

Using the information from the intervening chapters, we should reexamine the various system configurations to improve our understanding of how audio systems function and how components are configured into systems.

Public Address Systems

The simplest system is the basic public address (PA). It consists of a microphone, an amplifier, and a loudspeaker. The microphone is the transducer that collects an acoustic waveform and converts it into an electrical analog. The microphone output is connected to the amplifier input. The amplifier increases signal level and raises electric power to the strength required to drive the loudspeaker. The loudspeaker converts the sound from electrical to acoustic energy, and sets the surrounding air into motion at audio frequencies.

The fundamental purpose of the public address system is to pick up a human voice and return it immediately to the same environment at a greater sound power level than the voice alone could produce. One of the simplest forms of public address system is the bullhorn. All three parts of the PA system are incorporated in a single element. One end of the unit serves as the sound collector for the microphone, and the other end of the case forms the radiator for the loudspeaker. The amplifier is usually mounted either in the rear of the case near the microphone or in the handle with the switch that turns the system on and off.

More complex public address systems utilize a small mixer. They can accept several microphones and, possibly, the output from a record player or a tape recorder. Loudspeaker systems for larger public address installations usually involve more than a single radiator, and the amplifier itself may be mounted in a location that is accessible to the operator but out of sight of the audience. Figure 13-1 shows a diagram of a public address unit that has multiple inputs and mixing capability.

Acoustic Reinforcement System

The difference between public address and acoustic reinforcement is the difference between brute force and subtle persuasion. Otherwise, the purpose is the same: to pick up a live sound and return it to the listening environment. PA systems simply amplify the signal and project sound over the widest possible area. A reinforcement system should deliver sound quality comparable to direct audition. Reinforcement uses focused or distributed loudspeakers to cover zones of the

FIG. 13-1: PUBLIC ADDRESS SYSTEM

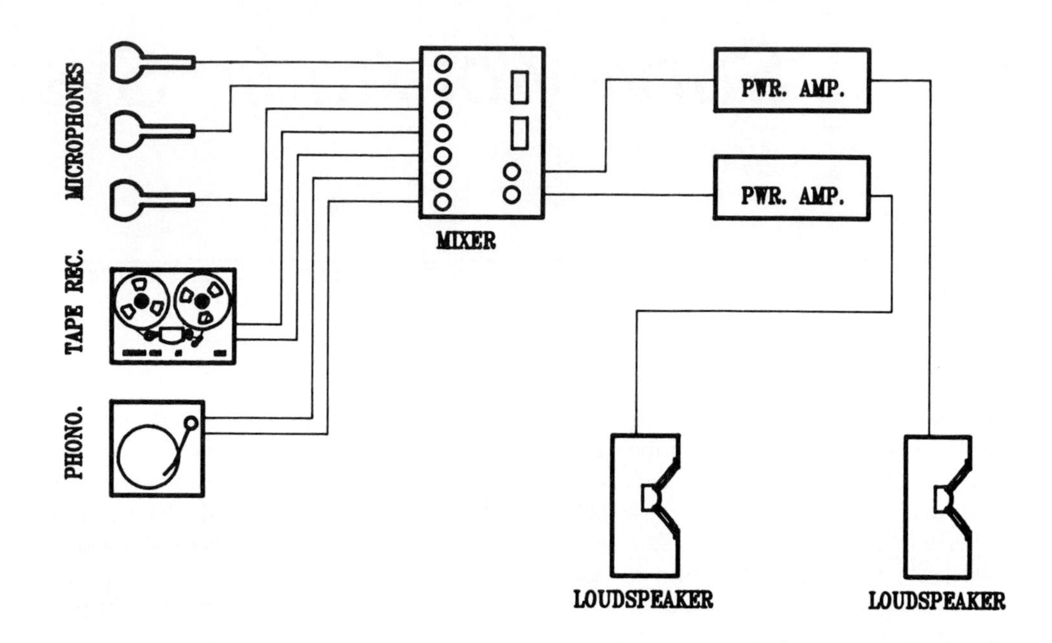

listening area. Sound pressure levels from reinforcement systems are generally lower than from public address systems.

A reinforcement system is either permanent or portable. A permanent system is designed for the acoustics of a particular auditorium. Portable systems must also match the acoustics of each space in which they are used; therefore, portable systems contain variable elements that facilitate adaptation to various environments.[1]

Acoustic reinforcement uses a large number of microphones. Microphone signals are combined into a composite output by the mixing console. The console output may be directed to a number of target destinations; however, the primary delivery point is one or more loudspeaker systems that return the sound of the performers' voices into the general audience area at a suitable sound pressure level so that everyone can hear clearly. Other destinations for the sound may include feeds to stage monitor systems, feeds to remote listening locations, or feeds to tape recording facilities for storage of the performance. Satisfactory reproduction of live sound from a stage requires signal processing functions such as equalization and delay. Different balances of EQ and delay may be needed for various, different outputs.

Acoustic reinforcement requires banks of power amplifiers to drive the number of loudspeakers that must be used. Highest quality sound reproduction usually involves bi- or triamplification, meaning that the signal is divided into separate frequency bands *before* power amplification. Each band is then routed to its own power amp, which, in turn, drives a particular section of the loudspeaker system. Electronic crossovers are used to divide the signal into bands just ahead of the inputs to the power amplifiers. Figure 13-2 shows a diagram of a reinforcement system.

Stage Monitor

Stage monitor systems are similar to acoustic reinforcement except that their purpose is not to amplify sound for the audience but for the performers. Stage monitor systems are tailored to permit individual performers to hear their own sound in relation to the general overall sound. Monitor systems may use microphones other than those for the primary reinforcement system, or may take one or more feeds from the reinforcement mixing console. Input controls of a mixing console specifically designed as a monitor mixer are usually rotary knobs rather than sliders, as, once set, input levels seldom change. Rotary knobs hold settings better than sliders and also reduce the cost of the system. Large banks of power amplifiers and a great many loudspeakers are also part of a monitor system. Figure 13-3 shows a diagram of a stage monitor system.

---

[1]Portable, in the case of reinforcement equipment, generally means that the equipment can be set into a road box, rolled onto a truck, and transported to the next engagement. The racks and consoles used in most traveling reinforcement systems cannot be carried like an attaché case.

FIG. 13-2: REINFORCEMENT SYSTEM

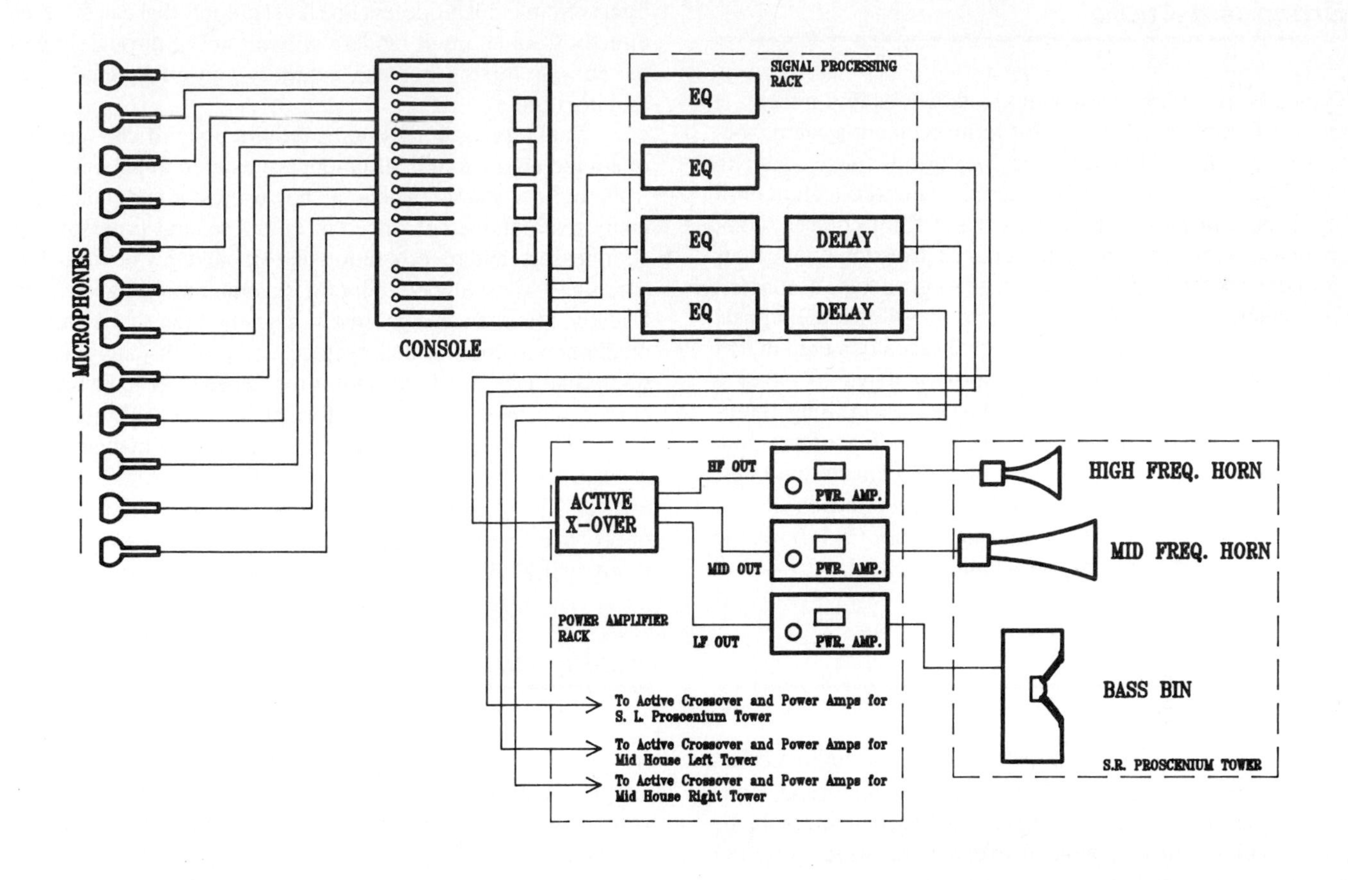

FIG. 13-3: MONITOR (FOLDBACK) SYSTEM

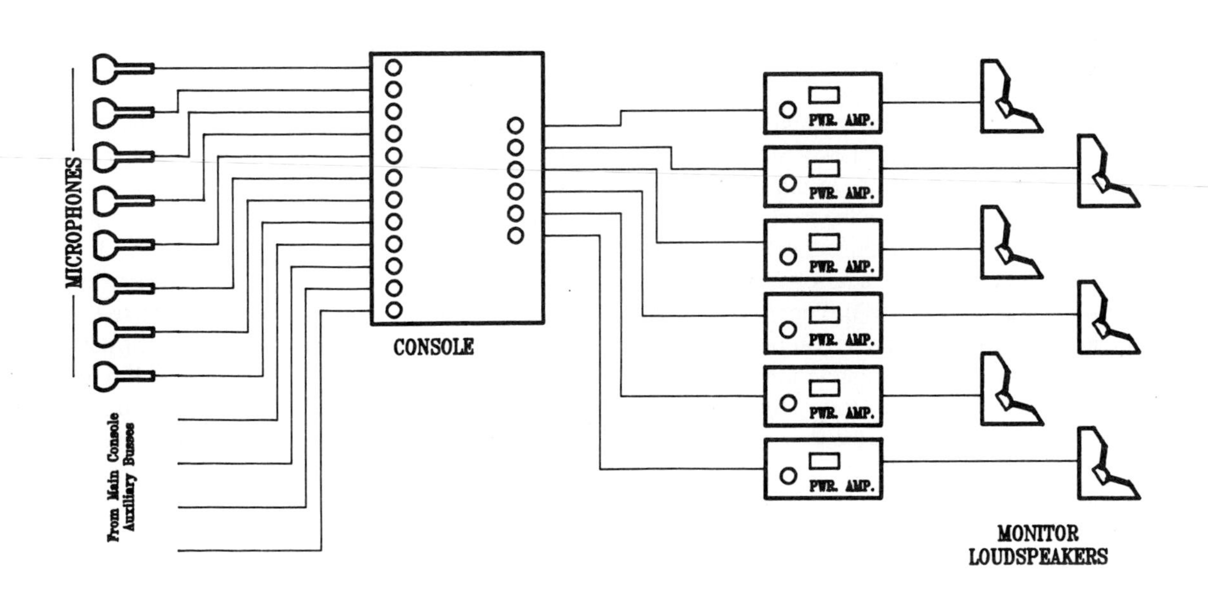

NOTE: The monitor system illustrated is rather simple. Especially for concerts, monitor systems may become extremely complex.

Broadcast Audio

Typically, most broadcast audio installations don't need the kind of flexibility necessary for reinforcement systems, because the range of functions they perform is fairly constant. Most broadcast stations don't attempt major production mixing. Inputs in most installations come from control room, announce booth, or studio microphones, from telephone lines (for network feed), and from storage sources (tape recorders, CD players, phonograph pickups, etc.). All of these signals are mixed down to a composite signal (usually stereo in FM broadcasting, mono for AM stations) and fed to either a transmitter or the input of a tape recorder. Processing equipment is minimal compared to recording and reinforcement applications. Except for EQ contained in console modules, processing is strictly outboard and is patched into the signal path after the signal leaves the console. Figure 13-4 shows the basic diagram of such a broadcast audio system.

Studio Recording

Explanation of the functions and purpose of a studio recording system requires a brief discussion of recording techniques. In modern recording practice, each instrument or singer is picked up by a separate microphone, which is routed to one track of a multitrack tape recorder. Original tracks are recorded "**dry**," which means that the sound is not processed in any way—no EQ or reverberation. Once all tracks are recorded, the tape is replayed and adjustments are made to provide correct equalization, reverberation, phasing, or other signal processing for *each* separate channel. When the best possible level and processing settings have been selected, the tape is mixed down to a stereo master tape.

The fact that so much of the work of making a recording sound balanced and clear takes place slowly, step by step, after the actual recording session demands system capabilities that aid in such work. Top quality recording consoles memorize and store sequences of control settings. Locator devices using SMPTE Time Code associate control settings with specific locations on tape. Using MIDI Time Code, synthesizers and other electronic instruments can be controlled automatically. Computer-controlled sound editors and processing devices can manipulate the pitch and tempo of sounds, either to correct problems or to create special effects characteristics.

Studio recording systems (see Fig. 13-5) usually stand somewhere between broadcast systems and reinforcement systems in complexity. Like a broadcast system, the principal signal destination is fixed (usually the inputs of a multitrack tape recorder), although recording systems are usually more complex than broadcast installations. Like reinforcement, subordinate functions require secondary outputs, such as directs from input modules (line level outputs that can be taken directly from an input module without going through one of the console mixing busses), auxiliary and effect sends, and monitor outputs.

Studio recording systems require a significant amount of equipment as a rule, although the exact components vary from studio to studio. In general, however, studios tend to have many microphones of different kinds, several types of EQ, compressors, and reverberation and signal delay devices. Recording studios always contain several recording devices. Usually, recorders are based on magnetic tape as the storage medium; but direct-to-disk systems will probably supplant the traditional tape machine within the decade. At least one recorder, whatever the storage format, will be multitrack (8-, 16-, or 24-channel). Another will be a two-track machine used to "master" the stereo mix-down of a recording session. SMPTE Time Code capability is essential, even if a studio uses only one recorder, because of the automated processes enabled through SMPTE.

Background Music/Paging

Background music/paging systems (see Fig. 13-6) tend to be simple in terms of form and function, but they are fairly complicated technically. Such systems have one main distinguishing feature: They have to drive a large network of loudspeakers distributed throughout a room or a building.

The front end of a music/paging system is uncomplicated. It contains a music source (usually a tape recorder), microphones, and a small mixer/amplifier. The mixer is designed so that pushing a keyswitch on the microphone overrides the music, either by interrupting the tape input completely or by dropping the level of the music and mixing the microphone signal over it.

The complicated part of a background music/paging system is its output. In a well-designed background music system, as in a restaurant or department store, the space to be served is zoned, and each zone requires at least one loudspeaker. Consequently, a great many loudspeakers are involved. A network of many speakers presents the problem of driving a large, distributed load economically. One typical solution to the problem uses the same principle as power transmission: Distribute high voltage at low current, transforming the power to necessary voltage and current levels at the point of use. The technique, called a **constant-voltage distribution system**, employs a transformer at the power amplifier to step up the voltage output and a transformer at each loudspeaker to tap off the power required for that particular unit.

Some theatres have adapted the constant-voltage principle along with switching matrices to permit signals to be directed flexibly to many loudspeaker units, but most theatres do not use constant-voltage output systems. Direct emulation of background music techniques is often used in furnishing

FIG. 13–4: BROADCAST AUDIO SYSTEM

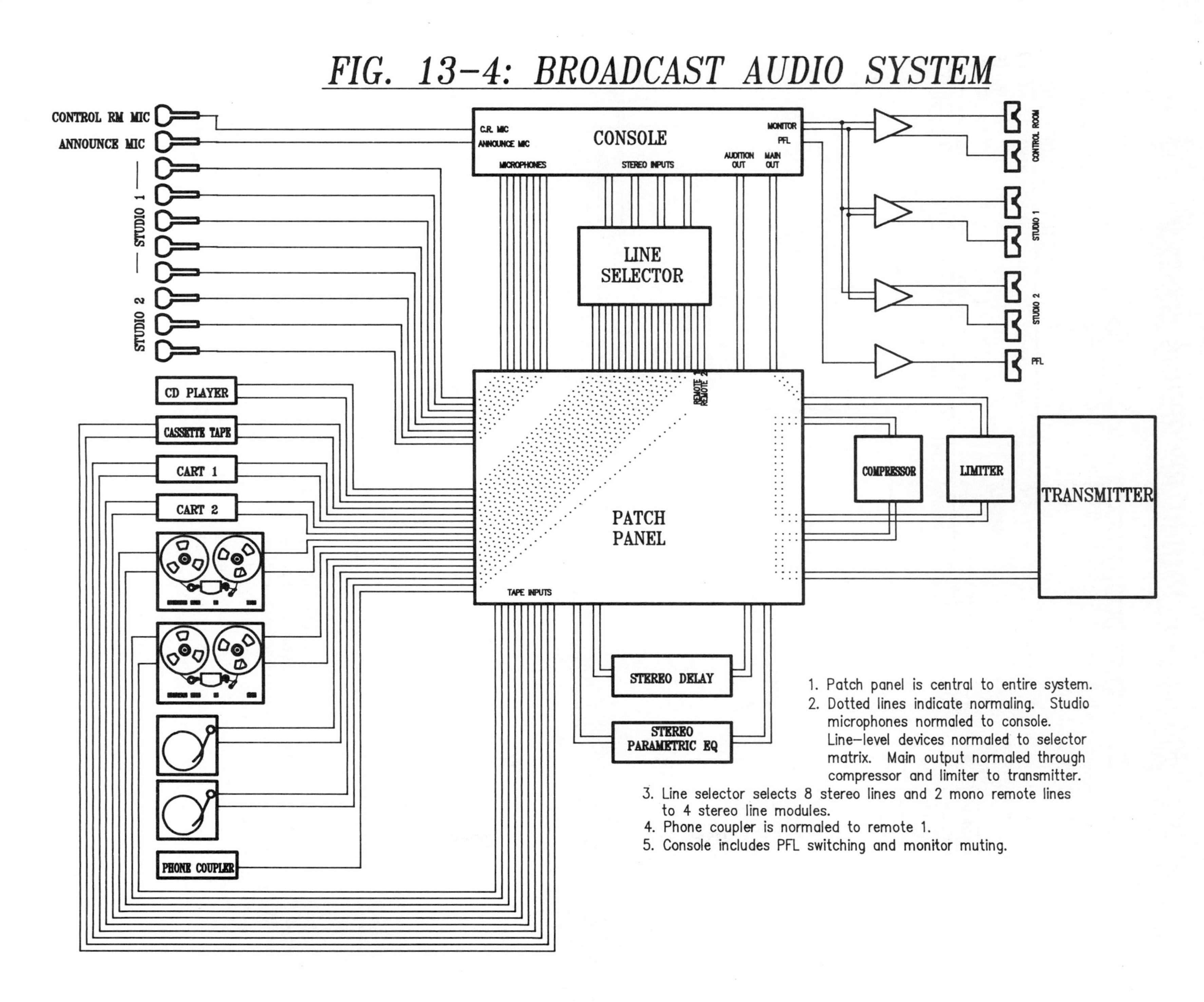

1. Patch panel is central to entire system.
2. Dotted lines indicate normaling. Studio microphones normaled to console. Line–level devices normaled to selector matrix. Main output normaled through compressor and limiter to transmitter.
3. Line selector selects 8 stereo lines and 2 mono remote lines to 4 stereo line modules.
4. Phone coupler is normaled to remote 1.
5. Console includes PFL switching and monitor muting.

FIG. 13-5: STUDIO RECORDING SYSTEM

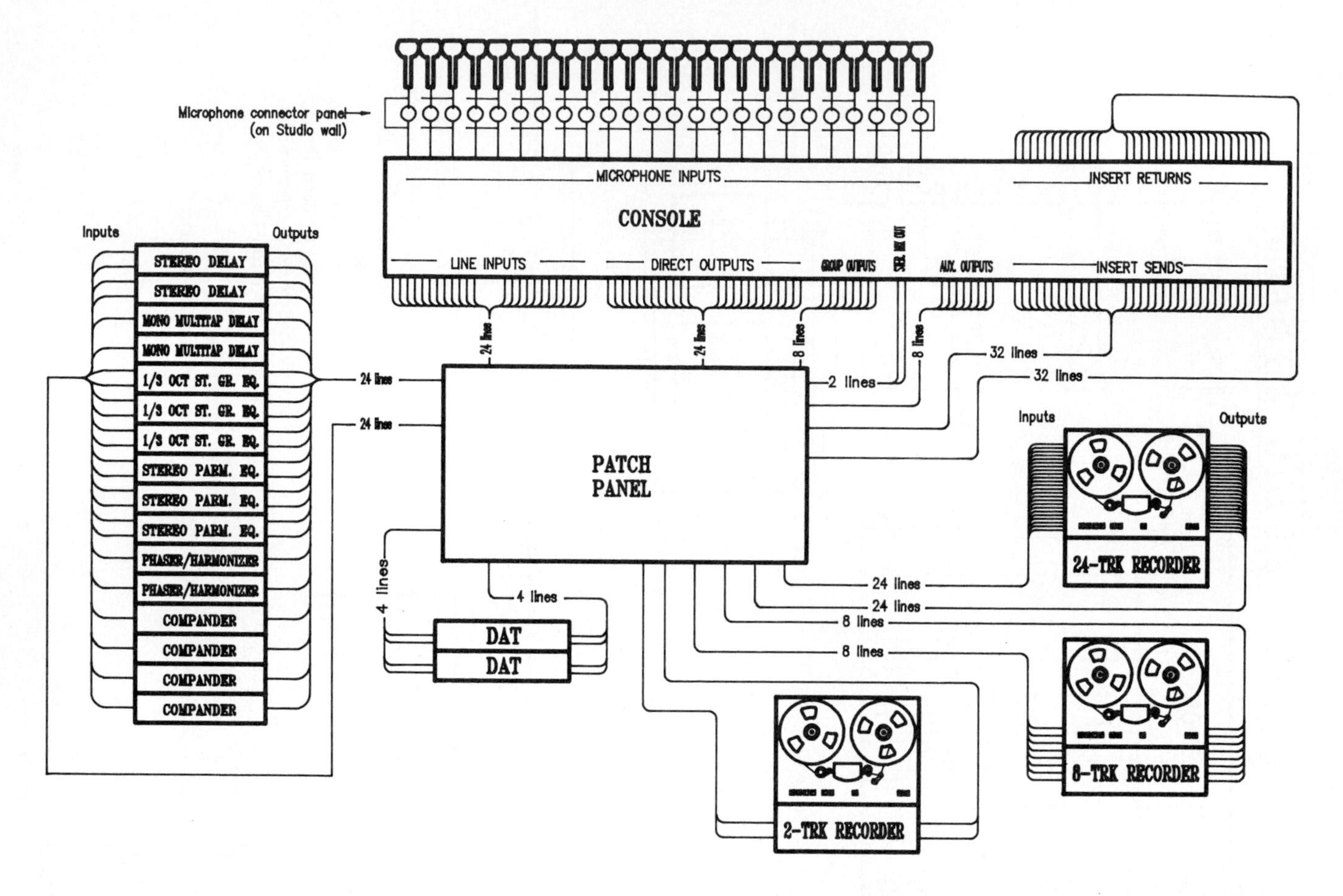

lobby music, however. In lobbies, we are essentially providing background music and the possibility for paging, so the constant-voltage idea is appropriate.

Domestic Music and Entertainment Systems

Domestic music/entertainment systems range from the very simple to the complex, though none duplicates either the complexity or the features of a professional audio system. Essentially, the purpose of a domestic music system is to provide ready access to a number of audio sources and to deliver the program to the listening space (or spaces) at whatever level and tonal balance the listener desires. Standard program sources for domestic music systems are record players (either tape, compact disc, or phonograph), radio (FM or AM), TV audio, microphones, and sometimes electronic musical instruments.

Domestic music/entertainment systems, while they may consist of a number of components, are fundamentally two parallel channels of audio mounted on a common chassis. (See Fig. 13-7.) One channel is stereo left and the other stereo right. The front end of the system provides loudness level and tonal balance and a means to switch from one audio source to another. The output is usually a very simple arrangement of power amplifier and loudspeakers. Such systems may be integrated, meaning that all components—phono player, preamplifier/input switching, power amplifier, etc.—are packaged as a single unit, sometimes with loudspeakers included. Other systems are made up of separate components so that the user can pick a specific group of desired components.

The control elements of domestic music/entertainment equipment are not suitable for theatrical use, and should be considered only where budget is the most pressing concern. The fact that home entertainment control systems do not permit mixing is the chief problem. Other items of home entertainment systems are frequently used in theatre sound—CD players, phono players, and power amplifiers. Tape cassette recorders used in home entertainment systems may be usable in theatre sound for some restricted tasks, but no other form of home entertainment tape recording system should be considered, even under severe budgetary constraints. Home entertainment tape recorders are simply too limited to be useful. Some do not have separate record and playback heads;

FIG. 13-6: BACKGROUND MUSIC/PAGING SYSTEM

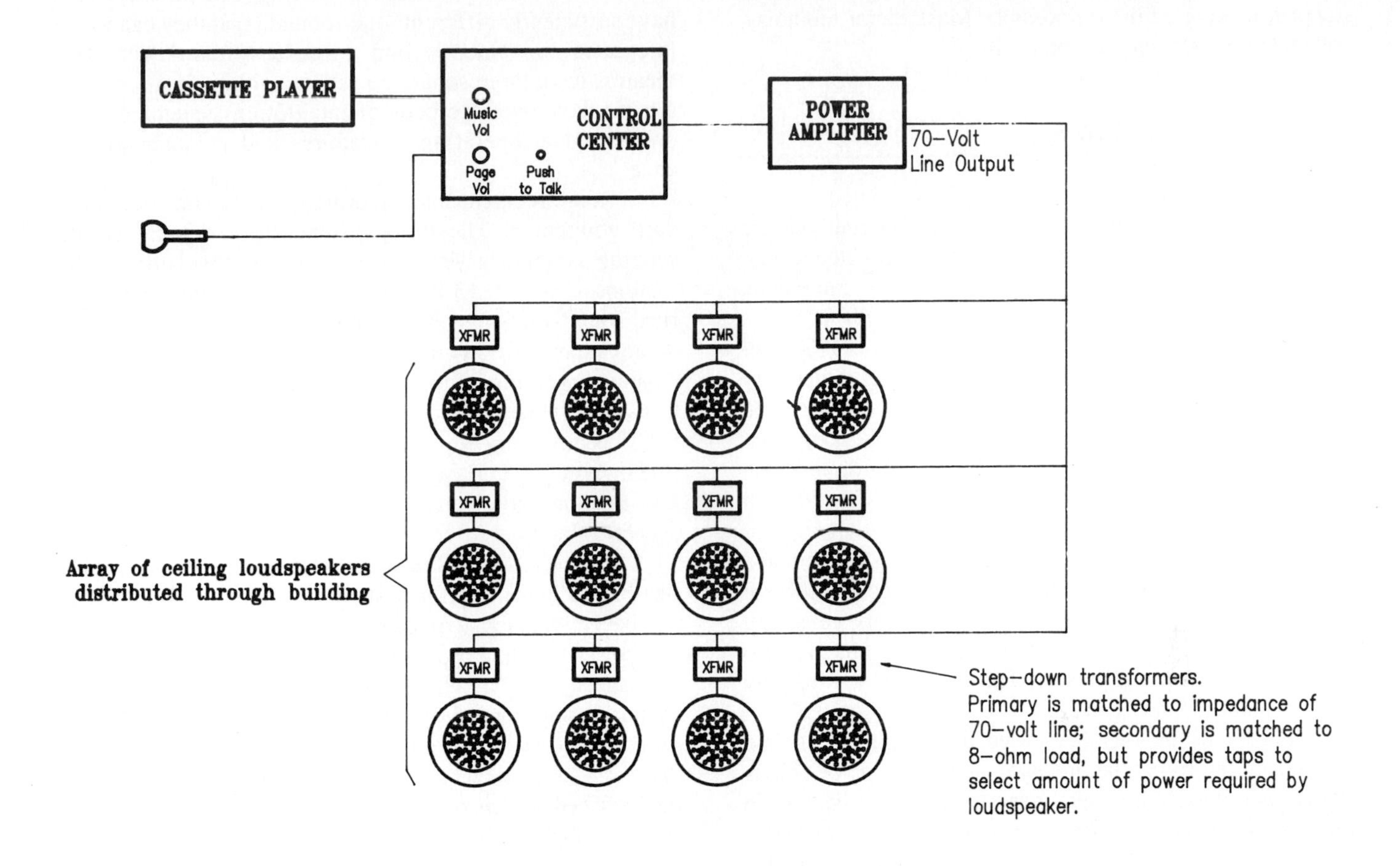

FIG. 13-7: HOME MUSIC SYSTEM

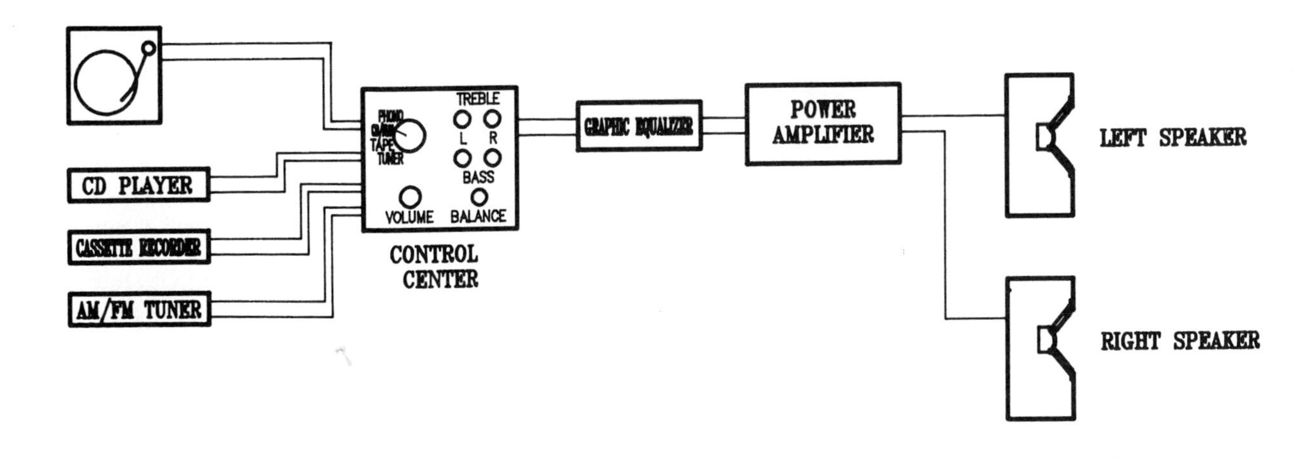

others disable playback amplifiers when tape motion is stopped, making editing impossible. Most are far too noisy, both electronically and mechanically.

Film Sound Systems

Film sound is in a state of change. Traditional methods are based on the technology of the vacuum tube era of electronics. Newer methods are based on the existence of integrated, digital circuitry and multitrack recording techniques.

Film sound mixing from the 1940s to the 1960s depended very heavily on fast, facile, live mixing under very stressful conditions. Mixers had to mix all of the sounds that might be included in a sound track: music, dialogue, environmental ambience, and Foley effects. Mixing was done in real time, often including live operations. The job was difficult and stressful, as one mistake could mean having to redo all or part of a section of the sound track. By the 1970s automated consoles had improved conditions. Operators could rehearse control changes, store those changes in memory cells, and recall them on demand. The process was analogous to memory storage of lighting presets. The final mix was usually made in real time, however.

Recent technology has eased the job of film sound mixing and improved the precision and the range of effects that are possible. Most operations need not be done live. In fact, addition of sound to film is customarily done in a process called **post-production**. Effects, music, and environments are processed and recorded individually. Using multitrack recorders and SMPTE Time Code, the sounds are placed at specific time points related to film frames. Finally, with audio recorders and film locked together under control of SMPTE Time Code, the master sound track is recorded and processed onto the film.

A film mixing system consists of multiple audio sources (microphones, specialized pickup devices such as vibration contacts, tape recorders, synthesizers, etc.); the mix-down console (which is usually very large); equalization, reverberation, compression, and other forms of signal processing; control switching equipment for starting, cuing, and stopping tape machines; and one or more "mag-stock" (magnetic film) or multitrack tape recorders and SMPTE Time Code generators/distributors. Consoles for film mixing systems are often custom-built. New consoles almost always incorporate computerized storage and control features.

Theatrical Sound Scoring Systems

Theatrical use of audio is unlike the use of audio in any other medium. In order to meet all of its audio needs, theatre really requires three different audio systems: reinforcement, recording, and sound scoring. The three systems have sufficiently different requirements that they cannot be integrated conveniently into a single system. Very few theatres have three separate systems, although some theatres do have two. Methods of integrating systems depend on individual operating procedures and personal preference.

Reinforcement and recording for theatre uses standard equipment. The third system, the theatrical **sound scoring** system, is very different from most other audio systems. (See Fig. 13-8.) Among audio systems, the theatrical sound scoring system is unique. At present, very few theatres have true sound scoring installations, although the concept is gaining in popularity and the number of installations is increasing. True sound scoring systems are rare, first, because very few manufacturers make such systems (the market for such systems is small), and second, because the functions involved are so unlike those of most other aspects of audio.

In effect, the theatre sound scoring system is a highly flexible audio distribution system, much like the output side of a background music/paging system except that the number and power-handling capability of the loudspeakers cannot be permanently fixed as with a music/paging operation. A theatre sound scoring control system should have a configuration *exactly the opposite* of most of the other consoles we have discussed so far in this chapter. If a typical audio console can be described as a *mix-down* system, a theatre sound scoring console should be classified as a *fan-out* system.

To create a satisfactory auditory environment for a dramatic production, a sound system needs to be capable of creating any kind of sound and making that sound appear to come from any distance and/or direction with any desired timbral quality—by no means a simple job. Because one of the stipulations is that sound should be able to come from any given direction (which must include the possibility of all directions simultaneously), the system requires a lot of loudspeakers. The system must be capable of sending signal to any or all of those loudspeakers at any time, regardless of what other activity is in progress in the system at the same time. The playback system must be capable of moving (panning) the sound from speaker to speaker smoothly in any order.

Sound scoring consoles usually accept any kind of line-level source as an input—CD players, phono disc players, synthesizers, tape recorders, signal processors, and other audio consoles. Usually, input is taken from one or more tape recorders. Signal processing is not normally included in sound scoring consoles. Most designers build equalization, level compression, or reverberation/echo onto the tape during cue recording and processing. For those occasions when it is needed, processing equipment can be inserted in the system, either in the input path, the output path, or as part of a return loop from output back to input.

FIG. 13-8: THEATRE SOUND PLAYBACK SYSTEM

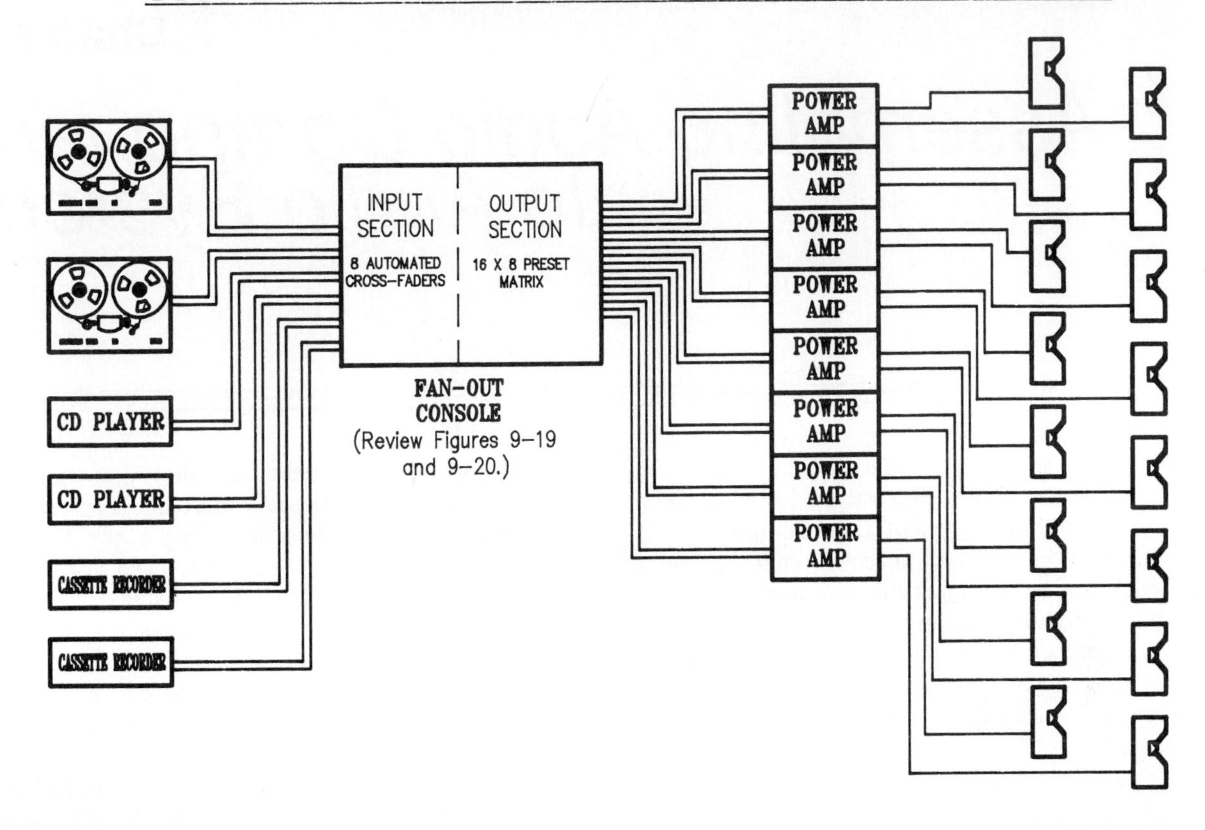

Summary

The overall discipline of audio contains many different system configurations. Most of these standard configurations are based on the principle of mixing several input signals together into a composite output. Two standard systems are frequently associated with theatre spaces: public address and reinforcement. Many theatres contain audio systems based on PA and reinforcement system design concepts. Reinforcement is appropriate where needed for the purposes of reinforcement, but is limiting for control of environmental effects and incidental music. Broadcast audio uses a very characteristic system format based on a number of fixed functions. Older theatres often have systems built around broadcast audio concepts. Studio recording utilizes a very flexible mix-down format. Recording consoles can be very flexible and will allow a great deal of creativity in building spot cues and environmental effects for theatre. Recording consoles, like all mix-down formats, are limiting for playback purposes. Film mixing systems are mainly recording systems adapted to handling a potentially large number of input sources.

Home entertainment systems and background music systems are very specialized configurations that are so specifically designed for their intended purposes that they have very little applicability to theatre.

Theatre requires three systems: reinforcement, recording, and sound scoring. The last is unique in almost all of audio. A theatre sound scoring system is a fan-out system that is designed to distribute sounds very flexibly so that an effect can seem to come from anywhere or everywhere.

Having gone this far in a discussion of typical audio configurations, we have not yet covered everything relative to the design and assembly of systems. Talking about the functions of systems and their typical signal paths provides little more than a meager knowledge of system interconnection. We still need to discuss matters relative to assembling components into systems. Those topics are covered in Chapter 14.

Chapter 14

Assembling Audio Components into Audio Systems

Chapter 13 gives a general overview of audio systems. In order to integrate the information gained in the detailed examination of individual types of audio equipment, this chapter will again concentrate on entire audio systems, directing attention to the characteristic ways in which equipment is configured to form systems. The first step in this process requires a look at one aspect of audio systems that was not given detailed consideration in previous chapters, namely, the methods of interconnecting pieces of audio equipment.

Patch Systems

A small system consisting of only two or three pieces of equipment needs only the connectors built into each unit and the appropriate mating cables, as replugging a small system to handle various tasks is not particularly onerous. Larger systems, however, can require hundreds of cables, and unless some means is provided to facilitate the job, replugging becomes a nightmarish occupation that takes far more time than it could ever be worth. An audio system without any simple means of restructuring the signal flow is essentially locked into a single function.

The customary means of providing flexibility in signal routing is a matrix of plug points called a *patch panel.* Figure 14-1 illustrates an audio system configured with a patch panel. Every input and every output in the system appear in this matrix. To connect one piece of equipment to another, the operator need only plug a patch cord from the appropriate output jack to the appropriate input jack. Consider, however, what would happen in a large audio system in which even the most rudimentary signal path had to be patched together every time it was needed. The system would be painfully frustrating and tiring to use. In order to confer ease of use without restricting flexibility, patch panels usually incorporate **normaling**.

Normaling is a method of allowing a signal to flow *through* a patch point when no patch plug is in place. Fig. 14-2 illustrates the way in which normaling works. The jack contains not only the contacts designed to mate with an appropriate patch plug; it also contains an additional set of spring contacts that close against the plug contacts when the jack is empty. These contacts are called *normaling contacts.* Normaling contacts of one jack are wired to those of another, permitting the signal to flow *as though a patch cord were plugged between those two points.*

Insertion of a patch plug into a jack pushes the plug contacts away from the spring contacts and breaks the normaling at that point. If the jack receives the output from a device, the patch cord will carry that output signal to some point other than the usual destination. If the jack serves an input, the signal arriving through the patch cord will replace the signal path usually accessed through the normaling contacts.

Patch jacks and plugs are made in a variety of contact configurations. The appropriate configuration for any given application is based on one or more of several factors: configuration of transmission lines, patch density, frequency of repatching, and cost of installation, among others.

Transmission line configuration is an absolute determinant of jack configuration in that balanced lines require two signal contacts plus a shield contact. Figure 14-3 shows the ways in which a balanced line can be accommodated in a patch panel. Note that a balanced line may be served by either a single plug or a double plug. The double plug uses only tip and shield contacts but has two of each; the single plug uses tip, ring, and shield to accommodate the two signal conductors and the shield contact both in a single plugbody. (Without exception, tips and rings carry the signal and shield is the ground or earth contact in audio work.) Unbalanced lines require only tip and shield in a single plug.

Large patch panels require a large number of jacks, and use of double plugs for balanced lines effectively doubles size and cost. Clearly, if either physical size or expense are limiting factors, tip-ring-shield, single-plug format is the obvious

FIG. 14-1: PATCH PANEL INTERCONNECTING SYSTEM

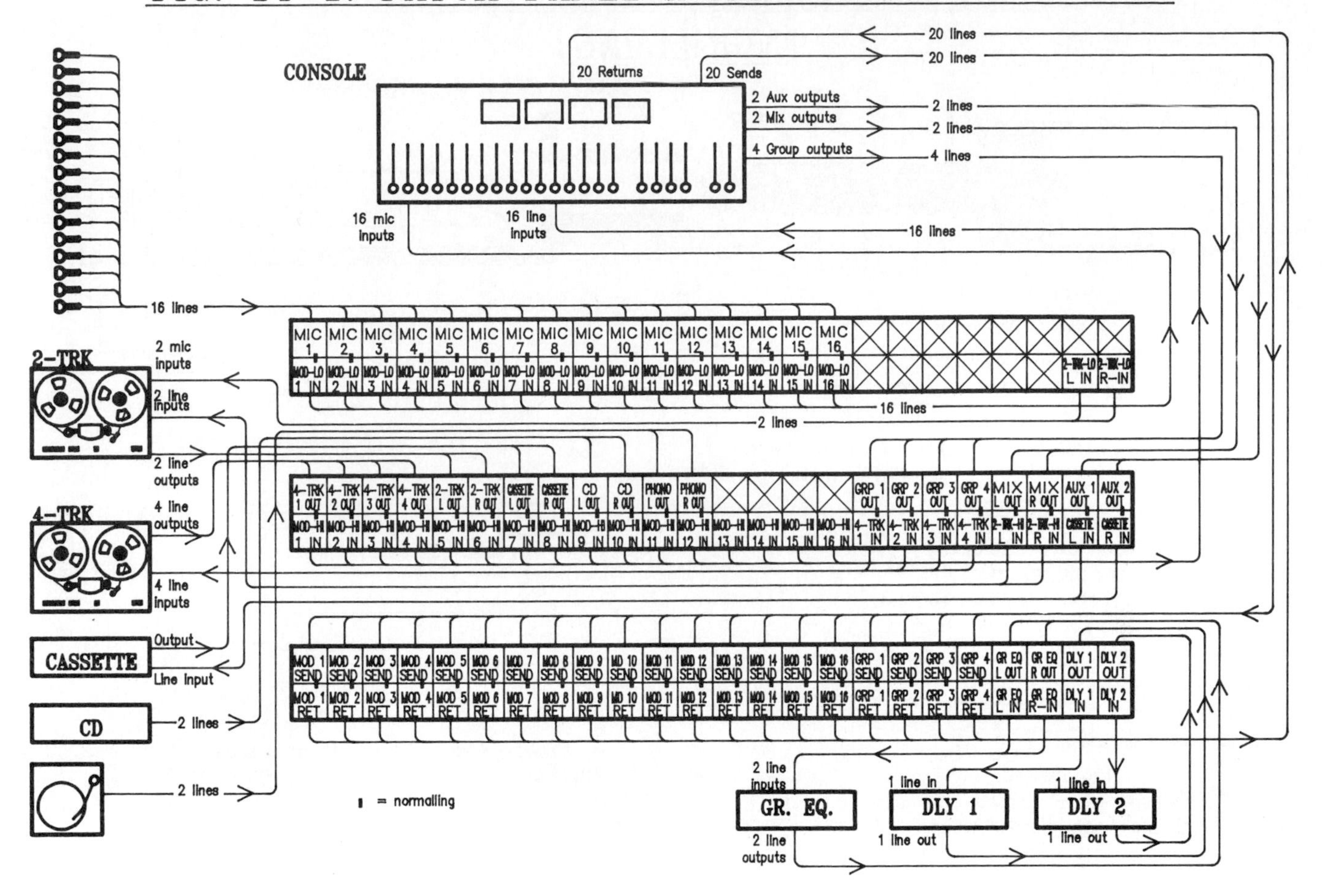

choice for balanced lines. Similarly, jacks with normaling contacts are more expensive than jacks without them; therefore, the amount of normaling used in a patch panel contributes directly to cost. On the other hand, the time required to patch for each use of the system cannot be ignored as a factor in what the system, once installed, will cost to use.

Organization of Patch Panels

Patch panels are arranged in whatever configuration best suits the particular installation, and few rules exist to govern placement of functions. One fairly common organizational approach, based on the idea that microphones, record players, and CD players are exclusively signal sources, is to locate outputs at the top of the patch panel aligned with console inputs in the row immediately below. Subsequent outputs and inputs follow in the succeeding rows until components with only electrical inputs are reached at the bottom row of the patch panel.

Transmission lines for all forms of equipment—microphones, mixers, signal processors, power amplifiers, and even loudspeakers—can be combined in the same generalized patching system; however, several potential problems result from this practice. First, the danger of connecting a power-level output to a mic-level or line-level input clearly exists. Granted that where only experienced hands are likely to use the system this danger is minimized, but it is always present. Second, the larger the system, the greater the chance of mistake in locating a particular patch point; and the time spent in finding patch points will necessarily increase with size, even if one is familiar with the system.

A practical solution is to divide the patch panel into separate groups: mic lines and their appropriate inputs in one panel, high-level outputs and inputs in a separate panel, and, possibly, connections between power amp outputs and loudspeaker lines in a third panel. The separation serves the purpose of making inappropriate or dangerous interconnections more difficult, as well as of organizing the patch system into somewhat homogenous groupings.

FIG. 14-2: PATCH JACKS AND PLUGS (NORMALING)

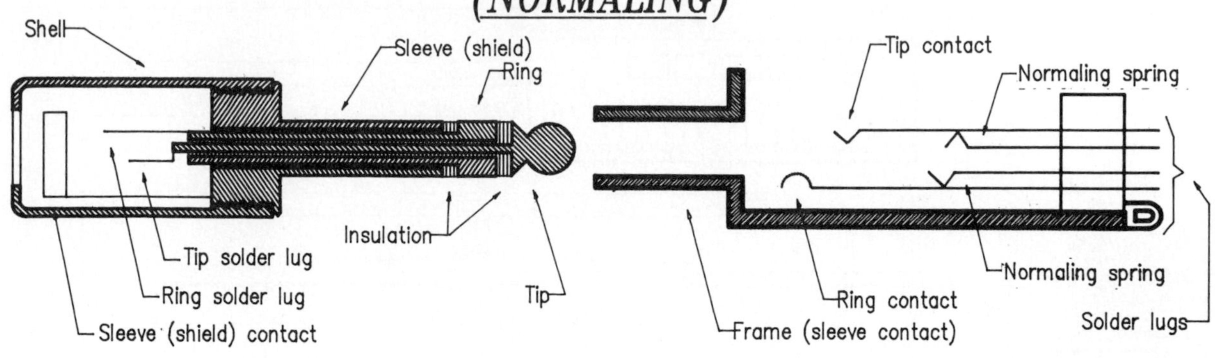

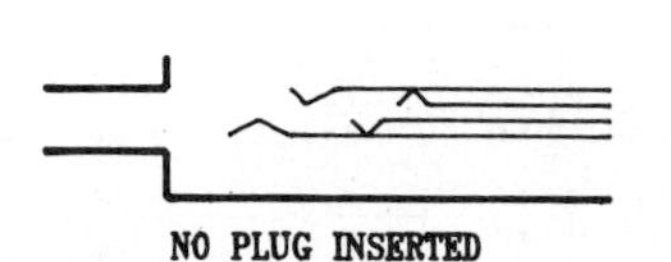

NO PLUG INSERTED
Signal flows between contact springs and normaling springs.

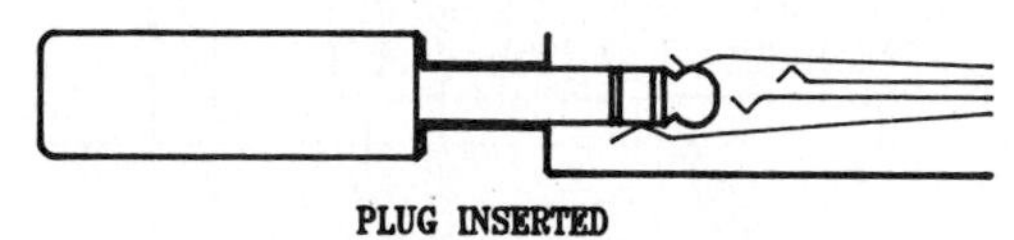

PLUG INSERTED
Connection between contact springs and normalling springs opened. Signal flows between contact springs and plug.

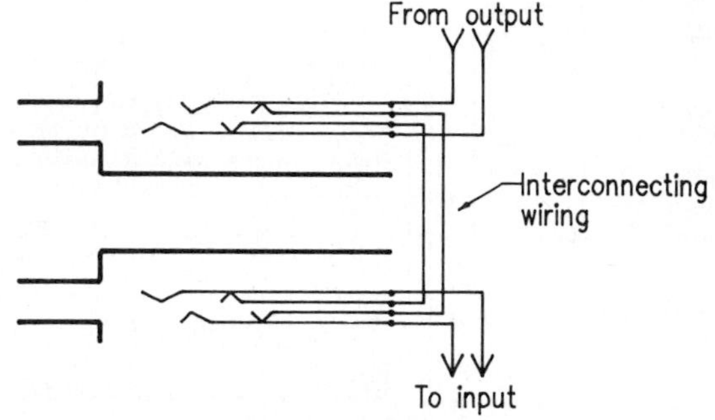

NORMALED WIRING BETWEEN TWO PATCH JACKS
Normaling contacts conduct signal from device A output to device B input with no patch cords used. Inserting a patch cord into either jack will open the contacts and interrupt normaling.

Interfacing System Components with the Patch Panel

Cabling from a patch panel to system components involves only running appropriate (usually two-conductor, shielded) cabling between the patch panel and the various system components. The cables are wired to appropriate connectors at the component end and soldered to the patch jacks at the panel end.[1] Although it sounds simple enough, a few significant concerns do exist. One is to ensure that when a connector is plugged into a piece of equipment, some internal function of that equipment is not changed or disabled. Another is to earth cable shielding properly so that it does not create partial short circuits or hum-producing ground loops. Further, we must always be concerned with the pattern of normaling that makes a most-common configuration possible without patching.

An example of preserving normal equipment function is illustrated by the process of bringing module inserts out to a patch panel. Inserts are simply localized, *normaled* patch points—send jacks normaled to receive jacks inside the module. If a plug is placed in either jack, signal flow through the module is broken, as with any normaled patch point.[2] Of course when we use the console's insert connectors to carry those functions to the system patch panel, we break the internal normaling; therefore, if we replicate a module's inserts at the patch panel, we must also provide the necessary normaling in order to maintain signal flow through the module.

[1] Soldering directly to the patch jacks makes modifications to a system difficult. A more common practice is to wire patch jacks to an interface connector—either to the so-called "Christmas-tree" connectors or to barrier strips. Moving connections on barrier strips or in a Christmas-tree harness is not as difficult as trying to move or replace individual connections directly at the patch jacks.

[2] Some consoles provide break in normaling only when a plug is inserted in the receive jack. In this way, the send can be used without disturbing the normal, internal feed through the module. Only when the receive is plugged does the signal through the output half of the module originate from an external source.

FIG. 14-3: BALANCED LINES IN PATCH PANELS

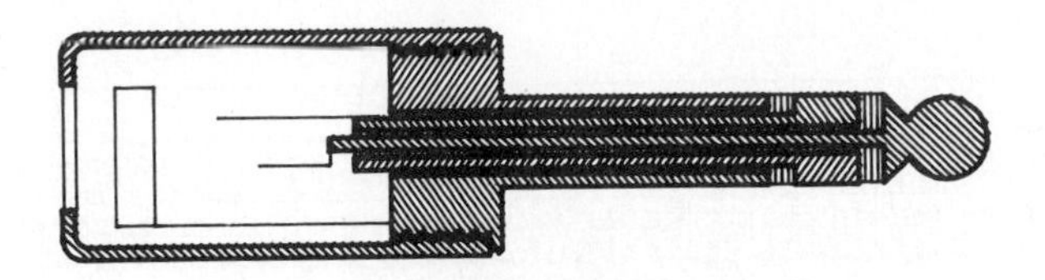

A. SINGLE PLUG
Balanced lines handled in a single-prong plug. Ring and tip used as hot lead connectors; sleeve (shield) used as neutral connector.

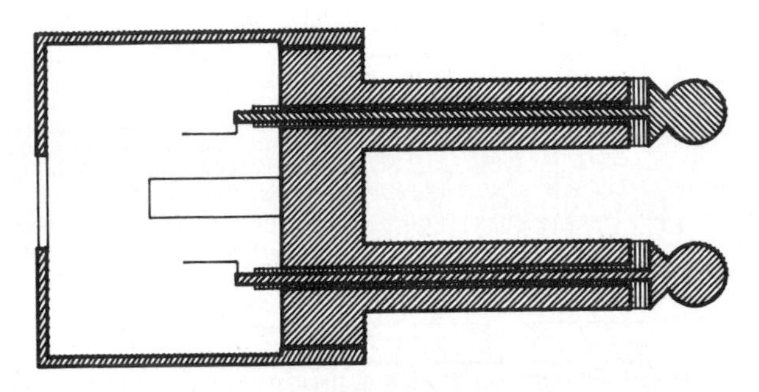

B. DOUBLE PLUG
Double plug provides two tips for hot connections. Sleeves (shield) used for neutral. Wider shell provides better grip to insert and pull plug. One edge of shell is sometimes knurled to help identify polarity.

Shielding is of concern because the more cable a system uses the more likely the system is to pick up noise. Because earthing shields is of concern to more than just the patch panel, it is covered in detail later in this chapter.

Whenever signals from any source or group of sources are commonly fed to specific positions on a console, those outputs are generally normaled to the particular console inputs. For example, if a studio habitually monitors the output of a multitrack tape recorder by connecting the track outputs to console line inputs, those outputs and inputs would be normaled together to facilitate such practice without the use of patch cords. Other possible uses for normaling occur at the interface between a distribution console and a group of power amplifiers, or between the tape recorders most often used in performance and the inputs to the distribution console.

Normaling is used when a line must feed through one point on the way to another. For example, in Fig. 14-4 we see an auxiliary patch panel used to facilitate the connection of processors to module inserts. The processor inputs and outputs, however, also appear in the main patch bay. Normaling contacts in the auxiliary patch are used to feed lines through to the main patch. A patch cord inserted into a processor plug point in the auxiliary panel will break the line to the main patch, so that one cannot put two inputs into the processor at the same time or split the processor's output to two different loads.

Complex System Interconnection without Patching

Patch panels are items to be used in fixed studios. They are not readily adaptable to portable systems that must be carried from theatre to theatre and applied to a variety of use conditions.

In a large, portable system, no way exists to avoid a long, and possibly tedious, hook-up procedure. The process can be facilitated somewhat by the use of multiconductor cables. Multiconductor cables—often called "mults," "multis," or "snakes"—are just what their name implies, a lot of individually shielded pairs of audio cables housed inside a common jacketing. Mults commonly have a plug box at one end that provides a connector (usually a female type XLR-3) for each shielded pair in the cable. At the other end—the "breakout"—the jacketing is cut back 18 inches to 2 feet, and each shielded pair is individually jacketed (either with heat-shrink tubing or electrical tape). Strain relief and a connector (usually

FIG. 14-4: NORMALING USED AS FEEDTHROUGH BETWEEN RACKS

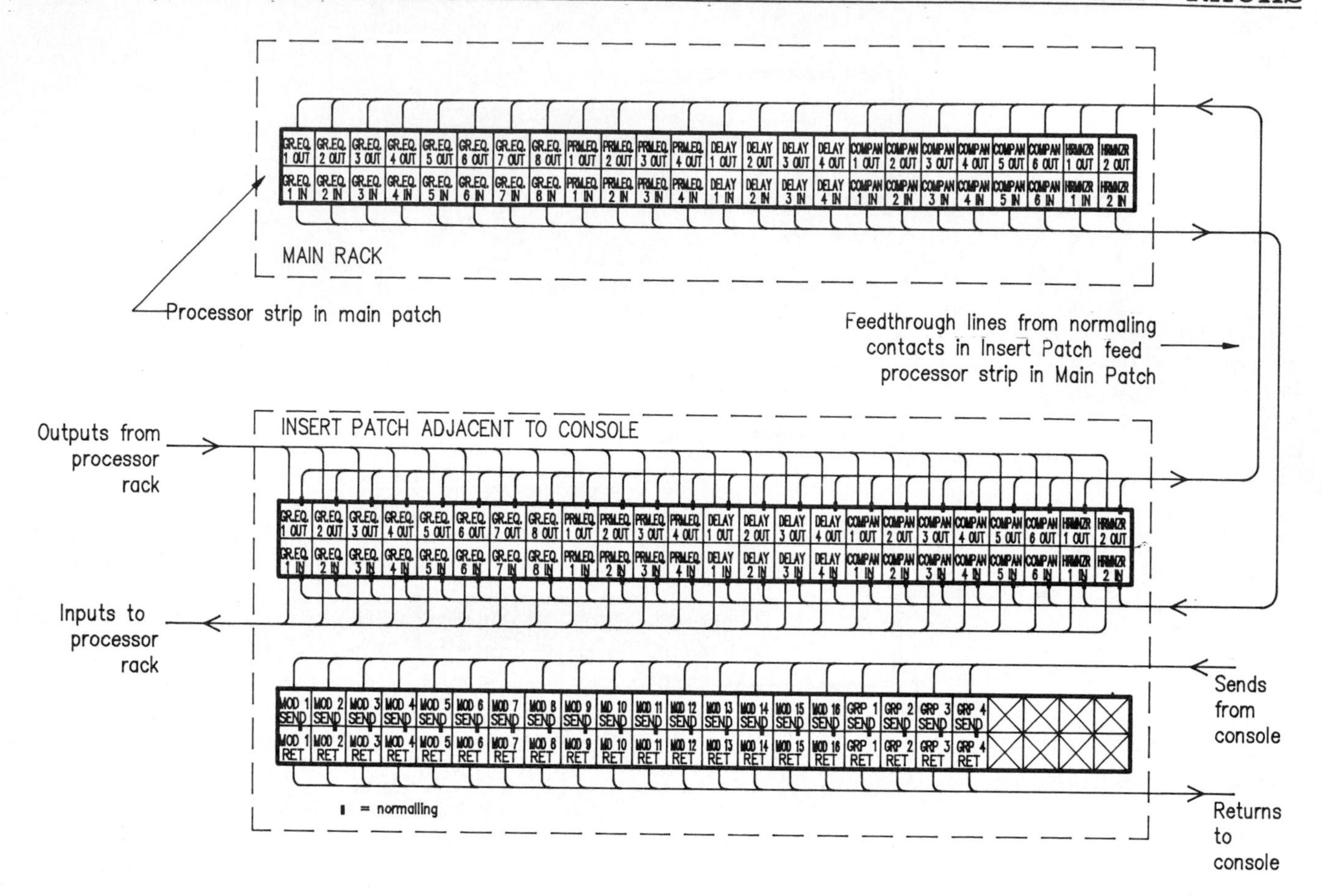

a male type XLR-3) are provided for each pair. A more expensive but more flexible procedure is to terminate the cable in a multipin connector. The plug tails are wired to a mating connector. Terminating a multiconductor cable in a plug helps extend the life of the cable, reduce wear on plug tails, and makes possible gender or type substitutions in the connectors placed at the ends of the individual pairs in the breakout.

Multiconductor cables are frequently used in reinforcement systems to carry microphone lines from stage to mixer and to carry line-level signals from group and matrix outputs to power amplifier racks. Mults may be employed to carry signals from the primary mixer to a monitor mixer, or from a mixer to its processor rack (if one is used). Individual jumper and adapter cables are used to connect from plug boxes to equipment input and output connectors. (Careful labeling of *All* connectors and cables is critical to retaining any degree of sanity when working with such hookups.) For companies on the road, use of mults makes life a great deal simpler. If all connectors at each end of the mult are appropriately labeled, or if the connections to a rack terminate in a multipin connector, setup time is reduced by several orders of magnitude and troubleshooting becomes much simpler.

Interconnections and Shielding

Whether interconnection is through a patch panel, multiconductor cabling, or individual cabling, a number of general considerations must be kept in mind regarding **shielding**. Shielding is the braid or foil covering wrapped around inner signal conductors in an audio cable. The purpose of the shield is to earth inductive field energy from power level or radio frequency sources in order to keep the corresponding noise that either kind of interference can produce out of the audio lines. Examine a length of microphone cable and you will observe that it really contains three conductors: the two signal lines and the shield. The shield may be a surrounding braid of wire, helically wrapped wire surrounding the inner conductors, or a casing of metal foil with a bare third wire conductor as part of the group of inner conductors.

A common mistake that even relatively experienced operators often make is to confuse the shield with the transmission line's neutral. The two are not necessarily identical and should not be presumed to be interchangeable. If single-conductor shielded cable is used, for example, to run an unbalanced line, then the shield necessarily serves as neutral (one reason why single-conductor cable is not good practice). Using a two-conductor shielded cable for the same purpose, the shield no longer serves as neutral but only as a shield against inductive field energy.

To accomplish its purpose, a shield should be tied to earth *at only one point*. If one attempts to earth the shield at more than one point, earth itself may provide sufficient resistance to develop a significant voltage at the noise frequency. This phenomenon is known as a *ground loop*. Recall that

$$E = IR$$

Power frequency or radio frequency energy fields induce current (I) into the shield. So long as the resistance (R) in the shield circuit is very low the developed voltage (E) will be very small. If the shield circuit resistance increases, so does the noise voltage.

Determining the point at which to connect a shield to earth is extremely important in minimizing system noise, and several factors, mainly related to the relative configurations of input and output, are essential to the determination.

Inputs and outputs in modern audio equipment may be of three kinds: unbalanced, balanced, or differential. These terms should be familiar by now, but let's review them quickly.

An unbalanced configuration consists of one neutral conductor and one conductor under electrical tension. A balanced line is composed of one neutral conductor and two conductors under electrical tension, one positive and one negative. A differential configuration consists of two floating conductors that respond differentially (in opposite polarity) around a virtual neutral. The surest way to avoid trouble is to connect a line to an input or output of similar configuration—balanced with balanced or differential, differential with differential or balanced, and unbalanced with unbalanced. Notice that differential and balanced lines mate easily, but balanced and unbalanced lines do not. Balanced and unbalanced lines can be mixed—*providing that various system components do not share a common earth and that the shielding in the interconnecting transmission lines is not connected in such a way that it shorts out one half of a balanced or differential input or output.* (Review Fig. 7-2 and the associated explanation.) Table 14-1 shows where to connect the shield for every possible combination of input-to-output configuration.

Connectors and Adapters

Several forms of audio connectors are used in audio in the United States and Canada, the most common being XLR-3 male and female, phone plugs and jacks, and RCA pin plugs and jacks. (Plugs are male; jacks are female.) Sometimes audio equipment will provide only barrier strip terminals for connecting inputs and outputs.

XLR-3 connectors are almost always used for professional quality microphones and for the majority of balanced transmission line connections. Differential inputs and outputs also typically use XLR-3 type connectors. The wiring patterns for these connectors are shown in Fig. 14-5. Note that pin 1, if used, is always shield, but an alternate shield connection exists. The difference is that pin 1 does *not* connect to the plug's case, whereas the earth connection does. The alternate connection is used in those cases where the connector must earth the shield without connecting it to the local neutral.

Phone plugs and jacks, like patch plugs and jacks, are made in balanced and unbalanced forms. The unbalanced version has only tip and shield, whereas the balanced version has tip, ring, and shield. RCA pin plugs are used only for unbalanced applications. They are most commonly found on home entertainment devices and on some pieces of semi-professional equipment.

Because two pieces of equipment, selected at random, are likely to have different connectors, adapters are often needed in interconnecting an audio system. Even in a permanently wired studio, pieces of equipment that come in temporarily (instrument amplifiers for guitars, basses, keyboards,

Table 14-1. Shield Connection Points

Source (Output)	Input	Shield Connects To
Unbalanced	Unbalanced	Input
Balanced	Unbalanced	Input
Differential	Unbalanced	Input
Unbalanced	Balanced	Source
Balanced	Balanced	Input
Differential	Balanced	Input
Unbalanced	Differential	Source
Balanced	Differential	Source
Differential	Differential	Input

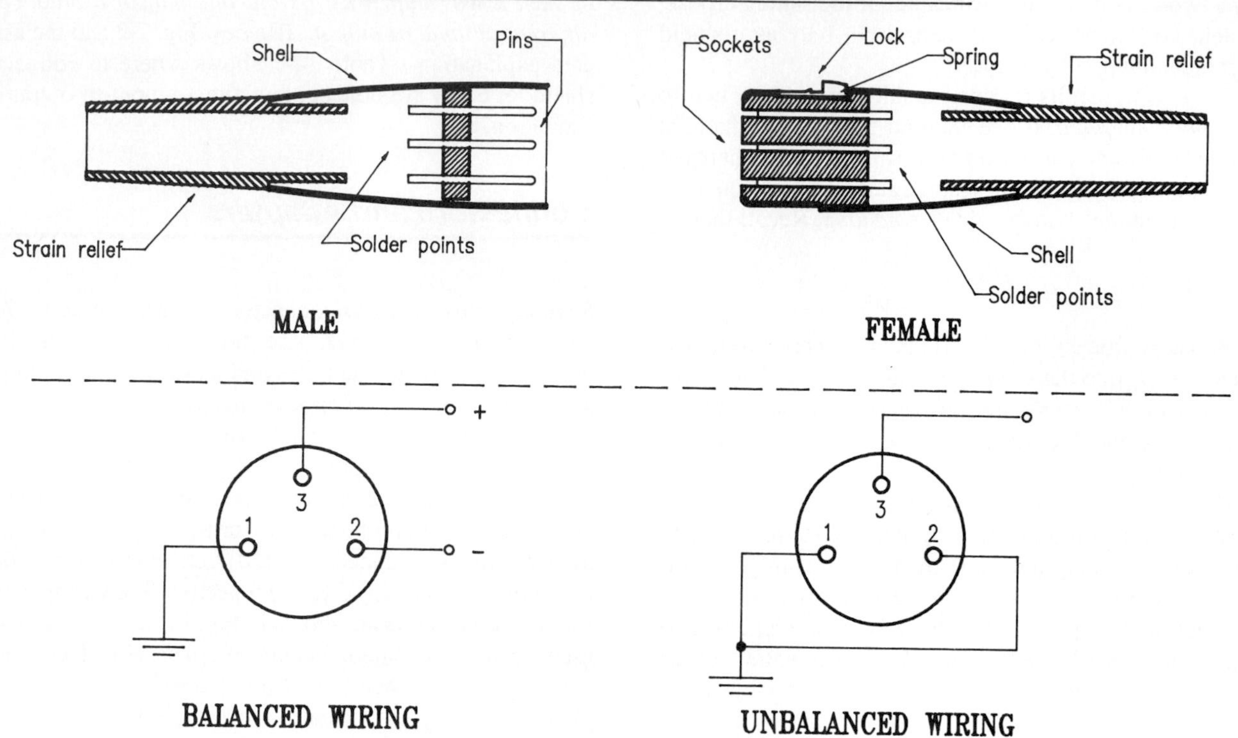

NOTE: Disagreement exists over whether pin 3 should be positive or negative. Some companies wire connectors with pin 2 positive and pin 3 negative. In an unbalanced configuration using pin 2 as positive, pin 3 is shorted to pin 1.

synthesizers, etc.) must be connected into the system for the time they are used. Thus, no studio can afford to be without at least a few adapters. Adapters can be fabricated in-house or can be purchased. Commercially made adapters have the advantage of being compact—often constructed within a shell not much larger than that of one kind of connector.

Adapters are designed to meet at least three problems: to interconnect pieces of equipment with different plugging formats (e.g., XLR to 1/4-inch phone plug), to change "gender" within a single format, and to branch one connection to two (sometimes three) others.

Adapters that convert one plug format to another are the simplest. They consist of a small length of cable with one kind of connector at one end and another kind at the opposite end. (As noted earlier, professionally made adapters usually take the form of a single shell housing both forms of connector.) Variations and permutations are necessarily many. For instance, a given situation may require a male of one kind of connector to a female of the other kind. Other situations could require that both types of connectors be either the male or the female of its particular type; and, though XLR, phone, and RCA plugs are the most frequently encountered, many other kinds of connectors exist and will be met from time to time.

"Gender changing" means altering the sequence of male-to-female connectors within a given type of connector. For example, microphone cable provides a female XLR at one end and a male XLR at the other. Occasionally, some unusual circumstance will present the need to cable between two connectors of the same gender. To do that, one needs an adapter having connectors of the same sex at each end (male to male or female to female). Such an adapter is called a gender changer or a turnaround.

Branching permits one source signal to be split to two load inputs or combines two source signals into one load input. (Sometimes a three-way branch is used, but not often.) Branching adapters are more commonly called "Wyes," "Y-connectors," or "twofers." Usually, branching adapters have the same kind of connector at each end, but they may also have different kinds of connectors at each end. Branches may be hermaphroditic or may be gender changers.[3]

[3]Using contemporary equipment, branching is a reasonably safe practice; but some caution must always be used in branching. The immutable fact is that placing two loads on one source output or mixing two sources into one load input alters the impedance relationships in the circuit—something that can potentially affect signal transfer, frequency response, and distortion.

System Configuration

The remainder of this chapter provides examples of typical system hookups illustrating the connections of individual pieces of equipment into total working systems. The examples use typical theatrical installations—a reinforcement system, a recording installation, and a sound scoring system.

Reinforcement

Reinforcement systems can be part of a permanent installation, but frequently are assembled as needed for particular productions. For the purpose of illustrating a system assembly without a patch panel, our reinforcement system example will be a rental package, specified for a particular production.

The reinforcement system will use 28 microphones (6 wireless, 7 foot mics, and 5 hanging shotguns onstage and 10 mics in the orchestra pit) plus 1 contact pickup and 1 feed from a direct box. The wireless microphone system will require a receiver for each channel, two diversity antenna systems (one diversity unit can only accommodate four microphone channels). The system will drive a central tri-amplified loudspeaker cluster, four side-fill speakers, and a group of four monitor (foldback) speakers. The power amplification rack includes nine power amplifiers and one three-way electronic crossover. Processing includes two digital delay units, three one-sixth octave graphic equalizers, and two parametric equalizers. The console provides 32 input channels with an 8 x 8 matrix output. (A block diagram of the system is shown in Fig. 14-6.)

To interconnect this equipment, we need a 27-pair multiconductor cable 250 feet long to run the stage and pit microphones and also the signals from the contact pickup and the direct box to the console. (Note that the wireless microphones are not included in this group, as their receivers can be stationed at the operator's position.) We will need 25 individual microphone cables ranging in length from 10 feet to 150 feet to connect the stage and pit mics, the contact pickup, and the direct-box output. The direct box will require an adapter having one end in an XLR-3 male and the other in a two-conductor (tip-shield), 1/4-inch phone plug. We will also need eight microphone cables 10 feet long to connect wireless mic receivers into the console. An additional 250-foot, 12-pair

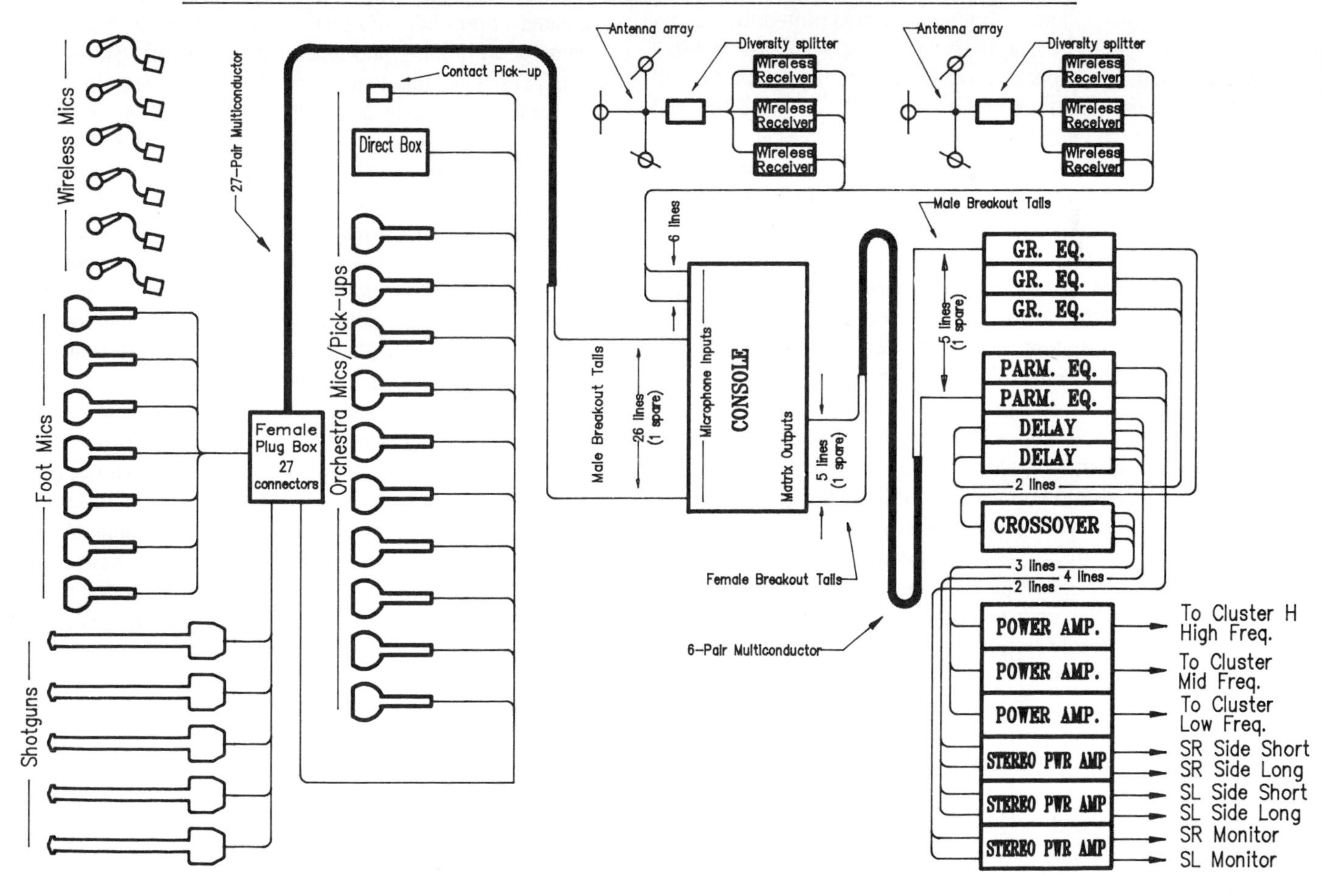

FIG. 14-6: TOURING REINFORCEMENT SYSTEM

multi will be used to run outputs from the console to the power amplifier rack. To connect from the rack to the speakers, we will use three 20-foot cables, two two-fers, one 50-foot cable, five 100-foot cables, and two 150-foot cables (three cables for the cluster, one for each side fill, and one long run plus a twofer and 20-foot jumper for monitor at each side of the stage). We will also require nine cables of 10 feet each to connect power amps to the multi to the power rack. The diversity antennas will require three coaxial cable runs each of 150 feet to connect the antenna mixers to the individual dipole antenna units—a total of six runs of coax cable.

On arrival at the theatre, all cases and boxes are unpacked and checked against the original order and the packing slip. Any important differences between the two are noted *in writing*. The equipment is positioned and cables are stretched according to length. We find that both multis come with a multiconductor male plug at the breakout end. The breakouts themselves drop out of mating female plugs. The breakout for the 27-pair cable provides 27 male XLR-3 connectors, and the breakout for the 12-pair multi uses female connectors to mate to the console outputs.

Each end of each individual cable, each adapter and twofer, each plug box connector, and each dropout plug is explicitly labeled to show what its function is and the point to which it should connect. The multiconductor runs are then pulled into position and dressed for minimum visibility. Mic mounts are placed and individual mic runs pulled from plug-box to mic position. The individual cable runs are bundled and dressed. Loudspeakers are hauled up into position, and cables are strung to them, dressed for minimum visibility, and connected to power amp outputs. The jumpers from the 12-pair plug box are connected to power amp inputs. These connections require gender changers, as the inputs to the power amps and the outputs from the mult plug box both are XLR-3 females.

At the control desk, the breakout tails are connected to the multis and the drop plugs inserted into the appropriate connectors on the console. The processor rack (which may also contain the wireless receivers) is positioned and plugged to the appropriate inserts on the console. (The digital delays are connected into the lines to the power amps driving rear side-fill speakers; the graphic equalizers are inserted in the group modules driving left and right fills and the cluster. The parametrics are inserted in two group modules used as sub-masters for the pit orchestra.) Because the inserts on the console are 1/4-inch phone jacks and the inputs and outputs on the digital delay units use XLR-3 connectors, adapters are required.[4] Wireless receivers are connected to mic inputs on the console. Diversity antenna splitters for the wireless receivers are set up and coaxial cable is pulled to the sites for the dipole antennas for each diversity unit.

Finally, when all cables have been plugged, the system is powered and an initial check is run to test all cabling and all pieces of equipment. At this point, problems in cabling, whether plugging errors or cable discontinuities, must be discovered and fixed. Finding problems is always simplified by meticulous and careful labeling.

Recording

A recording system is usually permanent, which means that it needs a patch panel. Planning a permanent installation is somewhat different than specifying a rental system for a single production. The most important difference is that flooring, walls, conduit, and cabinetry all help to protect and conceal cable runs.[5] Because such protection for cabling is available, and because two-conductor shielded cable is much less expensive than multiconductor cable, interconnection in a permanent installation is usually done with individual cable runs rather than with multiconductor cable.

Obviously, considerable time must be spent in planning the system layout for optimum flexibility and workability. The most used pieces of equipment need to be readily accessible from the console. The patch panel needs to be as centrally located as possible and positioned so that patch changes can be made as quickly and as easily as possible. To use a patch panel easily, its labels must be clearly visible, which implies placement as nearly at eye level as possible, and in some location that can be easily lit. Once position layout is decided, the number and length of cable runs must be worked out.

The system in this example uses a 16-input console with four group outputs, two auxiliary outputs, and stereo mix outputs. The system includes two two-track stereo, open-reel tape recorders, one eight-channel multitrack tape recorder, and two stereo turntables (locally preamplified). The system processing equipment complement has two digital delays, two sixth-octave graphic equalizers, and two compressors. A stereo, 50-watt/channel power amplifier is used to drive two wall-mounted monitor loudspeakers. A second 50-watt/channel power amp is available to drive loudspeakers placed in an adjacent studio room or on the stage. The control room receives 10 microphone lines from the studio and 10 microphone lines from the stage. A pair of loudspeaker lines runs from control room to the studio, and another pair runs to the stage.

The patch panel for this system will need to accommodate 20 microphone lines from stage and studio and 16 micro-

[4]Technically, the cables required to connect the equalizers to the console inserts are gender changers because both ends of those cables must be in male 1/4-inch plugs; however, pieces of equipment that use this style of connector always use jacks, not plugs, so that male-to-male phone plug cables are standard and do no constitute any kind of adapter.

[5]A note of caution: Electrical Code regulations apply to the installation of audio equipment. Especially in first-time installations in new buildings, every possible attempt should be made to conform to Code; but planning should always consider the need for future expansion by oversizing conduit and minimizing the number of right-angle turns in conduit runs.

phone inputs to the console; 16 line-level inputs to the console, 16 direct output lines from each console input module, 16 inserts from each input module (a total of 32 places because an insert consists of one send and one receive); four group output lines and the group inserts (one send and one receive per group for a total of eight places), left and right stereo output lines, three lines from each delay unit (one input and two outputs), and two lines each from the compressors and equalizers (a total of eight). Inputs to the tape recorders require 12 lines; and outputs from tape recorders and turntables add another 16 lines. In all, there must be 140 places in this patch panel.

We will decide on principle to separate the mic patch from the high-level patch, and we will use double plugs for the microphone patch. Single, tip-ring-shield plugs will be used for the high-level patch. Jack strips are made in single rows of 24 or 26 places, and double rows of 48 or 52 places. 24/48 place strips are specifically intended to be used in double-plug installations because their spacing does not permit a patch plug to be inserted accidentally into two places (half of the plug into one line and half into the adjacent line). The microphone patch will use two 48-hole strips. The 24/48 hole configuration allows 12 double plugs per row. Figure 14-7A shows how the mic patch will be arranged. Note that the studio mic lines are normaled to the first ten console inputs, and the first six stage mic lines are normaled to the remaining console inputs.

The high-level patch will require 104 places, so two 52-place jack strips will provide exactly the number of places that we need. Figure 14-7B shows the arrangement of this patch panel. The turntable and tape recorder outputs are normaled to the 16 console line inputs, and the four group outputs are normaled to the inputs of the two two-track recorders. The various sends and receives for input module and group module inserts are normaled at the patch panel to provide the usual feedthrough so that the modules will behave correctly with no outboard processor connected to the inserts.

All of the cable runs from both patch panels are dressed and bundled; then they are pulled through duct space beneath and behind cabinetry. At one end, the cable runs are soldered into the patch panel (or wired into interface connectors for the patch panel, e.g., Christmas-tree connectors or barrier strips); at the other end, the appropriate connector is soldered to the cable and then plugged into the piece of equipment it will serve. (Where a piece of equipment presents input/output connections in the form of barrier strips, forked spade-lugs are crimped onto the cable ends and are then secured to the barrier strip connectors.) When all cables are connected, the system is checked out for errors and discontinuities.

FIG. 14-7: PATCH PANEL FOR RECORDING SYSTEM

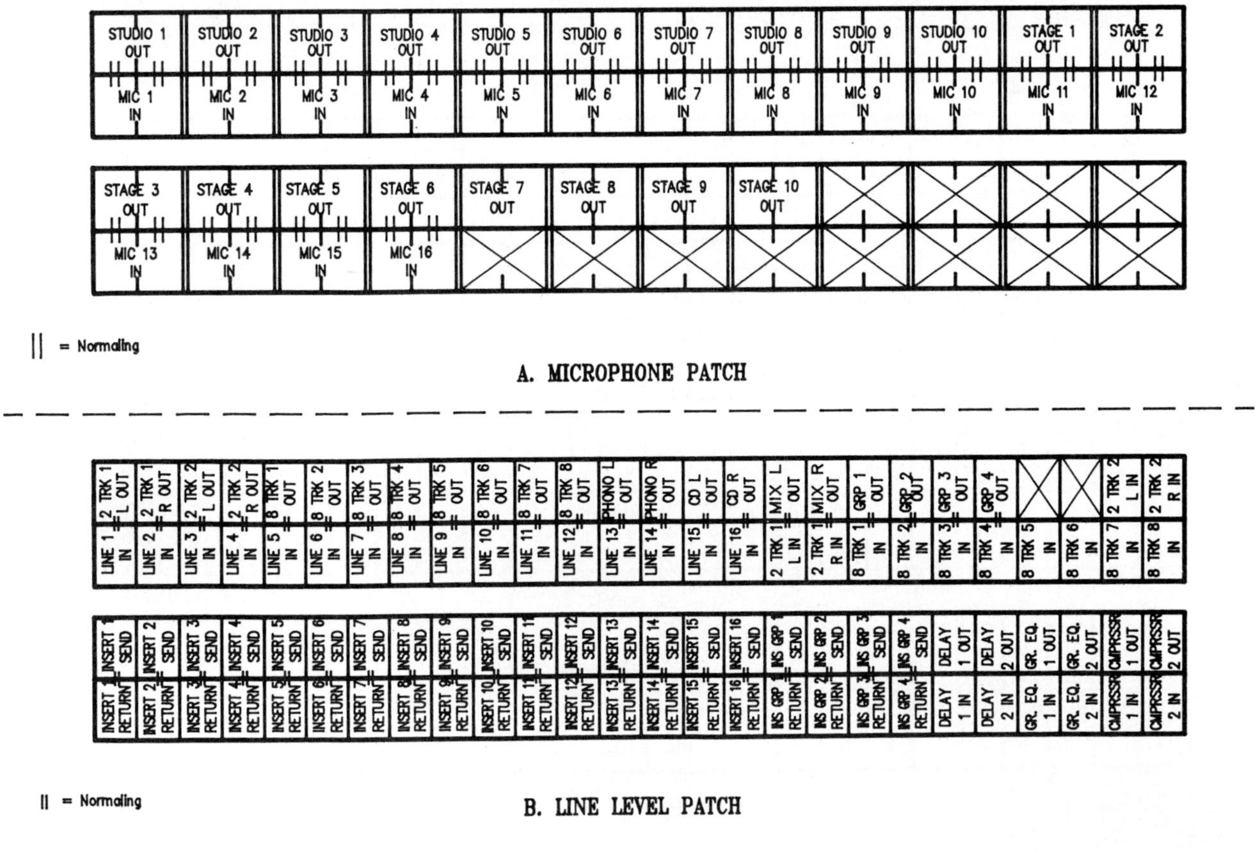

|| = Normaling

A. MICROPHONE PATCH

|| = Normaling

B. LINE LEVEL PATCH

Note the essential differences between the permanently installed recording system and the rented reinforcement system. First, the reinforcement system is set up to handle *only one configuration of inputs and outputs*; the recording installation, *through its patch panel*, can receive inputs from the stage, the studio, its turntables or tape recorders, or any combination of those inputs, and it can drive any combination of tape recorder inputs. The recording system does have a basic configuration, as revealed in the normaling pattern wired into the patch panel, but that pattern is not fixed as the only functional arrangement that the system can handle. Next, the reinforcement system utilizes multiconductor runs fitted with plug boxes and multiconductor plugs to ease the job of quick installation and removal. Clear labeling of all plug ends of cables is essential to prevent loss of sanity. The recording system uses individual cable runs soldered to the patch panel. Plug labeling is necessary, but it is not necessary to go into as much detail or make the labeling as readily visible as was done with the reinforcement system. Finally, note that adapters were not mentioned at all in the configuration of the recording system. Cables from the patch panel terminate in the kind of connector needed to mate to the input or output of the particular unit addressed. (If adapters are needed in such an installation, it will be for interface with equipment brought in for special purposes.)

Note, also, that the recording system does not offer accessible power amplifiers and loudspeakers. The only power amps and speakers are those involved in the control room and studio/stage monitor. These outputs are available only through the console monitor outputs. A selector switch is provided to let the operator direct monitor routing to the appropriate destination. This lack of available power amp/speaker outputs is consistent with the philosophy that the job of the recording system is only to prepare the tapes that will help to create the dramatic auditory environment during rehearsal and performance.

Theatrical Performance Playback System

The playback system, like the recording installation, is permanently installed, but it is much smaller and much less complicated. It does use a patch panel, but needs only a relatively few inputs and outputs.

The playback system is constructed around a console providing eight high-level inputs (two of which can be adjusted to accept feeds from microphones) and sixteen outputs. There are also three stereo, two-track, open-reel tape players, one stereo cartridge player, six parametric equalizers, and eight stereo power amplifiers. Ten microphone lines run from the stage to the booth housing this system, and 20 loudspeaker lines run from the booth to various parts of the stage and auditorium.

FIG. 14–8: PLAYBACK SYSTEM PATCH PANEL

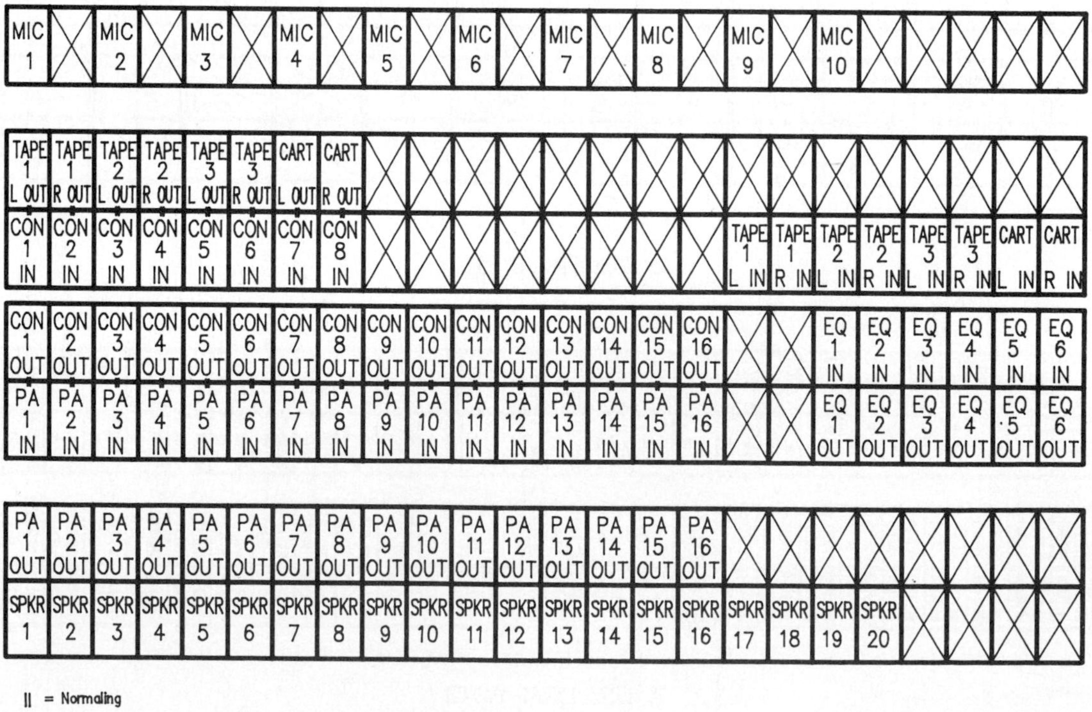

MIC 1 | X | MIC 2 | X | MIC 3 | X | MIC 4 | X | MIC 5 | X | MIC 6 | X | MIC 7 | X | MIC 8 | X | MIC 9 | X | MIC 10 | X | X | X | X | X

TAPE 1 L OUT | TAPE 1 R OUT | TAPE 2 L OUT | TAPE 2 R OUT | TAPE 3 L OUT | TAPE 3 R OUT | CART L OUT | CART R OUT | X | X | X | X | X | X | X | X | X | X | X | X | X | X | X | X

CON 1 IN | CON 2 IN | CON 3 IN | CON 4 IN | CON 5 IN | CON 6 IN | CON 7 IN | CON 8 IN | X | X | X | X | X | X | X | X | TAPE 1 L IN | TAPE 1 R IN | TAPE 2 L IN | TAPE 2 R IN | TAPE 3 L IN | TAPE 3 R IN | CART L IN | CART R IN

CON 1 OUT | CON 2 OUT | CON 3 OUT | CON 4 OUT | CON 5 OUT | CON 6 OUT | CON 7 OUT | CON 8 OUT | CON 9 OUT | CON 10 OUT | CON 11 OUT | CON 12 OUT | CON 13 OUT | CON 14 OUT | CON 15 OUT | CON 16 OUT | X | X | EQ 1 IN | EQ 2 IN | EQ 3 IN | EQ 4 IN | EQ 5 IN | EQ 6 IN

PA 1 IN | PA 2 IN | PA 3 IN | PA 4 IN | PA 5 IN | PA 6 IN | PA 7 IN | PA 8 IN | PA 9 IN | PA 10 IN | PA 11 IN | PA 12 IN | PA 13 IN | PA 14 IN | PA 15 IN | PA 16 IN | X | X | EQ 1 OUT | EQ 2 OUT | EQ 3 OUT | EQ 4 OUT | EQ 5 OUT | EQ 6 OUT

PA 1 OUT | PA 2 OUT | PA 3 OUT | PA 4 OUT | PA 5 OUT | PA 6 OUT | PA 7 OUT | PA 8 OUT | PA 9 OUT | PA 10 OUT | PA 11 OUT | PA 12 OUT | PA 13 OUT | PA 14 OUT | PA 15 OUT | PA 16 OUT | X | X | X | X | X | X | X | X

SPKR 1 | SPKR 2 | SPKR 3 | SPKR 4 | SPKR 5 | SPKR 6 | SPKR 7 | SPKR 8 | SPKR 9 | SPKR 10 | SPKR 11 | SPKR 12 | SPKR 13 | SPKR 14 | SPKR 15 | SPKR 16 | SPKR 17 | SPKR 18 | SPKR 19 | SPKR 20 | X | X | X | X

|| = Normaling

The patch panel for the system is semi-integrated in that it is all in the same rack space; but the microphone patch points are clearly segregated from the main high-level patch, as is the power amp output/loudspeaker patch. The microphone patch (a single jack strip of 24 places, only 10 of which are filled) uses tip-ring-shield jacks (utilizing single plug patch cords instead of double plug cords), so that adapters will not be needed when a microphone must be patched to either of the two inputs that will handle mic-level signals. The loudspeaker patch consists of one double jack strip of 48 places. The top row is filled to 16 places (for the 16 power amplifier outputs), and the bottom row contains the 20 loudspeaker lines.

The high-level patch is composed of two double jack strips of 48 places each. The top row contains the eight tape recorder outputs at the left end with the eight console inputs in the row directly beneath. Normaling exists between the tape outputs and the console inputs. At the right-hand end of the second row the tape recorder inputs are available. The third row holds the 16 console outputs at the left side with the equalizer inputs to the right. (Note break in pattern of output over input. It is done to facilitate patching equalizers into the output lines without cross-plugging.) The fourth row presents the 16 power amplifier inputs at the left, directly beneath the console outputs and normaled to them, and at the right are the equalizer outputs. (Figure 14-8 illustrates the system and its patch panel.)

The wiring and interconnection process is very much like that for the recording system; the cabling is done in individual runs rather than in multiconductor, and cabinetry can provide concealment and protection for cable bundles. The cables are soldered to the patch panel at one end and wired to the appropriate connector at the other. The system is interconnected and checked out.

Operation is very simple and straightforward. No patching at all is necessary unless an equalizer must be inserted in one of the output channels (between console and power amp), or unless the relationship of tape outputs to console inputs needs to be rearranged, or unless microphones are required. Note that the output normaling directs a console line straight to a power amp input. If an equalizer is needed in the chain (or if one needs to reroute an output back to one of the console inputs, as one often does), the patch cord *breaks* the normal flow, permitting a redirection of the console output for that line. Similarly, a patch cord plugged into one of the power amp inputs could receive a signal from an equalizer output, or could permit a signal from a tape machine to bypass the console entirely. Figure 14-9 shows the layout of the playback installation.

FIG. 14-9: THEATRICAL PLAYBACK SYSTEM

BLOCK DIAGRAM

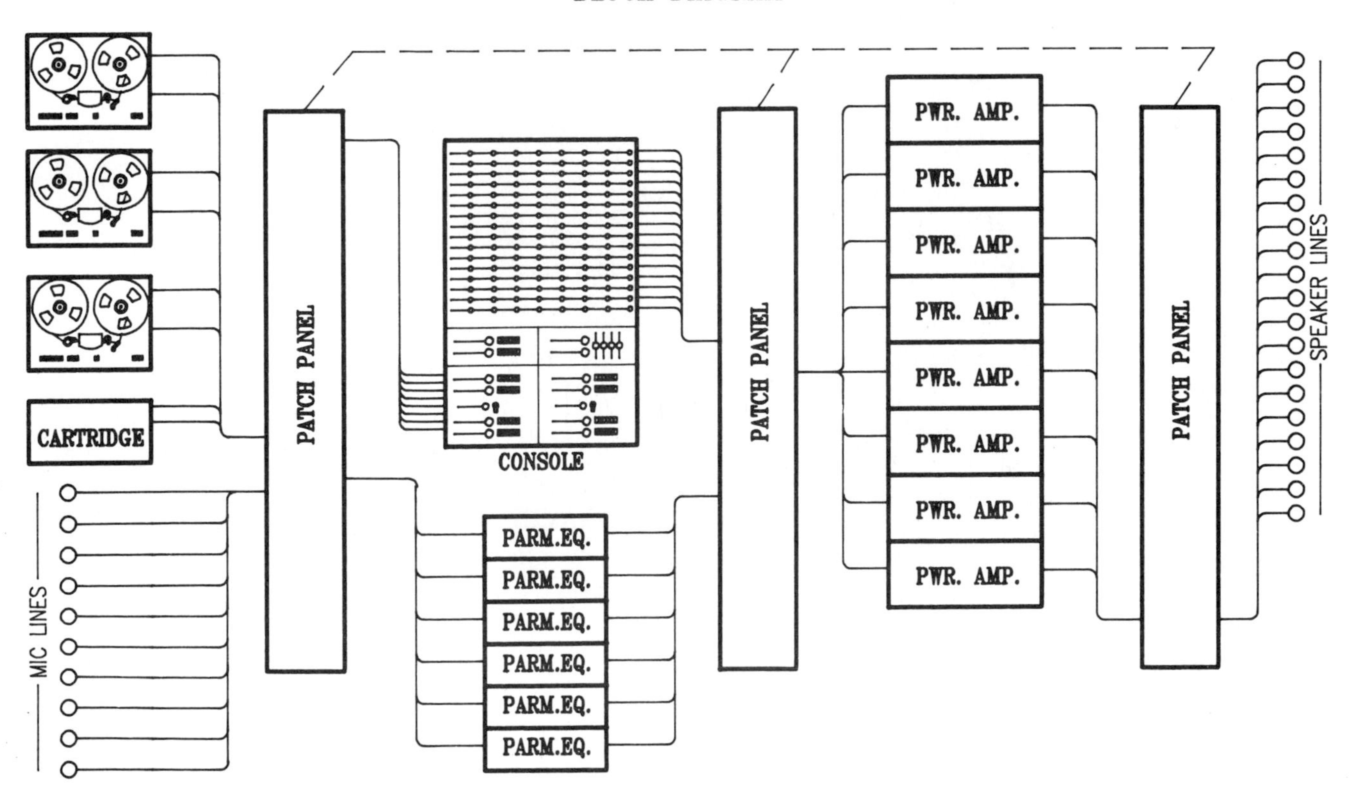

In each of the three examples, understand that interconnection is always a matter of planning signal flow from the first input to the last output. Planning a specific installation is, in many ways, the easier job because the functions required are clearly known. Specifying a flexible interconnection scheme for a system such as the recording installation is more difficult because system hookup must be as flexible as possible with every possible access point available through the patch panel, while at the same time a standard configuration must be preserved through the normaling wired into the patch. In all phases of planning the interconnection of a system, rules concerning electrical level, impedance, and correct shield earthing must be kept in mind. The placement of the individual system components within the overall installation can be planned only if one thoroughly understands the characteristics of both the overall system and those of the individual components. Without this understanding, serious mistakes and problems can develop that will waste time in troubleshooting that should be spent either on creative activities or in rehearsal.

Chapter 15

Basic Characteristics of Human Hearing

With this chapter we begin an investigation of an aspect of sound that is an important determinant of the choices a designer makes and of how those choices should be implemented. Sound as a physical phenomenon is a media-borne energy system in the environment; but sound as a subjective experience is based exclusively on the response of the human ear to acoustic energy and the way in which the human brain processes the neural output resulting from activity within the ear.

Sound as an information system depends on (1) human ability to group sequences of transient energy into event structures, (2) to classify those perceived structures and relate them to other experiences, (3) to use the analyzed percept as part of one's projected hypothesis of the character of the external environment, and (4) to form an emotional and attitudinal response to external conditions. In simpler words, sound helps us understand and respond to our immediate environment. The former sentence describes the process; the latter describes the result. The process is complex; the result is a familiar mode that all of us in the hearing world use without much conscious thought.

The business of the sound designer is to use sound to help a dramatic production achieve maximum effect. To do this, the sound designer must know how to shape sound to the purpose. Understanding the nature of human hearing is essential.

Sensory Information and the Environment

For human beings, sight and hearing are the dominant modes of perception. We derive most of our information about the environment from these two sensory modalities. The importance of these two senses arises from a simple fact: They are the only two senses that can extract detailed **distal** information (information for which the source is distant from our bodies) from the environment. All other senses essentially yield **proximal** information (information for which the source is close to or in contact with our bodies).[1]

Vision is easily the more important of the two primary senses; hearing is secondary. The expression "Believe half of what you see and none of what you hear" summarizes the skeptic's assessment of the reliability of visual and auditory data. Vision still wins. The primacy of vision is based solely on the structure of our brains. We are wired (for lack of a better term) to deal with the three dimensions of space as objective[2] elements, and vision is primarily a spatial sense. Vision is stable, simultaneous, and unidirectional; that is, the visual field appears to be a pattern of relatively stable objects, all simultaneously visible and distributed within a limited angle of view. Objects within the field of view possess the perceptual qualities of height, width, and depth. Visual objects appear to

[1]The sense of smell brings us information about things not directly in contact with our body surfaces, but olfaction does not provide sufficient detail to qualify as a major information source for humans. If smell were as important to us as to many other species, we would hardly try, as we do, to mask or even eliminate from the environment certain kinds of smells. Rather, we would need them for primary data, as dogs and other animals do.

[2]The terms *objective* and *subjective* for the context of this discussion specifically refer to those things that we can perceive as outside ourselves (objective) and those things that we can only experience through ourselves (subjective). Space is perceptually objective. Both visual and auditory experience tell us that there is a world outside of ourselves that stretches away from our bodies in all directions. We abstract the qualities of height, width, and depth. We generalize experience to apply these abstract characteristics to the immediate perception of spatial orientation. Time, by contrast, is purely subjective. We are trapped in time. We cannot use eyes or ears to perceive either the future or the past. We cannot move as we will in time—that is, take a step back or forward in time. Time, for us, is a **unidirectional continuum**. In a sense, even our very cognizance of time is due more to our ability to abstract and generalize than to any ability to discern by comparison and contrast. We can truly discern time only through the use of memory.

maintain a rather stable orientation within the perceptual field based on distance and direction from our bodies.

Perception of time is accessible to vision only by relative change of position in space. Audition, by contrast, is a temporal sense. Acoustic data occurs as a stream of stimulus energy extending through time. (So does luminous energy, of course, but the visual cortex is not set up to perceive it as such—a good example of how deeply our conception of our world is constrained by the structure of the brain.) Visual objects appear substantive and relatively permanent; auditory "objects" appear to be transient. Perception of sound is, essentially, a manifestation of time.

Sound has spatial components, as the propagation of acoustic energy requires a spatial medium, but the spatial aspects of sound are secondary to the temporal. The spatial elements of sound are not unimportant, however. Where vision is a unidirectional sense, hearing is omnidirectional. Hearing provides information about all points within the environment and provides that data continuously. Hearing can provide an assessment about the size and character of space; and hearing gives clues to the distance and directionality of objects that help to confirm and reinforce the report of our visual sense.

The two distal senses, vision and audition, work together to provide the data necessary for us to be able to act and react within the world around us. Vision provides precise, detailed data but is limited to perception of only a part of the environment at a time; hearing is less precise but brings in data constantly and from all directions.

Sound (like our cognizance of time) is a highly subjective experience. Try to relate to someone your experience of a sound and you will find yourself relying on vocal noises and words that simulate the sound you are trying to describe. Musicians have a written language to communicate information about matters such as pitch, loudness, speed, and duration; but only the musically proficient can translate those symbols into something akin to the experience of hearing musical sounds. Even proficiency in reading music cannot convey the experience of a particular performance. Handing someone the music for Moussorgsky's *Pictures at an Exhibition* is not the equivalent of hearing a recording of Vladimir Horowitz performing the piece. Further, musical symbols cannot describe the nature and character of nonmusical sounds. For example, annotating all of the sound qualities of a steam engine is virtually impossible.

The Anatomy of Human Hearing

The human auditory system can be divided into three basic subsystems: the sensory receptor (the outer and middle ear), the neural encoder (inner ear and eighth cranial nerve), and the auditory data processing system (auditory cortex and, roughly, whatever parts of the brain are required for derivation of identity and extrapolation of meaning). The first two parts of the system serve the functions of receiving and encoding the stimulus, adding some but not a great deal of processing to achieve meaning. The derivation of meaning is primarily carried out in the brain itself.

The parts of the ear are shown in Fig. 15-1. The external parts of the human hearing system are responsible mainly for collecting acoustic energy. The visible part of the human ear, the **pinnae**, tends to favor pickup of sound from the front and sides of the head, creating a slight sound "shadow" for sound arriving from the rear of the head. The head itself is also somewhat involved in a part of the hearing process, as it provides both a shadowing and a diffracting obstacle for the ear located on the side facing away from the sound source. The pinnae funnel sound into the **external meatus** (auditory canal)—the channel leading from the surface of the head to the **tympanum** (eardrum). The tympanum is a flexible membrane that can be easily set into motion by slight changes in the pressure of the external air. The pinnae, external meatus, and tympanum form what is known as the **outer ear.**

Connected to the tympanum is a train of small, articulated bones called the **ossicles**. The ossicles are located in a small cavity that is connected to the throat through a channel called the **Eustachian tube**. The Eustachian tube serves to equalize pressure between the ossicular cavity and the external atmosphere. The three ossicular bones carry Latin names describing their shapes: **malleus** (hammer), **incus** (anvil), and **stapes** (stirrup). Ossicles and Eustachian tube comprise the **middle ear**. The tympanum connects through the ossicles to the oval window of the inner ear. Together, the ossicles form an impedance transforming network, coupling the relatively compliant pneumatic system of the outer ear to the much stiffer hydraulic system of the inner ear. Within the region of the ossicular cavity a small muscle connects to the tympanum. This muscle, called the **tensor tympani**, reacts quickly in response to large increases in sound level to stiffen the tympanum in order to make it less susceptible to rupture from sudden, large changes in air pressure.

Housed in the mastoid bone of the skull is the **inner ear**. The primary structure related to hearing is the **cochlea** (Latin for "shell"), a helically coiled structure about as big as the tip of the small finger. The cochlea is divided into three chambers, two fluid-filled tubes called canals connected to each other by a small opening near the apex of the coil and a central duct that houses the input end of the working neural system that senses stimulus energy and passes it on to the auditory cortex. The third ossicular bone, stapes, connects directly to a membrane called the **oval window**. Immediately below the oval window is a second membrane called the **round window**. These two flexible membranes cover the two fluid-filled canals, the vestibular canal (oval window) and the tympanic canal (round window). Between the two canals is the central **cochlear duct**. (See Fig. 15-2.)

Inside the central cochlear duct nerve endings emerge in hairlike projections (cells) that grow out of a structure called the **organ of Corti**. The sensory hairs extend upward and into a structure called the **tectorial membrane**. Underlying the

FIG. 15-1: STRUCTURE OF THE EAR

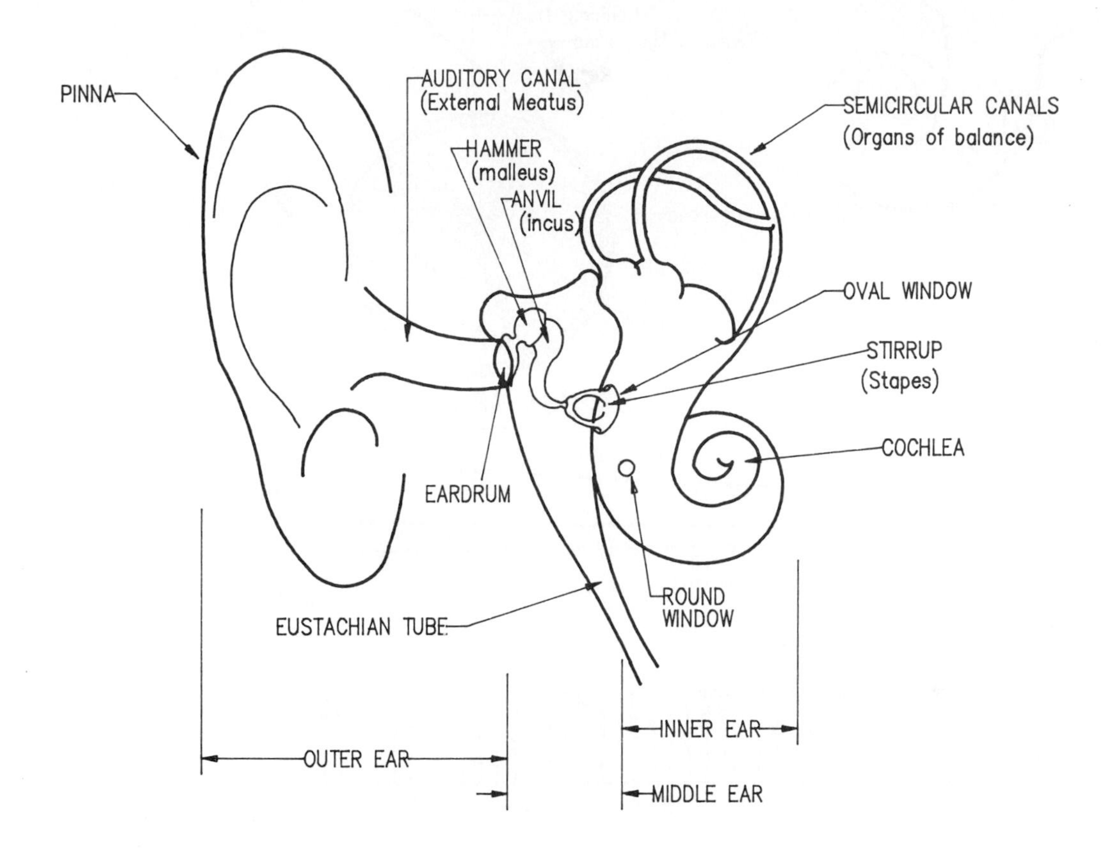

organ of Corti is one wall of the central duct, known as the **basilar membrane**. The opposite wall of the duct is called **Reissner's membrane**.

The hair cells of the cochlear duct are the lower extension of the **eighth cranial (auditory) nerve**, one of the major nerve trunks of the brain. The nerve rises from the cochlea to the **auditory cortex**. The auditory cortex contains the major processing centers that are responsible for primary interpretation of acoustic data. Along its path to the cortex, the nerve passes through several branch points where neural interconnections carry portions of the auditory signal to secondary processing areas. The various branches from the eighth cranial nerve provide signal input to areas of the lower nervous system that are responsible for defensive mechanisms such as startle response, the involuntary eyeblink after a sudden, loud noise, and the contraction of the tensor tympani.

This overview of the anatomical aspects of human hearing scarcely reveals the complexity of the auditory process. We know relatively little about specific activity beyond the cochlea. Our understanding of biological data processing is not sufficiently advanced. We do have a fairly good catalog of the behavioral characteristics of the auditory system, however, and the information is extremely useful. The study of the behavioral characteristics of human hearing is called **psychoacoustics**.

Psychoacoustics

Sound reaches our ears in the form of local variations in atmospheric pressure. These variations have measurable properties: frequency, intensity, relative phase, complexity, envelope, distance, direction, and duration. In response to external acoustic stimuli, the auditory system yields the subjective perceptual characteristics of pitch, loudness, timbre, volume, duration, attack, decay, reverberance, persistence, and localization. Some of the physical parameters can be related to psychoacoustic parameters, but none are direct correspondents. For example, frequency is generally interpreted as the physical basis of the subjective phenomenon of pitch, but both loudness and complexity can affect perception of pitch.

Occasionally, we try to liken the ear and the auditory cortex to an electronic audio system. The comparison is anything but exact. Performance characteristics of an electronic audio system can be specified in terms such as frequency

FIG. 15-2: INNER EAR

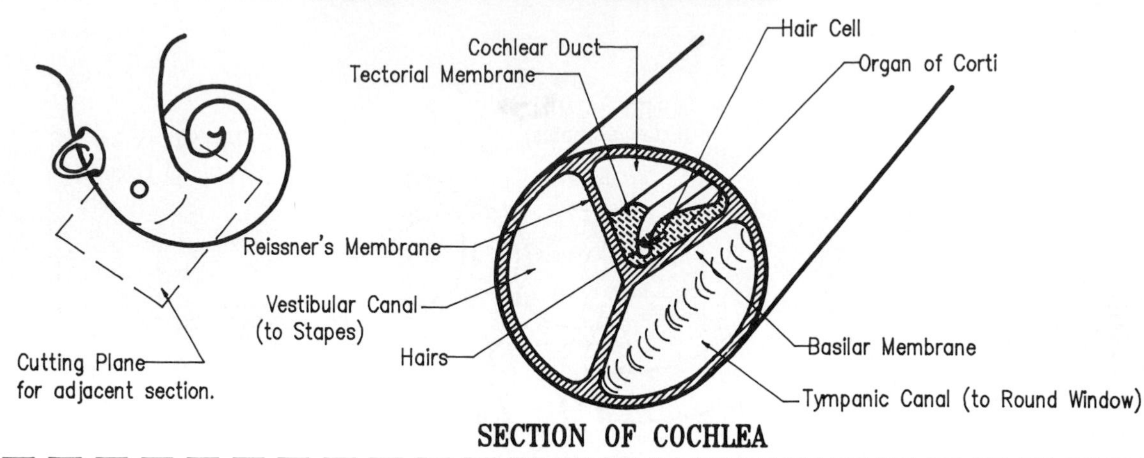

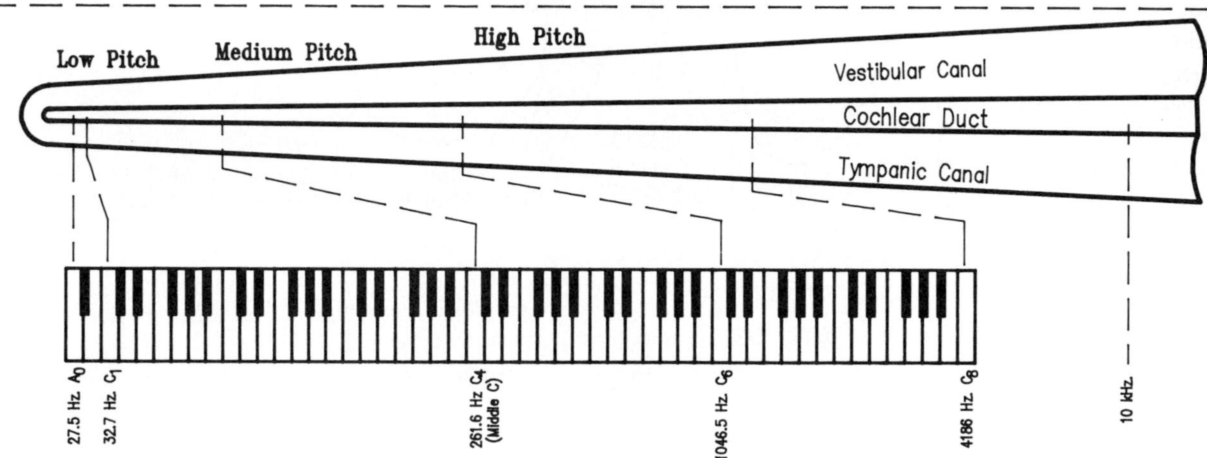

response, system gain, amount of distortion, plus a number of other important parameters. With an audio system, we can measure and quantify matters such as the input to the amplifier and its corresponding output. We can even look inside the amplifier to find out what is happening at specific points in the system. Specification of performance characteristics for the human auditory system is not as easy. In the first place, its output—if the auditory system can be said to have output—is difficult (if not impossible) to retrieve by direct means; and the relationship between input and whatever passes for output is by no means either simple or direct (as with frequency and pitch). Matters such as distortion are relatively meaningless. In fact, the brain uses distortion normally produced by the auditory system as a source of information. Moreover, the purpose of an audio system is to store and repeat acoustic information; the purpose of the human auditory system is to derive meaning from acoustic information.

Psychoacousticians can catalog most of the characteristics of hearing. These characteristics include spectral response limits, overall dynamic range and response to intensity, binaural phenomena (aspects of hearing based on use of two ears), and susceptibility to error. All of these characteristics are important to the sound designer as well as to the psychoacoustician, because these characteristics define the way in which human hearing will respond within a given environment, and how the ear can sometimes be fooled into perceiving something not actually present in the acoustic stimulus.

Response of the Ear to Frequency

The frequency of a sound produces a subjective response in human hearing that we know as pitch. A high frequency gives us the sensation of a high pitched sound. Flutes, clarinets, violins, and trumpets are all high pitched instruments. Low frequency gives us the sensation of low pitched sound. Cellos, tubas, and bassoons are all low pitched instruments.

The inner ear is responsible for detecting difference in frequency information in sound stimuli. The cochlea is filled with an incompressible fluid that occupies the vestibular and tympanic canals within the cochlear structure. The fluid-filled canals lead from the oval window to the round window, and are connected by a small passage at the apex of the cochlea. Acoustic energy flexes the tympanum, which, in turn, sets the ossicles into motion. At the end of the ossicular chain, stapes exerts a reciprocating force on the oval window, setting the cochlear fluid into motion. The fluid being incompressible, the round window membrane is necessary to accommodate motion of the fluid.

Acoustic input forces the cochlear fluid to flow back and forth following variation in atmospheric pressure. (See Fig. 15-3.) Forcing an incompressible fluid into alternating motion creates whorls, points of turbulence within the fluid. The location of turbulence depends on the frequency of directional alternation. Whorls within the fluid place pressure against the basilar membrane, causing it to deform. Deformation in the membrane stresses the hair cells inside the cochlear duct (lower section, Fig. 15-3), causing them to send a neural signal to the auditory cortex. The message contains not an analog of the signal frequency, but an indication of the position along the basilar membrane where pressure was applied. Positional information is interpreted as pitch. Pressure near the stapidal end of the basilar membrane yields a sensation of high frequency; pressure near the apical end of the cochlear duct is interpreted as low frequency. Neural activity from the middle of the cochlear duct is sensed as mid-range tones. The limits of the audible spectrum are directly determined by the length of the cochlear duct: Frequencies below 16 Hz or above 20 kHz simply fail to produce any significant action within the cochlear fluid that can be sensed by the hair cells within the cochlear duct.

The ear can discriminate an amazingly large number of pitches, many more than normally used in traditional Western music. The smallest change in pitch that the ear can discern varies with stimulus frequency. Near the low end of the audible spectrum a change in frequency of approximately 3 Hz produces a discernible change in pitch—at 20 Hz that represents a frequency increment of about 15 percent. The percentage of change decreases with rising pitch until, at about 1000 Hz, the percentage change levels out to about three tenths of 1 percent, dropping only to about two tenths of 1 percent at the upper end of the spectrum. The number of hertz difference in frequency, of course, increases with increase in pitch.

Response of the Ear to Intensity

As the input portion of the auditory nerve trunk, the cochlea must also detect and encode the physical intensity of the acoustic signal. The mechanism is relatively simple: Low

FIG. 15-3: DETECTION OF FREQUENCY

INSTRUMENT SOUNDING "MIDDLE C"

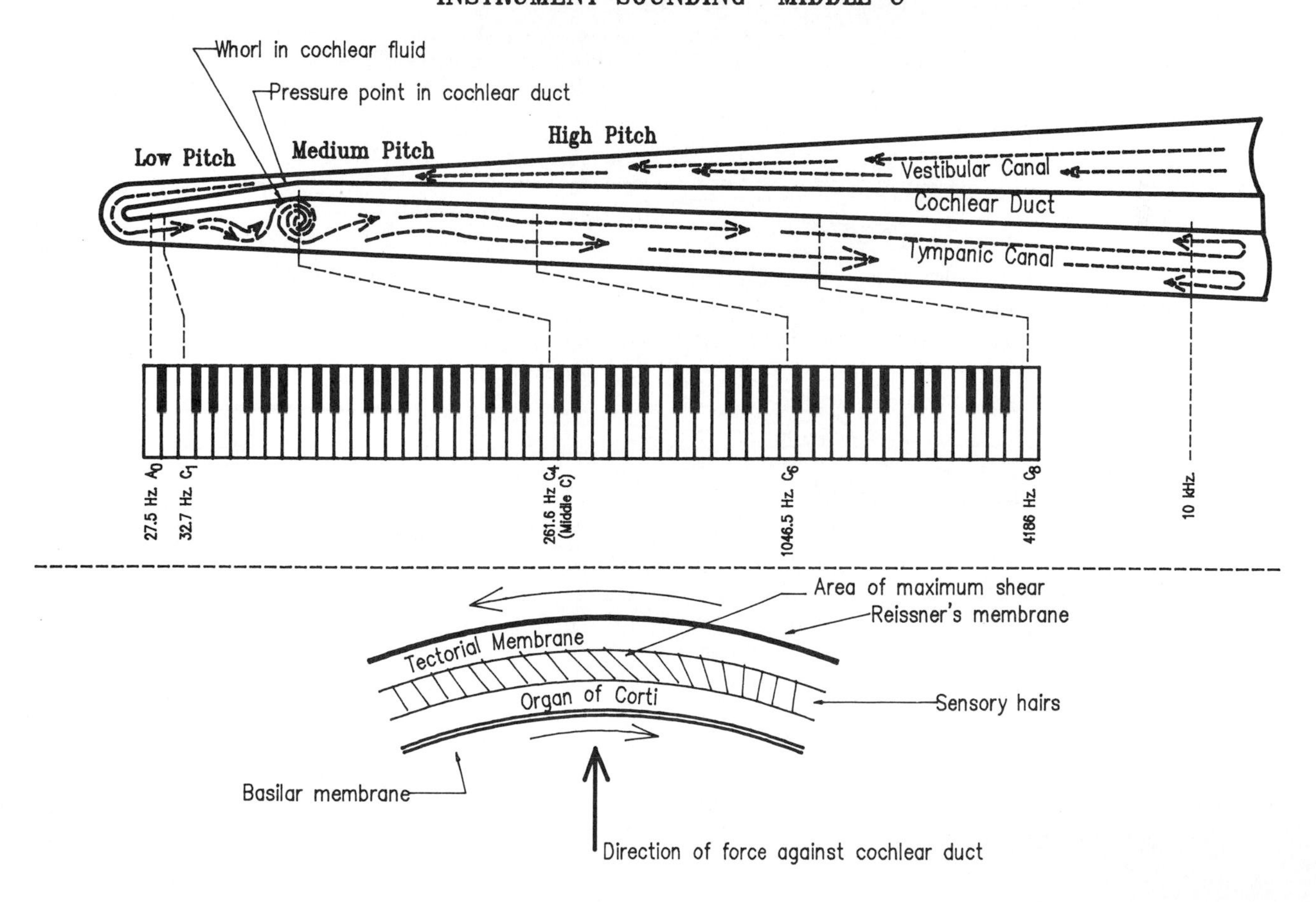

energy at any given frequency produces relatively little pressure against the basilar membrane; high energy produces correspondingly large amounts of pressure. A small amount of pressure on the basilar membrane deforms the cochlear duct very little with correspondingly small stress on the hair cells. A small amount of force exerted on the hair cells yields a small neural output signal. Larger amounts of energy produce a greater degree of stress, resulting in greater neural signal output.

Perception of intensity and perception of frequency are interdependent. Higher levels of intensity tend to stimulate wider areas of the basilar membrane than do low levels of intensity. (See Fig. 15-4.) Because perception of pitch is derived from positional pressure on the membrane, a widened area of stimulation essentially means that additional pitch information will be sent to the auditory cortex. In the middle of the auditory spectrum (i.e., in the middle of the basilar membrane), the expansion of pitch information is balanced symmetrically and we perceive relatively little if any displacement in pitch as a sound gets louder. At the extremes of the hearing range, however, the possibility of symmetrical displacement is limited by constraints at either end of the duct. Shearing effect will tend toward the more constrained end of the membrane. Consequently, increase in loudness in the higher frequencies tends to produce an upward drift in pitch (toward stapes); increase in loudness in the lower frequencies tends to produce a downward drift in pitch (toward the apex).

Response of the ear to intensity is a function of the amount of energy required to elicit a given sensation of loudness. As one might expect from the description of the structure of the cochlear duct, a certain minimum amount of stress on the hair cells is required to stimulate any response at all. The minimum amount of acoustic energy required to cause the cells to respond is known as the **threshold of perception**. (See Fig. 15-5.) Average normal hearing can detect sound pressure variations as small as those caused by a mosquito's wings several meters away. Note carefully that the energy required to produce threshold response is not the same for all frequencies. Much more energy is required to stimulate a hearing response at very low and very high frequencies than in the middle of the spectrum.

At the other end of the dynamic range, excessive stress on the hair cells ceases to yield increase in loudness and begins to cause a sensation of physical discomfort. The upper limit of intensity response is called the **threshold of feeling**. (See Fig. 15-6.)

FIG. 15-4: DETECTION OF INTENSITY

RELATIONSHIP BETWEEN INTENSITY AND PERCEIVED PITCH

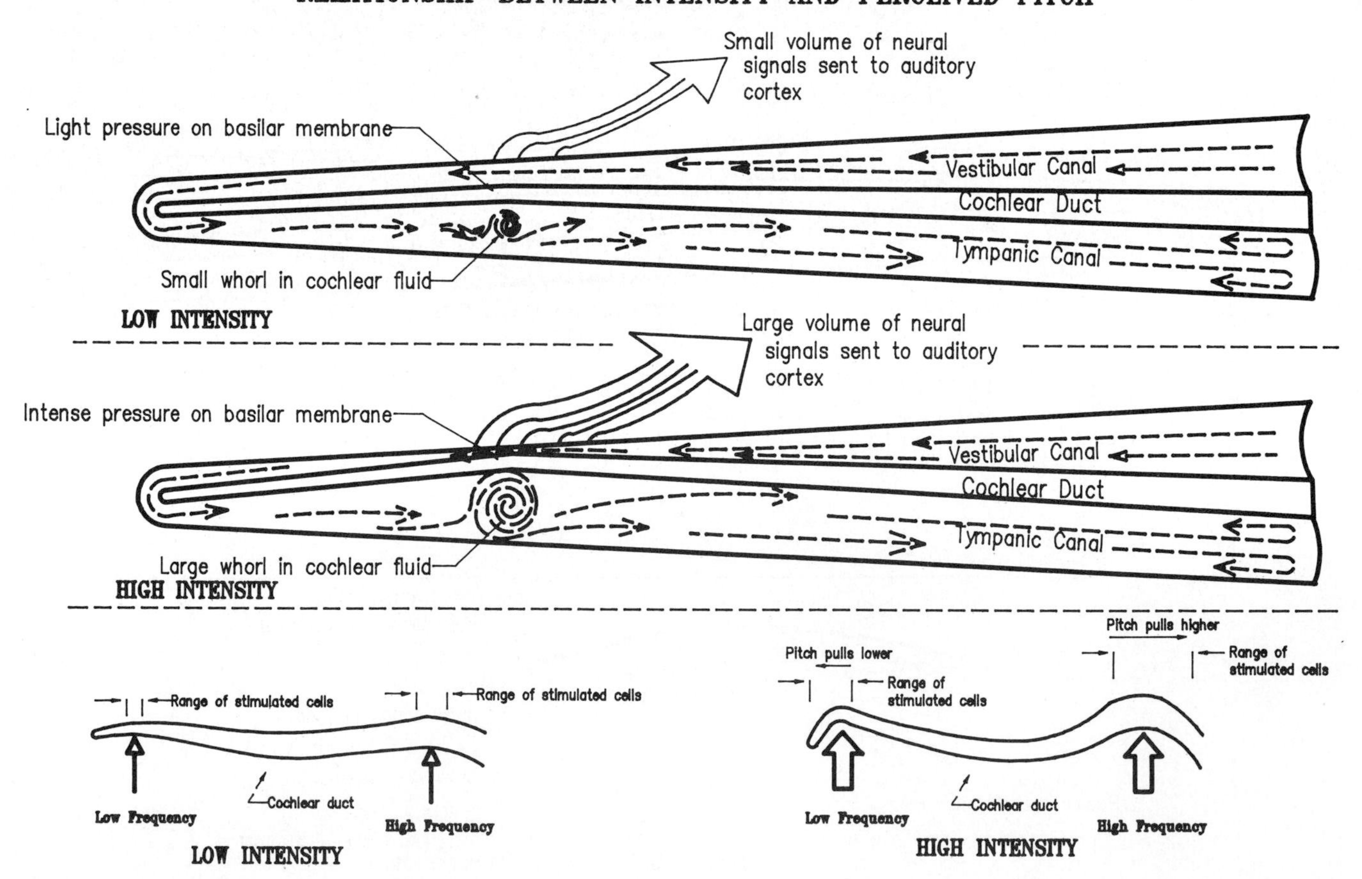

FIG. 15-5: THRESHOLD OF HEARING

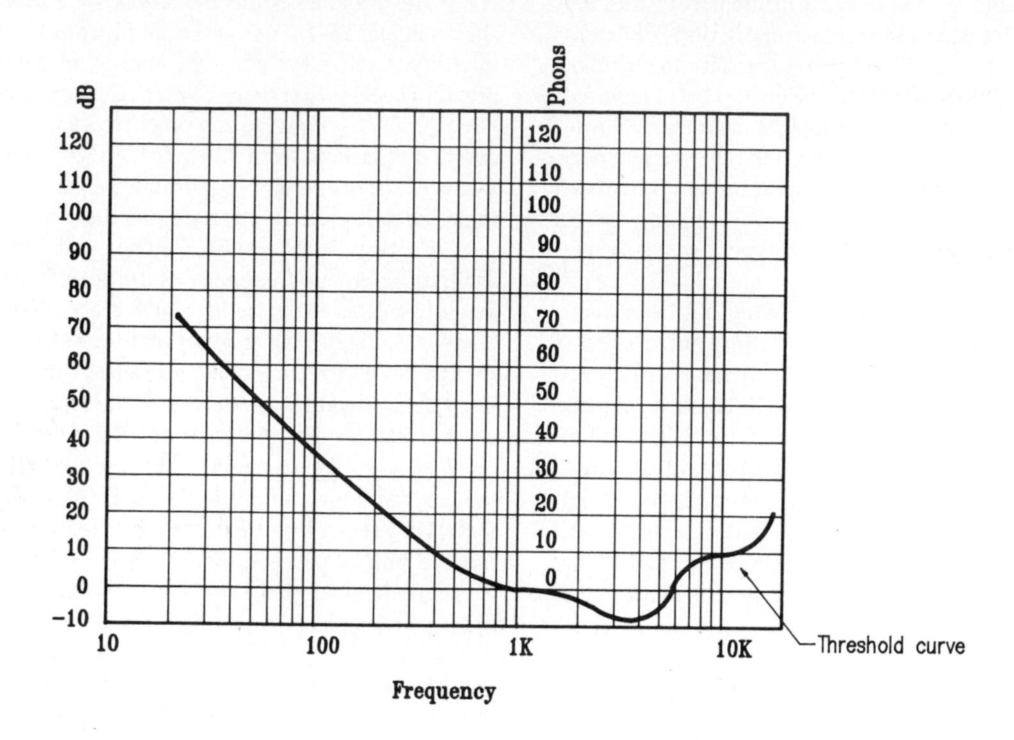

FIG. 15-6: THRESHOLD OF PAIN

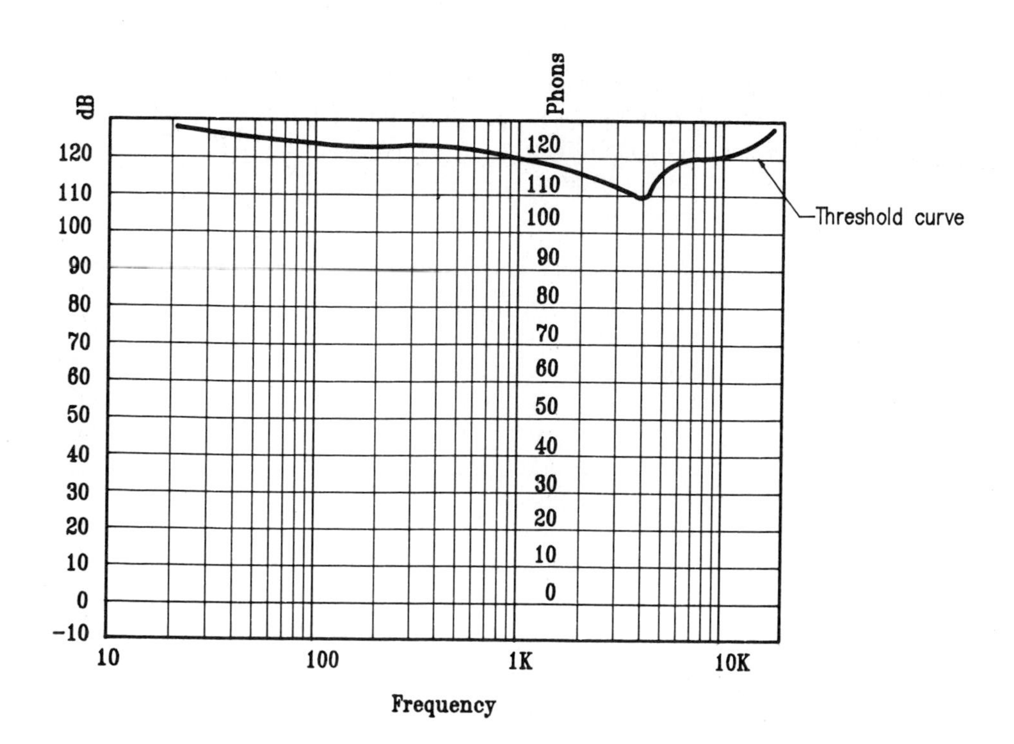

Beyond this threshold, further increase in intensity results, not in increase in loudness, but in increase in discomfort, resulting finally in physical pain. Note that, except for the band of frequencies from 2000 to 4000 Hz, the energy level required is approximately equal throughout the spectrum. From threshold of perception to threshold of feeling, average normal hearing has a dynamic range of approximately 120 dB at 1000 Hz. That means that the threshold of feeling represents an energy level approximately 1 trillion times greater than at threshold of perception.

The dynamic range of human hearing has been averaged over thousands of tests. The usual testing process seeks as the reference for the threshold of hearing the energy level where, given a stimulus frequency of 1000 Hz, test subjects report hearing a tone in 50 percent of the test trials and fail to report any sensation in the remaining 50 percent of the trials. This level represents an average sound pressure level equal to 1 trillionth of an acoustical watt and is designated 0 dB. The threshold of feeling at 1000 Hz, therefore, is 120 dB (one trillion times greater), equaling a sound pressure level of 1 acoustical watt.

Equal Loudness Characteristics

Given what we know so far about how the cochlea functions, we might suspect that the relationship between frequency and intensity involves complexities beyond those so far described, and that is, in fact, the case. Some parts of the audible spectrum require much less stimulus energy than others. A graph plotting frequency against the stimulus energy required for specified loudness levels produces a group of response curves known as curves of equal loudness. They are also called the **Fletcher-Munson curves**, after the discoverers of this aspect of human hearing.[3]

The equal loudness characteristic of human hearing, though it may seem illogical at first consideration, serves at least two useful functions: One, it provides a simple but efficient means of acquiring information about the distance of a sound source or about the presence of intervening structures, and two, it frees us from noise produced by our own body processes such as circulation of blood, action of skeletal joints, and respiration. The equal loudness function is one of the most important characteristics of hearing, so we shall discuss it in some detail. Of all aspects of human hearing, this is one that the sound designer must understand thoroughly.

The Fletcher-Munson curves of equal loudness are shown in Fig. 15-7. The curves are functions of frequency and intensity. Each curve plots the energy, in decibels, of sound pressure necessary for any given frequency to be perceived as having loudness equal to any other frequency lying along that curve. Frequency, from 20 Hz to 20 kHz, is plotted along the *x*-axis of the graph in a logarithmic scale. Relative intensity, in decibels, is plotted linearly along the *y*-axis.

To begin their study, Fletcher and Munson chose an arbitrary reference frequency of 1000 Hz. They tested numbers of subjects to determine the average amount of stimulus energy necessary to state levels for threshold of perception and for threshold of feeling at the reference frequency (1000 Hz). They determined that threshold of perception corresponds to a very small acoustic pressure of 0.0003 dyne of force acting on an area of 1 square centimeter (as we said earlier, about the loudness of a mosquito's wings at a distance of several meters). They found that the threshold of feeling is 120 dB above the threshold of perception.

As the next phase of the study, they began a sequence of tests to determine how much stimulus energy was required to make a tone of a lower or higher frequency sound as loud as a 1000-Hz tone. They tested starting at threshold of perception, increasing intensity in 10-dB increments. The curves shown on the graph detail their findings. They found that optimum sensitivity occurs at 3000 Hz and that we hear best in a band of frequencies between 2000 and 4000 Hz.[4] As stimulus frequency ranges above or below this optimum bandwidth, progressively more stimulus energy is required to make the input seem just as loud as a tone in the preferred range.

Fletcher and Munson created a new unit, called the **phon**, to specify levels of equal loudness. The phon is similar to the decibel, but decibels of sound pressure level and phons are identical only at 1000 Hz. A loudness level of 60 phons is equal to 60 dB at 1 kHz, but requires a physical intensity of approximately 80 dB at 100 Hz.

The major consequence of the equal loudness function is this: The lower the overall intensity level, the more hearing tends to focus into the central bandwidth between 800 and 5000 Hz. Although the loudness curves flatten as sound level increases, a great deal more physical intensity is required to make low or very high frequencies appear as loud as frequencies in the 2000 to 4000 Hz band over most of the dynamic range of human hearing. Only at the top of the dynamic range, near the threshold of feeling, does equal loudness result from more or less equal energy at all frequencies.

---

[3]Fletcher and Munson worked for Bell Laboratories during the 1920s and 1930s. Because Bell was interested in optimizing telephone voice transmission, the question of whether humans could hear one band of frequencies better than another was of significant importance. The curves of equal loudness resulted from tests devised by Fletcher and Munson to map the response of human hearing as a means of answering this question.

[4]Three thousand herz is higher than the arbitrary reference frequency of 1000 Hz that Fletcher and Munson selected as the basis of their tests. It happens to correspond both to the resonent frequency of the auditory canal and to the approximate frequency perceived at the center of the cochlear duct. Fletcher and Munson had no way of knowing that until after they and others had conducted numerous studies. Because 1 kHz was chosen as the reference, it has persisted as the coincidence of decibels and phons in subsequent measurement of equal loudness.

FIG. 15-7: FLETCHER-MUNSON CURVES

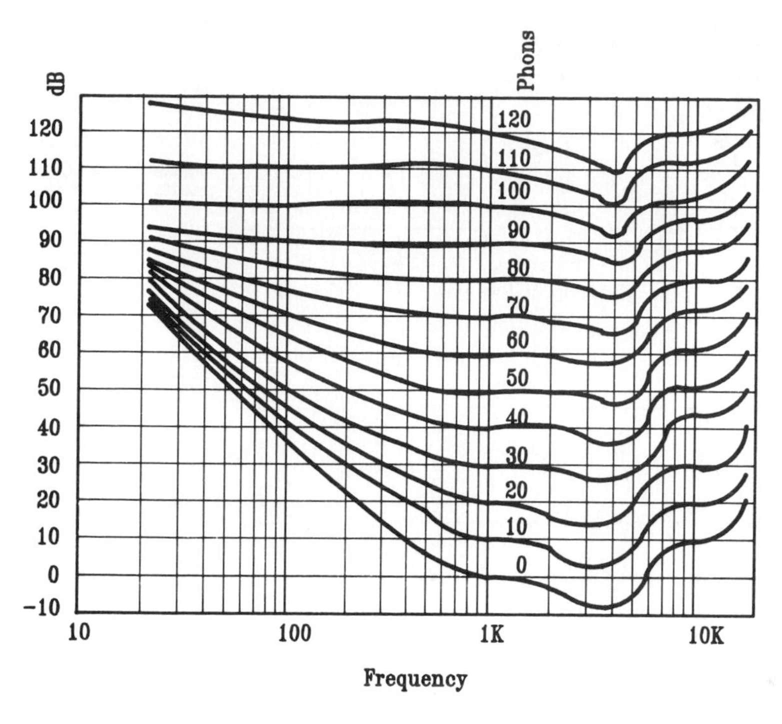

Curves define intensity required at any given frequency to match sensitivity at 1000 Hz. Curves are known as Equal Loudness Contours.

In practical terms, the equal loudness function means that we hear mainly mid-range frequency information in low-energy sounds. The auditory cortex uses this characteristic as one very important indicator of the distance of a sound source. As distance between our ears and a sound source increases, inverse square law acts to reduce the intensity of the sound quite rapidly. All of the energy that we perceive from the source focuses progressively into the central band of frequencies between 800 Hz and 5 kHz. Equal loudness also serves to preserve the intelligibility of auditory information, especially speech, at very low loudness levels. Most of the articulators that define sounds in general and words in particular fall into the mid- to upper-mid-range of the auditory spectrum. Later, when we study the phenomenon of masking, we will also see how the equal loudness characteristic helps to prevent distant sounds from obscuring nearer sounds.

The Fletcher-Munson curves are basic information for achieving any degree of realism in fitting an acoustic environment to a dramatic production, whether on stage or for film or television. Most recordings of sounds are made with the microphone very near the sound source, so that the microphone picks up all of the frequencies that the source produces—in other words, with a close perspective. Such recordings cannot be used to simulate the source at a distance for theatrical purposes without some modification. A simple reduction in intensity won't create the proper sound quality. The signal should be filtered to approximate the bandwidth appropriate to the intended distance. Usually, the designer relies on his or her ears to achieve the correct perspective, but understanding the way in which the equal energy function of hearing operates is essential to a successful job.

Loudness Perception and Time

Loudness perception involves a temporal component called **integration time**—the relationship between the duration of a sound and its perceived loudness level. The relationship based on experimental data is graphed in Fig. 15-8. Note that for all physical intensities of the acoustic stimulus, the peak in loudness level occurs at a duration of approximately two tenths of 1 second. In real life situations, however, the effect seems to involve somewhat longer duration, possibly up to as much as 1 second, depending on the loudness of the stimulus and on whether or not it is expected. Cognition and recognition of a sound event appears to be dependent upon a complex of variables, including both familiarity of the sound and conscious expectation of the appearance of the sound. At very low loudness levels, a sound lasting significantly less than 1 second in duration may well go unnoticed. Loud sounds of similar duration may gain attention but will not achieve full loudness.

FIG. 15-8: LOUDNESS AND DURATION

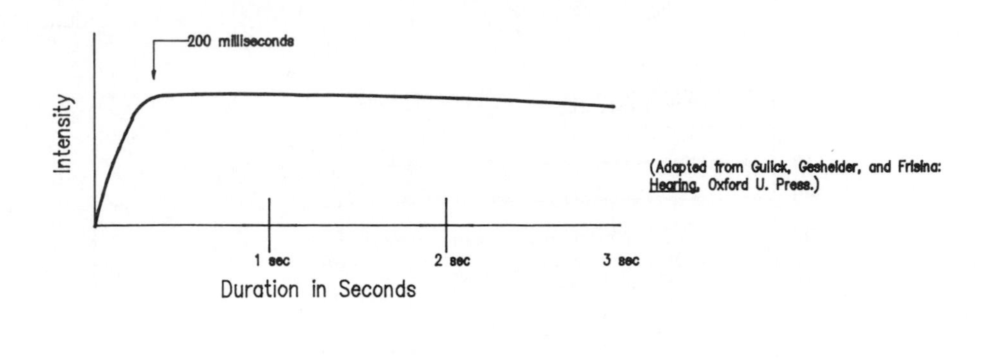

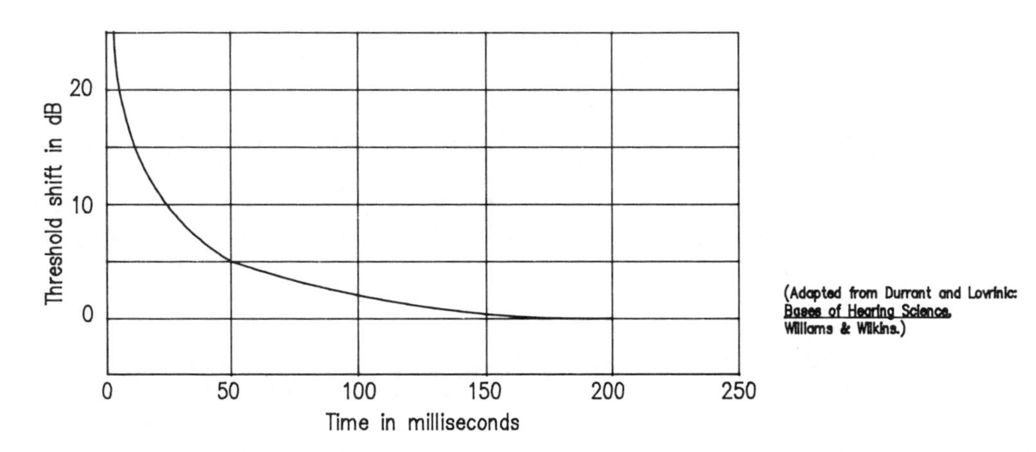

Integration time is a *threshold value in time* for the sense of hearing. It is the minimal duration necessary for the auditory cortex to recognize that an acoustic event is taking place, to identify and classify the nature of the event, and to relay that information to portions of the brain that filter important data for conscious process. When intensity level is low, more time is required for auditory processing to take place. A soft sound simply takes longer to establish itself as an event than does a louder one.

At durations less than a second, the brain may be able to recognize that some kind of auditory event has occurred, but may be totally unable to derive even a remote identity classification for the stimulus. Lacking the necessary minimum duration, the nature of the sound will be unclear. The minimum time required for the auditory cortex to recognize, identify, and classify the sound—that is, the time necessary for full integration—varies with stimulus intensity.

Integration time is another of the characteristics of hearing that is important to the sound designer because it determines the length of sound needed for adequate perception at a given level of loudness. Many sounds required in a dramatic environment tend to be short, sharp noises, such as pistol shots. When a pistol shot must be heard at a distance (low intensity), a real danger exists that the sound may fail to register upon audience awareness. Either the sound must be arbitrarily loud or else some way (such as generating extended reverberance) must be found to prolong the sound. (Another alternative, of course, is to rely on the help of the director or other designers to build a sense of expectancy that something is about to happen. When the audience anticipates an event that falls marginally near threshold, the likelihood that the event will be perceived increases significantly. Also, the possibility exists that reaction from actors following the sound can increase the likelihood of perception.)

Duration required for recognition seems to be related to the *familiarity* of a sound. In other words, the more familiar we are with a sound, the more quickly we will identify it. Comic sound effects can be particularly troublesome. Sometimes we use slaps and whistles and grinding noises to add humor to stage business. These noises, however, may not be sounds normally associated with a given action. Consequently, they need to be relatively loud if they are very short, or they need to be relatively long if they must be soft.

Recognition time also depends on the environmental context. We tend to expect sounds such as the noise of a power saw in a wood shop; we don't necessarily expect to hear the sound of a power saw during the middle of a theatrical production unless the dramatic context leads us to do so. If we were to introduce the sound of a power saw as a distant noise in the dramatic auditory environment, we would need to be sure that the sound was long enough for the audience to identify it. A quick cross-cut through a piece of 1 x 3 white pine would have so little duration that very few in the audience would be able to identify the sound, and a great many audience members might not even cognize that any kind of a sound had even occurred. Increase the intensity level of the sound and almost everyone would hear (i.e., cognize the presence of) the noise, but most people would be confused as to what it was and why it happened.[5]

Note that environmental context is the controlling factor. If the director can work a break into the dialogue (a long, wordless cross from one side of the stage to the other, for example) in which to present, in relative isolation, a sufficiently long, sufficiently loud instance of the sound, then subsequent uses of the effect can be much softer and much shorter. Obviously, an effect of this kind comes under the heading of an enhancement of the auditory component of the dramatic environment. As such, a valid, fully justifiable reason for the sound must be evident within the verbal and/or visual context.

Response of the Ear to Complex Stimuli

The ability of the human auditory system to detect intensity and to resolve small changes in frequency is by no means the extent of our ability to process acoustic information. We can discriminate subtle differences among sounds that contain multiple frequencies (complex tones). We can distinguish one tone from among a number of tones sounding simultaneously, or we can allow a group of simultaneous sounds to blend together perceptually.

Most of the sounds we experience are composed of multiple frequencies and are known as *complex sounds*. The cochlea senses complex sounds as multiple pressure points along the basilar membrane—a major pressure point corresponding to the dominant frequency of the sound, with subordinate points of pressure representing the presence of the most significant partials.

Our daily auditory experience consists of many sounds occurring simultaneously, all competing for attention. Our ability to process input from many separate sound generators at the same time, and to discriminate the character and quality of each sound at will, is much more difficult to explain than is our response to separate individual tones. Consider the way we hear a band or an orchestra, or just the normal mix of environmental sounds in everyday life: We are perfectly capable of listening to the total sound image or of selectively attending to one sound out of the group. If we want to concentrate on the lead guitar or the second cello, we just do it. The ability to perform this kind of selective listening is due to the auditory cortex and its capacity to correlate and use the difference information between the sound perceived by our two ears, a subject discussed further in Chapter 16.

Several psychoacoustic phenomena are involved in the handling of multiple, competing sounds. One is our ability to discriminate spatial depth through equal loudness processing; another is our ability to localize sounds in the auditory surround and to discriminate the identity, distance, and direction of a number of individual simultaneous sounds. (Localization is discussed in Chapter 16.) A third process that affects perception of multiple sounds is the phenomenon of masking.

Masking is involved whenever the cochlea has to respond to two or more sounds at the same time. The threshold of hearing is the amount of stimulus energy necessary to cause the hair cells of the cochlear duct to respond. Pressure at any point along the basilar membrane stretches the membrane, but if the membrane is already under tension, pressure from a second stimulus will have to exceed the normal required for its frequency in order to produce a detectable response. Because the basilar membrane and the cochlear duct are fixed at the stapidal end, the basilar membrane must stretch from the apex back toward stapes. Therefore, low frequencies, which are detected at the apical end of the basilar membrane, tend to affect the threshold for all higher frequencies, especially those near the masking frequency and at or near its integral multiples. High frequencies, which are sensed at the stapidal end of the basilar membrane, tend to mask only the frequencies above themselves, not lower frequencies. Figure 15-9 shows a graph of masking functions for two tones, one low and one high in frequency. Note that, as stated, the most severe masking occurs at frequencies immediately adjacent to that of the masker and its integral multiples.

Masking accounts for familiar situations in which hearing fails to give us complete information. Suppose that you are standing in the parking lot of an airport talking with several friends. During the conversation, a jet takes off, and for the time the jet is in its runout for takeoff, you have difficulty hearing what your friends are saying. You might say, "Well, of course. The jet engines are just too loud to talk over." But

[5]Confusion about the nature of a sound can result even when there is some kind of validating context within the action or dialogue of the play. For example, in Jarry's *Ubu Roi* there is a scene where Pere Ubu "tries" all of the noblemen, condemns them to death, and seizes their wealth. Death, in this case, means being thrown into the "disembraining machine." The nature of the machine is unspecified and the play is a dark comedy. As each nobleman is pushed offstage (thrown into a trap, or whatever the immediate action may be) the noise that comes from the "machine" needs to make some kind of humorous statement about the character of the person being "disembrained." The audience doesn't usually see the execution; so the sound is the only indication of what happens. A wide range of noises is possible, but invariably a sound of fairly long duration is required, and the staging needs to be managed to permit the sound to be heard at a fairly high level of loudness. Otherwise, the minima for satisfactory integration are not met, and the sound becomes confusing.

FIG. 15-9: THRESHOLD MASKING

MASKING TONES AT 20, 40, 60, AND 80 dB

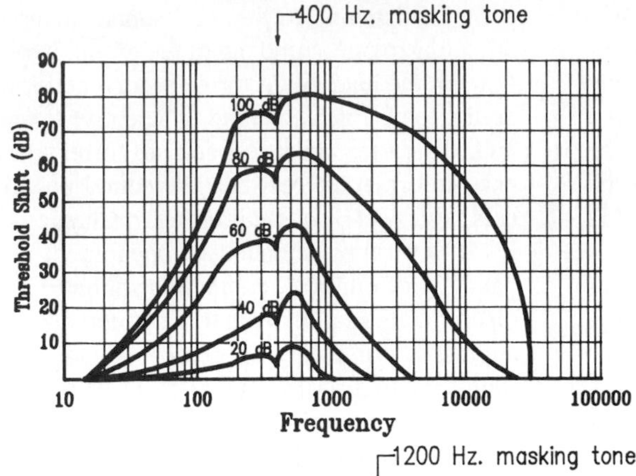

Low frequency masker shifts threshold for most of audible spectrum.

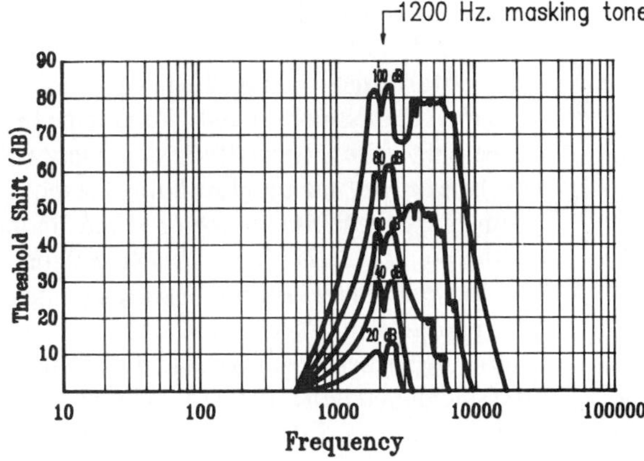

High frequency masker shifts threshold for frequencies above itself with little effect on lower frequencies. Most pronounced masking is for integral multiples of masking tone.

(Adapted from data by Wegel and Lane.)

think about it: The plane is probably a quarter to a half mile away, and your friends are standing within a radius of 3 to 4 feet from you. What is it that really makes it hard to hear your friends' voices? Masking, of course. The sound of the jet engine produces a large amount of low-frequency noise, and low frequencies mask high frequencies. If your friends shout, you can probably understand what they are saying. If one member of the group were a girl with a high pitched voice, she would be much more audible than the men in the group, because the effect of masking diminishes as second stimulus frequency gets progressively higher.

Masking is what makes underscore music for legitimate theatre difficult to do satisfactorily. The principle of masking—that low frequencies mask higher frequencies, but high frequencies do not mask lower ones—explains why, when underscoring does work, the music used usually has lots of high pitched instruments with very little bass, or has been equalized to reduce the level of the lower frequencies.

Consider, also, the effect that masking has on perception of timbre. First, masking explains why the lowest harmonic is usually the dominant frequency of a tone. Second, masking explains why the partials are not really heard as individual components. But the fact that partials are not heard individually also implies that a subthreshold stimulus can play some part in the hearing process, indicating that thresholds—at least those described by the Fletcher-Munson curves and by the masking curves—are thresholds of conscious awareness and not necessarily absolute thresholds.

Recall that low stimulus energy focuses sound into the middle of the audible spectrum because of the equal loudness contours of human hearing. Consider the relationship between masking and equal loudness. The equal loudness contours make possible facile perception of nearby sources without masking interference from the low-frequency components of distant sound sources, unless, of course, such noises have very powerful low-frequency components, as in the case of the jet plane. When we wish to underscore dialogue in a dramatic production, the adjustment we need to facilitate is to make the underscore music or sound assume a distance perspective so that its low-frequency elements do not mask the actors' voices.

The characteristics of hearing discussed in this chapter are the most fundamental aspects of audition. These characteristics are called the **monaural characteristics**—those that apply to hearing with a single ear alone. The auditory system is much more involved, and Chapter 16 continues the examination of the process of hearing and audition.

Chapter 16

Binaural Perception and Psychoacoustics

The characteristics of the auditory system that we considered in Chapter 15 would function even if humans had only a single ear. Human hearing is not a one-eared process, however. We have two ears, and, just as our two eyes do not receive exactly the same distribution of light energy, so the two ears receive signals that are slightly different in intensity, phase, time of arrival, and frequency content. Interaural difference is very important. The human brain is divided into two hemispheres. Each incorporates an auditory processing center that receives signals from the opposite ear. Using a process called **binaural summation**, the brain evaluates the difference in signal between the two ears and uses the interaural difference to synthesize a symbolic spatial model of the auditory environment. Interaural difference produces a sense of orientation in space. The details of that orientation include the proximity or remoteness of the sound, its horizontal and/or vertical location, the acoustic character of the environment (open or closed), and the effective spatial volume of the environment. Binaural hearing characteristics are extremely important to the theatrical sound designer, because the spatial distribution of sound in live theatre is always a critical element in the design of an auditory environment.

Interrelationship of Visual and Auditory Space

Humans frequently hear the direction and distance of a sound and then turn the eyes to confirm the placement and identity of the sound source. Thus, one function of hearing is to serve as a ranging and direction-finding system for vision. Although vision is more precise in localization and in discrimination of objects within the environment, its immediate scope is limited and reorientation times are relatively long. Vision operates through a limited field (roughly 180 degrees horizontally and 130 degrees vertically), which means that only part of the environment is visible at any one time. To bring the eyes to bear on any given part of the environment takes significant transition time, especially if not in the immediate field of view. Hearing, by contrast, operates through a full spherical field. Perceptually, audition senses the presence and direction of a sound source without any significant delay; thus, it is faster than vision in detection and primary localization of objects within the environment. Hearing is less precise than vision in recognition and identification of those objects.

Essentially, all of the cues to auditory spatial orientation result from the process of binaural summation. Loss of an eye does not completely sabotage the sense of visual depth or impair ability to perceive direction; loss of one ear, however, is invariably fatal to satisfactory perception of auditory space. Persons who are deaf in one ear can hear sounds, but lose most of the ability to detect depth or direction and to extract one sound from a jumble of noise.

Interaural difference enables the auditory system to detect the direction of the source of sound with respect to the orientation of the body in three-dimensional space. The perceptual result of this process is called **localization**. Localization is far more accurate horizontally than vertically, however. Therefore, we tend to discriminate directionality of a sound primarily according to its horizontal coordinates. Working with the monaural cues described in Chapter 15, binaural summation produces a sense of both direction and distance for sound sources.

Derivation of Direction

Perceptually, sounds reach our ears from four primary directions: front, right and left sides, and rear. Each primary direction possesses its own signature qualities that make it readily identifiable to the auditory cortex. Direct frontal and rear sounds essentially reach both ears with little or no interaural difference. Each ear receives exactly the same frequency content, phase relationships, intensity, and time of arrival. Sound from behind differs perceptually from that produced by a source in front only to the extent that the pinnae create a subtle but unique modification of frequency and phase characteristics that the auditory cortex readily identifies as "sound to the rear." For acoustic energy arriving from the left or the right, the signature character is marked by clear sound at the near ear, with obvious alteration and delay of sound at the opposite ear. When sound reaches the ear from the side, the head itself creates a rather dramatic sound shadow, producing significant interaural differences. The head mass reduces sound intensity and affects the equal loudness character for the shadowed ear, alters interaural phase relationships, and causes diffraction of energy within the middle and upper range of audible frequencies. As angle of arrival moves from direct side toward front or rear, the amount of interaural difference decreases.

The value of two ears in localizing sound resides in the fact that the brain can triangulate the coordinates of the sound source within acoustic space. The distance between the ears serves as the base of the triangle, with the sound source at the apex. The distance from each ear to the source form the triangle's legs. The fact that the ears are positioned along the horizontal axis of perceptual space accounts for our superior ability to localize horizontally.

Horizontally, we can spot the location of a sound to within an angle of 15 degrees. That's good enough to narrow the location of a singing bird to a particular tree, but not adequate to find the exact branch on which the bird is sitting. Vertically, we can discriminate placement of a sound to an arc of only 60 degrees. That's effectively adequate to locate a sound source only when it is almost directly overhead, and lateral reverberance easily defeats even this degree of vertical localization ability. (Apparently, our primitive ancestors didn't have many life or death threats hanging over their heads in the form of flying or tree-dwelling predators.) Once we do figure out that the sound source is above our heads, all we have to do is to tilt the head back (i.e., point ears upward) and use our more efficient horizontal localizing ability to spot the source of the sound. (Perceptually, "up" is always above the head, and the horizon is always at eye level.)

Human ability to localize rather accurately horizontally but somewhat less accurately in the vertical plane affects the sound designer in at least two significant ways: First, if we have to dub in the sound for an actor or a prop such as a radio, the loudspeaker that provides that sound must lie along or very near to a line through the actor or the prop to each member of the audience. Otherwise, audience members can clearly tell that the location of the sound doesn't line up with the visible object that the dramatic context suggests should be the sound's source. The loudspeaker works best, therefore, if it can be placed under, over, or right behind the actor or prop. Second, because only the horizontal placement of the loudspeaker is really critical, placing the loudspeaker above the stage is usually satisfactory. Unless height exceeds 60 degrees above audience eye level, very few people will discern the fact that a loudspeaker is actually higher than the supposed source of the sound.

Ability to localize direction is affected by field characteristics, however. We do better both horizontally and vertically in an open field (outside in open air) than in a closed field (indoors in a reflective room). We have less trouble localizing the sound of the bird in the tree than we would have locating the precise position of a friend calling to us from inside a large gymnasium. Reflections from the barrier surfaces enclosing a field create ambiguities that tend to confuse our ability to localize. The more reflective the surrounding surfaces, or the larger the number of boundary surfaces, the more we are forced to rely on vision to find the source of the sound.

The tendency of a closed space to blur localization cues often makes placement of loudspeakers for sound effects tricky. The stage house and the auditorium frequently have relatively reflective surfaces. Reflection times are usually short, so sounds from loudspeakers don't normally produce much reverberance; but directionality can easily be blurred by the reflectivity of stage house or auditorium walls. As soon as a set is placed on stage, however, it's a new playing field. Different configurations of scenery and the various different kinds of scenic materials that may turn up in a set all create unique acoustic effects. The designer needs to be aware that loudspeaker placement can seldom be finished until the set is completely mounted and masked.

Derivation of Distance

Localization primarily detects direction of the sound source. Estimation of distance relies on several additional factors: equal loudness contouring (which is the most important cue), presence or absence of peripheral noises, and reflections produced by boundary surfaces in the environment. Each factor deserves some examination in detail, as controlling the illusion of distance is a primary concern of the sound designer.

Equal Loudness

The equal loudness contours (Fletcher-Munson curves) affect the degree to which we hear the entire spectral output of a sound source. Most sound generators produce a fairly wide spectrum of frequencies. If the sound source is relatively near, we will perceive the entire bandwidth of the spectrum; but move the source some significant distance away and perception of the lower and higher frequencies will decrease markedly. Recall that intensity of a sound will decrease as the square of the distance from source to listener—that is, twice the distance, one quarter the intensity; three times the distance, one ninth the intensity, and so on. (Review Chapter 1, particularly Fig. 1-1C, for an explanation of inverse square law.) This logarithmic diminution of source intensity with distance enhances the apparent mid-band focusing of sound that is the characteristic of equal loudness contouring.

Assume that a machine produces wideband noise with a significant number of low frequency components, and creates a wideband sound pressure level of 100 dB at a distance of 2 meters. (Figure 16-1 shows the equal loudness relationships for this signal.) Intensity is relatively uniform throughout the auditory spectrum; therefore, loudness is nearly equal at all frequencies. Loudness ranges from approximately 180 phons (at 3 kHz) down to approximately 30 phons at 35 Hz. At 16 meters (Fig. 16-2), the sound pressure level drops to approximately 30 dB. On the Fletcher-Munson graph, however, an intensity of 30 dB equals a loudness of 30 phons at only two points in the spectrum: at 1 kHz and at approximately 4 kHz. Frequencies between those points will be audible as a loudness of 30 phons or better; above or below those points, frequencies will appear to diminish in loudness. Comparatively, a much narrower band of frequencies will be audible at 16 meters than at 2 meters. Clearly, an observer moving from 2 to 16 meters' distance from the machine experiences a much more pronounced drop in loudness at frequencies below 700 to 800 Hz than at frequencies between 800 and 5000 Hz. As a result the sound of the machine at 16 meters will be "thinner" or "lighter" than when the observer is only 2 meters distant. Apparent decrease in low pitched or very high pitched components of a sound is characteristic of increase in distance from the sound source, and is one of the most important cues that the brain uses to estimate the range of a sound source.

Figure 16-3 adds an additional factor that affects perception of sound at a distance, namely, the attenuation of high frequencies by the atmosphere. Notice that the highest frequencies diminish to a greater degree than equal loudness contouring would predict. Thus, with distance, any sound is naturally focused into the mid-range of the audible spectrum.

Normal environments contain many sounds. Those nearest us appear louder and timbrally more complex; those further away, softer and thinner. A part of a sound designer's function is to determine the necessary distance of a sound and to adjust spectral response of the system to enhance the illusion of distance.

Peripheral Sounds

One other cue to distance resides in the peripheral components of a sound that the ear can detect. Almost all sound generators produce incidental, operational noises, and proximity to a sound source usually makes those noises more noticeable. For example, if you are sitting a few feet away from a bassoon player, you will inevitably hear the instrument's keys clicking, and you will hear the player release and inhale air. Seated at the back of a concert hall you will hear only the tone of the bassoon, not the peripheral noises made by the

FIG. 16–1: PERCEIVED LOUDNESS OF HIGH LEVEL STIMULUS

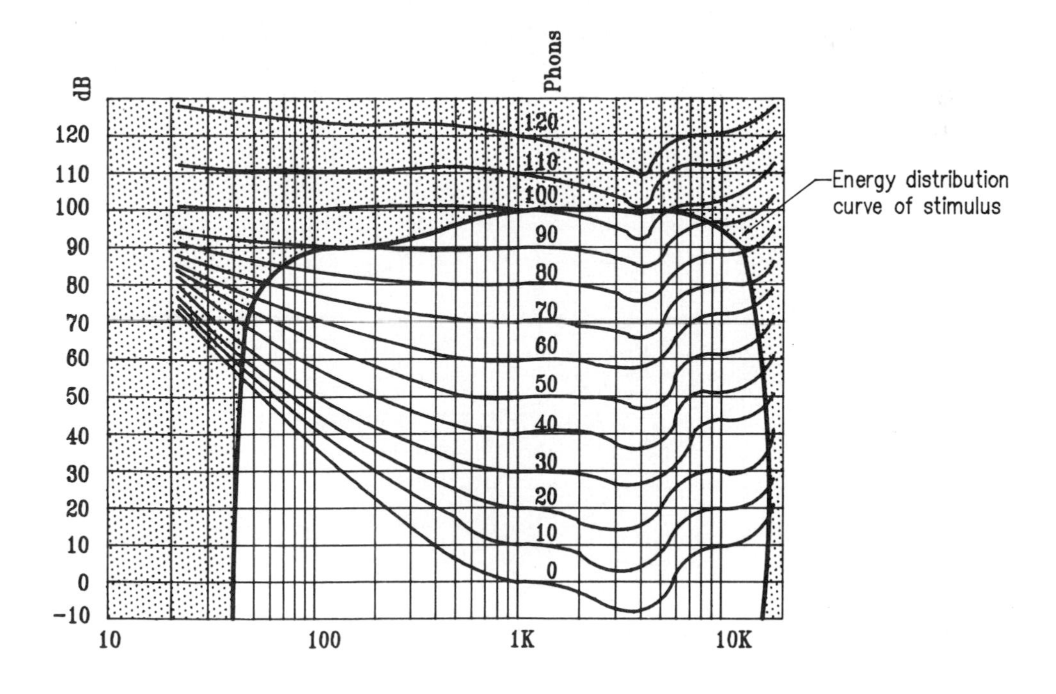

Perceived intensity with wide-band stimulus energy at high sound-pressure level.
Loudness is never less than 70 Phons for any significant frequency.

FIG. 16-2: PERCEIVED LOUDNESS WITH LOW LEVEL STIMULUS

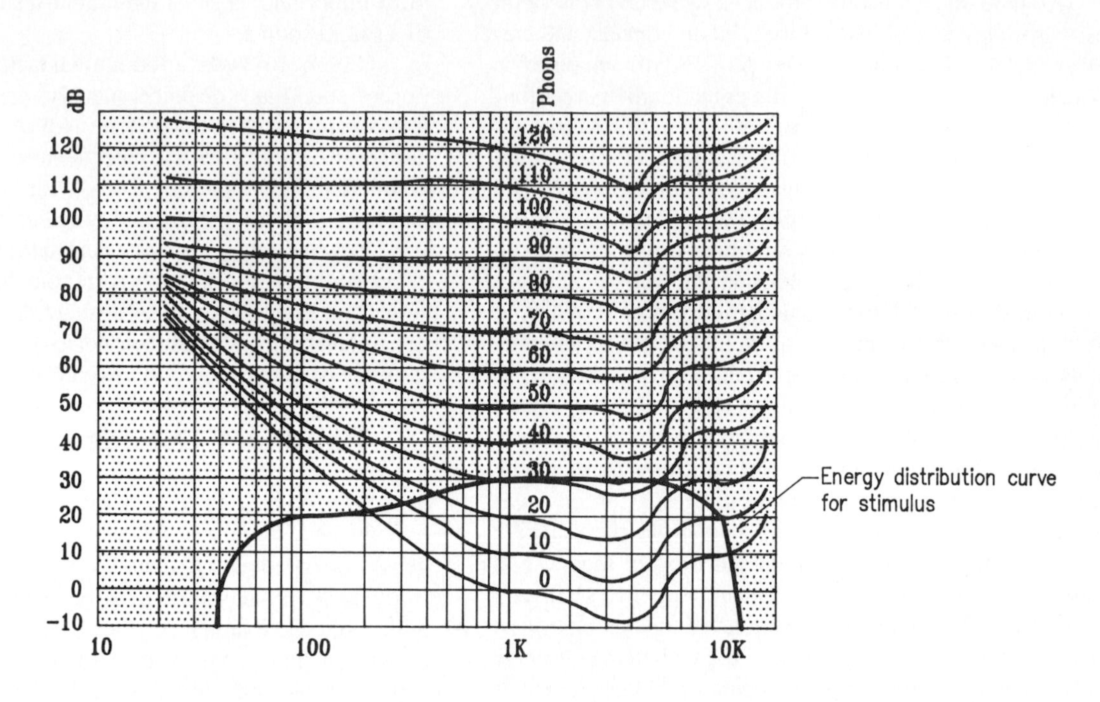

Loudest frequency components are between 500 Hz. and 10 kHz.

FIG. 16-3: PERCEIVED LOUDNESS OF LOW LEVEL STIMULUS WITH EFFECT OF ATMOSPHERE ON HIGH FREQUENCIES

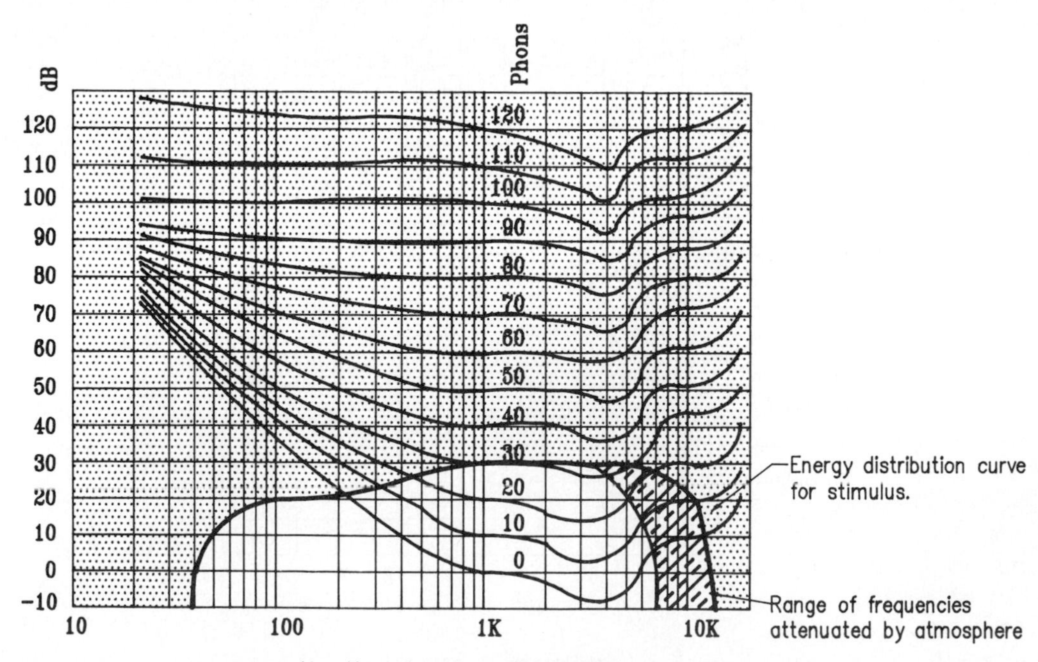

Air attenuates upper frequencies.
Major audible components are between 500 Hz. and 5 kHz.

instrument and the player. The brain uses perception of peripheral sounds as a cue to proximity even though we normally ignore the extraneous noises and attend only to the music.

One frequent problem for the sound designer is that recordings of various sources are usually made with the microphone fairly near the source. The microphone invariably picks up all the peripheral sounds. Peripheral sounds become a serious problem when the designer must make the sound appear to come from a distance. Because the ratio of desired signal to peripheral noises is fixed at the time of recording, reducing the intensity of the sound doesn't eliminate the peripheral sounds. Peripheral noises are very difficult to remove. Sometimes it's possible to find a more suitable recording, but most often the designer has to find a way to disguise the peripheral sounds or to compensate for their presence.

Environmental Ambience

Sounds always occur in particular physical environments, and the environment affects perception of the sound. We judge the nature and size of the environment mostly by the way in which it reflects sounds from sources distributed throughout the environmental space.

Boundary elements in an environment produce echoes of acoustic energy radiated in directions other than toward the listener. The auditory cortex assumes that direct sound from a source will arrive before reflected sound. The cortex estimates the size of a space by comparing the time delay between arrival of the direct sound and the arrival of the first reflection. Reflections may arrive too quickly for us to discern them as discrete echoes, but the auditory cortex is quite capable of recognizing reflections arriving as early as 3 milliseconds after the direct sound. Depending on the configuration of its boundary surfaces, a given environment will produce acoustic reflections ranging from simple to very complex. The nature and kind of reflections are extremely important to the brain's construction of a symbolic spatial model of the auditory environment.

If we are out of doors in an open space (no large trees, no buildings, no mountains or other large boundary surfaces nearby), and some distance away, someone is working with a hammer, we will estimate distance exclusively through the effect of the equal loudness contours. The acoustic model will give us only a sense of open space. Add a single, major boundary element to the environment, and the hammer will produce a single echo. We have added an element to the model, providing more detail about the shape of the space and the location of the noise within it. If both the hammer and the boundary are some distance away, both direct and reflected sound will sound relatively thin due to the effect of equal loudness contours.

Consider a slight rearrangement of the situation. Let's say that the person with the hammer is only a few feet away. Intensity of the hammer stroke is great, and equal loudness contouring introduces minimal low- or high-frequency loss. The reflective boundary surface, however, is still just as far away as it was in the previous example. The echo will have a noticeable delay, but the echo will also undergo timbral change, relative to the direct sound, due to equal loudness effects. The direct sound of the hammer will be significantly stronger and more powerful than the weaker, thinner sound of the echo, even though both sounds are produced by the same event. Localization and equal loudness contouring now clearly provide a foreground/background configuration that gives significant distance information about the source and spatial information about the environment. If we add a few more boundary surfaces to the environment, the number of reflections increases, but each reflection serves mainly to add detail to the model of spatial geometry.

When boundaries effectively enclose a space, however, auditory conditions change materially. Out of doors the hammer sound at a distance was thin and not very loud. Put the hammer inside a building, even in a space that is physically just as big as that outdoor field, and the hammer will sound louder no matter what the distance. Equal loudness becomes less of a factor because low-frequency energy is sustained for long enough within the environment for integration to give those frequencies greater loudness. Cues to distance now depend much more heavily on reflections than on the effects of equal loudness. (Note the implication that reverberance affects perception of loudness.)

Reflections inside an enclosed space can be quite complex, much more so than the few echoes that are likely to happen out of doors. First of all, the indoor situation provides potentially reflective surfaces on all sides of the source. Some of those sources are almost certain to be relatively close to the source, the observer, or both, meaning that some echo paths will be very short. Others may be fairly long. Reflections along the shorter paths will reach the observer quickly, perhaps only a millisecond or so after the direct sound. Early reflections within enclosed spaces usually tend to be strong. Later reflections are generally less intense. Because the entire space is enclosed by potentially reflective surfaces, the reflections themselves may generate reflections. One hammer stroke can set up a reverberant sound that takes a long time to die away. Overall reverberance forms a very powerful cue to the size and surface character of a room. The dominant cues to distance of the sound source for indoor environments are ratio of direct to reflected sound and delay between the arrival of the direct sound wave and the arrival of its first reflection.

Presence

Distance cues taken altogether constitute a psychoacoustic characteristic known as **presence**. The perceptual quality of presence is expressed as a feeling of intimacy or of remoteness.

The actual distance from sound source to listener is a major determinant of presence, meaning that equal loudness contouring is an important component of presence. However, reverberance also plays a significant role. Increase in reverberance usually decreases presence; increase in reverberance—again, because of the effect of integration—is interpreted by the brain as an *increase in loudness.* Increase in loudness usually involves an enhancement of frequencies affected by equal loudness contouring. Thus, reverberance and its interaction with integration tend to moderate the effect of equal loudness contouring.

Reverberance has subjective concomitants based on modulation of reverberant character through time and on manipulation of the perceived loudness of the direct sound in relation to its reflections. A sound that has full spectral range accompanied by significant reverberance can give the subjective impression of a huge object located in an immense space. Coupled with an appropriately suggestive context, this effect can produce a sense of threat or menace. Reducing the amount of reverberance gives the effect of a source moving from a distant background to the immediate foreground. Electronic musicians make very effective use of modulation of reverberant character, and manipulation of direct to reverberant sound level has been used successfully in the sound tracks of a number of science fiction and horror films.

Cocktail-Party Effect

Hearing with two ears accounts for the ability of the auditory cortex to perform a useful trick known as **cocktail party effect**. Binaural summation provides our ability to localize, but it also enables us to listen selectively—to pick out a single sound source from competing sounds in the auditory environment.

Consider what happens when you are in a relatively large group of people. You may be talking with one or a small number of persons, while the general murmur and laughter of the rest of the group forms a background around you. Assume that you are interested in the immediate conversation with your group, and you are not really conscious of anything else. Suddenly, from some other part of the room, you hear a familiar voice, or you hear your name or, perhaps, a phrase that catches your attention. Instantly, you can "focus" your ears on that conversation, albeit some distance away from where you are standing. If you choose, you can continue to attend to that distant conversation, or you can switch back to the one going on in your immediate group.

Cocktail-party effect is possible because we have two ears, each presenting the brain with slightly different aural images. The auditory cortex uses symbolic spatial modeling based on interaural difference to filter out extraneous sounds and to isolate the sound that we are interested in hearing. Cocktail-party effect is exclusively a two-eared phenomenon. Experiment has shown that humans cannot satisfactorily isolate one component out of a number of competing sounds using only one ear. To catch our attention and to permit us to comprehend, one sound within a busy environment, heard with only one ear functioning, must be at least 12 dB louder than when we listen with two ears.

Cocktail-party effect works mainly in situations where many sounds are present. When only two or three sounds are present in an environment, auditory attention is usually divided, with the result that attention to any one sound becomes difficult. When many different channels of sound are present, the brain simply accepts the energy as noise without trying to derive symbolic configurations from it. We can then focus on a single one among the competing sound sources.

Cocktail-party effect is useful to the sound designer in situations like underscoring, or at any time when sustaining a sound under a scene would be useful. Every sound designer is tempted, for any number of reasons, to play background sounds or music monaurally, perhaps even through a single loudspeaker. Reasons may include economy of loudspeaker placement or simply reluctance to expend so much effort on a background sound. Monaural reproduction, however, acts as a single sound source, and a single sound source is a strong competitor for auditory attention. Remember that the essence of cocktail-party effect is a surround of sound, where each source occupies a slightly different position in space and presents different information from other sources. Two or more loudspeakers spaced around the stage and the audience provide multiple sound sources and make the job of separating the actor's voice from background sound easier for the audience. (Be warned, however, that more is involved than merely splitting a monaural sound to numerous loudspeakers. Each loudspeaker needs to source a slightly different signal than its neighbor in order for this technique to work successfully.)

Psychoacoustic Illusions

Perception is not exclusively the result of stimulus energy reaching sensory receptors; it is also based on what the brain knows about the status of muscles and nerves, about bodily orientation, and about previous experience. Such internal components of perception can be extremely powerful, sometimes to the point of overruling the actual data of immediate experience. To use a relatively simple example based on the sense of vision, bodily orientation is an extremely important element in processing visual information. Place a subject in a specially prepared chair in a darkened room, then subtly tilt the chair so that the subject is sitting at a slight angle to the vertical. Presented with projected images that are aligned with true vertical, the subject will invariably report that the images are tilted. Lacking other cues, the brain assumes that the body is upright; therefore, the images appear to be tilted.

Loudness Illusions

Like vision, audition is also subject to illusions. The simplest of these illusions have to do with perception of loudness. We have already noted one of them—that the ear interprets increase in reverberance as increase in loudness. A similar illusion occurs when a very small amount of harmonic distortion is added to sound. Again, the ear interprets the change as increase in loudness, not as distortion of the sound. The probable reason for the distortion illusion has to do with the fact that, in the presence of very intense acoustic input, the tensor tympani stiffens the eardrum, making it less compliant. Stiffening the eardrum, however, induces a small percentage of distortion into the waveform transmitted through the ossicles. The auditory cortex is aware of the action of the tensor tympani and assigns the increased distortion to increase in intensity. Apparently, the situation works in reverse, allowing an increase in distortion to fool the cortex into believing that there has been an increase in intensity resulting in contraction of the tensor.

The simplest element of stereophonic sound reproduction is based on a loudness illusion—that the loudest sound indicates source direction. If we make one channel of a stereo system louder than the other, the stereo image shifts in the direction of the louder channel. That's one reason why stereophonic reproduction doesn't work as stereo in most auditoriums. Only a small portion of the audience is anywhere near the optimum position equidistant between the left and right channel loudspeakers. To make stereo reproduction in all areas of an auditorium possible, large and complex loudspeaker arrays are usually required. The auditorium space is carefully zoned. Precisely beamed horns are aimed at each zone for the left and right channels. When the zone lies farther from one channel than the other, the distant channel requires a narrow-angle horn and somewhat more intensity than the near channel, which needs a wider-angle horn and less intensity.

Problems of loudness as an element of localization plague reinforcement as well as stereo reproduction. If reinforcement loudspeakers are placed to each side of the proscenium in a theatre, sound will tend to localize to the loudspeaker rather than to the actors for audience members sitting relatively near one or the other of the speakers. If we use a single cluster directly over the center of the stage, the situation may be marginally better, but an error in localization develops when actors speak from extreme stage right or left.

Phase and Time Illusions

Because time is the primary dimension of hearing, manipulating the time-based variables of the acoustic signal can change the nature of the perceived event. As explained earlier, the brain uses interaural differences, both in onset of sound and in the interaural phase relationships of the signal, for information regarding horizontal placement of the sound source. If we place loudspeakers in a normal stereophonic configuration, we can use a digital delay to retard the signal to either of the loudspeakers by a few milliseconds (which is equivalent to an adjustment in phase). The effect will be a perceptual shift in the apparent location of the sound source. This phenomenon is known as the *precedence effect* (or *Haas effect*). For example, if we delay the signal in the left channel, the signal will appear to shift to the right. In other words, the apparent source location will shift toward the signal that arrives first.

Precedence (Haas) effect is commonly used in reinforcement to overcome problems in localization. The technique is simple: Place a delay in the signal fed to the reinforcement loudspeakers so that direct, unamplified sound from actors' voices will always reach the audience first. As long as this direct sound arrives first, the audience will hear even the amplified signal as originating from the actors rather than from the loudspeakers. A secondary benefit is that such use of the Haas effect also serves to improve gain before feedback.

Localization based on precedence is modified by localization based on loudness, of course. If some part of an audience is very near a loudspeaker, the loudness cue may overbalance the precedence cue, so that the perceived source of sound remains the loudspeaker rather than the actor.

Pitch Illusions

Localization is subject to a number of illusions based on frequency of stimulus. One that suggests potential uses in theatre sound is known as the **octave illusion**. The octave illusion occurs when two musical tones separated by an octave are played simultaneously to opposite ears, then switched. At one instant, the higher octave is delivered to the right ear while the lower is played for the left ear. At the next instant the high tone is played to the left ear and low tone to the right ear. We might tend to think that we would naturally hear octaves bouncing up and down in contrary motion in each ear, but that's not what happens. Instead, we hear the high tone in the right ear alternating with a low tone in the left ear. In other words, the localization subsystem of the auditory cortex tends to favor the higher pitch as an indicator of directionality. The ear that receives the highest pitch is perceived to be directed toward the source of the sound. The illusion is that the lower tone is perceived as being located to the left, when it really is being delivered to the right ear. Why doesn't the illusion work as a low tone switching ears? Why, instead, do we hear a high tone in the right ear followed by a low tone in the left ear? The answer seems to have to do with handedness. Most people hear the illusion as described, but a significant percentage of left-handers hear the illusion reversed—high tone in left ear alternating with low tone in right ear.

The importance of octave illusion for theatre is that, like precedence effect, it suggests a good way to control the apparent directionality of a sound. The perceived alternation of the experiment would rarely be useful for theatre sound as an effect, but its implication that high frequency is a cue to direction most certainly is useful. For example, a slight boost for frequencies in the favored 2000- to 4000-Hz band can often help to overcome directional ambiguity when loudspeakers must be placed in difficult locations.

Octave illusion suggests a solution to a well-known problem in reinforcement called (among other names) "sibilance bounce." Sibilance bounce occurs when an "S" sound suddenly gives away the location of loudspeakers. Even when the Haas effect is applied, sibilance can be strong enough so that loudness will override all other cues to localization. The sound will clearly appear to originate at the loudspeaker location, making the actor's voice momentarily appear to bounce up to the position of the loudspeaker. Octave illusion suggests that the cause of sibilance bounce is the tendency of the auditory system to use high frequency as a cue to direction.[1] Attenuating the band of frequencies associated with "S" sounds can reduce the effect of sibilance bounce. Signal processors frequently used in broadcasting, called de-essers or sibilance suppressors, are very helpful also.

One other potentially useful illusion of localization is the **drone effect**. If we take a melody, say, one that has a number of repetitive note patterns, and break up the notes, feeding each separate note, or possibly groups of two or three notes, alternately to each ear, the hearing system will fuse the stimuli into one melody line. Adding a second line, a lower frequency drone, to the pattern does not materially change the condition, provided that the drone is always fed to the ear *opposite* the one to which the melodic notes are played. However, if the drone is in the *same* ear as the melody notes, the illusion suddenly disappears and the switch between ears becomes very clear. One possible use for this illusion in theatre involves those times when we need to make a sound appear to fly rapidly and randomly around a space. Getting the sound to appear to jump from point to point is usually quite difficult because the auditory cortex does tend to try to make a single event of a discrete group of sounds. Use of the drone can sometimes facilitate the desired effect.[2]

Temporal Masking

Masking (see Chapter 15) has already been mentioned as a characteristic of hearing. The kind of masking discussed earlier is known as *spectral masking*. Another kind of masking, called **temporal masking**, is also possible. Temporal masking occurs in two forms: forward masking and reverse masking. Both concern the ability to reconstruct the memory of an acoustic event following the occurrence of the masking stimulus.

Forward masking is the situation in which a very powerful sound immediately precedes a less forceful sound. (In point of fact, the masker doesn't even have to be a sound. It can be any event that the brain interprets as requiring immediate and undivided attention.) Let's say that you are attending a piano recital. Just as the pianist sets his hands to play, you hear a loud crash from the back of the hall. The pianist, undaunted, starts the performance, but you are momentarily distracted by the noise and turn to see what happened. When you decide that there's no serious problem and turn back to the front, you realize that you can't remember hearing the pianist start to play. Try as you will, you cannot bring up even the slightest trace of a memory of hearing the piece of music begin. The noise from the back of the room masked the start of the music not because it stressed your basilar membranes and raised intensity threshold to a point where you did not perceive the energy from the piano, but because your brain focused attention on the event at the back of the hall and excluded any data that did not directly bear on the identification and assessment of that event. (Note the similarity of forward temporal masking to cocktail-party effect.)

Reverse masking is similar to forward masking except that the masker takes place immediately after the onset of the masked signal. Depending on the severity of the masking event, the displacement in time may even be quite long. Let's go back to the piano recital and change the conditions. The pianist sets his hands and starts to play, but shortly after he begins the pillars under the balcony give way and the whole balcony crashes down. Fortunately, there was no one under the balcony at the time, but, of course, the recital stops. Later, someone asks you if the pianist had begun to play before the balcony fell. Suddenly, you realize that you don't know. You have no memory of what happened just before the balcony fell. The noise of the collapse and the resulting shock masked the perception of auditory events immediately prior to the event (and may have masked perceptions from other forms of sensory cognition as well).

In both forward and reverse temporal masking, the brain appears to execute something like a "buffer dump" (to use a

---

[1]Note that when loudspeakers are located above the center of the stage, sibilance bounce is partly a matter of *vertical* localization. This situation implies that vertical localization may have more to do with frequency than with intensity, phase, or simple directionality. In fact, at least one study suggests that, deprived of other cues, subjects will tend to indicate a location above horizon for high-frequency tones and a location below horizon for low-frequency tones, even when the actual sound source is directly in front of them. If this characteristic of hearing exists, it could also have useful applications in theatre sound. That is, if you want a sound to appear to come from above, use a sound that has lots of high-frequency components.

[2]Note that most of the frequency illusions described are based on experiments done using either headphones or an anechoic chamber. These experiments suggest characteristics of the auditory cortex, but no guarantee exists that the illusions will work as described in a live field—one that contains reverberance or echo. Solutions to problems based on the experimental illusions may or may not work, depending on the characteristics of the immediate auditory environment.

computer metaphor) whenever an event of potentially critical significance occurs. Memory that holds incoming data pending recognition and identification/classification is emptied and readied for data pertinent to the emergent event. Thus, in both of our examples, you were unable to remember anything about the performance.

Temporal masking can sometimes be used in the theatre for special effects, but it is not the easiest phenomenon to control for artistic purposes. For example, one could create a loud sound as a distraction, so that some action onstage happens unnoticed. On some few occasions, something so disturbing as a sound loud enough or sudden enough to cause temporal masking may be justified, but extreme caution must always be used, as high impact noises are stressful and potentially very dangerous. Knowledge that temporal masking can occur, however, can certainly help the sound designer keep from interfering with the perception of other elements of the production. Temporal masking does not result exclusively from loud sounds. We should always remember that *any* distracting element in an environment, whether loud or not, can produce temporal masking. The need to figure out why an unexpected event occurs where and when it does is precisely the condition under which a buffer dump takes place. Thus, a sound effect that doesn't work within the context of the production can easily upstage actions and dialogue—to the point of making the audience lose track of a whole section of the play. Unless one has good reason to do so, and the event is planned into all aspects of the production, avoid making the audience wonder, "What is that sound, and why is it happening?"

Auditory Perspective

Auditory perspective is the term by which we refer to all of the perceived spatial characteristics of an auditory field. The word *perspective* is a visual term. The Latin roots of the word specifically mean "to look through or into." So, in effect, auditory perspective is a misnomer, but no other suitable term exists by which to refer to the spatial information that we derive through our sense of hearing.

With eyes closed we can estimate the size and surface quality of a space, we can tell the direction and approximate distance of sound sources within the space; and we can estimate the positional arrangement of sound sources and boundary surfaces. We can certainly tell the difference between being in an enclosed space and being out of doors. All of these things constitute auditory perspective. We use auditory perspective daily, but we do it so automatically that we are scarcely conscious of the process.

Auditory perspective applies to dramatic environments as well as to those we encounter in daily life. Within the dramatic environment, auditory perspective is extremely important because sound offers one of the more powerful means of building symbolic imagery and leading emotional response. When auditory perspective is incorrect and fails to agree with the visual elements of the scene, much of the symbolic and emotion-leading power of the sound is lost. Auditory perspective, therefore, is one of the most important aspects of sound that a designer must learn. Sound in dramatic production fails more often because of improper auditory perspective than for any other reason. Furthermore, auditory perspective applies to everything that happens in a dramatic production, including reinforcement.

In order to meet the needs of dramatic auditory perspective, a designer must understand both acoustics and psychoacoustics. To supplement this basic knowledge, the designer prepares an inventory of the major characteristics of the particular dramatic environment, an estimate of the distance of each potential sound source from the central locale of the setting, and an idea of the sound quality of each particular sound source according to its function in the development of the dramatic events. Each sound that the designer prepares must be adjusted according to its relative position within the dramatic spatial environment.

One additional but important factor that enters into the process of designing dramatic auditory perspective is the particular spatial relationship that the audience itself has to the stage and the auditorium. The relationship is both auditory and visual and includes the presence of the auditorium and the audience itself as well as the details of the stage setting. No matter how willing an audience may be to suspend disbelief, as dramatic theorists term the observers' resolution of apparent discrepancies between known real location and the presumed locale of the dramatic action, audience members never truly lose awareness of their surroundings—the boundary surfaces of the auditorium and the presence of other audience members. Nor should they have to reconcile a sound that doesn't fit properly into the visual scene. Auditory perspective must integrate comfortably into the audience's sense of spatial ambience, or else the designer's sounds will become distracting.

In the real world, sounds occur at random within an environment. In a stage environment, we usually design the timing of sounds to keep them from overriding or distracting from dialogue. Sometimes, however, playing a sound under dialogue is desirable. In such situations, the sound must not be allowed to mask any part of the dialogue. Using multiple sound channels to spread the sound field, we use cocktail-party effect to reduce the possibility of masking, if not by a full 12 dB, at least by a significant amount. Also, providing interaural difference makes a sound less recognizable as a recording. We can also use our understanding of the characteristics of simple masking to process the sound in ways that further reduce the possibility of interference. We know that low frequencies mask immediately adjacent and higher frequencies, so we can attenuate frequencies in the same range as the fundamental frequencies of the actors' voices. Usually, this distortion of the spectral character of a sound is not terribly noticeable, because the resultant spectral character tends to accord with equal loudness contouring for distance.

One might well ask if we should use multiple channels when a sound must appear to originate from one specific place, and if so, why and how. A short spot effect that does not compete with dialogue probably needs only a single channel, but with more complex or subtle effects at least two channels should be used. One channel should provide "direct" sound consistent with the direction of the intended sound source; any others should carry echo or reverberance information. Both direct and reverberance channels should be equalized appropriately for distance and environmental conditions.

Equalization and filtering of sound spectra is frequently necessary to auditory perspective, because recorded sounds rarely come balanced with equal loudness contours appropriate to the distance required by the dramatic environment. Recording technicians usually set out to get maximum fidelity and signal-to-noise ratio at all frequencies in the recording. To get the effect of a machine 16 meters distant, as in an earlier example, we do not simply move the microphone 16 meters away from the sound source. Microphones register decrease in intensity when moved further away from a sound source, but do not necessarily register the change in spectral character that equal loudness contouring would suggest. Distance from the sound source also increases the possibility of picking up undesired noises from the surrounding environment, degrading signal to noise ratio. In order to give the perceptual impression of a sound 16 meters away, a good quality recording (one with good signal-to-noise characteristics) must be processed to conform to the equal loudness contours, meaning that the low and very high frequencies must be reduced in intensity by an appropriate amount.

Psychoacoustics and Artistic Responsibility

The few theatrical applications of psychoacoustics and auditory theory that we have suggested in this chapter only begin to touch the possibilities. Sound exerts a wide range of control over both the emotional and physical life of humans, and phenomena exist that have not yet been adequately explored. For example, the possibility exists that subliminal and superliminal forms of acoustic energy may affect the human organism, even though we may not be able to perceive such stimuli through normal hearing. Even audible energy that falls short of the necessary integration minima may be perceptible at a level well below consciousness and may have a subjectively unrecognized but important effect on behavior and emotional response.

Because the possibility of subconscious control of emotional response exists, the use of certain forms of sound in theatre presents a considerable moral and ethical dilemma. The dilemma opposes the right of an audience member to retain control of his or her own behavioral and emotional response against the right of an artist or group of artists to use the best available means to create a desired emotional response. The essence of art is to manipulate emotional response, but whether the use of methods that percipients cannot readily detect, even if such manipulation heightens their aesthetic experience and increases its value, is open to question. Understanding of the subtleties of human perception yields powerful but dangerous tools. Care must be taken, and thought must be applied to ideas involving the use of these tools. Use of an effect for its own sake, however fascinating or emotionally stimulating, is not sufficient reason. Chapter 17 presents a more comprehensive discussion of the psychological foundations of auditory aesthetics.

Chapter 17

The Psychological Basis of Auditory Aesthetics

The discipline of aesthetics is the philosophical attempt to understand why humans respond emotionally to certain kinds of events, why they derive a sense of value or meaning from them and develop long-term feelings and attitudes about them. Why, for example, should a particular kind of event produce a feeling of joy or sadness in one context but seem emotionally neutral in another? Why, for some kinds of aesthetic events, is impact greatest at first experience, diminishing in intensity with repetition, while for others, the richest emotional involvement comes only with repeated experience? The history of aesthetics is filled with interesting and complex speculation about such matters, but speculation is not scientific fact. Although recent research in neurology and psychology has produced some tantalizing bits of information, a full understanding of the human brain is still remote. We are still limited to speculation (albeit with considerably more data), especially about our aesthetic responses; but that speculation is useful to the understanding of artistic processes.

In psychological terms, **aesthetics** might be described as the attempt to correlate stimuli of certain kinds with the mental states that we experience as emotion and to determine how emotional experience is transformed into meaning. We have learned a great deal about the behavior of biological intelligence through our efforts to produce artificial intelligence. By devising efficient ways to store and process data in electronic circuits, we have discovered what appear to be some general principles of information processing. Both brains and computers require both long-term and short-term memory, specialized sensors to acquire external data, and internal strategies to process large amounts of data quickly and without having to handle each individual datum repeatedly. Unlike electronic computational devices, however, biological brains are associative and heuristic. Humans "feel." We are all quite certain of that fact because we are humans and we experience feeling.

Sensory Input and Emotion

One fact is centrally important to our purpose in this chapter: *sensory input affects change, however slight it may be, in emotional state.* Nothing enters our awareness to which we do not have some form of emotional response. (Even emotional neutrality is consciously felt as an emotional state.) Ontologically, the purpose of emotion is to motivate—literally, to move. Emotion is the subjective state that moves us to respond to perceived conditions, to accept or seek to acquire things that appear desirable and to retreat from and attempt to ward off things that appear harmful. The response is not necessarily overt; in a great many cases the response is some form of mental processing rather than a bodily action.

Viewed as a necessity to the primal drive for survival, emotion cannot be exclusively a function of conscious awareness; and it is not. Emotion lies deep in preconscious portions of the brain. Emotion is reflexive, delivering a state of feeling to conscious awareness almost as a fact of the external world rather than as part of our internal response to that world. Like most reflexive responses, emotion is very fast. Speed of actuation is necessary to survival.

Emotions serve the need of biological intelligence to survive in a constantly changing physical and social environment, and the primary means of gathering information about the environment is through sensory experience. Therefore, sensory input must directly affect emotional state.

Note carefully the use of the term emotional state. We tend to think of emotion in terms of identifiable subjective states: as anger, or joy, or fear, or contentment, or any of the various conditions of subjective experience that are frequent enough and definable enough to have earned names. Emotion, however, is a *continuum*, the ongoing subjective experience of mental information processing. In most animal intelligence, change in emotional state usually results in immediate *overt* action. In human intelligence, overt action is often suspended. Such suspension of action in response to emotional experience is known as reflexive delay. We tend to think of reflexive delay as "maintaining objectivity," that is, the opposite of acting on subjective impulse. Change in emotional state that does not or cannot result in overt action (for example, what specific overt action exists for an emotion such as contentment?) is stored as feeling about the experience and used as data for processing other experiences. Change in emotional state experienced as the result of perceiving an artistic event such as theatre, a painting, or a piece of music is among the forms of emotional experience that are usually stored. The emotional impact is the primary experience of the event; the feelings that result act as factors that subsequently transform the experience into aesthetic meaning.

Consciousness, Awareness, and Meaning

Emotional state does not derive exclusively from external data, of course. Human intelligence is an extremely complex affair. Our brains constantly track internal body states, track the flow of data from the sensory organs, compare that data with information from memory, adjust the balance of attention and the focus of consciousness, and integrate the sum of these processes into a generalized conception of the external world and our relationship to it. We combine these concepts into new perceptions and decisions, relate them to past experience, and project hypotheses of both immediate and long-term future states. The sum total of mental life is **awareness**.

Awareness is not to be confused with consciousness, for they are not one and the same thing. Our brains are aware of our orientation in space, our need to breathe constantly, the processes necessary to activate muscular and glandular operations, the customary locations of objects in our habitats, the habitual operations needed to drive a car, and so on through a very long—perhaps endless—string of daily activities. We are not always *conscious* of these processes, certainly not of the glandular balances, muscular adjustments, repetitive motions, and the like, required to accomplish any of the thousands of things we do each day.

Consciousness is a very specific form of processing that enables the specialized judgments necessary to exist within complex social orders that transcend the immediate conditions of space and time. Consciousness allows us to select particular pieces of data and experiment with them creatively for a wide range of purposes. Consciousness enables what we normally call thought and imagination. Consciousness allows us to recall and rehearse past experience, to derive meaning from it, and to reshape it and consider the possibilities of alternate choices. Consciousness focuses present experience and projects possible future developments. Again one fact is important: *Conscious examination of experience, whether recall and rehearsal or creative imagination of future possibility, produces a shift in emotional state just as real sensory experience does.* After all, one necessary element of human survival is the ability to decide among preexamined alternatives. In order to do that, we must have feelings about which are desirable and which are undesirable alternatives.

The chief concern of human intelligence is **meaning** (so much so that we sometimes try to determine meaning on the basis of insufficient data). The capacity to derive meaning from experience is the cornerstone of human ability. We use sensory data to assess relationships in time and space. We sort the information for its contingency to present subjective state, its relevance to past experience, and to anticipate possible implications for the immediate future. Awareness experiences this process as a kind of "time corridor," a constant melding of past into future. Consciousness abstracts particular events, those with special significance, and uses them as "building blocks" to erect an architecture of reality, a symbolic structure that forms our understanding of the world and our relation to it. Within this symbolic framework of experience we derive two sorts of meaning: logical and aesthetic. Logical meaning derives from that symbolic architecture of objective reality that consciousness creates; aesthetic meaning comes from the transformation of emotional experience—the subjective states that accompany all mental processes (including sensation)—into personal feelings and attitudes. The need to *express* meaning is common to both modes. Words, numbers, and logical symbols are the common means of expressing logical meaning; art (in any of its manifestations) is the attempt to express aesthetic meaning.

Interrelationship of Vision and Audition

The quality of emotional experience is, in part, a function of the sensory modality involved. The human brain does not treat data from all senses equally. Sight and hearing are "distance" senses; they bring in information about the world at large. The other senses—smell, taste, touch—can only supply data about proximate conditions, those near or directly in contact with our bodies.[1] Accordingly, the distance senses rank higher in importance than the proximate senses in supplying the data from which the symbolic structure of reality is put together. Although the two senses complement each other, vision is without doubt the chief of the senses for humans. As a species, our world is predominantly visual. Our eyes provide us with a sense of a complete outlook on a field of relatively stable objects and orientations, of predictable patterns of mo-

[1]Smell, obviously, is to some degree a distance sense. The object that emits an olfactory stimulant need not contact our body surfaces directly. However, compared to othe species, humans derive very little environmental information through olfaction. For us, smells must be either very strong or proximate in order to produce a significant effect.

tion, and of a comfortable arrangement of depth and distance. This is so because of the way in which our neurological "wiring and programming" has developed; but it also means that the brain is most apt to credit visual data over data from the other senses, including audition.

Vision may well be the chief of the senses, but it has its limits. Vision is unidirectional; the view to the rear is available only by turning the head or even the entire body, a slow process, relatively speaking. By contrast, hearing is fast and omnidirectional. Sound is a kind of "all-points, early-warning system," but its information is frequently ambiguous. Many sounds are similar in nature; others merely fail to agree with our stored conception of them. In either case, visual confirmation is required for precise identification of the sound's source. (Clearly, not all sounds are ambiguous, but try keeping your eyes closed while other people are moving around you doing various things. Test your ability to identify objects and actions from sound alone. The results will probably surprise you, both in what you perceive correctly and what you do not.)

The power of sound to produce an immediate emotional response probably outweighs vision, however. Sound can startle a person into motion or immobilize, induce curiosity, send chills up the spine, or give a feeling of safety and shelter. (The reason for this characteristic should be obvious; after all, if sound is an "early-warning" sense, it should be able to make one move quickly, and emotion is the subjective condition that induces movement.)

The chief characteristic of sound is time—duration in time and change through time. Sound is the primary sensory embodiment of the time corridor through which awareness moves. Sound is like "touching" time (not surprising, as the sense of hearing is, evolutionally, an outgrowth of the sense of touch), a perceptible and almost palpable marker of the flow of events. Consciousness tends to symbolize time in terms of sound. Sound also has spatial characteristics, but they are secondary; they are, in fact, a measure of space in terms of time. The spatial characteristics of sound are directionality, reverberance, and component structure. Directionality is the apparent position of a sound source in space as revealed by interaural differences. Reverberance is the acoustic behavior of sound energy reacting with spatial boundaries. Component structure is the number and interaction of separate voices and frequencies (tones) operating simultaneously within an acoustic environment.

Hearing is, perceptually, *sequential*; vision, by comparison, is, perceptually, *simultaneous*. The physical energy variations that translate into our sense of sound or vision arrive one pulse after another in constant succession; but where vision is sensed as a field pattern of simultaneous events, sound is perceived as a sequence of data. Sound structures are perceived as temporal events changing in space. Visual events occur as spatial relationships changing through time. The temporal characteristics of sound are primary; those of space are secondary. For vision, spatial characteristics are primary, whereas those of time are secondary. Together, sight and hearing provide us with a complete flow of sensory data in four dimensions.

The four-dimensional character of experience originates in the perception of simultaneous relationships in "space" and of sequential connections in "time." Recently, we have begun to understand that our perceptions of such relationships derive from the structure of our brains. In effect, we have two data processors—one specialized for sequential relationships and the other for simultaneous relationships. The former controls our use of language and our ability to perceive and use logical structure; the latter organizes recognition of patterns in space and association of related events. The two processors, occupying, respectively, the left and right hemispheres of the brain, are connected by a large trunk of nerves through which they exchange information. The division of processing into sequential and simultaneous modes is an elegant strategy for a complex data handling problem: Specialize one processor for causal, stepwise relationships, the other for comparative/associative judgments, and mix the two to form the overall output.

Sound holds a unique position among the senses in this bilateral data processing scheme, because, more than any other human sense, sound tends to activate both processors. The structural and spatial characteristics of raw sound are simultaneous relationships, but the inflow of acoustic energy is sequential and the "content" of sound (speech and music, e.g.) is a sequence of related symbols.

Memory and the Integration of Sensory Experience

Memory is essential to both sight and hearing. We experience both the relative permanence of spatial (visual) objects and the integrity of sounds largely through the agency of memory. For sound particularly there would be no sense of "event" in any form without memory. Acoustic energy is evanescent; it is gone (as are all "present" events) almost before it can be consciously noticed. Yet, in spite of the transient nature of the physical stimulus, we think of sounds as "solid" events having extension in time and some degree of spatial embodiment. All sounds appear to be complete objects having something like form, line, and texture. Memory holds sounds and accumulates them; other processing centers assess their qualities and meaning. The processing of discernible auditory events is a means of articulating the time corridor into manageable units. In other words, sound provides a system of "markers" in time. (The term *marker events* will become significant when we begin to discuss the use of sounds as a component of drama.)

Earlier we noted the ambiguous nature of sound, pointing out how difficult could be identification by sound alone. Ambiguity, however, is useful in that the attempt to identify a sound instigates a search through memory associations for related auditory structures. In the process, other memory events come to mind. This process makes sound an excellent means to trigger wide-ranging memory. In point of fact, for humans all sound is associative—that is, it derives an identity because we associate it with a particular being or object. Association is one of the primary means by which we categorize sounds. (For bats, porpoises, and other creatures that use echolocation, the acoustic component of the world may well be a very different thing than we can even begin to imagine.)

The associative power of sound hardly explains the intricate involvement of audition in aesthetic experience, however. We all know, at least intuitively, that sound is a powerful element of experience. We enjoy sound. We enjoy manipulating sounds, listening to sounds, playing games with sound, experimenting with the subtle changes in sounds of words. We fill time with sound. (In fact, some of us are almost traumatized by lack of sound.) The human brain contains regions that experimenters have come to call reward centers. These centers impose on all aspects of experience a quality of either pleasure or annoyance. (Note the similarity to the earlier definition of emotion.) The sensation of sound appears to stimulate the reward centers very directly, and most sound appears to be pleasurable; hence our affinity for pastimes such as word games and musical entertainment. Sound is, in a sense, its own reward.

Sensory Modalities and Artistic Expression

Art is the attempt to communicate aesthetic meaning in the way that written language communicates logical meaning. (Note the dual status of poetry and literature within this context.) As grammar and syntax form the structure of language, design forms the structure of art. Design is the attempt to arrange patterns of sensory events in ways that will create an emotional response and, subsequently, the development of aesthetic meaning. Because the design, as raw data of percipient experience, can rarely (if ever) lead back to the original events that produced a significant emotional experience for the artist, the process of art is to devise a metaphor for aesthetic meaning within the modalities of perception of sensory data (visual, tactile, auditory, etc.).

Some arts rely on a single sense and some on the conjunction of senses. Qualitative differences exist between single-sense and multiple-sense arts. A single-sense art must achieve impact exclusively by stimulation of brain areas that process the data received by that particular sense. Only through a generalized search for meaning can other aspects of experience be called into association. Single-sense arts place a heavy demand on the skill of the artist to devise and execute suitable metaphoric symbols, and on the interpretive abilities of the person who perceives the work to call up a full range of associations to derive aesthetic meaning. Single-sense arts, therefore, tend to engender highly subjective, personal feelings that are often difficult to communicate to others. Multiple-sense arts offer a wider range of stimuli to trigger associations, both for the artists in creating metaphoric symbols and for percipients in deriving meaning.

Sound serves single-sense aesthetic expression in the form of music and participates actively in the conjunction of sensory data for multiple-sense arts. Sometimes that participation is in the form of music (underscore for film and theatre) and sometimes as an environmental element of dramatic structure (natural sounds, sound of human activity, etc.) (Sound is automatically involved as the medium of speech, but the verbal elements tend to assume a role in their own right in which sound is only a minor player.)

Musical Expression

Music, particularly instrumental music, may well be the most difficult art to explain. We know that music elicits emotional response very easily, but what is the *meaning* of a piece of music? There is no satisfactory answer for that question, but a great many people are absolutely convinced that musical meaning does exist. In the history of aesthetics, the attempt to determine musical meaning is a virtually insoluble problem. The cause of the difficulty is that musical meaning lies exclusively in the changing emotional states produced by the changing structural relationships of auditory stimuli over time. Visual art has an objective component through which we can find verbal associations, but music is purely a result of subjective experience—the enjoyment that humans derive from playing games with sounds.

Music used as a component of a multiple-sense art is a different matter than music in its own right as a single-sense art form. First of all, the musical meaning must in some way be shaped to match the overall aesthetic meaning of the larger work, which brings the artist directly back to the old aesthetic problem of how to equate musical meaning with other forms of experience. Here the associative power of sound does supply some amount of assistance, however, as music is used for particular purposes (e.g., military marches, fanfares, dances) and has very distinctive stylistic characteristics depending on the national or ethnic origins and historical period from which the music is drawn. The "objective" associations through common usage and historical origin can be aligned with visual and verbal components of the other modes of experience forming the larger work; but there must still be a certain amount of "intuitive" choice on the part of the artist, because a wide range of musical characteristics can adhere to music for particular applications and to music from any given style period.

Sound as a Component of Dramatic Experience

Sound articulates the time corridor of experience, but so much of sound is not relevant to immediate concerns at any given time that we suppress or ignore it. Suppressing or ignoring is not the same as not hearing. Awareness is usually cognizant of the total inflow of auditory data even if consciousness is not, and the instant an important auditory event (at least from the standpoint of immediate perception) occurs, we are instantly aware and attentive. Given that all sensory inflow alters the balance of emotional state, even very slightly, and that sound directly affects the status of the reward centers of the brain, the auditory component of experience clearly serves as a powerful force in shaping emotional response and aesthetic meaning. Sound is the sensory embodiment of the dimension of time; but while the normal condition of auditory sensation is an undifferentiated background of noise, the significant data to which we pay conscious attention serve as markers to articulate the time corridor into units of experience, ordered through derivation of meaning into a hierarchy of importance. Marker events that stimulate some degree of recognition, whether conscious or merely aware, are always slightly different to the prevailing character of surrounding sound, and they tend to trigger noticeable change in emotional state.

Two conditions, primarily, allow us to ignore or suppress conscious attention to sound: repetitiveness and lack of differentiation. The roar of city noise is continuous; the chirp of crickets on a summer night is repetitive. We can suppress or ignore either, but let an owl hoot and we take note of it. If an emergency vehicle goes screaming past, klaxon blaring, we become conscious of it. These characteristics of auditory perception reflect the way in which the ear/brain system is programmed, and the way in which auditory processing relates to other aspects of human intelligence.

A *designed* auditory occurrence is always a different experience than is auditory experience in day to day life, even if we try to simulate realistic auditory conditions. Structured by design, sound is a means of shaping emotional experience through time, as visual design is a means of shaping emotional experience in space. The auditory component of a dramatic production can include both background data and marker events.

Film is a dramatic medium that has made use of a full auditory component for many years; theatre less so. Perceptually, film has at least two attributes that have led to its more complete use of sound: First, auditory balance is relatively simple in film, as all of the sound is contained within an entirely electronic medium; second (and possibly more important), film is capable of a much more detailed articulation of space and visual focus, from panorama to intimate close-up, so that the demand for a corresponding auditory perspective becomes extremely important.

In theatre, the visual element is always at the mercy of distance, and perception of detail is exclusively dependent on visual contrast of adjacent surfaces (both foreground and background) and the enabling (or defeating) balance of color and intensity in lighting. Where perception of facial or gestural detail is essential to understanding, attention is very tightly controlled, especially auditory attention. Under such circumstances, we need a "clear channel" for speech, and additional sounds only interfere. Anything that competes with understanding of dialogue is not permissible!

The auditory component of live theatre is not all of a piece as it is in film. Where the auditory component of film is entirely contained in an electronic medium, sound in theatre is a blend of the live and the electronic, a blend that is often difficult to achieve satisfactorily. Therefore, a prevalent tendency in theatre production has been to strip down the auditory environment of the drama to its bare essentials: the dialogue and the marker events (i.e., the sound effects) needed to articulate special aspects of the dramatic time corridor that words alone do not cover. With increasingly sophisticated technology, the problem of blending is not nearly as severe as it once was, nor is the problem of competing with the actors' voices. Sound can now serve in theatre as it does in film, as a means to shape and lead emotional response, a role that theatre has yet to explore fully. Nothing here implies that every dramatic production should be underscored by sound from beginning to end—not even the most poorly conceived film does that—but the natural use of sound as an additional means of structuring emotional experience in theatre is an art that deserves attention.

The creative processes of the sound designer, like those of visual designers, are largely intuitive. Nonverbal use of sound is a function, in large part, of the pattern (simultaneous relationship) processing activity of the human brain, and involves the ability to understand the emotional character of certain kinds of perceptual data. Most artistic decisions are, by nature, intuitive, and intuition is as much a form of thought as is logical reasoning. It is, in fact, the "logic" of the simultaneous processor. Intuition does not imply either irrationality or caprice. Intuition is the conscious manifestation of aesthetic meaning. Because the arts are attempts to express aesthetic meaning, intuition necessarily plays a significant part in decisions about the structure of emotion-producing components of the artistic product. The art of sound is the meaningful articulation of time and the ways in which spatial characteristics can be arranged to structure temporal events, to affect emotional response. Sound is a difficult art for many people because most of us visualize but cannot "auditorialize." The job of the sound designer or composer is to manipulate auditory background to help lead and direct aesthetic response to dramatic experience.

The chapters in the remainder of this book are intended to explain and illustrate ways in which sound design can be applied in the theatre, how the sound designer functions both in his or her creative and collaborative roles, and some typical processes and techniques by means of which the sound designer can integrate his or her creation into the theatrical event.

Chapter 18

Creativity, Craftsmanship, and Design

An understanding of aesthetics provides a theoretical basis for art, but the actual process of artistic creation relies on artistic insight, on the artist's ability to relate the qualities of a given medium to the communication of general ideas, and on his or her ability to develop the craft skills necessary to use the medium as a vehicle for communication. Design is the application of artistic insight to a specific purpose and rests squarely on the development of both creativity and craft.

Creativity

Creation refers to the process of making something. An implicit sub-meaning in the word suggests that the result of creation is something *new*—something that has not truly existed before. Refashioning something that is already in existence, therefore, would be merely clever but not creative. Is novelty a valid criterion for judging creativity?

Novelty

To find something that is truly new, something that has never been seen or known heretofore, is a rare occurrence. It happens so infrequently that each occasion, if publicly known, becomes a marker event in the history of humankind's existence on this planet. These kinds of *creative discoveries* are the seminal events that drive great intellectual leaps. So, if creativity is to have a more generally applicable meaning, we cannot impose so limited a criterion as absolute novelty.

Even in the cases of seminal discoveries the creative insight discerns a novel set of possible relationships *implicit* within the framework of physical or social reality.[1] Creative discovery is possible because, and only because, we do not receive reality directly; otherwise, we would simply know all aspects of reality and there would be nothing else to know. We receive only the energy transmitted by or reflected from relatively small regions of the world *through a fairly short span of time and only within the limits of our sensory modalities.* We turn that energy into patterns within the brain that we can process, shift, rearrange, recombine. (Note the necessity for memory to supply our own mental counterpart of objective time.) From the transposition and mental recombination of patterns, from the gaps that appear in what we feel should be a set of causal relationships,[2] we are forced to project the existence of things for which we have no perceived data. We are forced to speculate about the realm of the possible.

Projecting the existence of possible but yet-to-be-proven objects and forces is one of the most dangerous things that we do, but also one of the most valuable. At such a point in the search for understanding, all of our assumptions about reality (our symbolic architecture) are on the line. Notice the word "assumptions." It means that we trust that, either as individuals or as a collective society, all of the things we have

[1]Reality is used here to imply the objective world that we can never truly know. We presume that it exists, because we are able to build within our "minds" that construction which I have previously called a symbolic architecture. The terms *physical* and *social* are meant to imply that both natural process and the interactions of life forms are contained within that reality. In any case, observe that the possibility of any set of relationships is either implicitly in that reality or else cannot exist. The discovery of the possibility is the insight, and the knowledge gained is the thing that is new.

[2]Causality, of course, is an illusion—a conception that appears to explain local phenomena. A part of human demand for causal interrelationship is our limitation in time. To be able to experience reality *directly* would demand that we be able to perceive four-dimensional infinity. Our local temporal limits force us to speculate on the historical—how things got to be this way, in order to explain what we see and project how things may come to be. All of our assumptions are tacitly and unconsciously put to the test. What we have perceived or surmised incorrectly will limit the accuracy with which we can continue to perceive and understand.

thought and have come to believe are correct. If we have been wrong before, clearly we will be wrong in future projections (or, worse, we will perceive correctly, but for all the wrong reasons). Guesses about reality accepted either uncritically or without adequate evidence become our superstitions; but without the ability to postulate the existence of previously unknown things or forces, neither art nor scientific advancement would be possible.

Projection and speculation exercise the imagination—the ability to arrange symbols of reality in the abstract (i.e., without the aid of sensory data inflow). Without imagination, human life as we know it at this stage of our development is impossible. Neither awareness nor consciousness can exist without imagination. At its simplest, imagination is no more than the ability to store and replay the mental symbol correlates of sensory data, the ability to recall and rehearse. At its most complex it is the ability to assemble, to formulate, and to reshape experimentally whole systems of possibilities, attending to the emotional resonances within the mind as the possibilities shift form and shape.

Insight

The creative urge results from **insight**. Insight occurs when some confluence of events brings an individual into a profound confrontation with some previously unknown element of reality, or into a confrontation with paradox—with ideas or appearances that seem entirely contradictory and incompatible with previously accepted notions. The confrontation creates a surge of awareness that sharpens the senses and heightens perceptions—that engages perception in maximizing data inflow and tunes the processing centers of the brain to begin the process of opening the symbolic architecture of reality for revision and restructuring.

Science, often thought of as consummately rational, avoiding imagination, could not exist without the use of imagination. The imaginative leap to project forces or agencies as causes, and to construct explanatory theories, is the basis of scientific inquiry. Scientists are constantly involved in projecting aspects of reality for which no perceptual evidence exists; but the scientific search centers on the idea of the projection as *hypothesis*. The strength of a hypothesis is judged by its adequacy to explain relevant phenomena and its power to predict correctly the results of any actions involving those phenomena. Until contrary evidence is found, the hypothesis is presumed to be valid insofar as it provides a useful theoretical framework. Subsequent investigation seeks to find proof not that the hypothesis is correct, but that the hypothesis is incorrect. A hypothesis can never be absolutely proved, but it can be disproved.

The business of science is to build a public symbolic architecture of reality, one that is accessible equally to any and all members of the world population,[3] by specifying unambiguously the nature of objects and forces operant within "reality," the laws that govern those objects and forces, and repeatable procedures that will always, unerringly, demonstrate those objects and forces operating in accordance with the laws. Science is normative and descriptive—that is, it states laws and describes processes.

The business of art is to open public awareness to new ways of perceiving—to the examination of new possibilities that will, in some way, alter both public and private symbolic architecture of reality. By contrast to science, art is deliberately ambiguous. Artistic expression can only be made by the individual (or, as in theatre, by small groups of individuals, but even in cooperative arts, the statement within a given production area is still mainly that of the individual artist); the interpretation of the statement is solely a function of the life experience of the percipient and will vary with individual difference. Art is aesthetic, associative, and intuitive. It produces emotional response based on personal internal associations, and the applicability of the percipient's experience is based only on his or her ability to reintegrate the artistic statement into his or her personal symbolic architecture of reality.

Art expresses relationships among objects and events. The very simplicity of this statement belies its complexity, however. A photograph captures a set of spatial relationships, but a photograph is not necessarily art. Art results from *engaged perception*—perception that has become actively involved in heightened awareness, in awareness that is suddenly opened to all of the patterns of energy in the environment, that sees or hears in new ways.

Engaged perception both results from and produces some degree of shock[4]—the shock of recognition that heretofore unperceived possibilities lie within the accessible field of external reality. Shock is necessary to open the symbolic architecture of reality to admit additions and alterations. Shock is the agent that produces the state of heightened awareness, that tunes the brain to highest receptivity to all elements of sensory data inflow and to symbol recombinations within awareness. Engaged perception is concerned with finding implicit patterns among objects and events in space and/or time that are not obvious to everyday perception. All of us enter states of engaged perception periodically, but for the artist the state yields something that must be actively explored. (This state also applies to the scientist equally. The divergence

[3]Accessibility is, of course dependent on the ability of the individual to incorporate the public symbolic architecture into his or her own symbolic architecture. The idea of a relativistic cosmos or of subatomic particles and forces is alien nonsense to someone whose reality is limited to primitive, commonsense perception.

[4]The word *shock* as used here refers to what some branches of psychology call an orientation response. Orientation response is a situation in which an individual becomes aware that some conditions within the immediate environment are unfamiliar and require a reassessment. Orientation response disappears when circumstances become familiar and well understood.

between artist and scientist lies in the nature of the engagement and in the intended outcome.)

The result of engaged perception is insight. Insight usually comes as a sudden rush of understanding, of new perceptions. Because our survival depends on assessment of perceptual data, the emotions necessarily become engaged. We either like or don't like what we have found. Indeed, we may have mixed emotions: elation at the discovery and consternation or horror at its content.

What an artist attempts to express is *both the insight and its emotional concomitant.* Emotional concomitant of insight refers to the emotional reaction experienced when important mental associations take place and something new suddenly appears. That is, when the individual's symbolic architecture of reality is changed in some way. The emotional concomitant of insight is the initial aesthetic component of discovery.

Expression

In perceptions that relate particularly to matters of aesthetic insight, to the experiences that modify feelings and attitudes, the mental associations are not verbal. They do not deal with or rely on words. They are perceptions of pattern and form, of the texture of events, of large direction of movement revealed by immediate occurrence. But that discovery, that insight, is exclusively within the artist's symbolic architecture of reality. It is his or hers alone; no one else shares (or can share) the experience. Yet that situation is intolerable to the artist; the insight *must* be shared. The artist *needs* a way to say, "See what I have found! Share my understanding; share my feeling!"

Most immediately, the artist needs a way to objectify the experience—to hold it so that he or she may return to it to refresh the feelings generated by the insight; but for the artist, *ordinary* words turn the sublime into mockery. I emphasize the word "ordinary" to point out that, for an artist whose medium is words, words crafted through design can, indeed, express the emotions of the experience and the feelings derived from it. But recall the sentence from Chapter 17: "Because the design, as raw data of percipient experience, can rarely (if ever) lead back to the original events that produced a significant emotional experience for the artist, the process of art is to devise a metaphor for aesthetic meaning within the modalities of perception of sensory data." For anyone other than the artist, the *original content* of the experience is not reproducible, even though the feeling of the experience, and to some extent, the insight, may be. The best that can be done is to try to put together colors, or words, or motions, or sounds that relate to the nature of the experience and stimulate some of the same emotional responses that the artist felt in the presence of the insight.

The artist perceives something that is (for the artist) profoundly affective; a flood of emotions follows from that perception. The aftereffects of the experience continue to reverberate through branching ideas and associations. Feelings about the experience begin to form. Those feelings take shape in terms of words or colors or sounds or textures—whatever the artist's natural medium of expression may be. The terms of the medium become the *metaphor* by which the artist communicates.

Why metaphor? What is a metaphor? In language a metaphor is the expression of one thing in terms of another, to enrich the expression by association with some particularly affective characteristic of the metaphorical object or event. For example, the reference to a "flood of emotions" is a metaphor. Technically, emotions cannot be said "to flood." But the association of emotions with flooding water conveys a sense of the sudden upwelling (another metaphor) of emotions, tumbling over each other before any one can be identified, overrunning the normal courses of rational thought, creating temporary confusion, opening awareness to heightened perception, sweeping away barriers to insight.

Metaphors need not be verbal, however. Consider the music used to underscore action scenes in motion pictures. Underscore music usually has a point of view. It carries the internal state of the character with whom we, as audience, are most likely to identify, or it seems to originate within ourselves as percipients, both following and leading our emotional anticipation. What usually happens when a fight is in progress? The music is likely to be loud, up-tempo, and punctuated by staccato accents that reflect not just the fact that the combatants are striking each other, but how hard and, possibly, the debilitating effect suffered by one or the other of the combatants. If, to further the example, the character with whom we are meant to identify is severely hurt and falls down a long staircase, the music will reflect both his falling motion and his degree of injury, finally changing to a slower pace but keeping the mood of internal stress played against a deep echo of the fight rhythm reflecting the menace of the victorious enemy. The music is not the fight; the music *alone* would, perhaps, not even call up the image of a fight. In the context of the motion picture, however, the music becomes a metaphoric reflection of the action.

Another good example of auditory metaphor is the use of some natural sound, crickets, for instance, in a quiet scene before a crisis. In Piscator's adaptation of *War and Peace*, at the scene before the battle, Andre is seen sitting in his tent, thinking about Natasha, thinking about the things that are likely to happen on the following day, and realizing that he may never see her again. (He will be mortally wounded, and we have the presentiment that this will happen.) He begins his thoughts with the words, "Chirp. Chirp. The crickets chirp." The noise of crickets not only motives his line, but it suggests by association the freedom of these insects to live out their lives, to do what is natural for them to do without the press of responsibilities imposed by the human notions of duty and honor.

Metaphor, as an element of artistic communication, requires symbols that can be expressed in terms of an artist's medium. Before everything else, an artist absolutely must have a lifelong love affair with his or her medium of expression.[5] The medium itself must provide for the artist a source of fascination that cannot be ignored. The medium must be the artist's most natural mode of self-expression. During at least one phase of the artist's life, the medium must virtually become the entire center of engagement, so that all of its subtle turns and quirks become familiar ground. Without that love of the medium, there is no way to express the emotional character of an experience in terms of the medium—not unless one manages to make a very lucky guess.

The most glaring characteristic of inadequate engagement with the medium, and also of failed artistic insight, is *arbitrary association*—for example, using an instrumental version of a song as a bridge between scenes in a drama simply because a word in the lyrics of the song also appears in one of the two scenes. Nothing may be inherently wrong with the song itself, or the instrumental arrangement of it; but to work for the dramatic production, the *character of the music*—not just the arbitrary association of one word—must match one or both of the scenes and facilitate the audience's emotional transition from the one scene to the next.

Craftsmanship

The ability to use one's medium effectively is the essence of craftsmanship. Craftsmanship is the responsibility of artistic life. One may have deep and powerful insights, but without command of a medium no way exists to shape those insights into the metaphor that will carry both the meaning and the emotional character of the insight to public awareness.

Skill and the Medium

Every medium is unique and possesses its own particular characteristics in space and time and its own particular use of energy; every medium also applies itself to one or more of the human sensory modalities. Understanding of the medium requires comprehension, both of the ways in which the medium distributes energy through space and time and the ways in which human sensory experience receives and processes that energy.[6]

The initial infatuation with the medium is usually on the level of intuitive understanding of its power, very much like a child's attachment to a new toy—a complete absorption with the object for its own intrinsic appeal. We see colors or hear sounds or experience the dynamic power of shapes in space. The intuitive confrontation with the medium is one of the first shocks of heightened awareness that the artist usually receives—the perception of the expressive power implicit in shape or color or sound. The initial confrontation tends to be imbued with a sense of such complete intuitive understanding that there is an immediate feeling of being able to communicate instantly within the medium.

Unfortunately, the first attempts to use the medium as a means of expression are quite likely to yield two profoundly revealing experiences: a blunt collision with the intractability of the medium and a realization that the ideas one can initially find for expression through the medium are trivial. The adversity of the experiences will either destroy the infatuation or drive the novice into pursuit of competence. The former absolutely precludes artistic mastery; the latter does not guarantee mastery, but it does open the door to the attempt.

A metaphor that has often been used in the history of the arts for the struggle to attain competence within the craft is *gradus ad Parnassum*—the climb to Parnassus—likening the effort to master one's medium to the labor of climbing Mt. Parnassus, the mountain adjacent to Delphi, the site of the Oracle of Apollo. Because Apollo was believed to inhabit the top of the mountain, presiding over the oracle below, climbing the mountain was to ascend to the level of the god. The climb was treacherous and difficult, and only the best could succeed. *Gradus ad Parnassum*, therefore, has often been used as a name for exercise books, especially in music study.

Phases of Development

Craftsmanship proceeds through phases of maturity just as humans themselves do, and most practitioners of an art do not escape any of the stages. The stages themselves are necessary, however, to the attainment of competence in the use of the medium. The stages, roughly, are juvenile exploration, adolescent technical virtuosity, and mature competence.

At the juvenile phase, one is involved in discovering and exploring the medium. During this time, a period that usually requires instruction, one accepts the elements of the craft in a naive way. For the novice, the elements of the craft are not necessarily related in any formal way. They merely are, and have to be mastered. This is the period of practicing scales and arpeggios for the piano student, of learning how to scale and shape linear forms for the artist, and of learning about audio equipment and the acoustic properties of space for the sound designer. The juvenile phase is one of acquiring a basic grasp of technique and of achieving primary familiarity with the dynamics of the medium.

At some point, one will be able to execute with relative ease most of the elemental aspects of the medium, beginning a transition to a phase in which he or she is concerned, among

---

[5] Perhaps the expression should be love/hate affair, as artists sometimes feel that the medium and the need to express drive them like harridans and deny them fulfillment of expression.

[6] Understanding, in this context, does not necessarily imply the understanding of the scientist with regard to energy and to psychophysics of perception. A great many artists and designers understand the physical and psychophysical dynamics of their medium on a purely intuitive level, and that is just as effective, in most cases, as technical understanding.

other things, with increasing the sheer skill of execution leading to an almost athletic display of ability. This is the adolescent phase. Often, during this period, the ability to perform technically difficult exercises gets confused with artistic expression. The adolescent phase of learning the craft is often one of the most difficult, both for the artist and for those to whom he or she attempts to communicate—as is the case with physical adolescence; but, like physical adolescence, this period serves to lay the foundation for technical and emotional maturity. It locks the kinesthetic patterns of technique into the neural and muscular system and shapes patterns of symbols in the brain that the emergent mature artist will learn to use as agents of communication. For the sound designer, the adolescent phase is usually one of learning all of the tricks that can be done with audio equipment and with the interface of sound and acoustic space.

The mature phase of development in the craft is that of subordinating the craft to the service of communication. Craft is not creativity. Craft will not produce creativity. Technical ability within the medium merely guarantees that if one has something to say, the means of expression are available. But that does not imply that the medium has nothing to do with creative expression. Far from it. The way in which the elements of the medium interact in space and time are prime determinants of the ideas that can be expressed through the medium. One cannot express the meaning of Brahms' Fourth Symphony or of Michael Jackson's "Thriller" by drawing a picture, by sculpting stone or metal, or by writing an essay or a novel.

The craft that enables one to use a given medium is a physical and mental skill that requires practice, usually a regimen of fairly frequent (if not daily) practice. For a craft such as audio, daily exercise is useful but not necessarily essential, as the operation of audio equipment is not usually a skill requiring the precise and detailed actions necessary to painting or performance on a musical instrument. The aspect of the audio craft that is likely to suffer from periods of neglect is the mental ability to direct physical activity. Ontologically, the purpose of a brain is to direct and control physical activity; therefore, without physical exercise both the brain and the musculature degenerate in skill and ability.

Art and Design

What differences exist between art and design? Do designers require insight in order to function? What is the purpose of design? Obviously, designers use elements of the various arts, but they use them in the service of something other than "pure" art. This is especially true if we are speaking of design as a means of planning the form and appearance of such objects as buildings, clothes, furnishings, automobiles, and so on. Design is the quintessence of subordinating craft to communication; but design also subordinates the art itself to the communication of very particular matters.

For any designer, the understanding of the expressive possibilities implicit in the medium is one of the most essential aspects of the craft. A designer must know, insofar as possible, what combinations of elements within the medium are most likely to produce what forms of emotional response and how to create subtle shifts in the arrangements of the elements to build and lead emotional response.

The primary difference between art and design in general is that design is art limited by a function other than its own. The designer as an artist is not free to enter into a state of engaged perception with just any set of possibilities in space and time. The focus of perception and insight must adhere to one particular object—the thing to be designed. That object to be designed is usually specified by some function other than the designer's own immediate interests. A designer, therefore, needs to be a person who can easily become actively involved with objects or functions for the immediate challenge that those entities present.

The object to which the theatrical designer's creative engagement is directed is the idea or ideas expressed in the playwright's script together with the director's interpretation of what that script means. The designer's function is to create, within the range of ideas expressible within his or her medium, a set of conditions that will promote the insights and the emotional character implicit in the script and the director's interpretive purpose. Those conditions may be the color and shape of the clothing worn by the characters; the depth, texture, color, and form of the surroundings in which the characters act; or the qualities of the auditory environment and/or the musical analog of the characters' feelings and interactions.

For the sound designer, the object of an engaged perception with the playscript is the auditory world of the drama. In order to perceive that world, and to be able to focus a heightened awareness on the interrelationships of that world, however, the initial engagement must be with the entire drama, not just its auditory surround. Recall that this world in the drama is the playwright's creation, that everything within it is there for a reason, and that the changing relationships in the dramatic space operating through dramatic time imply the necessary characteristics of this imagined world. The auditory characteristics of this world are what the sound designer must intuitively find. He or she must explore the possible relationships of those characteristics and find the set of sounds (and musical qualities, if needed) best suited to enhance the meaning and emotional dynamics of the playwright's intentions.

The result of the creative process for the sound designer is, of course, the assemblage of music and/or sounds that will be used to realize the auditory world of the production. That auditory world will be successful, in part, insofar as it functions as a useful metaphor for what the designer and director believe to be the meaning (the original insight and its emotional concomitant) intended by the playwright. The sound, along with all other aspects of production, must contribute to building for the audience the emotional structure that will lead them back to insights and feelings similar (in kind, if not in fact) to those experienced and communicated by the playwright.

Chapter 19

Sound and Dramatic Art

Sound as a Craft of Theatre

For most of theatrical history, uses of sound have been a matter of providing the **effect** of a sound, a mechanical simulation that served more as an indication, a reminder of the character of the sound, than as a realistic reproduction of the sound. Accordingly, we have come to use the words *sound effects* as objects that, like costumes and properties, are brought in, fitted to the particular character of a dramatic production, and set up for use in a sequence of cues determined by the playscript. Given our contemporary technology in the recording and reproduction of sound, we do not (at least we certainly should not) provide merely the effect of a sound when we can provide a **facsimile** recorded directly from the real sound source. Accordingly, the term that is used in this text is **environmental ambience**, not sound effects.

Sound in the contemporary theatre has three aspects each of which could be considered a separate craft: environmental ambience, music, and electroacoustic reinforcement. Environmental ambience is an all-inclusive term encompassing atmospheric noises and the sounds of the dramatic environment (including the obligatory sounds called for by the script), vocal assistance for the actor (meaning alteration and/or substitution of voice, not acoustic reinforcement), and the symbolic shaping of sound to imply extensions of dramatic space and time. Music involves the determination of the character and kinds of music to be used as incidental music to provide introductions, bridges, and closures for scenes and acts, as integral music used within a production as part of the dramatic action (usually but not always specified by the playwright), and as underscore music to help enhance and reinforce the dominant mood and character of a scene or to provide dramatic counterpoint to the immediate action within a scene. Electroacoustic reinforcement is the use of audio to improve the audibility of actors' voices by simple amplification, by corrective action to overcome acoustic deficiencies of the theatre or auditorium, or both. Although each area forms a separate craft, all three may be controlled by a single individual or each may be under the control of a specialist.

Sound as an Element of Dramatic Art

The art of sound in dramatic production may be considered under three primary aspects: the functions of sound in theatre, the properties by which a designer can manipulate sound as a dramatic element, and the methods by which sound for dramatic production can be accomplished.

Functions of Sound in Theatre

The functions of sound in the theatre are classifiable in three groups: dramatic, aesthetic, and practical. These categories are potentially overlapping, so we need to understand precisely what is meant by each term. *Dramatic*, in this context, refers to those things that directly advance or condition the progress of the drama or the environment within which the dramatic action takes place. *Aesthetic* designates matters that have to do with personal interpretation of the immediate emotional character of the drama and with the long-term development of feelings and attitudes as modified by the dramatic experience. *Practical* refers to the necessary considerations for audience comfort in perceiving the vocal component of dramatic production.

Altogether, there are seven functions of sound in the theatre:

Audibility
Motivation (environmental ambience)
Music
Vocal alteration
Vocal substitution
Extension of dramatic space/time
Mood

The functions that are primarily dramatic in nature are motivation and vocal alteration/substitution. Articulation of dramatic space/time and mood are functions within the aesthetic category. Audibility is the only function involved in the practical category. Music overlaps the dramatic and aesthetic categories equally.

What do the various functions mean with respect to dramatic art? How do the functions of sound relate to other aspects of theatrical design? Some are relatively self-explanatory and not unlike similar functions in other design areas. Audibility is much like the lighting function of visibility. **Audibility** means making all aspects of a production fully audible to all members of the audience, just as visibility in lighting means making sure that all members of the audience can see the performance. Unlike lighting, audibility is not the exclusive function of one person.

Audibility is usually carried out through the process of placing microphones on and around the stage and on the actors' persons. The signals picked up by the microphones are mixed together into signal paths that eventually lead to loudspeakers placed so that they appear to reinforce, as faithfully as possible, the sound coming from the stage. Along the way from mixer to loudspeaker, several kinds of processing may be applied to the signal. Most usual is equalization, to compensate for acoustic problems of the space, and delay, to enhance localization to the actors' actual voices and to improve gain before feedback. Having equipped the stage with microphones, mixers, amplifiers, and loudspeakers, however, the job is only partly completed. The best signal processing equipment in the world cannot make intelligible that which is garbled as the actor speaks it. The best microphone in the world cannot pick up a line that is not adequately projected. Like many other aspects of sound, audibility rests as much on the actors and the director as on the sound designer.

Motivation, as in lighting, means that all sounds that occur within the dramatic environment must fit. A sound must appear to come from a proper location and direction, and it must appear appropriate to the character of the ostensible source of the sound. For example, if the sound of a gunshot clearly comes from stage right when the actor firing the gun is almost offstage left, the discrepancy is both noticeable and disturbing. Also, if the sound has more of the character of a massive explosion when the visible weapon is only a small derringer, the effect is unsettling or even comic.

Music refers to provision of any of the three kinds of music that may occur during a dramatic production: incidental, underscore, and integral. Incidental music is neither called for by the playwright nor indicated as a vital part of the production, but it serves to set the mood for an act or a scene. Incidental music occurs at the openings of acts, as a bridge between scenes, and as a means of closing (tagging) an act or scene, reinforcing the mood of the final lines and actions. Incidental music also includes marches or dance music that accompanies parts of plays when the source of the music is arbitrary and not seen as a part of the visible action. Underscore, as in film and television, is music used beneath action and dialogue that reinforces and enhances the emotional impact of the scene. Music becomes integral when it is specifically called for by the playwright or is performed as part of the action of the play by actors or musicians on the stage.

Vocal alteration and **vocal substitution** are functions that cannot be carried out without the use of audio equipment. Both involve helping an actor to achieve an otherwise impossible vocal character. In the case of vocal alteration, the actor speaks lines live on the stage, but audio equipment is used to amplify and modify the quality and character of the voice. In the case of vocal substitution, the actor probably does not speak the lines on stage. Instead, either a recording is used or another actor speaks the lines offstage. The offstage actor's voice is picked up by an isolated microphone. The signal from the microphone is processed into a sound that no human voice could possible produce. The sound is sent to loudspeakers positioned on the stage so that the voice will appear to come from a visible object in the scene, either the actor in character or an inanimate object.

Extension of dramatic space/time involves use of sounds in ways that refer to events outside of the physical space and/or time frame of the dramatic action. All of the references in previous chapters to using the ambiguity of ill-defined sounds to trigger identity searches on the part of the audience come into play for this function. Alternatively, the sound designer may choose to use a very well known sound if that sound will unquestionably trigger associations with some other time, place, and circumstance.

Mood, as in lighting, is the manipulation of all of the other functions in order to enhance the mood and emotional character of a play, of acts and scenes within the play, and of individual beats and moments within acts and scenes. For example, one is not likely to choose a bright, trumpet fanfare as lead-in to a dark, melancholy scene in a jail cell.

The Controllable Properties of Sound

The functions of sound are carried out by manipulation of the controllable properties of sound.

Intensity
Frequency
Duration
Envelope
Timbre
Directionality

Intensity should be self-explanatory after our discussions of physical amplitude in acoustics, magnitude of current flow in audio electronics, and loudness in auditory perception. Intensity is the perceived loudness of the sound. **Frequency** is, simply, the number of cycles per second of physical vibration and the perceived pitch of the sound. **Duration** is the period of time through which the sound persists, roughly the sustain period of the ADSR characteristic. The apparent duration may be lengthened by reverberance. **Envelope** is the variation in loudness of a single sound plotted against the duration of the sound. As described in Chapter 1, the elements of envelope are the attack, the decay, the sustaining period, and the release (ADSR). Each of these elements is a segment of time measured in seconds or fractions of seconds. Normally, we hear envelope as an integral characteristic of a sound rather than as a change in physical intensity. **Timbre**, as explained in Chapter 1, is the characteristic quality of sound produced by any particular sound source. Timbre is the result of the unique blend of harmonics that the generator produces. **Directionality** also should be relatively self-explanatory. Directionality refers to the apparent direction and distance of a sound source measured from the listener's position.

Together, the functions and controllable properties of sound form the basis of the sound designer's tools for creating the auditory environment for a dramatic production. Each sound could be specified in terms of each of the functions and controllable properties. Most sound designers, however, are not so meticulous that they write out such a detailed specification for each sound in a design; but, whether consciously done or not, all sounds are conceived in terms of the functions and properties. When imagination fails to produce the necessary shape and character of a sound for some particular situation in a dramatic narrative, creative block can sometimes be overcome by attempting to specify the functions to be served by the sound and the specific value of each controllable property.

Styles of Production

Applicable techniques in any aspect of theatrical production are essentially determined by the style of the production. **Style** is usually defined as the degree of conformance to realistic representation of the dramatic environment and of the actions of the characters of the drama. Five styles are usually listed by most texts on dramatic production:

Naturalism
Realism, normal and selective
Stylization
Formalism
Abstraction

Naturalism is the most faithful to real-world environmental conditions. Naturalistic decor and acting attempt to re-create every minute detail of an environment, all the way to such extremes as bringing real earth, trees, grass, junked machinery, even rotting garbage, onto the stage. In a naturalistic setting, environmental sound would attempt to capture every bit of wind noise, rustle of leaves, and miscellaneous animal and human noises at varying depths and distances from the immediate scene. Virtually anything that the director and sound designer can think of that might be a part of the surroundings could be used for the production. Nothing would be ruled out, even though the sound might be irrelevant or even unnecessary to the meaning and understanding of the dramatic narrative.

Realism is faithful to the real-world environment but not so slavishly as in naturalism. Realism admits artistic license in arranging the events and components of the dramatic surroundings. Irrelevant detail is omitted simply because it is irrelevant. *Normal realism* includes all useful aspects of the environment. *Selective realism* reduces the amount of detail even further, choosing only what is aesthetically and dramatically necessary to the understanding and dramatic impact of the production. Everything that is represented is faithfully realistic; but only the most essential aspects of the scene are included. Realism in sound is a matter of creating the sounds that normally belong to an environment in their natural auditory perspective. Sounds are sustained only for as long as they are needed. Irrelevant sounds are omitted, and in selective realism only the most essential sounds are used.

Stylization is realism with full artistic license. Distortion for the purpose of aesthetic emphasis is permitted. Visual artists are free to shape the contours, textures, rhythms, and colors of the environment to whatever suits the purpose. The sound designer has similar license, except that the actions, noises, and visual images that result must be identifiable—that is, they must bear a reasonable resemblance to the real object from which they are derived. Stylization is one of the more difficult forms of sound for theatre, because finding how to distort sound without destroying its identity or meaning is relatively difficult. Visual arts are spatially stable and we can take our time and compare various arrangements. Auditory arts proceed through time, and comparison relies entirely on memory. Further, once a sound is placed in a context, its potential identity and meaning are altered. Sounds that seem perfectly acceptable in the audio control room may turn out to be unusable when played in the theatre in the context of setting and action.

Formalism is the name given to dramatic production in which the elements are arranged in planned patterns. The patterns are more important in and for themselves than realistic patterns that would might be characteristic of an actual locale. For example, everything in the visual elements of the production might be arranged in terms of radial or bilateral symmetry. Productions of the classics and neoclassics are often done in settings that are formally symmetrical. Dialogue and blocking are executed to create patterns of formal balance either in space or through time. Sound in a formalistic production tends to be fundamentally realistic except that the arrangement and choice of sounds are structured according to ordered patterns of rhythmic and tonal balance. Auditory perspective, directionality, and other factors related to environmental realism are a secondary consideration.

Abstraction is a style of production in which all elements are reduced to their most fundamental shape or form. These basic shapes and forms can then be enhanced and/or rearranged to create new patterns of expression that reflect a dominant aesthetic theme to be carried by the production. Abstraction may well be the most difficult style to use successfully, especially for the sound designer. Abstraction requires the artist to be able to strip away all of the outer detail from an event, an object, or an action, to be able to reduce it to its simplest form, and to find novel ways to utilize the basic schematic pattern of the object. In sound, schematization is difficult to accomplish and often not very interesting. Abstraction in sound (as in the visual world) usually demands the juxtaposition of several elemental entities in a collage of elements designed to make a larger and more complex pattern. Sound collages can be difficult to do well, that is, so that they are meaningful or recognizable.

Application of Craftsmanship to Design

In every design, the functions and controllable properties of sound and the style of dramatic production become the underlying basis of creativity that serves the needs of the production. Style of production determines the range of available choices open to a designer. All research, both for period characteristics and for sources of raw sounds (the original sources and recordings from which sounds will be developed), must keep style of production firmly in mind.

As the design proceeds, the designer should develop a solid notion of how each of the functions applies to the particular production. What kind of auditory character will be believable? How should music be used (if at all)? Is any form of modification of an actor's voice needed? How does the auditory environment fit into the extended story line of the drama? Should the auditory environment attempt to suggest events that lie outside the spatial or temporal frame of the immediate scene? What is the overall mood of the production, and how can sound help to generate the feeling appropriate to that mood?

Consideration of the controllable properties help to determine specific choices including technical decisions: What is the general intensity range of sounds for the dramatic environment? What frequency range is appropriate (i.e., should sounds be bright or dark)? What envelope qualities are needed and at what points in the production? How long do sounds need to be to fit mood? How long do sounds persist (reverberance) within the dramatic environment? What timbral characteristics should apply to the sounds chosen? (Timbral character is somewhat different from simple frequency in that a sound can have a variety of different tonal characters, depending on circumstance. All oboes, for instance, have similar tones, but each performer generates a slightly different timbre.) From what direction do sounds appear to come? For each sound used in a production, answers to each of these questions must be given, either consciously or intuitively. As we go through the production example in the following chapters, we shall see how stylistic considerations, functions, and controllable properties are applied to the decisions and actions of the sound designer.

Chapter 20

Reading the Playscript for Sound Design

The inevitable problem, often the first and largest barrier to getting underway with a design, is to read the playwright's creation and to draw from it an auditory construct, one that helps to bring the play to life for the audience and deepens the experience of watching and hearing the production. The playwright provides only words, and those words are mainly vehicles for spoken dialogue. Occasional rubrics indicate time, place, atmosphere, mood, and sometimes specific sounds; but, unlike novelists, playwrights seldom describe the auditory environment.

In novels the writer usually expends a great deal of effort to draw the scene with all its atmosphere and feeling. The novelist may tell us that "birds were beginning to murmur vague recognitions of the coming light as Katerina stepped out into the chill mist of the still-dark October morning." But a novel is not a play; the novelist cannot expect to have a team of actors, a director, and designers to bring his or her work to life.

The playwright sets out mainly to give actors words to say and the director meaning to explore through those words; but if a playwright does not provide direction as to atmosphere and locale, then how are we to function who must create and build the environment in which the playwright's characters are to live? If the playwright has ability and talent, the dialogue may tell us far more than could ever be put in italicized rubrics or production notes. If the characters begin to come to life through their words, then the details of their lives and the background of their world emerge with them; and what cannot be discerned directly, research can usually provide.

Acquaintance with the Script

To feel the world of the characters of the play emerging, one must get as close to that world as possible, and the only way to do that is to read the playscript. Not just read it, but reread it, and then read it again many times over. By the time one begins to anticipate the next line, he or she may be surprised to find that the characters have taken on visible form, have identifiable voices, and are moving and speaking somewhere in the reader's mind-space. Moreover, vague shadows of the characters' world may have taken shape around their speech and actions. From this dim, unfocused background, one begins to derive images and feelings that will lead to constructive ideas about the play and its meaning and about the environment in which it takes place. These constructive ideas provide the basis for the actual music and sounds that will be created later on. The underlying concept also begins to form during this stage of exploration.

When ideas and images of the look, feel, and sound of the characters and their environment begin to form, that is the time one should also find a wealth of questions about the kind of world that the playwright is trying to evoke. Is it realistic? What would one see or hear if one were in sixteenth-century London, or in New York or Paris of the 1920s? Is it fantasy, and, if so, what does it suggest about modification of realistic detail? When such questions begin to surface in the mind, the time has come to seek factual information, to research the details of the geographical region, the social circumstances, and the time period.

During research, one may find that mental images, acquired while absorbing the playscript, have little in common with the actual look and feel of the historical period of the play. Having collected data that tells you that the class of people who inhabit this play would not have worn a certain type of costume, and that most of the streets were dirt rather than cobblestones (meaning that horses and carriages passing by would hardly have made those lovely sounds you heard so clearly in your imagination), what does the designer now do? Was the imagination wrong in discerning the kind of mood and feeling for the play? No. Usually imagination is on the right track. Don't forget that your images arise from a dramatic narrative and its environmental circumstances, so one can presume that the playwright has a reason for the images that seem to emerge from the play. Study accurate data and let the matter rest for a while; a related but more accurate set of images will usually appear on the next reading, possibly the next several readings, of the script. Also, one may find reasons why the playwright has deliberately violated the factual reality; and one may even find scenes and events of the play mixing with images drawn from research.

Forming the Concept

Somewhere in this phase of activity one should begin to form a **concept** to guide future creative imagination. Forming a concept involves crystallizing all of the information and the feelings about the meaning and intent of the play into a single statement. That statement may take the form of a few words, or it may be a group of assembled sounds that express the emotions and feelings of the play. Forming a concept is not always easy, just as any process of real generalization of many facts and images into a single, all-encompassing, overarching statement is not easy. Insight and understanding must be very thorough in order to accomplish the task.

A very tempting alternative usually appears at this point. One can very easily entertain the following self-dialogue: "I have lots of ideas and images and they're really good. I can create a lot of nice background for this play, so maybe I don't really have to state a specific concept. I've got everything I need already." Remember: Self-deception is the first step to a less than spectacular effort, if not outright failure. *Concepts are essential.*

The designer's concept is an aesthetic subset of the ethical concept established by the director. The concept provides focus and unity to the process of creation, and helps to keep imaginative activity within the channels that serve the overall meaning of the production as a whole. Recognize that the process of creating a design takes time, perhaps only a few days in some cases but possibly several weeks. During the time one works on the production, the imaginative process does not stop. New insights and ideas develop, and the effort can easily go offtrack, especially when decisions have to be made under pressure.

Nothing is inherently wrong with changing one's ideas, providing one is working alone. Many artists create their best works in this fashion. Remember, however, that theatre is done by a large group of artists all working together. Each artist states a particular intent. Those individual intentions are adjusted in conference to fit into a unified whole, and at some point the creative idea must be fixed. Any revisions to insight that occur after the primary conferences are finished are irrelevant. The stated intentions must be stable and reliable if collaboration is to succeed. No single designer has the right to change intentions in midproduction—at least, not without the approval and consent of the entire group.

The decision to go without a fixed concept also fails to consider the needs of the audience. To communicate effectively with the audience, a dramatic production must trigger two forms of basic mental activity: perceptual engagement with the emotional qualities of the characters and their environment and associative reference to the events and circumstances of the audience members' personal lives. Decisions about the operational aesthetic concept help to determine the weighting of each of these activities in the overall perception of the production. Normally, perceptual engagement and associative reference are, or should be, about equally balanced. Comedies probably tend to emphasize the former. In serious plays the second of these two forms of mental activity must be made the preeminent goal, especially if the director intends for the message of the play to be highly relevant to the lives of the audience.[1] To achieve the necessary communication, firm decisions about the emotional qualities and contents of each element of production are necessary. The concept establishes those priorities and guides subsequent production decisions.

Aesthetic concepts function most successfully when cast in terms of the medium in which the designer will work. A concept for the visual aspects of theatre is usually a visual image. Vision is a simultaneous sense and can process a number of components at the same time. People can look at a picture and see all of its components clearly juxtaposed. Audition is a sequential sense. Audition does not handle simultaneous information well. Hearing tends to perceive events in sequence. Sequential processing often makes auditory conceptual imagery difficult in that, usually, no single sound can form a complete concept and simultaneous multiple sounds are difficult to integrate. Meaning is almost always derived from the way in which a number of individual sounds are combined together through time within the context in which those sounds occur. Placement within the sequence affects meaning. Thus, the conjunction of sounds to produce a concept image requires both a very secure understanding of the meaning to be communicated and the skill to capture and apply the component sounds required to create the image.

Conceiving auditory images is the traditional business of the composer. Composition is often felt to be one of the most arcane of the arts, almost a kind of witchcraft, because

[1] Relevance does not mean forcing a play to carry a purely contemporary or near-contemporary form, like staging *Romeo and Juliet* in the period of the American Civil War, or treating *As You Like It* as a clash between environmentalists and big oil companies in the 1970s (although either plan could conceivably be used in such a manner). Relevance, in the sense, means providing evocative circumstances and ideas that lead people to relate the events of the play to their own experiences and to derive meaning from the connection. If the production team does not decide on some particular forms of relevance as a focus for the production, then the overall impression that reaches the audience is very likely to be haphazard. That is why a central concept is essential.

for most people imagining a sound, not to mention a whole sequence of sounds strung together, is difficult if not impossible. Used here, imagination refers to true creative imagination, that is, to the construction of sounds in the mind without having heard them before in that particular form. Most of us can play over a favorite song or a well-known piece of music in memory, but that is not what a composer or a sound designer does. He or she puts sounds together in patterns that, presumably, have not been assembled in quite that form previously, at least, not within the composer's known experience.[2]

Devising sounds for dramatic production is not radically different from musical composition, at least in the sense of constructing auditory imagery. The designer, like the composer, must find a sequence of sounds, related sounds, that fits and enhances the immediate purpose. The primary difference between the composer and the sound designer is that, in most cases, the dramatic context gives the designer a basis from which to begin, whereas for the composer the process of creation is more abstract. The composer, however, must still maintain thematic unity, and the designer must have a central image to which all other images relate. The central image is the aesthetic concept. The skill to understand the emotional value of sounds presented in certain orders and within a particular context is analogous to the skill of the composer in creating a sequence of musical images and events. Native talent sharpened by experience and practice is essential.

A Practical Example

Talking about evolving mental images and finding a concept image is easy enough; accomplishing the task is not always as simple. To help in understanding this phase of the design process, an example using an actual playscript is in order. The script to be used for this demonstration is Tennessee Williams's play, *The Rose Tattoo*. This script potentially offers a chance for a liberal emphasis on sound and the auditory atmosphere implicit in the scenic environment. Williams, as usual, gives basic direction for sound, but not too much. He gives us information with which to begin and ample room for creative expansion.

A script must be read before it can be used for any purpose; so the reader should obtain a copy of *The Rose Tattoo* and read it through. For purposes of continuity, however, a synopsis of the script follows.

SYNOPSIS OF *THE ROSE TATTOO*[3]

Act I.

Serafina delle Rose is a Sicilian woman who has immigrated to the United States with her husband, Rosario. They live in an Italian community on the Gulf coast between Mobile and New Orleans. They have a daughter, Rosa, who, at the beginning of the play, is in her early teens.

As the play opens, it is a hot, humid late summer twilight—*prima sera* (the first of the evening). From all around are sounds of children at play and mothers calling them in to supper. Serafina, sitting in her living room waiting for Rosario to come home, is visited by Assunta, the old *fattuchiere* (the word means, approximately, jack-of-all-trades, i.e., a person who can take care of all kinds of needs: midwife, herb-cures, personal counselling, etc.). Assunta admonishes that pride is deadly, and Serafina is very proud—of her husband, of his manliness, of his title of *barone*, and proud because she considers her life to be special, and because she is pregnant. She tells Assunta that she knows she is pregnant because, some weeks earlier after she and Rosario made love, she awoke with a burning on her left breast and saw there, momentarily, an image of the Rose Tattoo that Rosario has on his chest. Assunta warns, "Serafina, for you everything has got to be different. . . . You speak to Our Lady. You say she answers your questions. . . . Because you are more important? The wife of a *barone*? Serafina! In Sicily they called his uncle a baron, but in Sicily everybody's a baron that owns a piece of land and a separate house for the goats! . . . But here what's he do? Drives a truck of bananas?" Serafina confesses that, yes, Rosario hauls bananas, but underneath the bananas, contraband merchandise to make more money. But this load of contraband is to be the last. After this the truck will be paid for, and then they will live a better life.

Assunta leaves, and Estelle Hohengarten, the Texas-born blackjack dealer at the local bordello, arrives bringing sewing for Serafina. She has a piece of beautiful rose colored silk which she wants made into a shirt for a man who, she says, is "wild like a Gypsy." As Estelle leaves, a neighbor calls out that the Strega's black goat has broken loose and is in Serafina's yard. The "Strega" (as Serafina calls her—the word means "witch") is an old American woman whose cataracts make her eyes look white and clouded. Serafina screams to Rosa to get in the house—not to look at the Strega. Rosa,

[2]Presumably one could get into some debate about how original imagination must truly be in order to supply a dog bark or a train whistle as prescribed in the script. The unique and original part of creation in such circumstances, however, is the adjustment of the character and the immediate auditory perspective of the sound to fit and reinforce the mood and to further the dramatic motion of the production.

[3]Tennessee Williams: *The Rose Tattoo*. Copyright 1950, 1951 by Tennessee Williams. Reprinted by permission of New Directions Publishing Corp.

largely educated in the United States, recognizes cataracts for what they are, but to Serafina it is the "evil eye."

All night Rosario does not come home and Serafina sits up, sewing the rose-silk shirt as she waits. Then at dawn, Father Leo and the local Sicilian women approach the house. Serafina sees them, and realizes that they have come to tell her that Rosario has been killed. In a fearful voice she cries, "Don't speak," and collapses.

Later, Father Leo and Assunta are talking on Serafina's porch. The doctor, called to tend to Serafina, comes out of the house and tells Assunta that Serafina has had a miscarriage. Serafina is sleeping; the doctor gives Assunta a hypodermic and a bottle of morphine, telling her to use it to control Serafina if she tries to get up. The priest confronts the doctor, ordering that Rosario's body must not be cremated. The doctor insists that Serafina's wishes must be carried out in spite of the laws of the church. Estelle appears, asking to pay her last respects to Rosario's body. We learn that she and Rosario were lovers; it was Rosario who was the man "wild like a Gypsy" for whom the rose-silk shirt was ordered. The Sicilian women drive Estelle away.

Three years pass and Serafina has become a recluse. The house is run down and Serafina's personal appearance is slovenly. The Sicilian women are noisily gathered around Serafina's door demanding the dresses Serafina has agreed to make for their daughters' graduation. Serafina refuses to let them in because she has learned that Rosa met a young sailor at a recent high school dance. In a fury of protectiveness she has locked the girl, naked, in her room for the last week, refusing to let Rosa go out or to let anyone else near the house. Suddenly Serafina bursts from the house, however, screaming that Rosa has slashed her wrists. Miss Yorke, Rosa's teacher, arrives and takes charge of the situation. She determines that Rosa's injury is only superficial and demands that Serafina unlock the closet where Rosa's clothes have been placed. Rosa's grades, it seems, have been high enough so that she will graduate, even though she missed all her examinations. Rosa leaves, and the women crowd in to find their daughters' dresses.

Serafina, a short while later, is frantically trying to get dressed for the graduation. None of her clothes fit her now. She hears the band begin to play in the distance just as Flora and Bessie arrive. These two are middle-aged and man-crazy American women, here to pick up a polka-dot blouse Flora has ordered. The blouse isn't finished and Flora is angry because she had intended to wear it to New Orleans for a Legionnaire's parade. She insists that Serafina finish the blouse immediately. Serafina objects but is finally coerced by Flora's threat to "report you to th' Chamber o' Commerce an' get your license revoked." Flora and Bessie make snide remarks about Serafina as she sews. Outside some Legionnaires pass by and Flora and Bessie shout at them. Serafina objects that hers is a respectable house, a shrine to the memory of her husband who was hers and hers only. She insists that the two women leave. Flora coldly turns on Serafina, mocking her with the information that Rosario and Estelle Hohengarten were lovers. Serafina, enraged and crushed, curses them and drives them out with a broom; then she rushes back in, looks at the rose-silk shirt almost realizing now why Estelle never came to get it, and afraid to look at the name on the card. She collapses before the statue of the Virgin crying, "Oh, Lady, give me a sign."

Two hours later, Rosa arrives from graduation with Jack, her sailor, to introduce him to Serafina. They are on their way to a party. They find the house shut and the shutters closed and think that no one is home. Inside, as they are talking of their first meeting and Rosa is just about to tell Jack that she loves him, Serafina hisses from a darkened adjacent room, *"Stai zitta, cretina!"* ("Shut your mouth, idiot!") Rosa makes Serafina come out and meet Jack. She attempts to calm her mother, but before Serafina will let Jack and Rosa go to the party she makes Jack swear before the Virgin to respect Rosa's innocence.

Act II.

Act 2 opens during the afternoon of the same day. Serafina, miserable and upset, sits on her porch. It is oppressively hot and muggy, and she wears only a slip. The neighborhood children peek at her and giggle, and Father Leo approaches to chide her for her slovenly behavior. Serafina demands to know if Rosario ever confessed to having an affair with another woman. She accuses the neighbor women of starting the story as jealous gossip. The father insists that the secrets of the confessional are sacred. Serafina begins to pummel him with her fists, and Assunta and the women arrive to restrain her. Father Leo escapes and Serafina drives the women away.

A loud-mouthed, red-neck salesman enters the yard selling a new item that is "bigger than television." He is interrupted by the sound of a large truck stopping, and a male voice with an Italian accent calls the salesman a "road hog." Alvaro Mangiacavallo comes into the yard, protesting that the salesman has called him "dago" and "wop" and has run him off the road. The salesman knees him in the groin and goes off laughing as Alvaro, in pain, stumbles into Serafina's house. The salesman calls back, "I got your license number, Maccaroni. I know your boss."

Serafina, upset and distressed that a stranger is in her house, follows Alvaro in, demanding that he leave. Alvaro is crying, which he says he always does after a fight, so he came inside so no one will see him cry; Serafina cries, which she says she always does when someone else is crying. His shirt is torn and Serafina, by now sympathetic to Alvaro's pain, offers to mend it. When he takes the shirt off Serafina gasps; his body reminds her of Rosario's. "My husband's body with the head of a clown!" she murmurs to herself. She asks his origins and finds that he is Sicilian. He is an immigrant truck driver, the sole support of a drunken father, a sister, and an old grandmother, all of whom are addicted to the numbers game and spend all of the money he makes. The grocery company has garnisheed his wages, and now, for fighting, he will probably lose his job. Serafina tells Alvaro about Rosario, about his tattoo, and how, on the night she became pregnant, she saw the tattoo appear on her breast.

Alvaro calls his boss to try to forestall trouble, but too late. He is fired. Serafina decides that it's too dark to sew, and gives him the rose-silk shirt, telling him to throw away the tag pinned to it and not tell her the name. Again, the Strega's goat breaks loose and gets into Serafina's yard, but this time Alvaro catches the goat and boots it out. Serafina is drawn to Alvaro, not only because of the resemblance of his body to Rosario's, but because of his good humor and simplicity.

Alvaro suggests that he can come for supper that evening to return the rose-silk shirt. Serafina tells him that if the shutters are open and there is a light in the house, then he can come in. Otherwise, Rosa will be home, and she says she intends to set a pure example for her daughter. Serafina is happy, deciding that the events of the afternoon are a sign from the Virgin that Flora and Bessie were lying—that Rosario was faithful to her.

Act III.

Act 3 opens that evening. Serafina has dressed and straightened the house. Alvaro arrives. He has gotten a haircut and has rose-oil on his hair. The smell upsets Serafina, and she is beginning to worry about Rosa, who has not yet returned from the party. Serafina complains about sailors and their ways. Alvaro asks if Rosa's sailor has a tattoo, then he opens his shirt to show her that he has a tattoo. Serafina reels. Alvaro has gotten a rose tattooed on his chest. Serafina is violently offended. Trying to explain, he nervously plunges his hands into his pockets, and when he withdraws them a condom falls out. Serafina, now enraged, tells him to take his passion to the bordello. He protests that he has only been trying to please her. She regains composure but still insists that he leave. She explains that her day has gone badly since the incident with Flora and Bessie. As he is leaving, she asks if he remembers the name on the rose-silk shirt. He does; it was "Estelle Hohengarten."

Serafina again flies into a rage and demands that Alvaro drive her to the bordello where Estelle deals blackjack—that she is going to cut the woman's lying tongue out! Alvaro tries to calm her, but she tries to call a taxi. Alvaro insists that Estelle wouldn't lie and calls the bordello. Serafina snatches the phone, and we hear Estelle's voice through the receiver telling her, ". . . I said, for a man that's wild like a Gypsy. . . . If you think I'm lying, come here and let me show you his rose tattooed on my chest!" Serafina hurls the phone to the floor and collapses. Alvaro goes to get ice to put on her forehead. She gets up and goes to where the urn with Rosario's ashes stands beneath the statue of the virgin. She picks up the urn and smashes it against the wall.

Immediately, the atmosphere clears. It is, again, *prima sera*, just as at the opening of the play. Again, the sounds of children playing and mothers calling. Alvaro returns, and Serafina loudly tells him good night. "You make me go home now?" he asks. No, she whispers; just pretend to leave, then return by the back way, so the neighbors don't see. Alvaro's truck is heard driving away, and Serafina blows out the candle in front of the statue of the Virgin, renouncing the image as a "poor little doll . . . with paint peeling off." By now it is dark. Serafina opens the shutters and moonlight floods in. Alvaro is heard softly calling from behind the house.

Just before dawn Rosa and Jack return. From inside the house they hear Serafina moaning. Rosa says bitterly, "That's just my mother dreaming about my father. She wanted me not to have what she's dreaming about." Jack protests that Serafina only wanted to protect Rosa. Jack's ship sails from New Orleans the following day, and Rosa wants to know what Jack will do between now and then. He says that he could lie and tell her he would go pick daisies in Audubon Park, but he really intends to go get drunk then get. . . . He doesn't finish the sentence, but Rosa understands him and asks him to first check the bus terminal about noon—that she may be there waiting for him. They kiss and Jack leaves.

About three hours pass; it is just dawn, and Rosa is asleep on the sofa. Alvaro sleepily stumbles out of the bedroom and bumps into a seamstress's mannequin, apologizes comically, then sees Rosa. Still partially drunk, he bends low over the sleeping girl and blurts out "*Che bella!*" (How beautiful!) Rosa wakes and screams, and immediately Serafina is out of the bedroom flailing at Alvaro with a broom! From next door the Strega is heard cackling, "The wops are at it again. Had a truck driver in the house all night."

Alvaro leaves; Rosa goes behind the screen to dress as Serafina tries to stammer out excuses for Alvaro's presence. Rosa emerges wearing her best white dress. When Serafina demands to know why, Rosa sneers and says she simply feels like wearing it. The only thing worse than a liar, she tells Serafina, is a hypocrite. Serafina, resigned and now relaxed, simply says, "How beautiful is my daughter. Go to the boy." Rosa, in turn, assures Serafina that Alvaro did not touch her. Rosa hurries away to catch the bus to New Orleans as Assunta arrives. Serafina shows Assunta the broken urn. "There are no ashes." Assunta observes, looking at the shards; "The wind has blown them away."

From outside Alvaro calls. The neighbor women gather around to taunt Serafina that there is a man on the road without a shirt, calling her name. Serafina confides to Assunta that last night she felt the burning pain and saw the sign of the rose on her breast; she is pregnant again. Calling "*Vengo, amore!*" (I come, love!) she picks up the rose-silk shirt and runs out to join Alvaro.

The Content and Imagery of the Play

Like so many of Williams's plays, the subject is purging and release from a self-inflicted burden. Unlike most of Williams's heroines, Serafina escapes and is liberated.[4] She does not degenerate to the madness of Blanche duBois, to the deliberate self-debauchery of Alma Winemiller, or to the despairing resignation of Laura Wingfield. Having broken the urn with her husband's ashes and set her daughter free to live her own life, Serafina joins her lover with a lusty cry of "*Vengo, amore!*"

Serafina is not an uncomplicated heroine, however. She is highly religious and superstitious, but at the depth of her despair and anguish, she is willing to challenge the sanctity of the confessional, attack her priest, and even renounce faith in the Virgin Mary. So, in a sense, Serafina does what all of Williams's heroines have to do: She finds a way to compromise with the effects of the personal and social constraints that have previously both held her life in check and also given it meaning. Unlike the others, Serafina's transition is to freedom rather than to a deeper and inevitable bondage. In a sense her final "*Vengo, amore!*" is not just directed to Alvaro; it is her own recognition of love as the motivating principle of her life and her surrender to it, free of artificial constraints.

The atmosphere of the Mississippi Gulf coast is essential to Serafina's story. The events of *The Rose Tattoo* could not happen in the New Orleans of *Streetcar Named Desire*, the St. Louis of *The Glass Menagerie*, or the small, southern, protestant community of *Summer and Smoke*. In her native Sicily, social constraints would have been far too powerful to permit Serafina's rebellion. Serafina's liberation requires the steamy hot weather, the teeming life competing for survival space, the ethnic friction, and the almost-broken cultural ties, stretched thin by time and distance, holding Serafina to her Sicilian traditions, countered by the frontier mentality of local protestant mores.

The play opens with sound. Williams says in his production notes,[5] "As the curtain rises we hear a Sicilian folksinger with a guitar.... Cars passing on [the highway] can occasionally be heard." In the rubrics before the opening scene he writes, "The mothers of the neighborhood are beginning to call their children home to supper, in voices near and distant, urgent and tender, like the variable notes of wind and water. . . . The calls are repeated, tenderly, musically." All the way through the play, the dialogue and stage directions suggest sounds that compare and contrast the bucolic, feudal Sicilian world that drives Serafina's outlook on life with the realities of the racially and ethnically biased, primitively mechanized environment of the southern United States in the late 1940s. The Sicilian folk songs, the high school band, the goat and Serafina's parrot, the sounds of trucks and cars on the highway, and (although not part of a sound designer's responsibilities) the contrast between the Sicilian accents of the primary characters and the harsh, nasal drawls of the southern Americans.

Williams uses sound (sometimes artificially) to mark important points in the drama. When Alvaro appears for the first time in the play, we first hear his truck pulling up and stopping. Then Williams writes in his rubrics, "At the moment when we first hear his voice, the sound of a timpani begins, at first very pianissimo, but building up as he approaches, till it reaches a vibrant climax with his appearance to Serafina beside the house."

At the onset of Serafina's deepest crisis, the sound of the high school band playing in the distance marks the flow of time as it rushes past, threatening to swamp Serafina, who is desperately trying to pull herself out of her self-imposed exile from life. As she hears from Flora the fact of Rosario's affair with Estelle, the band is heard playing a counterpoint of Sousa's "Stars and Stripes Forever."

Before Rosario's death, the bleating of the Strega's goat signals the encroachment of Serafina's loss of control of her life and her descent into misfortune. During the first encounter with Alvaro, when the goat again escapes, Alvaro captures it suggesting that stronger hands have taken control of the situation, permitting Serafina to emerge from her pain and begin to live in the world again.

Turning Insights into Ideas

Once one sees the emotional threads and the symbolic structures running through a playscript, what does one do with that information? How does one change those insights into a concept and into ideas for a design? Are we obligated to provide every sound that a playwright calls for in the rubrics? Do we need all of the animal sounds and highway traffic noises? Are these merely going to be distracting? Are some other choices perhaps better for our production? Notice that the moments pointed out as indications of the importance of sound in the play are not all specific sounds demanded in the playscript by the author. For example, nothing other than lighting and actors' voices is called for as a part of *prima sera*, but the moment is important enough that some kind of auditory enhancement seems necessary. The sound designer needs to pay special attention to such circumstances, even if the script calls for no particular sound at the time.

What about music? Williams mentions only a Sicilian folksinger accompanying himself with a guitar. Is that done by a live musician functioning as an actor or is it on tape? And how frequently do we hear the singer or Sicilian folk music? How does the folk music interface with the high school band music? Need it do so at all? Are there any other points in the play that suggest use of music other than the band music and folk songs? And what about the music mentioned in connec-

[4]The notion that Serafina is liberated represents an optimistic view of *The Rose Tattoo*. An alternate opinion is that Serafina simply plunges right back into the trap from which she had an opportunity to escape.

[5]Williams, *The Rose Tattoo*, p. 9.

tion with the salesman's and Alvaro's entrances? Is that too artificial, too obviously symbolic? Or is it right for the purpose? If it is right for the purpose, what do we do to make it sound reasonable and not arbitrary?

Conflicting ideas need some unifying theme to help in making choices. The unifying theme in a design is the concept. So how do we derive a concept for *The Rose Tattoo*? We begin to ask questions of the script and, possibly, of other research documents that we have available. For example—

What human values is Williams exploring?

What is he saying about people and their behavior?

What is the point of the play?

What are the dominant images in the play?

What are the dominant themes?

How is time represented in the drama?

How does time affect the characters and their situations?

What are the spatial limits of the scenic environment?

What lines of ingress and egress are implicit in the spatial environment?

Obviously, not all people will come up with the same, or even similar, answers to these questions. Some potential answers that could be given to these questions are the following:

What human values are explored? Love, fidelity, loyalty, truth, falsehood, hope, despair.

What aspects of life and behavior does the play examine? That people cling to familiar systems of values, however inappropriate they may have become for a given situation; or that perceived injury or injustice becomes a driving motivation to change; or that inflexible personal standards and ideals eventually come into critical conflict with emotions and desires.

What is the point of the play? Possibly, that the person is as good as dead who lives for a dead love; or that only by accepting that those we love are separate individuals from ourselves and interacting with their freedom can we assert our own happiness and individuality.

What dominant images appear in the play? Sunset, night, dawn, sunlight, heat, humidity, moonlight, shadow, artificial light, animal noises, mechanical noises, music, odors. These are a few of the most evident images. Many others could be found.

What are the dominant themes of the play? Love, life, death, conflict between nature and convention, conflict between cultural traditions and values, conflict between generations, belief and despair, rekindled hope, new life.

How does time affect the characters and their situations? Williams specifies that there is a three year lapse between the first two scenes and the rest of the play. Rosa clearly grows from girlhood into adolescence. Serafina degenerates over that time from a proud, self-respecting woman into a slovenly, despairing creature fixed on the memory of her dead husband. Over the course of the play, from the wedding-dress scene to the end, Rosa achieves a measure of independence; Serafina goes through a major life crisis and emerges renewed and ready to live again. Nothing really changes with respect to her system of beliefs, since she becomes involved with another Sicilian man. She will still believe that the Strega has the evil eye (if that is of any further importance); she will still believe in special events, and she may even come back to faith in the Virgin.

What are the spatial limits of the scenic environment? The environment is limited to Serafina's dwelling, but constant reminders of the impinging world outside appear throughout the play: the highway, the sounds of traffic, the noise of the high school band, and the sounds of other human voices in the distance.

What lines of ingress and egress are implicit in the spatial environment? The highway, the path to and from the house, and the broken fence giving access to Serafina's land from the Strega's adjacent property. Alternatively, the cane brake and any suggestion of surrounding woodland imply return to a more primitive, more animalistic existence.

For *The Rose Tattoo*, time is one of the most important elements. One of the most important moments in the play is the reprise of *prima sera* after Serafina smashes the urn containing Rosario's ashes. Almost as in musical sonata allegro form, Williams brings us back to the feelings of the plays beginnings—the feelings of warmth and integration with a live, comfortable world. Again we hear the children's voices and mothers calling them home. The final rays of the setting sun break through into a quiet and tranquil early twilight. The moment is immediate and vivid, clearly marking a transition in Serafina's life, clearly marking the beginnings of her release from her self-imposed penance, and it is followed in the play by a fierce return of sexual desire. Williams uses it to mark the completion of a cycle in time—the ending of a process of working through the psychological effects of a major life catastrophe. The moment is observed in lighting and in the action, as well as by the sound of the actors' voices, so it is of unquestionable importance to Williams.

Questions and considerations such as these are the basis on which a design is founded. The designer does not necessarily achieve final answers to these questions, and certainly he or she does not do so apart from the views and opinions of the other collaborators on the production, most especially the director. From such questions and their resolution in pre-production planning discussions must emerge the specific design outline that is to be expanded and realized.

Chapter 21

Developing the Design: Deriving a Concept

Chapter 20 began with a strong statement advocating the necessity of a concept as a guiding element in creating a design. So, now that we have read *The Rose Tattoo* and have acquired some degree of insight into the nature of the script, what do we do for a concept? The time has come to be specific. To begin, we need two things: a statement of our own perception of the play and a directorial concept. The statement of personal perception sums up our own feelings about the play and what it means in aesthetic terms. The personal statement is usually written before hearing the director's ideas, and ensures that we can approach the initial design conference with a point of view and an understanding of the play.

The designer's personal summation about *The Rose Tattoo* could be expressed as follows: "The play carries a feeling of sadness, loneliness, and loss of motivation. It involves a painful breakthrough to an understanding and acceptance of a harsh reality, followed by the promise of a renewed vitality and reason to live. It presents images of light and dark, of both man-made and natural cruelty, of superstition, of ignorance and knowledge, and especially, of animal 'being' versus human 'becoming'."[1]

The Director's Concept

The director's concept is the statement of what the play means to him or her in social and human terms, and what he or she intends to focus on as the central point of the production. Let us say that the director chooses to use as the central thematic element of the production the idea that personal standards and ideals, when too closely held, eventually come into critical conflict with emotions and natural desires. He embodies the concept in the statement

> Serafina delle Rose has chosen to die with her husband. She tries to imprison time but finds that she has only succeeded in imprisoning herself. She can only regain her life by an act of desecration.

The director's thematic and concept statements stress the element of time. For sound, that should bring up immediate possibilities: the articulation of marker events, possibly, or the reinforcement of the stream of emotional awareness by the choices of underlying background sound elements, or both. On the basis of the director's concept, we decide that the auditory aesthetic concept must embody some reference to time. The director's statement refers to imprisonment and violent rebellion, and the notion of freedom as something to be purchased at high cost; so we must also conclude that the our aesthetic concept must bear some relationship to the nature and value of freedom.

Is there a conflict between the designer's statement of feeling about the play and the director's concept? Probably not, as both talk about loss and recovery of a life orientation. If no conflict exists, then we can proceed to look for a satisfactory aesthetic concept by which to construct the auditory design.[2]

What form of auditory aesthetic concept can we derive from the director's thematic and ethical statements? Recall

[1] Obviously, the author's statement will not agree with every reader's idea of *The Rose Tattoo*, but consider the statements regarding meaning as useful for illustration in the overall context of the chapter.

[2] What happens when the designer's ideas and the director's stated concept do not agree? One of the two options are available: the first, to attempt to convince the director of the merit of the designer's view in the hope of persuading the director to modify or change his or her concept to one more compatible with the designer's own idea; the second, to accept the director's concept and begin to find a way to modify one's own thoughts and feelings about the play.

that anything aesthetic has to do with feeling and emotion—with the pure, sensory elements of experience—and we have already determined that our own feelings about the nature of the play involve quite a range of feelings and emotions. We need to find sounds and sound structures that evoke associations with emotional experience and help those who see and hear the drama to associate their own feelings and experiences with those of Serafina and Rosa. To do this, we need to find sounds that are specific to the dramatic environment but also common to general experience.

First of all, the sounds used in a production cannot be independent of the atmospheric and geographic environment of the drama. We know that the time is summer and that the place is the Mississippi Gulf coast. The locale suggests immediately that there could be a large number of bird and insect noises. Farm animals are also likely to be in the vicinity. We also know that there is a highway immediately adjacent to Serafina's house. The scenic conditions mentioned in the playscript talk about a palm tree that stands near the house, and there would certainly be both pines and oaks. The oaks would probably be draped in Spanish moss. Williams also mentions a nearby canebrake. The rich variety of vegetation suggests that the sound of wind moving through several different kinds of leaves is a good possibility. So we have a wealth of natural and man-made sounds on which to draw.

Association of Ideas with Sounds

How do we associate certain kinds of sound with particular ideas? Lets start with the idea of freedom that is expressed in the director's ethical concept. Freedom superficially appears to be the right to do what one wants to do when one wants to do it, with no constraints on one's ability to act. Nature appears to humans to be a state of freedom. Birds and animals seem to do as they wish without the constraint of human rules and laws. Alternatively, other people frequently seem to be more free than we ourselves, especially when we are in a state of crisis, or when we have been denied some element of our own freedom. Using nature as a symbol of freedom, therefore, sounds of birds and animals could be used to contrast to Serafina's sense of her own imprisonment. Similarly, the passage of the cars on the highway could suggest a measure of human freedom that Serafina does not enjoy.

Freedom, mentioned in the abstract, always sounds like a desirable thing; however, freedom can also be threatening. In a sense, freedom means the right to do as you wish, but it often means sacrificing the known and familiar for the unknown and the strange. Freedom in nature is paradoxical: As humans we tend to see animals as free—free of our worries and social constraints, doing as they please when they please. But animals really have no choice; they do what they do out of instinct and of necessity. The freedom of nature is purely a human concept, and one that is something of a romantic oversimplification. Use of nature sounds as a counterpoint to Serafina's suffering, therefore, would potentially call into association both sides of the paradox of freedom. We are looking for ways to enhance the emotional impact of the drama, however, so the idea seems promising.

Human freedom involves the possibility of a certain amount of conscious choice, both personal and social. Nonetheless, we are what we are because of our mental structure and our modes of perception. Serafina chooses to create artificial constraints for herself, but in so doing she is following her instinctive mode of behavior. However, her attempt to hold onto Rosario by enshrining his remains begins to reduce her chances of survival. The irony is that her marriage, as she conceived it to be, was over before Rosario was killed; and she refuses to accept the evidence of it by forgetting who spoke the words "wild, like a Gypsy" and by not wanting to hear the name on the claim card for the rose-silk shirt. Then, when she does permit herself to remember, her first reaction is to destroy Estelle—to restore the purity of her world by destroying the evidence of its impurity; her second reaction, and the step that frees her, is to smash the tangible memory of Rosario. Sounds that focus on and contrast animal and human elements call up associations of the constraints to freedom and action. After Serafina smashes the urn, the pure sounds of wind, the light, musical voices of children at play, and the voices of the mothers calling the children in to supper form a marvelous contrast, enhancing the freedom that Serafina gains as she overcomes her self-imposed confinement.

In the designer's preliminary statement of feeling about the play, we mentioned sadness and loneliness. The sounds of the highway always stand as a constant reminder to Serafina (and to ourselves as spectators) of how Rosario died. From the viewpoint of Serafina's self-pity, both highway sounds and nature sounds give the impression of life going on all around without any care for her hurt and loss. The association of external life against internal struggle fits particularly well in the context of the graduation-dress scene, where Serafina is literally hounded to distraction by the pressures of life outside her own immediate concerns. As an added element, the highway sounds also reinforce the fact that the place is not Serafina's native Sicily. So does the intrusion of sounds characteristic of the American South. The proximity of farm animals carries reminders of Serafina's and Rosario's Sicilian origins, but also emphasizes the differences between the southern United States and Sicily.

All of the various sounds mentioned so far would work well as elements to convey a background stream of time. All have "light" and "dark" variants that can be used to obtain an enhancement of the emotional progression through time and to achieve the contrast necessary to enhance the urgency of Serafina's need to escape from her self-imposed entombment.

We always have the possibility of using music to motivate and reinforce the emotions of the play. Williams asks for a Sicilian folksinger accompanied on a guitar, and he asks for underscore music to enhance the events that lead to the meeting of Alvaro and Serafina. The Sicilian folk music could serve as a symbol of the inbred values and human nature that Serafina cannot escape, that will dictate her actions in spite of her conscious choices.

Notice that we cannot really speak actual auditory images of sounds in words. We can only try to describe them. In order to use auditory images as an aesthetic concept we need to single out the most essential characteristics of sounds in terms of tonal quality, spectral character, duration, and so on. If we select representative sounds, we can meld them into a palette of "colors" and "textures" that can serve to guide what we intend to do with the application of sound to the production.

Devising an Auditory Concept

Let's take a group of bird calls, mix the calls with wind in Spanish moss and cane fronds, add into that some occasional truck and car noises (tire noises on the road, engine sounds), fade all of that into a distant chorus of frogs, night birds, and soft winds; then blend into a dawn chorus of birds followed by farm animals, and finally back into daytime birds. Through all of this we will intermix fragments of Sicilian folk music. This mixture of fragments will be less than a minute and a half long, but it will provide a palette of sound color and texture that will communicate the essence of what we want to do and will guide our choices in expanding sounds into actual background for the play.

Why does this mixture of sounds form an auditory concept, and why do we form an aesthetic concept using sounds? Don't we need to be able to state a concept in words? Nothing is wrong with a verbal concept statement—if a verbal statement is what an individual as artist needs in order to do his or her work. In fact, we shall try to state an aesthetic auditory concept in verbal form somewhat later, just to explain how that can be done. First, we need to examine the nonverbal concept described above, to see just how it works and why it forms a concept and how the concept leads to the design.

How does the mixture of sounds form a concept? First, the chosen sounds create an audible image of time passage through the large structure of the play, in its change from day to night to day, from light to dark and back to light, from stability to despair through a return to some measure of stability. Second, the mixture introduces both the natural and the artificial as contrapuntal elements—emotional marker symbols for the components of Serafina's self-alienation. Finally, the group of sounds includes the narrative element of music as a thread tying the various elements together, and unmistakably inserting human values into the structure of the auditory aspects of the drama.

The concept is formed in the manner described because auditory images permit us to state relatively unambiguously the emotional character and the auditory color and texture that we, as sound artists, will try to add to the production. The auditory concept gives the director something to which he or she can react directly. The director can say clearly that the sound designer's concept fits or does not fit with his or her feelings about the nature of the production. (Remember that the director is going to have an aesthetic orientation to the play, even if he or she does not explicitly form an aesthetic concept, and the director's aesthetic orientation to the script is going to have a great deal to do with how that person accepts a given designer's ideas.) The director can listen to the concept mixture repeatedly, using it as a planning element in shaping the production, and the mixture serves as a working tool in discussions about how sound is to be used in the production. If the director has this kind of concept statement to work with, anything that he or she comes up with to enhance a scene is more likely to be compatible with our own ideas instead of conflicting with them.

Devising a Verbal Concept

As we said, we should try out a verbal concept statement, just to illustrate how such a concept is formed. What would we have to change in order to make the concept work in words? Essentially, nothing relative to the kinds of sounds in which we are interested. We just have to try to capture in words the ideas and feelings contained in the sounds we have chosen. We might state something like the following:

> Serafina delle Rose's world is never free of sounds. Sometimes they annoy her, sometimes they mirror her feelings, and sometimes they seem to function as omens for her life. She lives between the natural sounds of wind and creatures and the man-made intrusions of the sounds of trucks and cars on the highway. The birds, the wind, and the animals mirror her feelings and reflect her fears. The highway brings her husband's infidelity and his death; but it also brings Alvaro.

This statement embodies the same ideas as the mixture of sounds described earlier. The major difference is that the word concept does not actually show any of the intended colors and textures, and it cannot really show the unifying effect of the Sicilian folk music. We could mention the folk music, of course, but in such a verbal statement, we begin to lose some of the economy and the focus of the concept if the statement becomes simply a description of all the elements that we might want to use in a design. The focus of the concept is the power of natural and human sounds to support, enhance, and frame the emotional elements of Serafina's story. Notice that, similarly, the sounds of the high school band were not included in the auditory concept. The omission of the band music by no means implies that we won't use it; we will, because it is an important element.[3] As part of an auditory concept statement, however, the band music would be confusing, as would a reference to folk music in the verbal statement.

We must find sounds that will be potentially useful in order to begin putting together both the concept mixture and, eventually, the design itself. But how does one know what sounds will be potentially useful? Remember that (theoretically) we are still in the stage before the auditory concept has been formed and stated; we know what we want to do, but we are in process of looking for material from which to build the concept. The obvious answer is that there is, necessarily, an indispensable element of *auditory imagination* that cannot be omitted from this process. In fact, parts of the process run together and overlap. That's why the designer prepares a preliminary statement of his or her feelings about the play. The designer, as we said earlier, must find that the world of the drama has taken form and shape in his or her mind during the process of familiarization. That form and shape must focus on the color qualities of sound in the imagined dramatic world. *This creative predisposition directs the search for raw sounds from which the designer will shape both the concept and the elements of the design itself.*

Nothing prevents the designer's original ideas from changing under the influence of research. One may very well find things that had not occurred to the imagination but which are so obviously right and so compelling that they cannot be ignored. This is a very real part of the creative synthesis. Some of the best ideas happen by chance and by accidental discovery. Such discoveries need to be found as early as possible, however. Remember that, after the primary production conferences are finished, any major change requires the consent of the entire production staff. Great revelations late in the production process are invariably more damaging than they are beneficial.

When new possibilities are discovered through experimentation with sounds, then a standard of judgment is needed to be sure that simple fascination with the power of a sound or of a new idea is not misdirecting the focus of the search. (The danger of seduction from the basic objective is the primary reason for the establishment of a concept in the first place.) In order to make a decision, one must sometimes create several trial concept mixtures to hear exactly how the various elements fit together. The trial mixtures must then be tested against the play and against the directorial concept. Remember that, like colors, the perceptual qualities and characteristics of any sound can change radically in the presence of other sounds and in the presence of various different contexts. A part of the business of putting together a concept mixture is to hear how the various elements work together and to work out the correct balance among the auditory elements.

[3]One question that is likely to come from almost any director about the purely auditory concept is, "Where is the band music?" The sound of the band is relatively important, and any director will probably have this sound fixed as a central element that must be accounted for. One of the things that sound designers must always be prepared to do is to assure a director that a concept mixture is not a catalog of sounds to be used—it is a means to establish emotional qualities to be evoked. The submission of rough cue list in conjunction with the concept mixture—a cue list that includes all of the major required marker sounds—can be reassuring to a director.

Chapter 22

Developing the Design: Organizing Ideas

Charting the Play by Scenes

To develop a design for any production, the first thing to do after we have achieved familiarity with the script and have decided on the dominant themes and images is to take stock to determine what resources we are going to need. One way to do this is to go back through the script and establish categories of sound by act and scene, even by French scenes[1] if necessary. With each grouping, we make a few notes about the action and primary emphasis of the scene. We note causal relationships from previous scenes, and look for events within the scene to serve as causal forces in succeeding portions of the drama. All of this is mechanical and somewhat arbitrary; but it does help to reduce the play to its skeletal events and action lines and helps us compile a catalog of the raw materials that we will need for the design.

To continue the process of working by example, let's set up a chart by act and scene for *The Rose Tattoo.*

General Locale: The Gulf coast between Mobile and New Orleans. Hot summer weather. High humidity.

Act I/Scene 1

Time: Early dusk. Children playing, mothers' voices. (These noises onstage, made by actors.)

Characters: Serafina, Assunta, Rosa. Later, Estelle. Briefly children and the Strega.

Actions/Causal Relationships: Establishment of Serafina's belief in signs and miracles and of her pride. Establishment of pregnancy. Establishment of Rosario's contraband activities. Introduction of Estelle as possible interfering force. Introduces Serafina's superstition and fear of the Strega.

Sounds: Called for in script: Folksinger; cars passing on highway. heavy truck approaching, passing. Serafina's first mention of conceiving pointed with phrase of music. Another truck passing. Goat bleat, jingling harness; crash of splintering wood. Goat sounds repeated several times.

Possible design additions: Evening bird sounds; wind; nearby domestic animals.

Act I/Scene 2

Time: Just before dawn the following morning.

Characters: Serafina, Father Leo, Assunta, women.

Actions/Causal Relationships: Serafina sewing on the rose-silk shirt. Approach of women and Father Leo to announce Rosario's death.

Sounds: None called for in script.

Possible design additions: Night sounds—wind, a cricket.

Act I/Scene 3

Time: Noon, the same day.

Characters: Doctor, Assunta, Father Leo. Later, Estelle, Rosa, the women.

Actions/Causal Relationships: Miscarriage; Father Leo objects to cremation of Rosario's remains as "pagan idolatry." (Establishment of Serafina's intent to enshrine Rosario's remains.) Estelle comes to pay last respects

[1] French scenes divide a play by the entrances and exits of the major characters. For example, at the opening of *The Rose Tattoo*, the few lines between Serafina and Rosa would constitute one scene. When assunta enters, we begin the second French scene. Assunta's exit would mark the beginning of the third French scene, and the entrance of Estelle starts the fourth French scene. French scene division is mainly used by costume designers as a means of organizing what costume is required when.

and is beaten and driven away by the women. (Establishment of the affair between Rosario and Estelle.) Rosa reacts to her father's death.

Sounds: None called for.

Possible design additions: Distant birds (crows, jays), wind.

Act I/Scene 4

Time: Morning, June, three years later.

Characters: Women, Serafina, Assunta, Miss Yorke, Rosa.

Actions/Causal Relationships: The women want the graduation dresses for their daughters. They gossip about Rosa's being locked up, naked, in the house. (Introduction of Jack Hunter as factor in Rosa's life.) Serafina's scream is heard inside, and she comes out crying that Rosa has slashed wrists. Miss Yorke determines no serious injury; takes Rosa to graduation. Women get dresses.

Sounds: None called for.

Possible design additions: None needed except scene bridge music in and out.

Act I/Scene 5

Time: Same as I/4.

Characters: Serafina, Flora, Bessie.

Actions/Causal Relationships: Serafina trying to get dressed; Flora insists on having blouse. Serafina, angry at their brazen behavior, says she and Rosario had a perfect relationship. Serafina leans about Rosario and Estelle.

Sounds: High school band in distance, car with Legionnaires driving by, squawks from Serafina's parrot.

Possible design additions: None needed.

Act I/Scene 6

Time: Two hours later (afternoon).

Characters: Serafina, Rosa, Jack.

Actions/Causal Relationships: Rosa, with Jack, stops to change for a beach party. Serafina confronts Jack, accuses him of taking advantage of Rosa; Serafina makes Jack swear before the Virgin to respect Rosa's innocence. Rosa and Jack exit to join students.

Sounds: Car arriving and leaving (twice in act—at beginning and end). Car horn.

Possible design additions: Wind, birds, animals, traffic noise.

Act II

Time: Two hours after I/6 (mid-afternoon).

Characters: Serafina, Father Leo, briefly women and Assunta, salesman, Alvaro, briefly children and the Strega.

Actions/Causal Relationships: Serafina's point of deepest despair. Father Leo chides Serafina for slovenly dress, and she demands to know if Rosario ever confessed to an affair; flails at Father Leo with fists when he refuses to answer. Women and Assunta rescue him. Salesman enters, followed by Alvaro; Alvaro is kneed in groin. Alvaro and Serafina become acquainted; likeness of Alvaro to Rosario established, Alvaro learns details of Rosario's appearance and mannerisms which he later uses to try to impress Serafina. Alvaro suggests possible breakage of Rosario's urn as means to freedom for Serafina. Strega's goat escapes; Alvaro catches it. Alvaro and Serafina agree to meet for dinner so he can return the shirt.

Sounds: Children's voices, salesman's car, Alvaro's truck, goat.

Possible design additions: Wind, road traffic, birds, animals.

Act III/Scene 1

Time: Evening, same day (similar to I/1, just a little earlier.)

Characters: Alvaro, Serafina.

Actions/Causal Relationships: Alvaro returns for dinner. He makes advances; she finally orders him to go, but business over the shirt leads to recall of Estelle's name and relationship to Rosario. Serafina wants to go to the bordello and cut out Estelle's "lying tongue." Alvaro calls Estelle; Serafina hears her voice over phone confirming her affair with Rosario. Serafina smashes the urn; agrees to spend the night with Alvaro.

Sounds: Truck arriving. Later truck leaving. Estelle's voice over the telephone. Night animal sounds, shrieks of laughter, goat, wild night birds.

Possible design additions: Wind, evening bird sounds, children's voices, animal noises, road traffic.

Act III/Scene 2

Time: Predawn, following morning.

Characters: Jack, Rosa (Serafina heard offstage).

Actions/Causal Relationships: Serafina heard moaning, making love, in bedroom. Jack and Rosa return. Rosa wants Jack; Jack, in respect of his promise, resisting with difficulty. Agreement to meet at bus terminal in New Orleans. Jack leaves; Rosa beds down on sofa and cries.

Sounds: Rooster crowing.

Possible design additions: Predawn breeze.

Act III/Scene 3

Time: Three hours later, first daylight.

Characters: Alvaro, Rosa, Serafina; later Assunta and women.

Actions/Causal Relationships: Alvaro, drunk and sleepy, finds Rosa. Rosa's scream brings attack from Serafina which Rosa stops. Alvaro, in confusion leaves; Serafina miserable and apologetic, trying to pretend she had no knowledge of Alvaro's presence in the house. Rosa condemns Serafina as hypocrite. Serafina relents, tells Rosa to go to Jack. Assunta enters, finds the urn smashed; women outside set up mocking shriek about Alvaro without a shirt. Serafina tells Assunta about new pregnancy, goes to take shirt to Alvaro.

Sounds: "Bleating duet" (*Che bella*) between Alvaro and goat, cockcrow, music.

Possible design additions: Dawn chorus of birds, highway traffic, wind.

Plotting the Tension Curve

In the outline, all of the primary, causative actions are emphasized. Note that there is a progression that traces a process of development and change in Serafina's life. A secondary story line of Rosa's growth into early maturity and independence also takes place beneath and intertwined with Serafina's primary story line. Figure 22-1 plots the curve of tension and climactic points in the course of the drama. The main trace (the heavy line) is Serafina's; the secondary trace is Rosa's.

Note the breakdown of acts and scenes as percentages of the total length of the play. The percentages are useful only as a means of showing the fundamental rhythmic structure underlying the overall drama. Observe that the first act is highly articulated and episodic, a group of short scenes that quickly sketch the conditions of Serafina's life. The first act constitutes approximately 40 percent of the play's length, and is used to introduce most of the characters, set up the circumstances by which the main conflict of the drama is conditioned, and bring us to the central developments of the play.

The second act is not divided into scenes; however, four subscenes are implied, dividing the act into rhythmic steps.

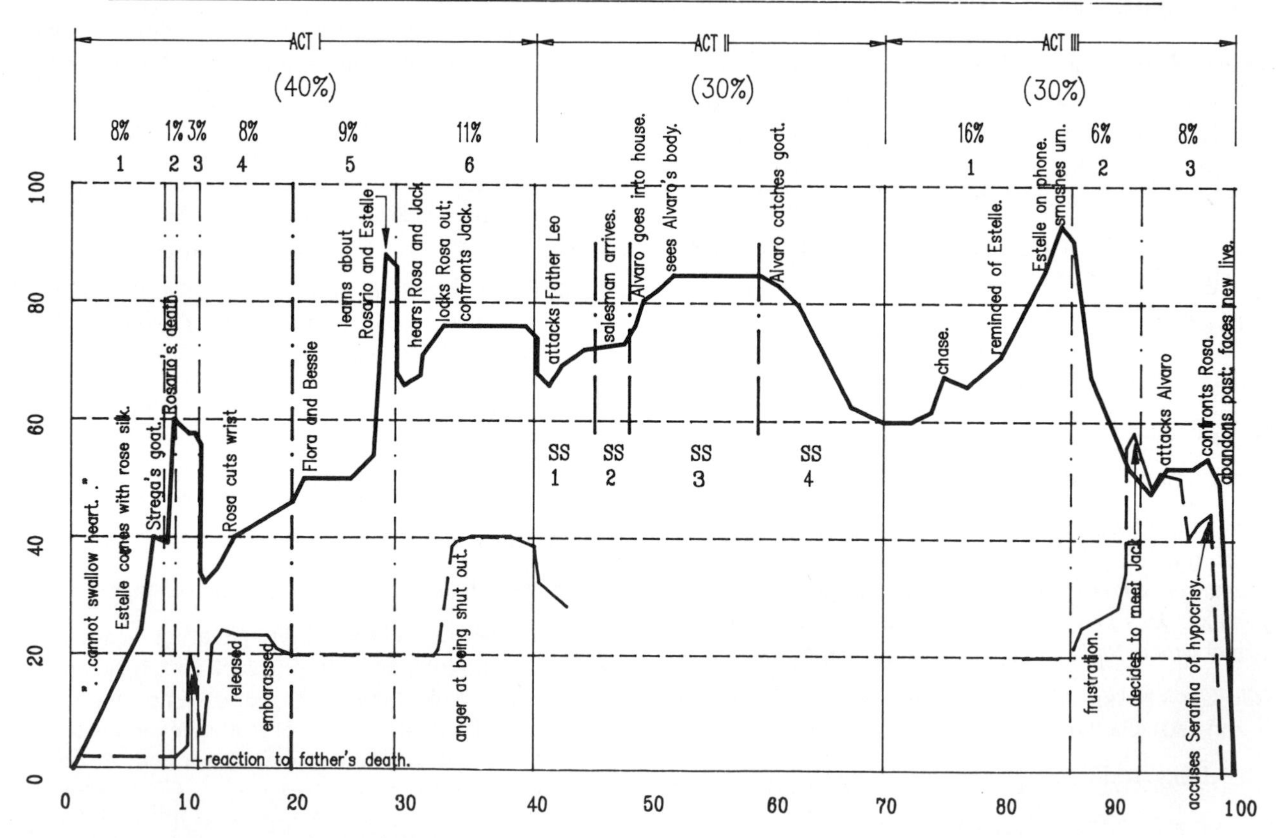

FIG. 22-1: TENSION CURVE FOR THE ROSE TATTOO

The steps are necessary to decelerate the action from the frenetic pace of act I to the semi-intimate interlude that passes between Serafina and Alvaro, and then to quicken the action in preparation for Act III.

The first of the rhythmic episodes is the scene with Father Leo; the second, the salesman's appearance and the initial appearance of Alvaro. Both of these rhythmic elements tend to retard the pace of the action, even though both have their flurries of activity. The Father Leo scene finds Serafina apathetic about anything except her internal misery. When the Father upbraids her about her behavior, he triggers a brief flurry of rage that quickly subsides to confusion and frustration under the badgering of the salesman. The one-sided battle between the salesman and Alvaro is the last, short burst of frenzy before the relatively tranquil portion of the act during which Alvaro and Serafina become acquainted.

The second act has to serve as introduction for Alvaro, give us his basic qualities, and establish reasons for Serafina's attraction to him. Most of the act is taken up by the first meeting between Alvaro and Serafina, and gives them (and us as spectators) ample time to let the relationship begin to take root. Most of the act, therefore, needs a somewhat slower, more relaxed pace. Finally, the business from the second escape of the Strega's goat to the end of the act begins to quicken the pace again in preparation for the events of the third act.

The third act begins with another long scene between Alvaro and Serafina, starting with more frenzy and energy and quickly building in tempo to the smashing of the urn when we have a brief reminder of the opening pace and feel of the play. Then we quickly return to the frenetic energy of Serafina's still outflowing rage. The second scene transfers that energy to the second thematic element of the play, the developing passion between Jack and Rosa, letting it build in an echo of the tension curve of the previous scene. Finally, in the last scene there is a brief opening flurry of Serafina's rage and a frenzied attempt to maintain her old familiar patterns. Then this gives way to a complete emptying of the soul (so to speak) and the regeneration of new life.

The tension curve starts low, primarily because we, as spectators, don't know anything about Serafina or her problems at the outset, but it quickly begins to rise. The first significant marker point in the tension curve is the discussion with Assunta about "not being able to swallow my heart." Assunta's comment that "a woman should not have a heart too big to swallow," that she has to go attend to a woman who took rat poison "because she had a heart too big to swallow," presages the oncoming catastrophe of Rosario's death. Estelle's appearance with the rose-colored silk and the order for a shirt for a man "wild like a Gypsy" is the second marker in the tension curve. This is a point that potentially needs delicate enhancement by some element of production other than the action and dialogue to make sure that it carries its necessary weight. The audience should have some feeling that Estelle and the shirt are somehow going to engender difficulties for Serafina, but we shouldn't find it too obvious that Estelle and Rosario are lovers.

The incident of the Strega's goat marks a large jump in the tension curve, because it is pure omen—a symbol of things to come. The black goat is loose—the goat of the adversary that Serafina believes to have "shaken hands with the devil." The incident symbolizes that, although she doesn't really know it, the events of Serafina's life are being carried beyond her control. Again, the treatment has to be delicate, or it will become comical and tasteless. (A good rule to follow with such symbolism is to let the dialogue and action carry the entire burden if at all possible.)

Act I, scene 2, is the shortest in the play and carries the second highest sudden increase in tension as Serafina learns of Rosario's death. Scene 3 of act I is the only scene in which Serafina does not actually appear onstage. The tension curve remains high, however, as we are told of the miscarriage and of her intent to enshrine Rosario's remains. The tension level drops at the outset of scene 4, although, with the clamor of the women for the graduation dresses, the energy remains fairly high. The tension curve climbs quickly with Rosa's self-injury and Serafina's accusation that the high school provides opportunities for lewd and indecent activities to take place. Serafina's frustrating attempt to pull herself together to go see Rosa graduate carries the tension level as scene 5 begins. The entrance of Flora and Bessie increases tension, and finally, as Flora tells Serafina of the affair between Estelle and Rosario we get another steep rise in the curve, to a level that constitutes the first major climactic moment of the play.

Typical of Serafina's response to catastrophe, the tension level has dropped somewhat by the time Jack and Rosa arrive to get ready to go to the party at Diamond Key. Her frenzy of fear that Jack is going to seduce Rosa leads to a small but significant increase in tension that holds throughout the scene to the end of act I.

At the beginning of act II the tension level has subsided a bit with Serafina's lethargic response to having to endure her misery. The petition to Father Leo brings a flurry of energy, and then the appearance of the salesman primarily serves to increase the confusion and annoyance for Serafina. The salesman's fight with Alvaro heightens that state, leading to her fear and confusion when the stranger, Alvaro, enters her house without permission. Tension increases to its peak and sustaining point for the act when Alvaro removes his shirt and she observes the resemblance of his body to Rosario's. Under the influence of this observation, she begins to unburden some of her pent-up feelings, so that when Alvaro takes control and kicks the Strega's goat out of the yard, she begins to feel as if a change is about to begin, and the tension curve starts to decline.

Act III again starts with somewhat reduced tension, but, like act I, scene 5, Serafina is again experiencing frustration trying to dress in clothes that no longer fit. Tension escalates with the revelation of Alvaro's rose tattoo and momentarily peaks as Alvaro drops the condom and precipitates her rage again. The curve drops slightly but only momentarily. When Serafina confirms that Estelle's was the name on the claim card for the rose-silk shirt the tension swiftly rises through the point where we hear Estelle's voice confirming the affair with Rosario and culminating at the main climactic point of the play, when Serafina smashes the urn.

Although Serafina does not appear on stage in scene 2, we can hear her moaning offstage as she and Alvaro make love. With the release of a new intimacy, Serafina's tension curve drops markedly through this scene. At the beginning of scene 3 the curve rises briefly as Serafina tries one last time to uphold a pretense of purity for Rosa's benefit, then it peaks just a little more with the final confrontation before Serafina lets Rosa go to pursue her own life. Finally, with Serafina's realization that Rosario's ashes have blown away, that she is no longer bound to old memories, the tension curve drops, ending this story, and Serafina exits to begin a new life pattern.

Observe that Rosa's tension curve largely echoes that of Serafina. Rosa has her reaction to Rosario's death precisely one scene later than Serafina's reaction. Rosa's tension curve in act I, scene 4, rises more steeply than Serafina's and momentarily leads it. In act I, scene 6, Rosa's tension rises in direct response to Serafina's actions in questioning Jack and making him take an oath before the statue of the Virgin.

Again, in act III, notice that Rosa's tension curve is almost an exact duplicate of Serafina's from the previous scene. Finally, in the last scene, the two curves peak at the time when Alvaro awakens Rosa, she screams, and Serafina attacks him with the broom. Rosa's curve drops in tension as she determines that she will go to Jack after seeing that her mother has had Alvaro in the house all night. There is a brief increase in tension at the final confrontation between Rosa and Serafina, followed by a steep decline in Rosa's tension curve when Serafina releases her to go, with Serafina's blessing, to Jack.

Using the Statistical Data

How does all of this outlining, percentage taking, and curve plotting help in constructing a sound design for *The Rose Tattoo*? First of all, we have created a list of the required sounds and possible added enhancements within the context of each act and scene of the play. We now have a clear picture of the rhythmic motion of the play—where its pace is rapid and where it is slower. Finally, we have designated the climactic points in the drama and we have a pretty good idea of the underlying schema of the play's structure.

Let's look at the graph again, this time with the required sounds (those which Williams specifically mentions) added in. (See Fig. 22-2.) Observe the placement of sounds. At the start of the play we get a small helping of atmospheric sounds that are directly related to the action and the symbolism—the truck's passing and the bleating of the goat. Two quick, high-tension scenes and the graduation dress scene pass without any sounds called for; but with the Flora/Bessie scene, a number of sounds are used in quick succession to heighten Serafina's tension and frustration. There are no effect sounds in act I, scene 6, until at the very end we hear the car horn and the voices of friends calling to Jack and Rosa.

In act II, there a number of effect sounds at the opening: wind, goat noises, perhaps the salesman's car, then Alvaro's truck; we probably hear the salesman's car leaving. With the salesman's and Alvaro's appearances, Williams asks for music, and he also calls for music when Alvaro removes his shirt. By this point in the act, however, the frequency of effect sound has dropped considerably. There is another small group of sounds just before the second episode with the Strega's goat.

Act III opens with the sound of Alvaro's truck arriving and with the sound of children playing nearby (which may be live rather than produced by the sound system). Most of the scene is devoid of effect sound; then, near the end, we hear the goat, Estelle's voice (through the phone), and finally Alvaro's truck driving away from the house.

Act III, scene 2, is punctuated by a rooster crow, adding to the urgency for Jack and Rosa to part. In the last scene, we hear a rooster again, signifying dawn, then the Strega's goat—the omen of trouble. This time the goat sound is made comic by the "duet" with Alvaro—his saying the first syllable of the word *bella* in a manner that echoes the goat's bleat. Finally, near the end of the scene, we hear a train whistle, which increases Rosa's haste to leave.

While the placement of the called-for sounds in the context of the tension curves and the rhythmic structure of the play is instructive, the charts become extremely useful when we begin to consider the possibility of atmospheric enhancement. For example, we can clearly see that any of the sudden increases in tension and the run-ups to those points become obvious parts of the play on which to focus attention. The two main climactic points are of extreme concern in terms of atmospheric enhancement.

The outlining and charting processes discussed in this chapter are not necessarily carried out by a designer every single time he or she sets about to create a sound design, at least, not overtly. Rather, the examples given here are illustrative more of the kind of thinking that goes into a design than

FIG. 22-2: MARKER SOUNDS ADDED TO TENSION CURVE

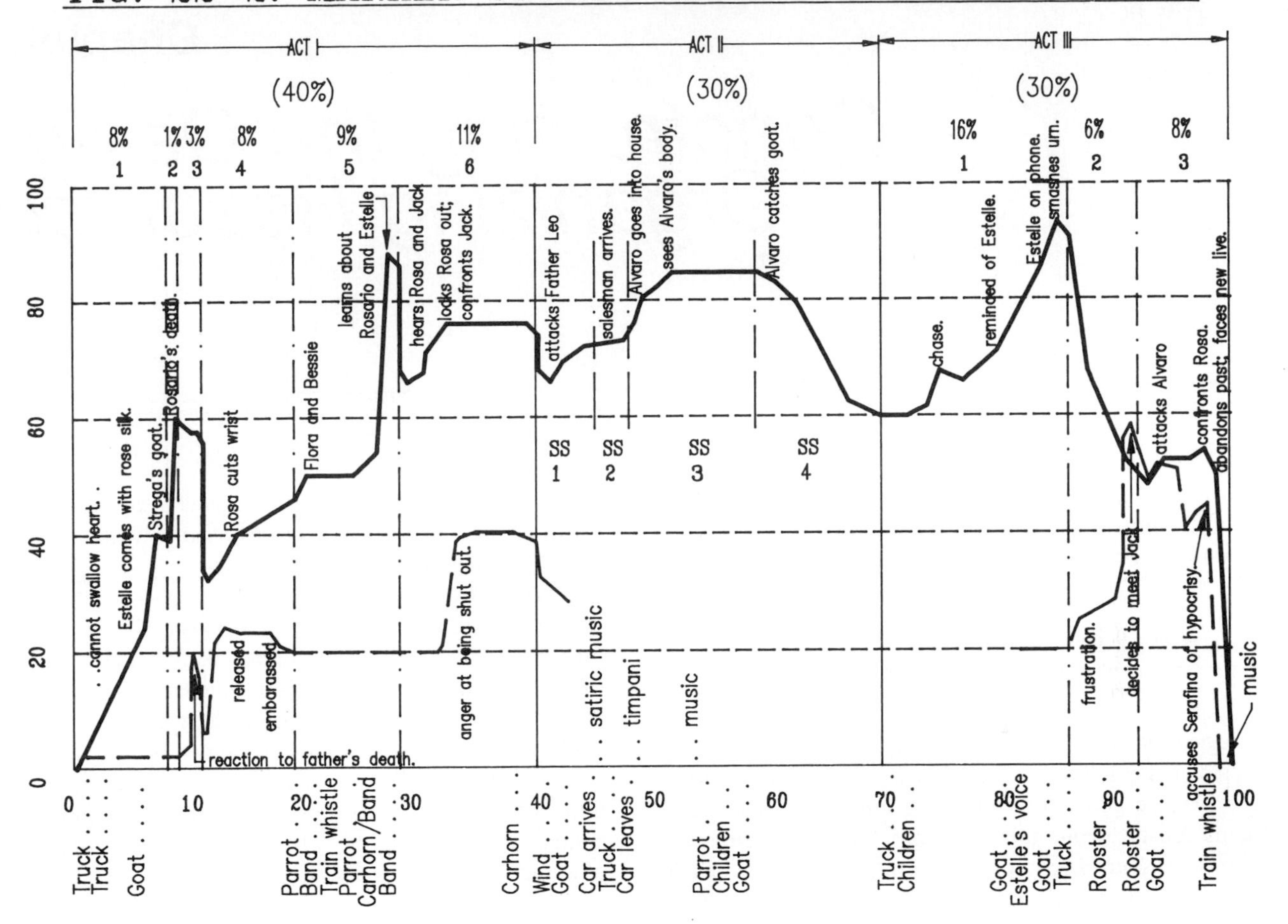

of actual processes (although there are times when use of the full process is not only helpful but essential).

The purpose of the outline is to summarize the dramatic conditions through the play in a scene-by-scene manner, together with a catalog of the required and possible sounds. The purpose of the chart is to provide a graphic illustration of the tension curve and of the climactic points in the play, so that we know where the focal points of the dramatist's efforts are directed. Often, the outline is reduced simply to a list of sounds, both required and possible. The tension chart is frequently a matter of one's intuitive grasp of the play, understood but not made formally explicit. The more designs one has accomplished, the easier it is to perceive the tension line and the climactic focus of a play. In other words, experience brings insight. The novice should try to go through the full, formal process at least a few times in order to learn what a satisfactory level of insight into a play feels like and what such insight can accomplish toward realizing a design.

Chapter 23

Constructing the Design: Organizing Resources

Now that we have a directorial concept, a design concept, and a set of preliminary ideas with which to work, we shall proceed to construct the design for *The Rose Tattoo.*

Cataloguing the Required Sounds

The concept has committed us to an attempt to flesh out at least part of the auditory environment of Serafina's world. That is, we are going to provide a number of sounds that Williams doesn't mention in the script: birds, domestic animals, insects sounds, wind, various night creatures, and a number of man-made sounds of the American South. We will need some ideas to guide this undertaking, and we will probably find that some of the best ideas develop as we go through the process of creating sounds for the production. Very often the act of finding, making, and shaping the required sounds and music for a script gives rise to new ideas and inventions that could add measurably to the emotional impact of a play. Also, attendance at rehearsals, to watch the growth and development of the characters as the actors intend to bring them to life, can bring about new thoughts and insights. Remember, however, that radical changes require the approval of the entire production team. Measure every new idea against the concept to be sure that it is in keeping with the objectives that we have set for ourselves in realizing the design.

Introductions to acts, scene bridges, and conclusions of acts will use the Sicilian folk music that Williams specifies, except that instead of using a singer, we choose to use instrumental music, primarily guitar.

We will provide all of the marker sounds that Williams calls for with one exception: We will omit the timpani roll that Williams mentions at Alvaro's first entrance. There may be another way to emphasize the event if that proves to be necessary. The timpani, however, seems overly dramatic and heavy handed. We will need to experiment with this part of the play to determine what feels right.

As a next step, we must try to generate a catalog of the various sounds that will be needed, putting this catalog in the form of a tentative cue list. The cue list is made as a table of rows and columns. Each row is a full description of a cue: its name, number, and location in the script and a mnemonic description. The first column in each row is the act and scene number. The second column is the tentative cue number; the third is the page in the script.[1] The fourth column is the cue name or function, and the fifth column contains a description of the kind of sound we need to find.

Finding the Sounds with Which to Work

Some of the sounds we need are not going to be easy to find. The Sicilian folk music is a case in point. Most of us don't keep libraries of Sicilian folk songs; in fact, most libraries don't keep collections of Sicilian folk songs, not on records or tapes, at any rate. Even in commonly available anthologies of folk music, Sicilians are not liberally represented. One needs a good music library with an excellent wide-ranging collection of both records and printed music. (If you can find only the printed music, then you must also have a way to turn the printed musical symbols into a performance of the music.)

Useful libraries of sound effects can also be surprisingly difficult to procure. Most of those that are available are protected by copyright, meaning that they should not be duplicated without permission, not even for transfer to tape for purposes of editing. Copyright problems aside, most public libraries do not keep anthologies of sounds; such collections are not something in which there is wide general interest.

[1]Taken from *The Rose Tattoo*, acting edition (New York: Dramatists Play Service, 1951).

TABLE 23-1. Tentative Cue List and Outline of Sounds

THE ROSE TATTOO

Act/Scene	Cue No.	Page	Cue Name/Function	Sound Description
Preshow	1	**	Preshow music	Sicilian folk music. Mainly guitar, but with other instruments included.
I/1	2	13	Show open	Main music theme (Sicilian, guitar). *Prima sera.* Evening birds, light wind.
I/1	3	15	Truck	
I/1	4	17	Truck	
I/1	5	18	Goat	
I/1	6	20	Bridge I/1>I/2	Sicilian music (guitar).
I/2	7	20	Cricket	Scene background.
I/2	8	21	Bridge I/2>I/3	Sicilian music (guitar).
I/3	9	21	Daytime birds	Random bird noises, different distances.
I/3	10	23	Bridge I/3>I/4	Sicilian music (guitar).
I/4	11	23	Birds, cars	Random daytime birds (jays prominent). Sounds of cars passing on highway.
***	**	29	***********	No bridge music. Continue bird/car background.
I/5	12	29	Parrot squawk	
I/5	13	29	High school band	Distant. Something that sounds like a school song.
I/5	14	31	Train whistle	Distant. Two blasts.
I/5	15	33	H.S. band	Distant. Playing what could be a graduation entrance march.
I/5	16	35	H.S. band	Distant. Sousa's "Stars and Stripes Forever."
I/6	17	36	Insects, distant birds	Light drone of small insects, occasional cicada; bird sounds very distant.
I/6	18	36	Car approach, stop, idle	
I/6	19	45	Car horn	Two short blasts.
I/6	20	45	Close act I	Sicilian music, guitar. (May also include other instruments.)
II	21	46	Open act II	Sicilian music, guitar. (May include other instruments.)
II	22	46	Dry, hot wind	Cane and palm fronds rattling in wind.
II	23	48	Goat	
II	24	50	Car approach, stop, with radio blaring	Car approaching, stops quickly with a short squeal of brakes. Radio is playing loud country music (late 1940s vintage). Radio keeps playing in background until car leaves.
II	25	52	Large truck approach, stop	

The Rose Tattoo (cont.)				
Act/Scene	**Cue No.**	**Page**	**Cue Name/Function**	**Sound Description**
II	26	53	Car starts and leaves	Motor revs and tires squeal as car takes off. Radio fades as car pulls away.
II	27	63	Children playing	Distant. Blends in but rises just above any othe noises that may be playing in background at this point.
II	28	63	Goat; wood breaking	
II	29	66	Large truck	Truck starts, drives away.
II	30	66	Close act II	Sicilian music, guitar. (May include other instruments.)
Note: Act II will also need some bird and wind background not specified in above list. Birds like act I; wind, hot and dry. Emphasis on rattling cane and palm fronds.				
III/1	31	67	Truck passing	Moving at moderate speed, slow enough to fool Serafina into thinking truck might be Alvaro arriving.
III/1	32	67	Truck arriving	Truck arrives and stops with squeal of air brakes. Engine shuts off.
III/1	33	75	Goat	
III/1	34	76	Estelle	Voice heard w/"distant" telephone perspective.
III/1	35	77	Birds, wind	Reprise of *prima sera*. Birds, soft wind, as at open of play.
III/1	36	78	Goat	
III/1	37	78	Truck leaving	Truck is heard starting, then pulling away, disappearing into distance.
III/1	38	78	Sinister night sounds	Parrot, then goat bleat. Gust of wind; bird cries and flapping of large bird's wings. Distant dog. Distant wild shrieking, laughter.
III/1	39	79	End III/1	Sicilian music, guitar only.
III/2	40	79	Open background	Periodic quiet breezes; occasional cricket.
III/2	41	80	Rooster crow	
III/3	42	81	Rooster crow	Distant but clear.
III/2	43	83	Bridge >III/3	Slow guitar chords, very soft.
III/3	44	83	Dawn birds, then rooster crow	Dawn chorus in distance. Rooster is relatively close.
III/3	45	84	Goat	Several bleats, long. Timed to Alvaro's "*Che bella*!"
III/3	46	84	Transition	Begin transition to daytime bird background.
III/3	47	86	Train whistle	Moderately distant; two long blasts, one short, one long.
III/3	48	88	Final Music	Sicilian music, guitar first, then other instruments. Hold under scene, then fade up for end of show. Begins as shirt moves to Alvaro.

The most readily available collections of recorded sounds are commercial ventures that can be purchased through record shops or by direct order from the manufacturer's marketing agent. Sound collections available in record stores tend to be aimed toward people who like to do sound backgrounds for their home movies and videos. Such collections usually have a smattering of every kind of sound—domestic animals, the more "ferocious" wild creatures, a few automobile sounds, and a motley assortment of "space" noises, buzzes, and crashes for comic effects. Most of these sounds are more like caricatures than serious reproductions of real noises.

Serious sound effect collections are usually advertised in magazines aimed at the broadcast, film, and theatre industries. The collections available are generally larger than the one- or two-record set provided for home video buffs in record stores; but they are also considerably more expensive. (Sometimes they are more expensive without being noticeably better in quality than the record shop collections.) The latest and best sound effect libraries are huge collections recorded on CD. Most of these productions include fully cross-indexed catalogs. A few sound contractors do exist for the professional motion-picture and legitimate theatre industries, and these shops will supply individual sounds on request. There is no guarantee, though, that the sound supplied will be usable. Sounds available from these industrial jobbers are taken from their libraries and are restricted to whatever they have collected under the particular category requested.

Special purpose sound libraries exist in some rather out-of-the-way places. You will need to discover and learn how to use these sources because they are often the best suppliers for certain kinds of noises. Model railroad magazines can often provide information about where to find the sounds of steam- and diesel-powered trains. Large university engineering and science departments sometimes keep recordings of sounds related to their disciplines. For example, one major university ornithology department maintains an immense collection of nature sounds—not just bird sounds, but anything that might potentially form a part of any bird's natural environment. One potential difficulty in specialized sound collections, however, may be the length of their sounds. Many disciplines do not really need protracted cuts of a given sound, as, for the purposes of identification and recognition, spectral measurement, and so on, short samples of the sound are as useful and much more convenient than long ones. Such short recordings can be difficult for the theatrical sound designer to use, though sometimes they may be the only available source.

For *The Rose Tattoo*, we find a music library that has a number of printed anthologies of various Sicilian folk songs and dances. We will have to arrange the music to suit our purposes (or find a musician to do the arrangement). We will also need to locate musicians to perform the arrangements. Fortunately, we manage to find enough music to take care of all of the preshow, act introductions and closings, and scene bridges that we have tentatively set out to create.

Band music proves to be relatively easy to come by except for one small problem: Performances available on commercial recordings are just too good. We have to assume that a high school band in a small town would have a few intonation problems, a few players who weren't right on the note, and at least one or two hotshots who think loud is good. So, tentatively, we are going to need to find some way to make the band music sound just a little tinny. We don't want it to be terrible, just not professional-sounding.

Several of the marker sounds turn out to be rather easy to find. A rooster crow, a goat bleat (even including a very long one for the Alvaro-goat duet in act III, scene 3), and various domestic animal sounds prove to be available on a sound effects record that is not restricted by copyright. Another, similarly unrestricted set of sound effects contains a usable car horn. The truck and car sounds are not so easy to find. We find one sound effect record set that has a cut of New York City traffic (totally unsuitable) but no other cut of general traffic noises. The specific sounds of a car or a truck pulling up and stopping, then starting and taking off just are not within easy reach. (We assume that if we could have all of the time we needed and could spend any amount of money to purchase special collections until we found what we needed, we would eventually find something worthwhile; but the time and financial resources don't exist.)

A number of nature sounds—bird calls, crickets and other insects, various animal sounds—we find in a major collection at a large university ornithology department. Getting sounds reproduced costs a moderate fee, but the results prove to be worthwhile.

So now we have a fairly large number of raw sounds, most of what we need, in fact. Some rather critical sounds we do not yet have—wind sounds, the important truck and car noises, sounds of children at play, the train whistles, the parrot, the sound of wood breaking (when the goat breaks loose the second time).

The Equipment Available

Before we can proceed with constructing a sound design, we need to know what equipment we have available. We are speaking now of equipment for recording and producing the tapes to be used in performance, not the performance playback system that will be used to integrate the tapes with the staging of the production. That system we will deal with later on.

Our recording system consists of the following components:

Mixing Console

- **16 input channels, each module provided with**
  - in-line faders
  - 3-stage, quasi-parametric equalization
    - shelving low- and high-frequency controls
    - peaking mid-range control
    - four switch-selected center-frequency steps per stage
    - 12-dB boost and cut
    - EQ stage bypass switch
  - 2 sends per channel
    - separate level control for each send
    - 1 send equipped with pre/post fader switch
  - 4-position buss switching
  - panpot designed to fade odd-left to even-right (In other words, any odd-numbered channel can be assigned to

the left side of the pot, and any even-numbered channel can be assigned to the right side of the pot.)

module on/off switch

cue switch (assigns module output directly to monitor)

direct outputs

4 output channels, each module provided with

in-line faders

3-stage, quasi-parametric equalization with same features as on input modules

2 sends per channel, as on input modules

panpot (no switching)

auxiliary input level control

output module on/off switch

cue switch (assigns module output directly to monitor)

1 effects receive module provided with

level control for each of 2 receive inputs

assign switching for each input to route effect returns to any of the four output channels or back to either of the send busses

cue switch (assigns effect receive directly to monitor)

1 slate/talkback/headphone module equipped with

built-in microphone

assign switching to direct mic to any output channel or to sends

switch to direct mic output to talkback system

headphone output and level control

1 stereo master module containing

split, stereo output faders

module on/off switch

monitor switching and level controls

Note: Input to stereo master module is received from panpot L/R busses.

2 high-quality stereo turntables, locally preamplified (The turntables have a vernier speed control that permits a small degree of pitch and speed variation.)

2 half-track, stereo tape recorders

1 eight-track tape recorder with synchronous head switching

1 good quality, portable (battery-powered) stereo tape recorder

2 broadcast type stereo tape cartridge recorder/reproducers

1 performance-type analog synthesizer (older model, MultiMoog or similar)[1]

2 third-octave graphic equalizers

1 two-channel parametric equalizer

2 digital delay units equipped with extra memory to provide delays up to 3 seconds, infinite-repeat switch, and modulation controls

2 compressors

2 filter units (4-stage, cascaded) containing

high-pass filter, sweepable from 20 Hz to 2000 Hz

2 notch/peak stages, sweepable in stages

20 to 200 Hz

200 to 2000 Hz

2000 to 20000 Hz

range selectors for the notch/peak sections

switches to select notch or peak and Q

notch, 10%

notch, 20%

notch, 50%

peak, 10%

peak, 20%

peak, 50%

low-pass filter, sweepable from 20000 Hz to 2000 Hz

Microphones

10 cardioid dynamic

5 condenser preamplifiers with interchangeable

wide angle cardioid heads

medium narrow angle cardioid heads

hypercardioid (shotgun) heads

4 omnidirectional dynamic

2 pressure zone microphone

Monitor

2 large, 3-way wall-mounted loudspeakers

2 near-field loudspeakers mounted over console

2 high-quality 50 watt/channel stereo power amplifiers (one each for wall-mounted and near-field systems).

2 patch panels

microphone patch (routes all mic lines to console mic inputs)

line patch (accesses all console output lines, directs, processor lines in and out, turntable outputs, tape recorder lines in and out, and synthesizer outputs)

---

[1]This kind of synthesis device is certainly obsolescent. However, older analog synthesizers, especially the oldest patchable units which antedate something like a MultiMoog, make signal and control routing very accessible and flevible, sometimes in ways that exceed more modern devices.

Time Constraints and Scheduling

We shall assume that the initial process of reading the script and forming first impressions began approximately ten weeks before the date the production is scheduled to open. Conferences with the director and the other designers took place about eight weeks ahead of opening. Completing the research and locating initial raw materials took another two weeks; so we are now approximately six weeks away from opening.

Technical rehearsals are scheduled to begin ten days before opening, starting with cue-to-cue rehearsals without actors. All technical elements are expected to be in place and operable by the start of technical rehearsals. We have one particular request from the director, however, that sound associated with the meeting between Serafina and Alvaro be ready to be used in acting rehearsals at least two weeks prior to opening.

Essentially, then, we have four weeks to complete the design, except that at least some of the music must be ready earlier. In that time, we must process and adapt the sounds we have, find or make those that are not available on prerecorded sources, select, arrange, record, edit, and process all of the music, and design and specify the playback system to be installed. We have a very full four weeks! The first step has to be to sit down and lay out a schedule of deadlines. We do this by means of a calendar with essential deadlines clearly marked. (See Fig. 23-1.)

Not everyone has the ability to arrange or compose music, so we shall assume for the purposes of this example that we can find a musician to arrange the Sicilian folk songs and dances. We know that we want a guitar as a very strong leading element in the music, so we will communicate this information to the arranger. We shall assume that we have anticipated the need and were lucky enough to find a musician who is an arranger and also a creditable guitarist and who is interested in doing the music for us. After several conferences with the arranger we decide that the other instruments to be used are flute, bassoon, and percussion. The arranger also suggests that she can probably find a group of instrumentalists who may be willing to record music that we can use for the high school band cues. If this idea works out that would eliminate the need for the commercial band recordings.

All of the dates on the schedule related to taping music are set up in conference with the arranger. Note that the scheduled date for taping the music comes perilously close to the end of the rehearsal period. The reasons are, first, that the

FIG. 23-1: ROSE TATTOO PRODUCTION CALENDAR

SUN	MON	TUE	WED	THU	FRI	SAT
44	43 Work on general atmospheric sound.	42	41	40	39	38
37	36 Music arrangements complete.	35 Locate musicians. Schedule reh. and record sessions.	34	33	32	31 Atmospheric sound roughed in.
30	29 Plbk. system system designed. All raw sounds on hand. Marker cues on tape.	28 Begin locating plbk. system equipment.	27	26	25	24 Atmospheric sound complete.
23 Music rehearsal.	22	21	20	19	18 All plbk. equipment located and ready.	17 Marker cues complete.
16 RECORDING SESSION. All music on tape.	15 Begin system installation.	14 Act II internal music ready. Run Act II internal music w/ rehearsal.	13 Music cues edited. Run Act II internal cues w/ rehearsal.	12 Playback system installation complete. Run Act II internal cues w/ rehearsal.	11 Check out and tune playback system. Run Act II internal cues w/ rehearsal.	10 TAPES COMPLETE! 1st Q to Q! Act II Q's w/ rhrsl.
9 2 Q to Q! Act II Q's w/ rhrsl.	8 DRESS PARADE! Run complete sound during rehearsal.	7 FIRST TECH.	6 SECOND TECH.	5 THIRD TECH.	4 FULL DRESS/TECH.	3 FULL DRESS/TECH.
2 Tech. run through. (No dress.)	1 PREVIEW/DRESS!	0 SHOW OPEN!!				

arranger needs as much time as we can allow to complete the arrangements and write out the parts; second, that the arranger feels that, once the parts are written, she must allow the musicians at least a few days to become familiar with and to practice the music before they can reasonably be expected to record competently; and finally, it works out that Sunday afternoon is the only time when all performers and technicians concerned can be available, and the only free Sundays (one for a rehearsal and the second for the actual taping sessions) fall as scheduled on the calendar. So we know to begin with that we need to get as much done as possible on all other aspects of the design in order to clear the time to process and edit the music, especially the internal cues for Act II, which we have agreed to provide for use in rehearsals by the Tuesday following the taping session.

From here on, the process of completing the design essentially becomes a matter of working to schedule, day by day. Chapter 24 details the process of that work.

Chapter 24

Constructing the Design: Recording and Editing

Initial Experiments

43

We begin on Monday, the 43rd day from opening, with some experiments in combining background sounds. The first of these experiments is aimed at finding a satisfactory set of wind sounds. For this, we use the synthesizer's white noise generator and filter controls. After a number of trials, shaping the filter output to a number of different configurations, we come up with a number of caricatures of wind, but nothing so far sounds like real wind. So we put that experiment aside for the day and turn to other work.

A few trials with the synthesizer gives us some interesting sounds that bear such close resemblance to electric and air horns that they could be used as part of the highway background sounds, or even for the car horn cues in act I, scene 6. The synthesizer settings for these sounds are carefully written down and filed so that we will have them later, and notes are placed in the cue list indicating that we now have a potential source for car horns.

Next, we pull out all of the effects recordings that we have available and spot through them for sounds that can be combined into various kinds of background noises. Using the eight-track tape recorder, we put down several repeats of a cut of New York City traffic noise. On a separate track of the tape we add a couple of synthesized car horns, randomly spaced. On one of the effects records there is a cut of a blue jay call, which we add, randomly, on a third track. Playing this back all together, trying various balances among tracks, yields very little that is worthwhile. We begin to realize that putting this background together is going to require a collection of decent sound recordings from which to draw the components. The present combination, no matter how we adjust it, sounds contrived and not very much like the background ambience of a semirural southern community. We don't have very much else to work with yet, so we make notes and leave for the day.

Finding and Purchasing Sounds

42

Work time on the following day (day 42) is largely taken up with the business of acquiring the sounds that must be purchased from outside sources. While talking with personnel in the audio unit of the ornithology department that will supply the bird and natural sounds, we learn that they also have cuts of wind sound, including wind in pines, wind in palm trees, and even wind in a canebrake, so we add those items to an already large order. For insurance, we also ask for several cuts of roosters crowing, plus several other kinds of animal sounds. The tapes, they promise, will be sent out by the end of the week.

After ordering natural sounds, we start looking for ways to acquire the car and truck sounds that we need. We have decided, after some thought, that the most logical thing to do is to go out and record the sounds locally, if we can find the right sources. Several calls to garages that specialize in renovating old cars provide the name of several people who own cars from the late 1940s and early 1950s. By luck, we find one of these car owners who thinks it would be fun to have a recording of his car, so we agree to provide him with copies of our tapes in exchange for his time and the use of his car. We arrange to do a recording session on the following Saturday afternoon.

On a whim, we pack up the portable recorder, some cable and microphones, and go out to a local truck stop. Several trials there prove only that there's too much constant activity to permit a good take of a single truck's sound. The one time when we are lucky enough to have only one truck running and moving, an airplane flies overhead and spoils the take. The one thing that we do manage to get out of the field recording excursion is some fairly good general highway noise, with cars passing periodically, separated enough so that, with a little editing, we can get good individual car sounds. We even get individual sounds of large trucks passing by.

Building Marker Cues

41

On Wednesday (41 days to go) the director asks in a production conference if a tape of the goat can be provided for rehearsing Alvaro in act III, scene 3. We agree to get it to him by Friday, and go to work on marker sounds (there really isn't much else that can be done yet). By the end of the day, we have the following cues on tape:

Scene	Cue No.	Name	Notes
I/1	5	Goat	
I/5	14	Train whistle	
I/6	19	Car horn	
II	23	Goat	
II	28	Goat	(Tentative. part of a bigger cue.)
III/1	33	Goat	
III/1	36	Goat	
III/2	41	Rooster	
III/2	42	Rooster	(Tentative, part of a bigger cue.)
III/3	44	Rooster	
III/3	46a 46b 46c	Goat Goat Goat	(Tentative number of cuts. May vary with way the cuts work with actor in rehearsal.)
III/3	45	Train whistle	

We make a copy of cue 46 and get it to the stage manager for use in rehearsals. The need to be able to cue the tape precisely eliminates the possibility of using a cassette player to run the sounds in rehearsal. Instead, we arrange for the stage manager to use a small, portable reel-to-reel machine. The rehearsal tape is edited with white plastic leader tape between each cut and, of course, at the head and tail of the tape.

Note carefully the breaking of the cue into three separate parts, which we do because we assume that the director will want to be able to time the goat bleats and Alvaro's lines with some care. We surmise that timing would probably be difficult with three goat bleats on tape unmarked by leader. We shall have to wait for the results of rehearsals to find out what sort of timing will be involved. We decide that it's time to attend rehearsals more frequently to see and hear what kinds of dynamics are developing in the actors' realizations of the character, situation, and dialogue.

We should point out that each time we record anything associated with the show, a 1000 Hz. setup tone, recorded at 0 VU, will be placed at the head of each reel of tape. Each time we cue the tape, we will reference the setup tone, setting the playback level to 0 VU, and all equipment will be set to 0 VU using a reference tone generator in the mixing console. Also, each time we finally record a cut for inclusion in a show master reel, the settings for every piece of equipment used will be carefully logged and kept, so that if we ever need to repeat or rework anything, the original control settings will be available. Use of a setup tone is a rule that should be observed in all sound production work, no matter how many or how few cues.

Experiments with Environmental Cues

40

On Thursday (day 40), after some considerable thought about methods of synthesizing wind noises, we try another experiment. This time we run the output of the synthesizer through the filter unit and record onto the eight-track recorder. We build a pink noise band at a low level and use one of the notch sections of the filter to sweep a notch through the band. The result sounds a lot like wind in pine trees. Next we record about 2 minutes of the sound on track 1 of the multitrack recorder, varying the notch randomly. We move to track 2 and do a similar take which parallels track 1, but with the variations in the sound echoing those on track 1. Next we vary the sound quality from the synthesizer just a bit, and write a 2-minute take to tracks 3 and 4.

Four tracks should be enough to give a reasonable proof of whether or not this idea will work. Time to go haul out four relatively portable loudspeaker units and set them up, roughly in the corners of the auditorium. The four loudspeakers are connected through power amps to the eight-track recorder. Amazingly enough, the effect is almost that of wind moving through a surrounding stand of pine trees. We have to improve on the relationships among tracks and make the quality of the pink noise sound a bit more convincing, but it could work.

Following this, we get an idea for producing some leaf noises and go looking for various items such as a grass skirt (borrowed from the costume department), rather large clumps of foliage from several different kinds of trees, and a large piece of kraft paper, torn into strips. We set up a microphone and record several 2-minute cuts of the sounds these objects make; then we adjust the white noise from the synthesizer to a slightly higher band, again with filter sweeps. After several tries, the result played through the four loudspeakers in the auditorium produces the effect of very credible wind in deciduous trees and in foliage such as palm fronds and cane ribbons. In fact, we can even localize the sound of rattling cane and palm fronds to one part of the house, while a gust of wind moves through oak and pine trees from house front to house rear (or in any direction that we want). The problem of wind appears to be solved, even though we still have some wind sounds coming in from the ornithology department sound library. We will listen to the tapes when they arrive to find out if they provide either a better wind effect, or if adding them to the synthesized wind can improve the sound.

39

On Saturday afternoon (day 38), we take the portable tape recorder out to record the car sounds. By luck, the person also owns a large, stake-body truck of reasonable age (but in good condition). So we are able to get a variety of truck noises as well. We set up a pair of microphones to get a stereo recording of the sounds, and, checking them through headphones, they seem to be good. Back in the studio, we find a couple of unfortunate background noises that we didn't notice through the headphones, but the tapes are still usable. Before we get too involved and forget to do it, we dub a copy of the tape to send to the person who provided the car and truck to record, as per our agreement. The tape gets mailed on Monday.

Planning the Playback System

38

We spend part of the weekend working out a playback system configuration to be used for the production. Most of the equipment is included in the fixed system located in the theatre sound booth. Loudspeakers are a slightly different matter. Although the theatre has a central cluster (originally intended for reinforcement) located over the middle of the proscenium arch, most of the loudspeaker units must be gotten out of storage and installed to meet the requirements of each particular production.

Here is a list of standard production playback equipment:

The tape recorders listed in the recording equipment inventory

Playback console, providing

8 line-level inputs arranged as follows

2 stereo groups, each group fitted with dual input controls, output switching for each level control to route the input signals to output presets, and an automatic crossfader that smoothly fades routing from one preset to another.

2 inputs, each fitted with output switching to route the signal to output presets (no automatic fader)

2 inputs dedicated to presets 7 and 8, and each configurable for microphone or line-level signals; each has a bass/treble tone control.

8 presets of 16 output faders. Each preset will accept signal from one, all, or any combination of inputs.

Note: The console described here is not the mixing console described in the recording equipment. This one is a sound scoring console (similar to but slightly larger than that illustrated in Fig. 9-19) designed to fan a small number of input sources out to a potentially large number of loudspeaker locations.

Loudspeakers

10 heavy-duty, three-way systems with 90 degree horizontal field angle x 60 degree vertical field angle; 100 watt input rating.

18 small, narrow-beam, two-way systems; 50-watt input rating.

The permanent house cluster—a three-way, wide-angle system; 100-watt input rating.

Various microphone and loudspeaker cables, adapters, etc.

12 stereo power amplifiers, each rated at 100 watts/channel

The system we plan to use is shown in Fig. 24-1. It will include 7 of the large three-way loudspeaker systems, and 6 of the two-way units, plus the cluster. Fourteen power channels are required, therefore. The playback console, two 2-track tape recorders, two cartridge players, and the multitrack recorder will also be used as part of the production sound system. We have one small problem in that we wish to use the multitrack tape as the basis of the environmental surround, but if we connect all eight outputs from the multitrack machine to the console inputs, then we won't have input positions for the two stereo reel-to-reels and the two cartridge players. Momentarily, we defer the solution to this problem in favor of continuing work on the tapes.

One other problem needs some thought: placement of the operator. If the show is going to include the amount of environmental sound that we are planning, the operator must be able to hear what the audience hears. The best place to hear as the audience hears is from a position in the house rather than from a location inside the sound booth. Obviously, moving the operator outside the booth will have a number of ramifications. First, tickets for some seats will have to be pulled for each performance. Second, the operator's position must be as inconspicuous as possible. Third, no equipment that generates significant operational noise can be placed at the operator's position. Again, after some thought about the matter, we defer the decision.

36

On Monday (day 36) the arranger checks in with samples of completed music. She plays it for us, then for the director. The arrangements seems good, and suggested instrumentation (flute, clarinet, bassoon, two trumpets, two trombones, and assorted percussion) sounds as though it will work. She also reports that she has been able to borrow copies of several

FIG. 24-1: ROSE TATTOO SOUND SCORING SYSTEM

PRELIMINARY PLAN FOR HOOKUP

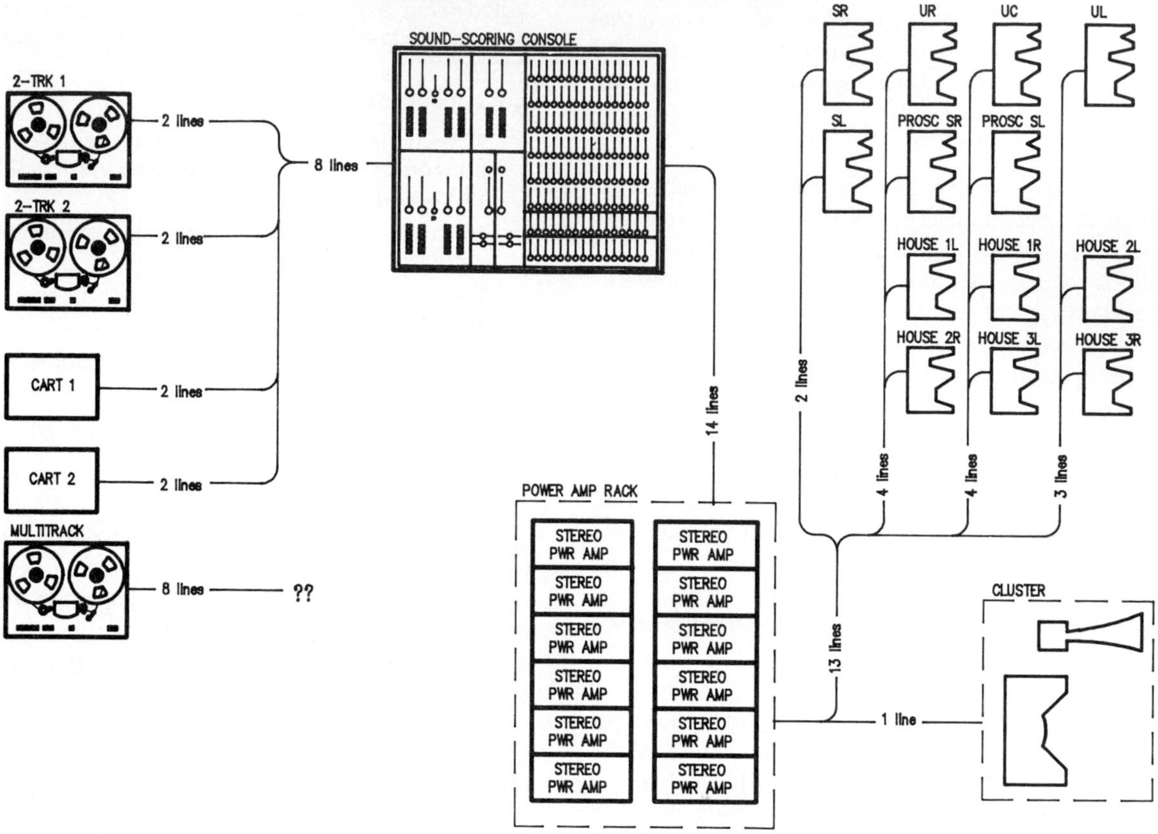

suitable band pieces, including "Stars and Stripes Forever," for use in recording the high school band sounds. Since we won't have the full band, she says she intends to make a few changes in the parts to bridge empty spots in the instrumentation. We ask if overdubbing the parts will help. She agrees that it would. The director wants to know if there will be enough time to record everything during one Sunday afternoon, especially if we add the additional burden of having to overdub to get all the band parts? (Good point.) The arranger suggests that she will be ready to perform some of the pieces for guitar alone well before the two scheduled Sunday dates, if that will help matters. To that we all agree and schedule recording sessions for the Wednesday and Thursday (days 27 and 26 before opening) of the following week. On Sunday (day 16), therefore, we need record only the ensemble music.

The tapes from the ornithology sound library also arrive on Monday, and we go through them to hear what we have. The various general bird recordings are quite good. (These are recordings of sound fields with lots of birds surrounding the microphone position.) Some of the individual cuts of specific birds are excellent, among these, fortunately, the parrot. Some of the individual sounds, however, are unusable, either because they are too short or because they contain far too much extraneous backgroud noise. (One tape even has muffled undertones of people talking in the background.) The tape also contains various sounds of cows, a goat (a better recording than we found on the effects record), dogs, horses, and chickens. The taped cuts of wind are not very good, but one cut does have sounds of wind through bamboo leaves that could potentially be combined with some of the synthesized sounds.

Building a Concept Tape

Recall that we still have not done the concept on tape, mainly because, up until now, we haven't had the necessary resources with which to accomplish that task. So the next step is to construct the concept tape. We have our notes written down about the nature of the concept, and, referring to them, we go through the ornithology tapes, looking for appropriate segments to use.

For reference, we'll review the concept idea here. We decided to take a group of bird calls, mix the calls with wind in Spanish moss and cane fronds, add into that some occasional truck and car noises (tire noises on the road, engine sounds), fade all of that into a distant chorus of frogs, night birds, and soft winds; then blend into a dawn chorus of birds, farm animals and birds, and finally back into daytime birds. Through all of this we will intermix fragments of Sicilian folk music. This mixture of fragments will be less than a minute and a half long, but it will provide a palette of sound color and texture that will communicate the essence of what we want to do and will guide our choices in expanding sounds into actual background for the play.

We decide to make one small change, using the sound of wind in pines in place of the wind in Spanish moss; otherwise, we can proceed to put on tape a wind background. We add some of the road traffic sounds we recorded, and, partly overlapping the road sounds, we bring in a few bird noises. The bird sounds fade into night sounds—mostly frogs and crickets. The night sounds cross-fade to a dawn chorus of birds with a cow and a rooster spotted into the near background. Finally we have the daytime bird sounds and sounds from the highway. Paralleling all of these tracks, we add snatches of guitar music taken from a record of miscellaneous guitar pieces. (The guitar music isn't Sicilian, but it's close enough for the purpose, and we aren't using long segments of the music.) We've done all of this recording on multitrack, so the final step is to mix the tracks down to a short segment of stereo tape. The remix requires balancing the various sounds so that what comes out in the final two-track tape is a smooth, well-blended progression of sounds. We now have our concept tape, and we place it on a separate reel, clearly marked, so that we can find it easily when we need it.

More Work on Marker Cues

Next we complete the taping of the marker spot cues called for in the script. We recut the various goat cues using the recording supplied by the ornithology department. We add the various car and truck arrival and departure noises, then find a suitable group of parrot squawks and cut in those cues.

Finally, we start to work on the sinister noises for act III, scene 1 (cue 38). We have gotten several cuts of various birds flapping their wings as part of the tape from the ornithology library. Unfortunately, none of these sounds can be used as recorded, primarily because the wing flaps are not readily identifiable as such. After several trials, however, we happen to run the tape at twice the recording speed, and this provides an interesting sound. Mixing this sound with a group of chicken and bird alarm squawks, plus the torn strips of kraft paper that we used to make the sounds of palm and cane leaves in building our wind sounds, produces a sound that could resemble some sort of animal predation taking place in tall, stiff grass. This gives us the idea of adding a distant dog howl at the first of this cue, just before the parrot squawk.

The one remaining component that Williams calls for is "a distant shriek of Negro laughter." Williams's purpose seems to be to add a particular kind of human element to the situation. Obviously, the specified sound is of considerable importance; however, it is not going to be easy to acquire. Williams's specific call for "Negro laughter" probably comes from the likelihood that, to persons familiar with that time and place, there would have been a very distinct and identifiable difference between laughter of whites and laughter of blacks. Just how important is absolute fidelity to this distinction 40 years later? Is the essential point some reference to the conditions of Negro life in the South at the time, or is just a sound intended to convey a particular emotional quality? We choose the latter, and decide to recruit one or more actors to try to elicit the sound we want. Because actors are not immediately available, we put off finishing this sound until later.

Several sounds, aside from the laughter, require participation by actors: the one integral cue associated with Estelle's voice heard through the telephone in III/1, and the women's voices calling children home to supper for the opening cue of I/1 and for the *prima sera* reprise in III/1. All of the cues involving voices are going to require outside help. They, too, must be deferred to another day. We will need to check with the director and the stage manager to schedule actors for recording sessions. We also need a recording of children at play (cue 27, act II).

We have several other marker spot cues left to build: the synthesis of the sinister noises for III/1, the wood crash needed when the goat breaks out in act II, and the salesman's car radio in act II.

Building the Auditory Environment

35

On Tuesday (day 35) we start to work on the background atmosphere tapes. In building the concept tape, some of the potential benefits and problems of an environmental surround began to appear. In an early production meeting, the director expressed a bit of reservation about extending the auditory atmosphere, so we decide that preparing a demonstration would be prudent. We set about to build a tape of extended examples to explore the kinds of sounds we can reasonably expect to create. When we have worked out the sounds to our own satisfaction, then we will ask the director to come in, listen, and react.

Using the sounds from the ornithology library, plus some synthesis, we have a sample of background atmospheric sounds ready to try out on the director by the middle of the week. For the trial with the director, we set up six loudspeakers—the four in the auditorium corners as for our wind experiments and one in each upstage corner. We have a taped sequence worked out that surrounds the audience with a whole panorama of sounds. These sounds work in a sequence following the times of day associated with each scene of the play: evening, night, daytime, hot evening, hot steamy night, predawn, morning. The final morning segment is generally brighter than all the rest because it will accompany the resolution of Serafina's immediate problems; the hot-steamy night segment is the darkest in tone color because it accompanies a portion of the play that is psychologically dark and tormenting for Serafina.

On Tuesday evening at rehearsal, we talk to both the director and the stage manager to set up, first, a time for the director to hear the atmospheric sample tapes that we have built, and second, a list of actors to participate in recording sessions and start the stage manager arranging times for the recordings to take place. The director agrees to listen to the atmospheric tapes the next day; the stage manager says that he will get us the list of names and recording times by the end of the week.

34, 33

On Wednesday the director hears the auditory atmosphere samples that we've prepared. He seems to like the sounds in themselves, but remains fearful that underscoring long portions of the play with these sounds will distract the audience from the dialogue and action. His first reaction is to rule out the background sounds altogether, but we talk him into waiting to hear the sounds in the context of the play, promising that if they don't work there will be no objection to cutting them out.

We go pick up some scrap wood from the scene shop and record several cuts of wood breaking and splintering. After some processing, one of the cuts works for the sound of the goat breaking through Serafina's fence (cue 28 in act II).

The remainder of Wednesday and most of Thursday are spent in working out as precisely as possible the sound progressions for the atmospheric cues, and the order of cue traffic in general for the show. By the end of the week, the specific structure of the cues has taken a clear shape, but there is a great deal of work to do. Also, we begin to realize, if the cues work as planned, the operator is going to be a very busy person.

Thursday afternoon we take the portable recorder out again, this time to a small park at the outskirts of town. We need to get a recording of children playing, but we don't want a large number of children. We anticipate that if we try to record children from a distance that a cardioid microphone can handle, we will attract attention to the fact that we are recording, and that will simply bring the children around, curious about what we're doing and asking to hear what they sound like on a recording. That kind of situation will produce no usable results. We decide to try to record from an inconspicuous spot as far away as possible. To do that, we need a hypercardioid ("shotgun") microphone. Fortunately, we manage to borrow one from a friend at a local radio station.

Naturally, when we get to the park, a few problems develop. At first, no children are around. Then, when some children arrive, they don't make very much noise at first. Finally, when they do start into a loud running game, there is so much automobile traffic that nothing we can record is of any use. Just at the time we decide we can't stay any longer, however, the traffic clears for about a minute and one half. It's not much, but it's all we can get. Back in the sound room, we listen to the tape and decide that we can use it. We have to do a bit of processing, making several overlapping dubs and using enough delay to create a pseudo-stereo recording,[1] but finally we have what we need, and we add the cut to our reel of finished spot cues.

32

On Friday (day 32) the stage manager delivers a list of names of actors, and asks if a recording session on Wednesday (day 27) would work. We have already agreed to begin taping guitar music on that day, but given that the recording session for the actors is early in the day and shouldn't take more than about an hour or so, we agree.

Friday and Saturday are largely spent working to meet the deadline to have the atmospheric cues roughed in. Since we have previously prepared samples for the director to listen to, most of the work is a matter of building extended versions of those sample cues. This takes time, although having already plotted general atmospheric activity for each scene helps. The plot shows patterns of wind motion, bird activity, and other incidental sounds (crickets, cicadas, animals, etc.). The atmospheric plot is shown in Fig. 24-2.

The general plan is to provide a surround of diverse wind and bird movement that approximates a natural outdoor perspective. This means that if a breeze starts to blow in from southwest, we will first hear the sound of air turbulence and leaf movement from that direction; but then the sound will reach trees immediately around us, then move on to the northeast. By the time the moving air affects trees to the northeast, the breeze may have died down in trees to the southwest, where this particular event began. Some of these events pass directly over us, and some move around the perimeter of the area. Similarly, birds answer back and forth, and occasionally, a particular bird moves from tree to tree. With blue jays, groups of birds tend to move together from time to time. At this point in time, we plan to use one track of the multitrack tape for each major compass direction, with the remaining four tracks for "fill points" between. So we have quite a bit of detailed recording to do, including some rather complicated crossfades from one track to another.

The most logical (and simplest) way to approach the construction of the auditory surround seems to lie in creating several individual sound elements, on reel-to-reel or on cartridge

[1]Pseudo-stereo: creating a small delay between two channels of otherwise identical monophonic sound. The delay gives the illusion of stereo separation.

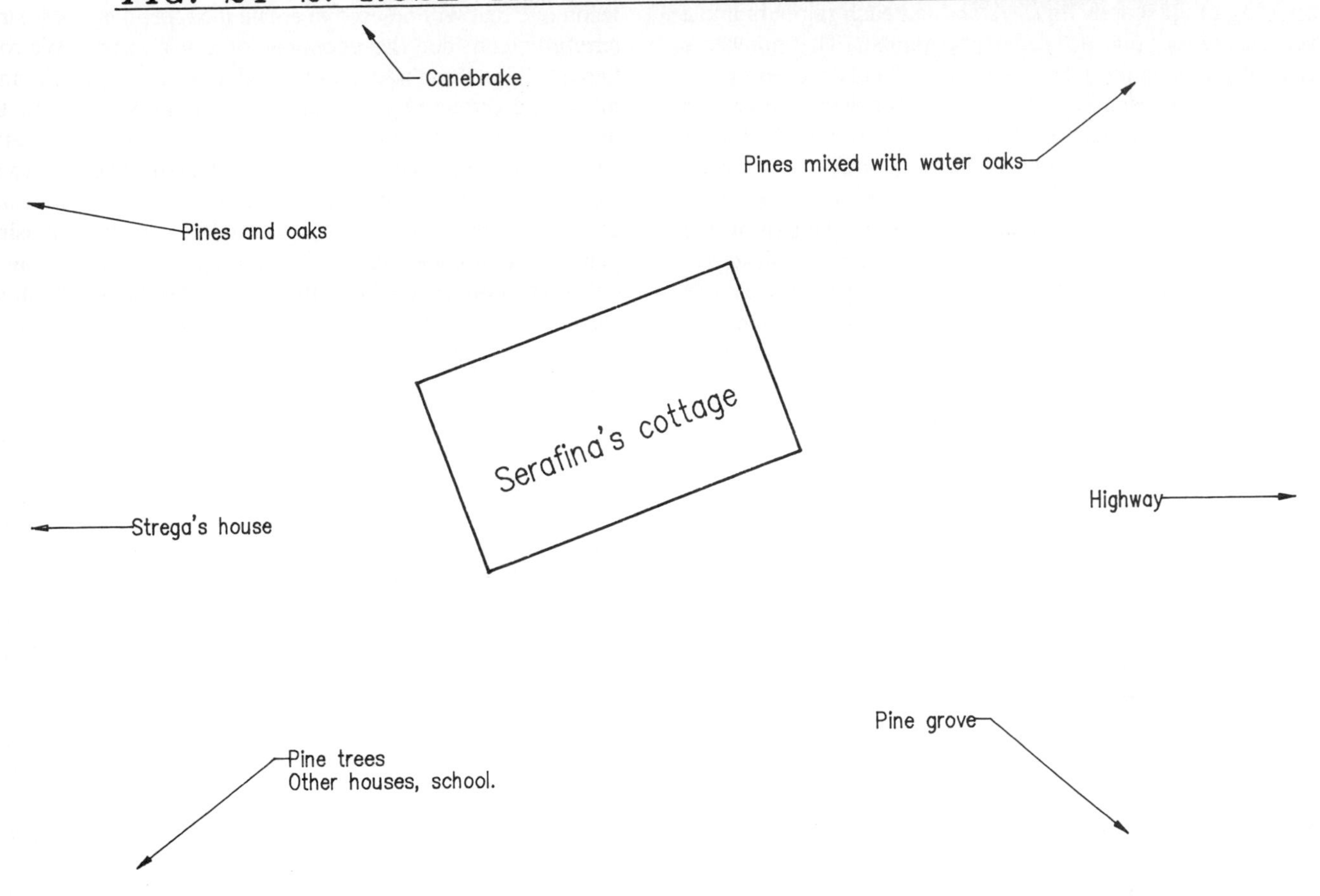

loops. As things turn out, we need both loops and open-reel segments. For optimum control of crossfades we need the precision of the open-reel machine, but for continuity to bridge parts of a track where the open-reel tape is momentarily inactive, we need the loop to provide low level fill sound.

We spend quite a bit of Friday standing in front of a microphone shaking various small tree limbs, recording the sounds of moving leaves. An assistant helps to add density to the sound for the heavier gusts of wind. We try to make as large a variety of random motions as we can in order to prevent noticeable sound patterns from developing. By Friday evening, when we begin trying to combine sounds, we discover that synthesized wind is going to play a far less significant role than originally anticipated. We use the synthesizer to provide distant wind noise, as of wind in trees considerably beyond the immediate perimeter of our acoustic surround. White noise patterns help to reinforce stronger gusts in nearby trees, however. By the end of the evening, we have most of the wind ambience on tape, but in listening to the sounds in the auditorium, we decide that, for the night scenes, we need to redo some of the tracks to get a "darker," more subdued quality.

31

The first part of Saturday's work session is spent reworking the wind ambience for the night scenes. We use very little wind for act I, scene 2, finally reducing the atmospheric sound to four cues: an opening breeze in nearby trees stage right (the direction from which Father Leo and the women enter), a breeze that moves from house right to house left, the sound of wind in pine trees beginning as Father Leo says, "We must go to the door," and a breeze that stirs the leaves as Assunta wraps her shawl around Serafina. For act III, scene 2 we make a stereo loop of wind in pines, then five cuts of breezes and wind gusts. We also decide to use small amounts of this wind-in-pines loop during act I, scene 2. Finally, we prepare a cartridge loop of crickets, which we will have the operator fade in and out periodically during the night scenes.

Setting Up a Work Log

By this time, the amount of tape we have recorded and cut is beginning to add up. To become confused about what we've done or not done, and where a recorded segment is, would be very easy. The chance exists of throwing away, recording over, or just losing an already completed sound. Time to stop, draw up a log of all sounds finished, and cut each segment into a storage reel, giving each segment a number. That number is carefully written onto the leader immediately preceding and just following each cut. The tape of finished segments is always stored *tails out.*[2] Segments recorded on cartridges are labeled and stored in a lockable cartridge rack.

After reworking the night wind ambiences, we move on to dealing with adding birds and other sounds to the atmosphere. We will not use birds in the night ambiences at all, but we still need a lot of bird sounds for all of the various daytime scenes. We record a general bird background on a long cartridge loop, getting it ready to use for making the multitrack ambience tapes later. We also make a variety of short bird "event" tapes—that is, individual bird sounds and sounds of groups of birds, including several different patterns of birds answering back and forth from tree to tree and moving from place to place. This group of bird events gives us a variety of details that we can play with in making up the ambience tapes.

After all of the source tapes are completed, cataloged, and labeled, we have to begin the actual job of building the auditory ambiences. We first concentrate on the opening of the play, the ambience for *prima sera* (which will also serve us at the reprise of *prima sera* in act III, scene 1). This must be as remarkable a moment as we can make it without letting it be too intrusive. One of the cuts we have asked for in our order to the ornithology sound library is of birds settling in to roost. This cut we listen to several times over. It is a monophonic recording, approximately 2 minutes long and fairly homogenous (i.e., there are no significant changes in the sound over the entire cut). This we make into a cartridge loop. Using the loop, we build three tracks (one for audience right, one for audience left, and one for the rear of the auditorium) on the multitrack recorder. The three tracks are to add depth (multiple directions) to the effect of the roosting birds. We start the loop and place the first (left) track on tape. The second (right) and third (rear) tracks are recorded by overdubbing, staggering the starting point for the loop for each track. This sends differing sound patterns to each channel and gives an illusion of multidirectional depth. This segment of roosting birds is going to have to fade under the opening scene, then gradually fade away completely under the evening wind in the pines.

Timing is critical, and one of the things that we have carefully done during attendance at rehearsals is to time acts and scenes, noting the time lapse to various particular points in the script. Also critical is the management of the tracks on the multitrack tape. Ideally, we should be able to build as many tracks as we need and mix them down either to four or eight tracks; but that degree of liberty requires the use of several multitrack recorders, and we only have one. The technique that will be used to create the tape, therefore, must carefully consider the economy of track usage. We will "bounce" the three tracks of birds that we have just finished to tracks 5 through 8 (splitting track 3 to tracks 7 and 8) and mix wind in pines onto tracks 5 through 8 at the same time. We will do this pass by pass, a slow and laborious process, but the only one readily available to do the job. Fortunately, we can shorten the process somewhat by the way we have prepared our two-track source tapes. Again, the object is to get a different auditory pattern on each directional track of the surround. We made the source tape by offsetting the sounds so that wind starts in one place and moves to another. We can gain an additional pair of information channels by offsetting the two-track tape when we record tracks 7 and 8, so that they have slightly different information than do tracks 5 and 6. Getting the balance for the mix requires a bit of trial listening to the sounds played back through the loudspeakers we have placed in the auditorium. (Remember that we cannot adjust the balance of the mix once we have recorded these tracks, because the ratio of birds to wind will be permanently fixed, so we have to get it right the first time.)

The wind-in-pines noise is recorded at sufficient length to underscore the entire act (even though it may not remain audible at all times.) Of the opening bird sounds, we recorded only enough to play through Estelle's entrance, so that they fade under and out. Finally, we record random car noises on tracks 1 and 2, then random animal and insect sounds on tracks 3 and 4. Toward the end of the scene, we also use tracks 1 and 2 for occasional insect sounds.

Next we record a general bird background for act I, scene 3. This needs four tracks. We overdub two tracks of wind in parallel with the birds, but are not sure yet if the wind will really be used.

For act I, scene 4, the pattern of sounds will be both more immediate and more random than in other scenes we've built so far. We start at the opening of the scene with nearby birds, especially blue jays, with squawking chickens briefly in the background. The chicken and random bird sounds are built only on two tracks, but the jays are built on four tracks with a pattern of motion that is somewhat like the treatment of the wind in act I, scene 1, but not as complex. Individual birds echo back and forth. Periodically, birds move to another point of the surround, a pattern that is continued throughout the scene. (The operator will control level on the track so that the sounds can fade and hold under the scene without being obtrusive. Periodically, the sounds will fade up to a higher level, then back down again.)

For scene 5 of act I, the ambience uses a general bird background (again, on two tracks), and adds random highway noises on another two tracks. The same pattern continues

---

[2]As a general rule, tapes should always be stored tails out. The reason has to do with "print-through," a condition in which magnetism on one lap of tape bleeds through to the lap above. Print-through seems to work from inside to outside on a reel of tape. Stored heads out, one hears an "advance echo" of the sound one turn before the recorded sound should begin. Stored tails out, the actual program tends to cover print-through.

through scene 6. Again, the plan is to give the operator specific times to lower the general level to minimum threshold and other times to raise the level to noticeability.

Act II needs to convey a somewhat different atmospheric feeling, reflecting the alterations in Serafina's own feelings. She is like a ship on an urgent voyage, becalmed in steamy, tropical seas. She desperately needs a resolution of her emotional conflict, but she has lost all means of orientation and mobility. The terrible, blazing sun, the heat, the oppressive humidity of the afternoon, all compound her dual anxiety about Rosa and about the charges by Flora that Rosario was Estelle's lover. The auditory surround for this act needs less wind and more animal noise than those for the scenes of act I.

We begin act II by putting down four tracks of random cicada sounds, most of them distant with an occasional nearby chirp. Another two tracks adds an occasional car from the highway, but there is much less use of automotive noises overall. Those road noises that are used are much more subtle and are forced well into the background, just above what we hope will be threshold level. Naturally, we check out the quality using the trial loudspeakers in the auditorium (although this has to wait until late in the evening when rehearsals are over and no one else is around, either to distract or be distracted).

We plan to use one loop of random, distant bird sounds as part of the ambience for this act. The loop only needs to be heard on random occasions, as neither the birds nor the cicadas are continuous; both come and go, varying in directional characteristics and distance.

30

Finishing the atmospheric backgrounds requires some work on Sunday (day 30), to complete act II and to build and complete act III. Act II has a time progression involved that requires that the atmospheric background change during the course of the act. Also, there will be times, we decide, when a particular kind of sound needs to happen in conjunction with particular lines and actions. (We have begun to see particular patterns of dramatic tension and emphasis emerging in rehearsal, and we feel that some of these call for augmentation in the auditory background.) The way to handle such sounds, which must be cued to dialogue and action, is to treat them as spot cues, either on open-reel recorders or on the cartridge players. Accordingly, we take several domestic and wild animal sounds that we need and place them on two-track tape, temporarily using the open-reel machines. (We can transfer these cues to cartridge later if we need to do so.)

Toward the end of the act, about the time that Serafina gives Alvaro the rose-silk shirt to wear, we should begin to hear some wind movement in the auditory surround. The time in the story is moving toward evening, and also, symbolically, life is beginning to lighten just a little for Serafina. We will accomplish this effect by adding tracks of wind sounds—wind in pines on one track, wind in deciduous trees on another. Just a bit ahead of the point where Alvaro gets the shirt, we will begin to fade in more of the wind track, starting with wind in pines. As the act moves toward its conclusion, we will add more of the second wind track, brightening the quality of the sound. Also, we will use a track of evening bird noises that will help to lighten and brighten the sound quality.

For act III, scene 1, we have to construct a real time progression, from early evening into night. Wind will play a large part in this atmospheric background, starting with evening breezes in the deciduous trees but with distant soughing of wind in pine trees in the far background. We develop a pseudo-stereo track of birds, much like the daytime bird tracks of I/6. This bird track is long enough to play to the point in the scene where Alvaro drops the condom, but no longer. We will have faded it out by that time. The sound of wind in the pines on previous tracks is now louder and more toward the foreground. We begin to hear a few random gusts of wind in the palm tree and the canebrake. Building up to the climactic point at which Serafina finally smashes the urn, we introduce a track of random blue jay and crow noises. Starting in the deep background, the sounds move into the midground as the critical moment of smashing the urn approaches.

We have a potential problem in making the transition from this daytime auditory surround into the *prima sera* reprise. We have only one multitrack player, on which we have loaded quite a few tracks. There will not be time to fast-forward and cue up the *prima sera* reprise cue. Somehow, we must find a way to accommodate the transition instantaneously and smoothly. We go ahead and build the reprise cue (we can't just copy the tape from the opening of I/1, again, because there is only one multitrack machine), and temporarily defer the decision on how to accomplish the transition.

The auditory background cues for III/2 are already done, so we move on to III/3. We start with a pseudo-stereo track of light wind in pine trees. Next we add sounds of a dawn chorus of birds in the mid-background. Gradually, random sounds of jays and crows begin to appear along with random gusts of wind and breezes through the canebrake and the palm fronds. Various domestic animals are periodically heard in the very far background. Finally, we end the tape with the kind of general bird background that we used in act I. With the exception of our problems with the *prima sera* reprise, the auditory surround is reasonably complete.

Adjustments to the Playback System Design

Over the weekend, we spend some time deciding just how to handle the problem of accommodating eight tracks from the multitrack recorder while still allowing inputs for the two-track open reel machines plus the cartridge players. Eight separate tape channels are not really necessary to handle the ambience effects. Essentially, we have built a four-track surround, with some details of the auditory ambience constructed as parallel stereo tracks. The primary need is the ability to control the distribution of the sound from these various tracks, which the presets on the sound scoring console can manage with no difficulty. The problem is insufficient inputs for all 16 channels that we will have to accommodate (eight from the multitrack, four from open-reel machines, and four from cartridge players). The solution is to run the eight tracks from the multitrack recorder into the 16 x 4 mix-down console (described in the recording system). The outputs from the two cartridge machines will also connect into the mix-down console. The four outputs from the mixer will then run to the first four inputs of the preset console. The outputs of the two-track, stereo, open-reel machines will go to inputs 5 through 8 of the preset console, as most of the cues run from the open-reel machines are spot, marker cues with no requirement for directional change or motion (none, that is, that cannot be taken care of by track-to-track panning when recording the two-track tape). The final configuration of the system will appear, now, as shown in Fig. 24-3.

The remaining problem is placement of the operator. Although having the operator in the same space as the audience would make the job of controlling levels much easier, the additional equipment and setup time that would be added to the installation makes an operator location inside the sound booth more practical. With regret, we conclude that we cannot place the operator in the house.

29

On Monday (day 29) the system diagram is handed to the sound M.E. (master electrician) so that he and his crew can begin locating the necessary equipment. All of Monday is spent placing the cues that we have built onto their appropriate reels, in show order. To do this, we have prepared a cue traffic chart (see Table 24-1). (Notice that, in setting up the cue traffic we found the solutions to the problems of transition from III/1 general atmosphere into *prima sera* reprise.) By the end of the day we have all of the open-reel cues so far recorded cut into the show master reels and the various cartridges prepared and labeled. The process of preparing the cartridges to conform to the cue traffic list required some rerecording to place several cues on a single cartridge with cue tones stopping each cut at cue point. The individual cartridges are numbered for ease of identification (see cues 4, 5, 48, 50, 51, 53, 56, 56a, and 56b). There are now five show masters: one for each stereo machine and three reels for the multitrack recorder (one reel for each act). The number of cartridges is now 17.

FIG. 24-3: ROSE TATTOO SOUND SCORING SYSTEM

FINAL CONFIGURATION

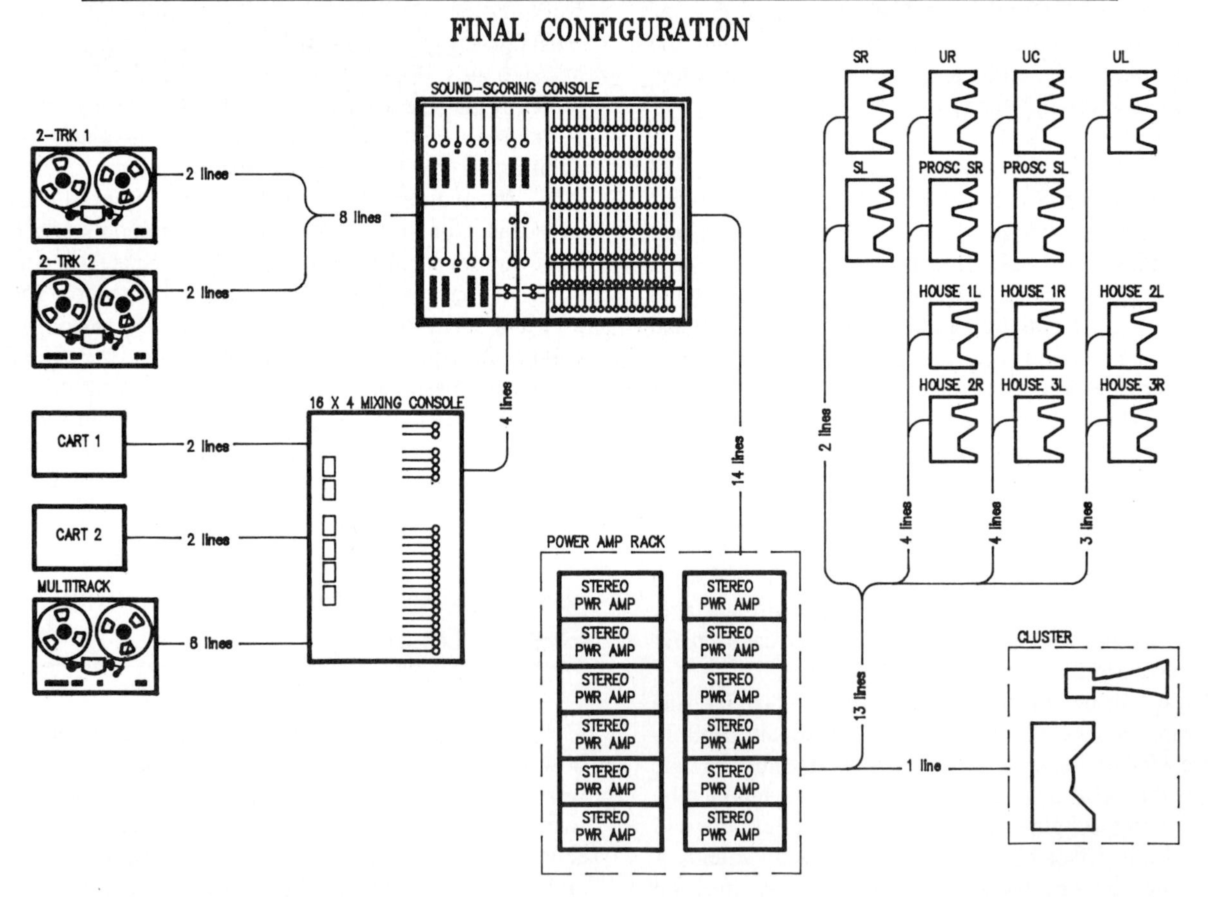

TABLE 24-1. Cue Traffic Plot

THE ROSE TATTOO

Cue No.	Cue Name	Source	Reel No.
1	Preshow music	2-track #1	A
2	Fade preshow	(2-track #1)	(A)
3	I/1 Open atmosphere	Multitrack	C
3a	I/1 Open music	2-track #2	B
4	Truck passes	Cartridge #1	1
5	Truck passes	Cartridge #1	1
6	Goat	Cartridge #2	2
7	Bridge, I/1 to I/2	2-track #1	A
8	Cricket loop	Cartridge #1	3
8a	Wind-in-pines loop	Cartridge #2	4
8b	Gust of wind	2-track #2	B
9	Gust of wind	2-track #1	A
10	Increase level—wind in pines	Cartridge #2	3
11	Long breeze	2-track #2	B
12	Bridge, I/2 to I/3	2-track #1	A
13	Fade bridge	2-track #1	(A)
14	Blue jays	2-track #2	B
14a	Fade jays	2-track #2	(B)
15	Bridge, I/3 to I/4	2-track #1	A
16	I/4 atmosphere	Multitrack	C
17	Parrot	Cartridge #2	5
18	Band	2-track #2	B
19	Train whistle	Cartridge #1	6
20	Band	2-track #1	A
21	Band	2-track #2	B
22	General fade		
23	I/6 atmosphere	Multitrack	C
24	Car arrives	2-track #1	A
24a	Car leaves	2-track #1	B
25	Car horn	2-track #1	A
26	Close, act I	2-track #2	B
27	Open, act II	2-track #1	A
27a	Hot wind gust Atmosphere	Cartridge #1 Multitrack	7 C
28	Goat	Cartridge #2	8
29	Car approach/radio	2-track #2	B
29a	Fade radio under	(2-track #2)	(B)
30	Alvaro's truck approach	2-track #1	A
31	Salesman's car leaves	2-track #1	A
32	Children playing	2-track #2	B
33	Goat breaks away	2-track #1	A
34	Alvaro's truck leaving	2-track #2	B
35	Close act II	2-track #1	A
36	Open act III	2-track #2	B
36a	Atmosphere	Multitrack	C
37	Truck passing	Cartridge #1	9
38	Alvaro's truck arriving	Cartridge #2	10
39	Goat	Cartridge #1	11
40	Estelle's voice	2-track #1	A
41	Background transition Fade/recue multitrack	Cartridge #2 (Multitrack)	12 (C)
42	*Prima sera* reprise	Multitrack	C
42a	Fade under and hold	Multitrack	(C)
43	Goat	Cartridge #1	13
44	Alvaro's truck leaving	2-track #2	B
45	Sinister night sounds	2-track #1	A
46	End III/1	2-track #2	B
47	Atmosphere III/2	Cartridge #2	14
48	Breeze	Cartridge #1	15
49	Rooster	2-track #1	A
50	Breeze	Cartridge #1	15
51	Breeze	Cartridge #1	15
52	Rooster	2-track #1	B
53	Breeze	Cartridge #1	15
54	Gust of wind	2-track #1	A
55	Bridge, III/2 to III/3	2-track #2	B
55a	Atmosphere (incl. rooster)	Multitrack	C
56	Goat	Cartridge #1	16
57	Train whistle	Cartridge #2	17
58	Final music & curtain call	2-track #1	A

28

Having caught up on all of the taping and editing that can be done, Tuesday (day 28) is spent getting ready for the recording sessions of the next two days. First, in order to make sure that we have complete control of the background ambience of the voice recordings, we take some old velour stage curtains and drape the room that serves as a recording studio. Next, we try out several microphones in the room to see which give the kinds of sound we want. (Obviously, no decision will be final until we hear just how the mics will treat the actual voices to be recorded.) The necessary mic lines are plugged, tape channels patched, and the whole mixing/recording system is checked out.

The selection of microphones is done with as much care as we can manage, limited by the choices available. We choose one of the dynamic cardioids for recording Estelle. For the voices of the women calling their children (part of the *prima sera* effect which we will add to the multitrack atmosphere), we tentatively select two pressure zone microphones (PZMs) to be hung adjacent to each other at a 90 degree angle. For the guitar, we decide to use two of the condenser microphones with wide-angle cardioid heads. The condenser mics will be carefully suspended in shock mounts on booms that can quickly be positioned for optimum pickup.

Initial Recording Sessions

27

We're in the sound room by 8:00 A.M. on Wednesday (day 27). Since the actors are supposed to be on hand at 9:00, we have to have everything ready to go. One or two people are late, so we get a late start. Then a microphone cable develops trouble. (An open shield producing hum. Stepped on, perhaps, by one of the actors? Who knows? It worked yesterday; today it doesn't.) So we take time to replace the mic cable.

Getting the women to call in just the right way takes a while. Using the PZMs to pick up this effect was the right decision, but we have to get the actresses the right distance from the microphones; and then we have to get them to use their voices to produce just the right sound. All of this takes some coaching and rehearsing. Finally, with one or two people starting to complain that they have to be somewhere else by ten o'clock, we manage to get a good take.

Most people leave, and we are ready to tape Estelle's voice; but just as we are starting, the arranger/guitarist arrives. She's willing to wait until we can get done; but in the middle of what appears to be a very good reading, some people come into the auditorium shouting about something. (Maybe they couldn't read the "Recording in Progress. Quiet, Please!" sign we left on all of the doors?) They're very apologetic, but the take is still ruined. By 10:30, however, we have what sounds like a usable recording, and the actress playing Estelle leaves. We don't have the laughter for act III, scene 1, yet, but that will have to wait.

Now we have to begin recording the guitar pieces. Naturally, we encounter an immediate problem. We provided a music stand, but we did not consider that the arranger would need two stands to place side by side to accommodate several pages of staff paper taped together. Getting a second music stand takes a few minutes.

The first part of the recording session is involved with listening to the guitar through various microphones, placing the mics differently each time. Because the arranger, who is also the performer, has a say in how the recording sounds, each of these trials has to be taped; then the tape has to be played back so that the arranger can hear each one. Finally, we and the arranger decide that we will use the condensers placed relatively near the instrument for the opening and also we will record from the PZMs for a more distant effect, in case it turns out that we need a less present sound. This means setting up the multitrack so that we will have more than two synchronized tracks.

By the end of the morning, after several takes of each one, two pieces of guitar music are on tape. We break for lunch. When we get back, we find that there is enough prerehearsal activity going on so that there really is no possibility of getting a clean recording. Accordingly, we abandon the effort until the next morning and hope that we can get everything done.

After the arranger leaves, we go to work putting the first two pieces into place—the opening section into the *prima sera* effect for act I, scene 1, and the bridge between I/1 and I/2. Both pieces seem to work quite well.

After the music cuts are in, we start on the voices. First, we process Estelle's voice (using the parametric equalizer) to sound as though it's coming through a telephone receiver. That doesn't take very long, though a lot of time is spent getting just the right equalization to make the effect work. (Remember that this effect is going to have to be audible to an entire house full of people, but should still sound as though it is believably coming from the telephone.)

Next we start in on the women's voices for *prima sera.* While we are listening through the various test takes and extraneous noise, we discover something fortuituous that we failed to notice during the recording session: During a test take, one of the women found something both embarrassing and very funny and, as a result, first gave a small scream, then a hysterical laugh. Hearing it through the monitors at a moderate level, it almost sounded right for the laughter in the III/1 sinister sounds. There was a lot of talking and other extraneous noise; but could it possibly work? We equalize it and run it at a level that we think might approximate its level in the show. Sure enough, it's very close. We'll have to try it in the theatre itself before we'll know, but it could work. (This lucky accident illustrates one reason why trials of microphone setups should be recorded.)

Adding the women's voices to *prima sera* is little more than processing for distance and reverberance and editing the sounds into the two-track masters. Later, when we can get quiet time in the theatre, we'll figure out just how we will dub it to the multitrack tape.

Finally, at the end of the day, we get some time alone in the theatre and can play with the level and quality of Estelle's voice, with the calling women, and with the laughter that we got by accident. Without having to stay too far into the night, we get good balance on Estelle and the calls in *prima sera*; and the laughter mixed with all the other sinister sounds works very well. (No doubt someone from Williams's generation and native locale would recognize that the sound is not the "Negro laughter" that Williams asks for, but it works acoustically with the rest of the effect. As long as the director doesn't say anything about it, it's not a major part of the show.)

26

Thursday (day 26) is the second day set for recording the guitar music, and the arranger/performer is back and ready to go. We have five more pieces to record, almost all to be recorded just as we did yesterday—by taking a close perspective from condenser cardioids positioned close to the instrument while taking a more distant perspective from two PZMs hung in a 90 degree angle. With the exception of minor technical problems (mainly working out a satisfactory way of avoiding excessive fret noise and noise of the performer's breathing in the close perspective), and a few difficulties with the performer's execution of a couple of tricky passages in the music, the taping goes smoothly. We get all of the pieces on tape by early afternoon.

After a break for lunch we get back to processing and putting the music into the tape. The major question that came up in the process of recording and working with the music is that of the most satisfactory auditory perspective for the guitar music. Should the guitar be a voice in the foreground, or should it be further back in the atmospheric surround? Or should it take either perspective, dependent on dramatic context? The last answer seems the most promising. Clearly, the opening and most of the bridges would seem to suggest a foreground perspective, perhaps fading under the scene. On the other hand, the use of the guitar in the *prima sera* reprise and in the chordal bridge from III/2 to III/3 seems to suggest that the guitar sound should be somewhere in the middle ground. After a few trials the pattern seems clear, and we proceed to process and edit the music into the master reels.

25/24

Friday is used to locate enough music stands and folding chairs for musicians to use on Sunday during the music rehearsal, to plan the microphone layout and hookup for the recording session, and to attend the evening's rehearsal. During rehearsal, we check with the stage manager to make sure that the theatre is still clear for the music rehearsal on Sunday. Also, we check to be sure that the arranger has left a list of the musicians' names for inclusion in the program. Saturday is spent in getting the control room and system ready for the music rehearsal. We will make a trial setup of the recording system, so that we can hear just how the mic placement is going to work, and to locate any potential problems that may arise.

Music Rehearsal

23

Sunday (day 23 before opening) is the day set for music rehearsal. So that we are ready to deal with any technical problems, the planned recording system is set up. Monitoring the rehearsal through the system will permit us to isolate potential problems and look for ways to solve them, so that no time is lost during the actual recording session.

As usual, when musicians are brought in to prepare music with minimal rehearsal time, tensions run high, the problems are large, and some niceties in both smoothness of performance and interpretation of the music get lost. Overall, however, the music sounds as though it's going to be pretty good. The high school band marches are rehearsed at the beginning of the session to release those musicians who will not be involved in performing the arrangements of the Sicilian music. The imitation of the high school band is going to be quite good; just the correct balance of right notes to sour ones. In fact, if it were not for some extraneous talking in the background, we could have used the tapes from the rehearsal to make the band cues. Were there more time available, we could really tape the band music today, but rehearsal time is at a premium. Best not to disturb the arranger's plans. She needs her time to rehearse and drill the performers.

At this point, we are well ahead of schedule, so we can take some time to go back through the tapes very carefully and listen for places where improvements might be made. We have approximately one week in which we can make revisions. The first thing to do is to go back to the concept tape and make sure that everything we have put on the show master reels so far has been consistent with the concept. This appears to be the case, but some parts of the atmospheric surround seem to be a little heavy-handed on second hearing; and some of the wind and bird events don't move from location to location quite as we had wanted. (We have made sure to listen to the sounds through our trial system in the auditorium and stage.)

Since some of the problems that we identified require reworking of the multitrack tape, do we really want to or can we afford to start revising? Two or three segments seem particularly offensive; so we'll revise these and leave the rest as they are. We approach the sections to be revised by remaking those portions of the atmospheric sound, and even then we only do parts that present the largest problem. In listening to the original tapes, we find that there are some spots before and after the offending passages where we might be able to cut and splice, so that we can take out the old sound and put in a revision. In doing this work, we prove the value of having setup tones on each tape and of carefully recording all control settings as we made up each cue originally. This kind of precision allows us to get very close to our initial sounds, so that splicing in a revised cut of the sound will not be noticeable when the tape is played as atmospheric background for the play. The revisions go smoothly, and we are finished with time left to check on progress locating equipment for the system installation and to plan for the recording session the following Sunday (day 16).

We still have one cue that has not been completed: the sound of the salesman's car and the car radio. The problem is finding appropriate music. We assume that the salesman would probably listen to country music; but country music in the 1990s is not what it was in the 1950s. One thing we have been doing, in part through the arranger, is trying to locate a source of 1940s and 1950s country recordings. We have discovered that a local disc jockey has such a collection, and he has agreed to tape a selection of these tunes for us to use. (We will acknowledge his assistance in the show's program.) This tape arrives on Tuesday (day 21) and turns out to be pretty good. (No major scratches and the quality isn't too bad.) Since the salesman's car stops but the radio keeps playing, we have to balance the sounds carefully. To some extent, we have to guess, because we don't have the playback system, and the trial system we've been using is not going to give us conclusive information for two reasons: First, the set is not in place, so the acoustic effect of the scenery cannot be determined; and, second, the loudspeaker placement is not exactly what we will have once the full system is installed. We put the cue on tape, and note that we may need to rework it once we hear the effect during final rehearsals.

By the end of the week (day 18), the sound M.E. reports that all equipment has been pulled from storage and checked out, and that he is ready to proceed with installation.

To be sure that everything is prepared, we make a quick check with the stage manager to be sure that nothing has been scheduled in conflict with the recording session on Sunday.

17

Saturday (day 17) is spent getting equipment ready for recording. Although we set up the system last week, we want to check everything out to be sure that no last minute surprises hold up what is likely to be a tight and hectic day. We also made quite a few notes last week about how various things sound, and on how microphones might be repositioned to improve the recording. We go over these, and make last minute modifications accordingly.

Music Recording Session

16

Sunday (day 16) starts at 8:00 A.M. setting up microphones and stands and rearranging the stage speakers so that the musicians can hear tape playback. The chairs we found for the rehearsal the previous week have been left stacked offstage, so these now are placed and the music stands set up. Just about 9:30 the musicians start to arrive. At 10:00 everything is ready to go. The arranger wants to play through each piece once, work on any rough spots, then tape. We will do the pieces requiring the most players, the high school band marches, first.

Since the high school band is to be heard as though in the distance, we use a stereosonic pair of cover microphones only.[3] The instrumentation is two flutes, four clarinets, one tenor saxophone, three trumpets, two horns, two trombones, baritone, tuba, and percussion. After we tape the first march and listen to a replay, the arranger says she isn't satisfied with the balance between the woodwinds and brass. We try a take using microphones placed close to the flutes, clarinets, the saxophone. We use four mics for this purpose. We develop two problems using this approach: first, phase cancellations. The brass, coupling into the mics for the woodwinds, is out of phase with the brass as picked up by the cover mics, so the brass winds up sounding dull and muffled. Second, the woodwinds now sound too close and squeaky. We suggest going back to using the cover mics, but physically moving the brass about halfway across the stage. A first placement doesn't work very well, but after one or two tries, we come up with an arrangement that records fairly well. The arranger isn't completely satisfied, but we assure her that for the purpose and the level of audibility, the recording will be fine. The rest of the marches go fairly quickly.

Once the band music is done, we take a break to allow the musicians who are no longer needed to pack up and clear out, and also to permit repositioning of microphones to record the smaller group performing the Sicilian music. Once the session resumes, some testing and repositioning are required to get a satisfactory sound. This time we want to record each instrument on a separate track of the multitrack recorder. The biggest problem proves to be isolation of the individual musicians from adjacent microphones. Separating the musicians doesn't work, because they tend to lose contact with each other and the playing suffers. Finally we go to the scene shop and borrow several sheets of plywood and improvise baffles between each instrument. The musicians can see over the baffles, but the individual microphones are shielded from other performers' sounds.

---

[3]Stereosonic microphone placement sets two cardioid microphones at right angles to each other, with their heads as close together as possible without touching. Naturally the left channel microphone will be on the right side of the pair, and vice versa. The arrangement delivers a very good stereo image with minimal hassle.

We get only one piece on tape before lunchtime. After lunch, however, things go much faster. Again, we take the pieces requiring the most people first, letting each musician go as his or her parts are completed. After the last piece is recorded, we and the arranger go over the tape to get some preliminary ideas about the balance of the final mix. We also set a time on the following day to mix down the multitrack tapes to two-track stereo. Meantime, the sound M.E. has collected all the mics, stands, and cables and put them away. The speakers are put back at their up-right and up-left locations, where we will use them until the full system is installed.

15

On Monday (day 15) we hear from the stage manager that cue 56 (the goat for the "duet" with Alvaro) doesn't work as three separate cuts. The director would either like to cut the cue or work it as a single tape somehow. We make a note, and then turn attention to the mixing session. A few troublesome moments develop trying to get the arranger to understand the problems of auditory perspective for some of the music that has to underscore scenes. Finally, when she understands what will have to happen when the music is played, we manage to agree on the balance for mixing the pieces of music in question. Most of the music for the preshow, scene bridges, and intermissions is easy to balance and mix down. By the middle of the afternoon, the mix is done and all the pieces are cut into the show masters.

Now we turn our attention to the problem of the goat and to getting the high school band into the tape. Goat first. Actually, it proves to be fairly simple; we make a tape with two bleats, dub it to cartridge, and that's it. No more 56a and 56b. The high school band is not too hard, but we wait until late evening, when we can get a quiet theatre and use the speakers in the house, to check the processing required to give the band the required perspective distance. By the end of the day, all cues are recorded, processed, and on tape.

Building a Rehearsal Tape

On Monday (day 15) while the electricians are installing the system, we make a rehearsal tape of the internal music and cues for act II. This includes the salesman's car, Alvaro's truck, and the music associated with the first encounter between Serafina and Alvaro. This tape will be used to rehearse timing, as requested by the director during initial production conferences.[4] Later in the afternoon, we meet with the operator to go over the way in which these cues will work. Also, using the temporary system on which we have been testing sounds, we set levels for the operator to use in running these cues for rehearsals beginning the following evening.

14-11

The rest of the week is spent at rehearsals making notes. The internal cues work reasonably well. The director is not happy with the way the salesman's radio sounds, but, again, is persuaded to wait until all of the system and the scenery are in place. Again, we assure him that if the radio makes a problem, it can fade out under the scene. By Wednesday, the system is installed and checked out. In spite of the fact that all of the scenery is not completely in place, on Thursday and Friday evenings, after rehearsal when the theatre is quiet, we and the operator work on levels so that not too much time will be consumed during cue-to-cue and technical rehearsals.

Everything is set. We are ready to go into final rehearsals.

[4] See Chapter 23.

Chapter 25

Rehearsing Sound

Technical and dress rehearsals are critical times. The design is really made or broken during these rehearsals. Managing the operation and execution of the design is one of the sound designer's primary responsibilities. In order to see how these rehearsals work, we will continue to follow our production of *The Rose Tattoo*, beginning with the first of the rehearsals to involve all technical elements.

Technical Rehearsals

10

Saturday (day 10 before opening) is the first of the technical rehearsals for our production of *The Rose Tattoo.* This first rehearsal is a cue-to-cue without actors, held in the afternoon, with the usual acting rehearsal in the evening. We have been running a very few of the act II cues since Tuesday. However, we have not used the system in the way we intend, even for these few cues. First, the system was not fully set up, and, second, the operator had not been taught full cue routing to use the system. The purpose of running the cues was primarily for timing and to accustom the actors to a few very critical sounds that were important to the scene.

Cue-to-cue is a chance to familiarize the operating technicians with their cue lines and their actions. It is the first chance for the stage manager to call cues and to get the feel of where in dialogue or blocking a cue must be called in order for the sound or lighting or other action to happen on time. Cue-to-cue rehearsals can sometimes be ego-crushing disasters, because nothing appears to work right; some cues are run again and again and again, never turning out correctly. Curtain raisers can be especially difficult, and so it happens here.

The scenery, with minor exceptions, is all in place, lighting is ready, and all of the sound equipment is installed. The show tapes have been finished for several days, and we have had a chance to go through the show several times with the operator. None of this guarantees a smooth run as we begin to try to integrate all of the production elements. We run about 20 minutes into the tape of preshow music, start it, let it fade with the house lights, then start sound just before the stage lighting comes up. Somehow the cue doesn't work; the timing is off. There is too much music before the lights fade up. We recue and go over it all again. This time the curtain takes far too long. Try it again; this time the sound and light cues get called before the curtain is fully opened. After several more tries, the sound and lights are working well together, but the curtain is still slow. Sitting in the house watching the rise, we get the feeling that there is darkness onstage when we really want to be able to see from the beginning; and, since all of the sound for the opening *prima sera* is played through speakers onstage, we first hear muffled sound that clears as the curtain opens fully. Several more tries vary the timing of the light/sound calls with respect to the curtain cue. Nothing seems to work. Finally, everyone decides to go on to the next cue and think about ways to solve the problem with the curtain.

The next several cues are all lighting cues. We have the operator stop the atmosphere tape in order to make it easier for people to communicate.

The next two cues are only sound, the truck passing noises. This is the first time we've heard the cues with the full set in place. The truck is quite a bit too loud the first time—partly overcompensation on our part trying to guess how much sound the set will absorb from a backstage loudspeaker and partly an error by the operator. Unfortunately, the mistake gets the director edgy, and he immediately wants to start lowering levels—on everything, the atmosphere, and the truck. We run the cues several more times until the director decides the level is right. We are pretty sure that, once the cue is put with dialogue and action on the stage, that the level is going to be far too low; but the director insists, so we let it go for the present. His choice will catch up with him soon enough; but we'll probably go through a small round of having to assure him that the levels are all set as he placed them during this cue-to-cue before he'll believe that the sounds need to be louder.

We really can't get any kind of fair idea about the atmospheric tapes during this cue-to-cue simply because we are skipping and stopping far too much to let the atmospherics play through, and also because we will never know how the atmosphere tapes really work until we hear them along with the dialogue. We do, however, try to keep some kind of atmosphere under the cues that we run.

The next cue, the first time we hear the Strega's goat, presents a problem that happens several times during the evening. In trying to keep up with everything on the cue sheet, the operator forgets to change presets and accidentally routes the sound to the preset used by the truck cues; so we wind up with a goat bleat coming from all over the stage. Well, part of cue-to-cue is to allow the operator to gain initial familiarity with the operations, so we run the cues again to straighten out the mistakes.

Although rough spots exist, and though some cues seem not to work as planned, most of the rehearsal goes fairly well for sound. Not too much can be judged without dialogue and action. Tomorrow's rehearsal is not going to be much different. Although we will have actors, we'll still be jumping from cue point to cue point through the play. So much of the sound in this production is based on continuity that it will be difficult for both designer and operator, and probably for everyone else. Mistakes are going to happen, and nerves are going to get frayed. Our best first trial of the sound will happen on Monday when we run sound with a full rehearsal for the first time. Inevitably, problems are going to come up, of course. When portions of sound for a production are virtually continuous, starting and stopping always throws the timing off somewhere; and there is a lot for the operator to do on this show. Sometimes three to four tapes can be running at the same time, each requiring routing to different presets. Timing of all operations is critical, and these rehearsals are as much for the operator to learn what has to be done as it is to integrate the sound into the production.

9

The cue-to-cue with actors on Sunday evening (day 9) is an unpleasant rehearsal. In fact, trying to keep up with the atmospheric sounds gets to be such a burden that we finally make an agreement with the director to omit all of the background sounds except for a few critical cues (such as *prima sera*). We simply run the spot marker cues. Unfortunately, all of this does nothing to improve the director's predisposition against the atmospheric underscore. Because sounds needed to make the design work are integrated into the atmosphere tapes, mistakes happen, and the director and a number of the actors are upset by the end of the rehearsal. In order to try to forestall problems, we schedule a practice run with the operator for the following day. This practice run is going to be a long one, because we will have to sit through most if not all of the atmospheric tapes in order to be sure that levels and operating sequence are correct. That means, naturally, that any mistakes are going to prolong our practice run considerably.

8

The practice run on Monday (day 8) proves to be quite difficult, as other work has to go on in the theatre at the same time. Sound does not co-exist well with other technical operations. The sound people need quiet, and the sound tends to drive carpenters and electricians crazy when they are trying to work on the set. (Lighting has similar problems, but there are often ways to compromise that are not available for sound.) We do manage to get enough practice to make the evening rehearsal much less likely to be a disaster, however.

Late in the afternoon, the director and the designers meet to talk about the problem of the opening curtain, which continued to prove difficult through the Sunday cue-to-cue. During this meeting the lighting designer proposes that upstage illumination be preset, so that the curtain opens on a lit set, frontal fill lighting to be added as the curtain flies out. The solution sounds workable, and we will look for adjustments to minimize the problem of initially muffled sounds as the curtain is rising.

The evening rehearsal gets underway with dress parade, which takes a significant amount of time to work through. There will also be other stops during the rehearsal to deal with costume problems. But this rehearsal is primarily for costumes to spot problems and difficulties so that the shop can deal with the alterations during the rest of the week.

The sound work-through earlier in the day proves useful. The operator is much more capable of handling the cues, and the levels are very close to right—not right as we think the levels must finally be set, but right as the director has decided he wants them. Unfortunately, most levels are, indeed, far too low. The director is becoming noticeably agitated as he listens. Sure enough, during act I, scene 5 he is complaining about sound levels being incorrectly set. We pick up headsets for a pro-forma check with the operator, but all levels are set as recorded during the previous rehearsals. We've been adjusting the levels of the atmospheric background during the evening's rehearsal, but we've scrupulously avoided changing anything that the director has set. Now, with the director's complaint, we strike an agreement with him: that we will trim levels of all cues during rehearsal, and he will give notes afterward on what cues were too loud or too soft. (He would do so anyway, of course, but agreeing to handle matters in this fashion permits each of us to concentrate and get productive work done during the rehearsal.)

Some cues are simply too short to adjust as they are running, but we don't ask to stop during this particular rehearsal. We simply work through, talking with the operator on headsets, correcting where we can, and making notes about everything else. A lot of productive work gets done during the evening, even though we had some rough moments because timing got thrown off when rehearsal was stopped to deal with costume problems.

7

Tuesday (day 7—one week before opening) is the first full technical rehearsal. The lighting designer's idea works, and the opening curtain, after we rehearse it a few times, is no longer a problem. Unfortunately, the stage manager makes a mistake and calls a stop after the curtain cue. He had not understood that the director wanted to keep going. The director does not want to go back and redo the opening, however. To compound the problem, the stage manager has instructed the sound operator to recue tapes to the beginning of the show, so the timing of the atmospheric tape gets thrown off again. Consequently, minor problems occur, first, when parts of the auditory surround change before they are supposed to, and then, again at the end of scene 1, when the tape runs out before the scene is over.

Generally, the evening goes fairly well. Lighting calls several stops, but we are lucky that none of these happen at points that seriously affect tape timing. Afterward, during notes, we get the first indication that the director may be losing some of his fear of the auditory surround. He indicates that he still may want lower levels, but that he was not distracted from dialogue and action by the atmospheric sounds as he had anticipated that he would be. We give notes to the operator, and make notes for the master electrician to move the placement of one of the upstage loudspeakers slightly. The present placement of the loudspeaker makes the auditory perspective of several cues seem incongruous.

6

The second technical rehearsal on Wednesday (day 6) goes fairly smoothly. During act II, however, the director expresses concern about the sequence of cues, beginning with the salesman's entrance. He doesn't like having the salesman's radio playing during the encounter between Serafina and the salesman and during the fight between the salesman and Alvaro. He calls a stop. He wants the radio music switched off when the salesman's car stops. Then he wants the radio back on when the car starts again. This, of course, causes a problem, because the tapes aren't set up to accommodate doing this. The radio is on one tape and the car leaving on another, meaning, or so it first seems, that the operator is going to have to perform an extra set of tape stops and starts. After a bit of discussion, however, we suggest that the operator can simply fade the level out, leave the tape running (as it would have in the original cues), then fade the radio tape back up again after the car start on the second tape. This solution doesn't really change anything for the operator except to add one additional fade. We try it, and the director seems satisfied.

During act III, scene 3, the timing of the "goat duet" causes a brief stop. The stage manager is having trouble calling the cue in a way that permits Alvaro to say his *Che bella!* without having to take an obvious beat to wait for the goat to bleat. Several first attempts at anticipating the cue only succeed in bringing in the goat noise well before Alvaro sees Rosa. Finally, we get a timing that works.

5

The following evening (Thursday, day 5) also runs smoothly. There is a stop for lighting, but otherwise the rehearsal plays straight through with no major hang-ups. One power amplifier starts making some very obnoxious sounds near the end of the evening. Evidently one channel is in trouble, but a backup is available, and the problem can be fixed during the day on Friday.

Dress Rehearsals

4

Friday (day 4) is the first time since the dress parade that actors will be in costume. Some problems will develop again tonight, as actors adapt to wearing the costumes. Before the rehearsal starts, the director comes over to say that he has now decided that the auditory atmospheric surround works quite well; but he wants the salesman's radio back into the show. He would, however, like to have the level a little lower than we originally set. This takes a few minutes on headsets with the operator, but causes no other difficulties.

During act III we notice that the upstage left loudspeaker now has a very strange sound and the directional quality seems to have changed. A trip backstage shows that costumes and props have been very busy rearranging the backstage geography. Where our loudspeaker was there is now a prop table. The loudspeaker has been moved behind a costume changing screen, and, in fact, the dresser has turned the loudspeaker around so that it is facing stage right and slightly upstage, aiming it squarely into a rack of costumes.

We go out to talk to the costume designer and the props master. A conference backstage after the rehearsal suggests that the prop table and the dressing area have been moved for fairly good reasons. The prop and costume designers apologize for the fact that sound wasn't consulted before moving the speaker, but explain that it was done hurriedly during rehearsal to take care of a problem that developed during the evening.[1] Unfortunately, there seems to be no way to handle the problem but to move the loudspeaker. The speaker in question, however, is the one that was repositioned during the rehearsal on Tuesday, and its placement is fairly critical. We

[1]This particular problem wouldn't take place (or, at least, shouldn't take place) in a professional situation, if for no other reason than that members of one union won't touch equipment that falls under the jurisdiction of another. It shouldn't happen in amateur and educational theatre either, but it does. Similarly strange things do happen in commercial/professional theatre as well. David Collison (Stage Sound [New York: Drama Book Specialists, 1976]) relates an anecdote having to do with finding an elderly actor using a loudspeaker as a bench. Unfortunately, the actor was wearing a very thick, heavy robe, which almost completely muffled the loudspeaker's output. The problem is included here as an illustration of the kinds of things that can happen.

decide, much to the master electrician's chagrin, that the upstage loudspeakers will have to be flown—a process that is going to mean a full afternoon of rigging spot lines to dead-hang the units. We don't want them more than 8 feet off the deck, and they don't fall at a location where they can be picked up by any free batten.

3

Saturday afternoon (day 3 before opening) the upstage loudspeakers are dead-hung, and we go through an hour of careful aiming and trimming to get exactly the right directional qualities in the auditorium.

During Saturday night's rehearsal, little reason exists now to maintain headset communication between sound designer and sound operator, other than to make minor adjustments in levels. The stage manager is just slightly off time with two cue calls; these mistakes are jotted down to be given as notes to the stage manager after the rehearsal.

In the intermission after act II, the director and the arranger indicate that they are not happy with the underscore music for the meeting of Serafina and Alvaro—the guitar music played as Serafina sees Alvaro silhouetted against the window. The arranger doesn't feel that the music is distinctive enough to create the needed effect (something with which we agree, by the way, but had not considered significant enough to make a point of it. If the director and the arranger wish to deal with it, however, that's fine.) What the arranger wants to do, though, is going to require a bit of rerecording and reworking of tapes. The effect that the arranger suggests, however, is appropriate; so we agree to a recording session on the following day, if the arranger can get the additional musician she wants. (This being a Saturday evening, reaching people on short notice could be a problem.) The arranger goes to try to contact the performer.

During act III, the arranger returns to say that she has found the musician (a flautist), and she suggests a time for recording the following afternoon. We have to check with other departments (lighting, props, scenery, costumes, etc.), but the time seems to be alright. The arranger then leaves to go rework the piece of music in question.

Rerecording Segments of Music

2

Sunday is only two days before opening, but at 2:00 P.M. we start recording a new version of the Serafina-Alvaro underscore. The music is essentially the same, except that, instead of using only the guitar, as in all the other underscore segments, the arranger has added a flute to this piece. Recording the piece doesn't take very long, but processing and editing the new music into the tape takes about two hours. In addition, we have to remove and store the old tape carefully, so that in case this new piece doesn't work for some reason, we can go back to what we had before. Testing the new piece takes just enough time so that we decide to order pizza for dinner. (Not enough time to go home before rehearsal.)

We have used our setup tones and logged control settings to make the new piece of music just as close in level to the old piece as possible, but, inevitably, there will be some change. Before rehearsal starts, while the actors are still getting into costume and makeup, we run a quick trial of the new piece with the operator. Several changes are necessary to get the correct level and the desired sense of distance and directionality. The revised music does seem to work well, however.

In fact, during the rehearsal, the revision to the music turns out to make a great deal of difference. The levels have to be trimmed slightly. Our estimated setting before rehearsal without actors on stage turns out to be a bit too loud. Also, now that the flute has been added to the piece, the quality of the auditory perspective isn't quite right. The sound needs just a little more reverberance to achieve the sense of distance that has been planned for the effect. That's going to mean a remixing of the new piece, unfortunately, and another adjustment. The problem, of course, is that the only remaining rehearsal is a preview, and there will be an audience in the house.

1

During the day Monday (1 day before opening), we rework the new piece of music and put the new tape into the appropriate show master. The director and the stage manager have kindly gotten the actors playing Serafina and Alvaro to agree to come in to run the scene before rehearsal so that levels can be set. The reprocessing is very satisfactory. The music now has a more distant quality and sounds as though it originates from somewhere in the neighborhood—music that fortuitously coincides with Serafina's recognition of Alvaro's likeness to Rosario.

The operator recues the tapes, the actors go back to finish their makeup, and we can only wait now to see how the sound behaves and feels with an audience in the house. Some inevitable changes always occur, no matter how well designed the auditorium is acoustically. Usually, all sound levels have to be very slightly higher. Sometimes a slight rebalancing is necessary, although this is much harder to accomplish. Rebalancing is to be avoided if at all possible, especially when it means changing settings for the operator. Overall level correction can be accomplished by raising the input controls to power amplifiers slightly. Power amplifier input controls are items that the operator should have to check only during setup each night. These settings should never be varied or used as operating level controls unless there is no other way to accomplish some desired effect. Increase in level also can be accomplished by a uniform increase in all input level settings or by an increase in a master control. Rebalancing, by contrast, means increasing some power amplifier input settings more than others to make certain all loudspeaker channels produce equal sound pressure at a majority of seats in the auditorium—a process that requires several hours of adjustment and testing.

Fortunately, with an audience in the house no real changes seem to be needed. The only trouble during this preview is a case of nerves—a line dropped by an actor starts the problem. The missed line throws the stage manager off, and he calls a cue early. In turn, the sound operator panics and tries to make an adjustment, which results in two missed cues before he can recover and get one of the two-track recorders back to the proper cue point. All we can do is to make notes and pass the comments to the stage manager and the operator after the rehearsal.

In general, everything is set and working well, and we are ready for the run of the production.

Chapter 26

Postscript on the Design Process

Some thoughts and ideas can only be shaped by looking back at a process and analyzing it. The process of design can be understood best by those who have experienced it and can, by review, see the patterns of imaginative and practical activity.

In the preceding chapters, we have talked through a rather large design. This "talk-through" certainly cannot substitute for firsthand experience of the design process itself, but it can point to some of the essential features of that process. We need to look back at the example to draw out the salient features of the process.

Not all plays need the amount of sound that we used for our production of *The Rose Tattoo* (and, of course, for some kinds of plays, sound designs can grow far beyond the amount needed in our example). The choice of a relatively large design was deliberate; it incorporated three very important areas of sound design: the usual sound effects (spot cues called for in the script); an example of extended auditory atmosphere; and a significant amount of music, both incidental and integral.

Spot Sound Effects

Spot sound effect cues occur in almost all plays. Sometimes a play may call for as few as one or two spot sounds, but most plays probably average around twelve. The fact that the majority of plays do not explicitly call for large amounts of sound should never lull one into minimizing the importance of sound to the production. In fact, the fewer the cues, the more important each one becomes. If a play has only one sound cue, obviously the playwright simply could not find a way to do without that one sound. That suggests that the sound has a very important motivational role in the action and overall meaning of the play.[1] That one sound, therefore, deserves all of the care one can give to understanding its purpose in the drama. Just what kind of sound did the playwright have in mind? Of course, no one can know precisely what the playwright imagined, but someone like Tennessee Williams clearly had very particular sound images in mind. The value of a sound designer to theatre lies in that artist's ability to understand intuitively the kinds of sounds required for the drama and to be able to produce them.

Atmosphere

Extended atmosphere, like musical underscore, is a device to help lead emotions and focus perception. Not all plays need the level of enhancement that extended atmospheric sound provides. In fact, most plays probably do not need this level of sound support;[2] but when the sound designer decides (and the members of the production team, especially the director, agree) that extended atmospheric sound is needed, the nature and character of that sound must be carefully considered and meticulously crafted.

The most essential concern in producing extended auditory atmosphere is to involve the audience completely in the world of the play—to support the emotional direction of the

[1]The converse of this proposition—that the more sounds a playwright calls for, the less important the sound is to the drama—is hardly true. One should assume that, whether one cue or many, if the playwright puts the sound into the script the sound probably serves a real purpose. This doesn't mean that we don't sometimes take out a cue, or go off on a totally different tack than that suggested by the playwright; but then that is what theatrical interpretation is about—finding meaning in something that the original author did not necessarily see. Significant variations in meaning don't happen often, but when they do, there is no reason not to consider them and pursue them if they are valid.

[2]One of the real values of theatre is its ability to abstract—to strip a dramatic situation to its minimal values, without the added baggage of a fully realized environment. Precisely this kind of economy makes theatre the single most effective vehicle for dramatizing intimate human conflicts.

drama in order to help the audience focus their perceptions on the central elements of the production. The problem lies in running frequent cues, sometimes even continuous background sound, without in any way distracting from the dialogue and action of the drama. To be able to do this, the designer must understand the nature of human auditory perception. After all, human beings do live every single day in a continuous auditory surround, and the sounds of our accustomed environment usually do not distract us from the things we have to do or want to do. That attribute, however, is not due to any ability to prevent the inflow of auditory energy.

As we noted early in this book, we can shut off vision by closing our eyes, but we have no way to shut our ears. In order to permit consciousness to attend selectively to some or even none of the auditory environment, our brains literally filter out sounds that do not signal something of importance to us. In other words, we have the ability to push sounds that are not of immediate consequence into the background. Human auditory selectivity rests on two properties of physical sound: multidirectionality and continuity. Random patterning of sound energy is also valuable. A sound that becomes rhythmic or obviously repetitious attracts attention and becomes annoying. The sounds we selected for the auditory surround in *The Rose Tattoo* were multidirectional, continuous, and as random as possible.

The most universally "negligible" sounds are those of nature. Almost any human can adapt to a background of birds, wind, and water. (Be careful with water sounds, however; they tend to promote concern for the bladder over interest in dramatic dialogue!) City sounds and some general human noises can also be filtered into the background.

Creating the effect of an atmospheric surround in theatre requires duplication of natural conditions insofar as possible. Several random tracks of sounds are essential. Monophonic reproduction of sound (i.e., one track through one loudspeaker, or even one track through multiple loudspeakers) absolutely will not work. Even stereophonic reproduction is not sufficient. The effect of an environmental surround requires at least three but usually more than three tracks. (The acoustics of the particular space determine the requirements.) The tracks must contain similar but not identical information. In *The Rose Tattoo* we used four tracks of surround. We started with monophonic recordings of bird and wind noises, but we recorded them so that no two speakers would ever have exactly the same pattern of sounds at the same time. Thus, although each track carried similar information, the instantaneous program of each track was different from all other tracks.

Use of many tracks rather than one helps to keep the auditory surround from interfering with perception of dialogue. The brain can filter input if it has a variety of signals from many directions (cocktail-party effect), but, given two or three very strong, competing sounds, it tries to attend equally to all of them. In the context of an auditory surround, actors' voices become one among many sound channels. The brain can filter all of the information and fasten onto the most important element, usually the dialogue, as speech tends to be the most important perceptual element in any auditory environment. Strange as it might seem to say that a greater number of sound tracks *prevents* the sound from distracting from the dialogue, that is, in fact, the case. (Naturally, proper balance of sound levels is also necessary. If the auditory surround is too loud, then simple masking takes over and we cease to hear the dialogue.)

Human voices should integrate naturally into an auditory environmental surround. There really were no major problems in blending the auditory environment with voices in *The Rose Tattoo*, but there was one situation in which we recorded actors' voices, if you recall. For *prima sera* the voices of the mothers calling children home to dinner were recorded. Usually, directors have actors call such lines from offstage; however, offstage dialogue (even having actors face offstage) cannot really match the quality of human voices in proper auditory perspective for the visual environment.[3]

Although not a problem in our *Rose Tattoo* production, the integration of all elements of the auditory environment into a unified whole is potentially a touchy area for the sound designer. Unification of the auditory environment requires some control over the sounds of actors' voices and the incidental noises that they make moving about the stage, and very few people in theatre at present have ever even considered the idea that the sound of the actors' voices and the sounds of the "stage machine" should be in any way a part of the sound designer's concerns. (The sounds of the "stage machine" include footfalls on the set, the sounds of the properties that actors use, and the sounds that the scenery itself makes in normal use.) When a sound designer needs to create a total auditory atmosphere, however, the sounds of the actors' voices and the sounds of the stage machine cannot be ignored. For example, if the scenery for a production of *Richard III* provides the illusion (even in some degree of abstraction) of a stone dungeon for the Tower of London scenes, the sound designer is likely to be tempted to make the voices and footsteps sound reverberant. Not only does the auditory character fit the visual character of the scenic element, but it adds to the sense of evil and threat and to the sense of bloody death that should permeate the Tower in that play.

---

[3]In an actual production situation, we might have had quite a battle to get the director to approve recording those voices. Most directors (except those who have learned about sound and understand how it can be used and what well-designed and executed sound can accomplish) feel that they have more control when a live actor speaks a line than when a technician starts a tape machine. In our *Rose Tattoo* example, we decided that the calls would be recorded, because by so doing they can be made to blend into the auditory environment, whereas, spoken by actors offstage, they would probably contradict the auditory environment.

Controlling the auditory ambience of the theatre potentially requires sound-deadening material on or under platform tops and hidden microphones to pick up voices and other sounds such as footsteps and property noises. If sound-deadening material is needed in the construction of platforms, then the need for it must be discussed during early phases of production planning. The purpose of the microphones is not to reinforce the actors' voices but to allow reverberance to be added to all of the sounds happening on the stage. Microphones can be hidden in various parts of the set or hung over the stage, as needed. Placement is not as critical as for reinforcement, because we do not need maximum gain before feedback; we only need enough sound level to permit a very small amount of atmospheric reverberance or echo to be added to the overall sound of the production.

Finally, an auditory surround should be developed from the spot cues that the playwright specifies in the script. The spot cues give some of the best clues as to the nature of the auditory environment. When the auditory environmental surround is completed, the spot sounds should work as specific time marker elements within the environment. The markers, of course, must be strong enough and commanding enough to act as demarcations in time, setting the start and/or end of particular temporal events in the dramatic action.

The idea of surrounding the audience with sound sources is a disturbing notion to many people in theatre. One school of thought feels that the stage, especially in proscenium theatre, is the focus of action, and that anything that pulls attention away from the stage is, simply, wrong. Perception is influenced by emotional predisposition, and people whose logic confines all production elements to the single direction of the stage may never be able to hear and feel the aesthetic impact that multiple sound directions can provide. Conversely, those individuals might say the same essential thing about people who do consider a multidirectional surround useful. The most difficult problems occur when the sound designer is of one school and the director of the other. As usual, compromise is the only possible resolution, although a larger compromise is usually required from the designer than from the director.

Music

The Rose Tattoo provides a good example of the problems involved in supplying music for a dramatic production. First of all, the music must necessarily be of a particular kind (unless, of course, the entire play is to be wrenched from the setting which Williams prescribes). Further, the playwright specifically calls for music at particular points in the script, some of it related to his notes on the auditory background of the scenic environment (meaning that it must fit into the auditory surround) and some of it purely symbolic of Serafina's emotions. Probably we could have found prerecorded music that would have served the purpose reasonably well; but we chose to try to provide original arrangements and performance for one particular and special reason: Music designed for the needs of a production potentially offers a more powerful enhancement to the emotional impact of the drama than does music recorded for its own ends. Further, music written especially for a particular staging of a drama simply fits the production better than would any piece of prerecorded music, however good that prerecorded piece may be.

Original music can never be left to the last minute. Even if the sound designer happens to be a musician and can write the music, the process must be started as early as possible—at least as soon as the acting rehearsals begin. Otherwise, the music may get lost in the last minute rush of the final phases of production, or there may not be sufficient time to find and record all of the musicians needed to perform the score. (In fact, one might be well advised to find out what instruments are available before starting to write the music.)

On the other hand, one might argue that prerecorded music, professionally performed, will offer better musical quality than would be available from local musicians working in a less than optimal studio situation. No doubt the quality available from a commercial recording would be better in some respects, but if reasonably good musicians are available, satisfactory recordings are possible. The balance of the decision, then, rests on the desirability of music specially written to integrate with the immediate production.

General Processes

The first step in designing *The Rose Tattoo*, as it should be with any design, was a careful reading and rereading of the entire script to form ideas about the dramatic meaning and about the relative significance of the sounds for which the playwright asks. We pursued this schema of insights by looking for an ethical, then an aesthetic concept. Once we had an idea of the symbolic importance of the sounds we could begin to develop some idea of just what auditory qualities each should have.

Shaping the Characteristics and Qualities of Sounds

Granted that for any given kind of sound a limited range of qualities exists, we usually have enough latitude for artistic choice. That is, we can alter the timbral, dynamic, and/or durational properties of a sound. If we alter too radically, we soon have a different sound, but for most sounds, component values may be modified without detriment. Those values are the objects of consideration in choosing the dramatic properties of sound. Sometimes the simple choice of the apparent distance of a sound will make all the difference in its dramatic meaning. For example, thunder in the distance *before* a storm implies the onset of danger; thunder in the distance *after* a storm signals release from danger.

All sounds derive meaning from their context, but for maximum efficiency in contributing to the meaning and impact of the drama, precision in adjusting the qualities of a sound to the context is absolutely essential. To use thunder as an example again, merely reducing intensity will fail to simulate distance. One also needs to adjust the spectral balance and the reverberance of the sound to achieve a convincing simulation of distance.

Control of sound to achieve effective dramatic expression requires subtle changes in sound quality. With the thunder approaching (threatening), we need more low frequency—more of the quality of a ferocious and threatening growl. As the thunder recedes (the threat passes), we need a lighter sound—more in keeping with a sense of relief at escape from danger. Without in any way diminishing the importance of the potential symbolic value of auditory imagery, we need to remind ourselves of the potential dangers in symbolism as well. We can never expect the audience, or even the rest of the production staff, to derive the same symbol significance from sound that we, ourselves, see in that sound. *Before all else, the sound must be a credible part of the dramatic auditory environment.* If it fails in this, then the sound must be changed or else removed from the production.

Housekeeping

Attention to detail such as the timbral properties of a sound is important; but the craft of sound design also consists of more mundane things. Recall that in the production of tapes for *The Rose Tattoo*, we spent time writing down lists of everything that we would need in the way of raw sounds; then proceeded to try to isolate sources that could fill those requirements. We spent time in research, looking for ideas specific to place and time to lend authenticity to the design.

Recall two things, especially, from the process of executing the design: careful logging of steps and processes, including control settings used to create each cue; and meticulous planning of cue traffic—of what cue would originate from which tape recorder—and how tapes would have to be constructed and arranged to provide adequate control over the sounds in each cue or group of cues.

Writing down everything involved in the creation of a cue makes it much easier if revision or reworking becomes necessary. Usually, we try to salvage the existing cue before we decide that we have to build a completely new one; therefore, the data on just how we made the cue is always useful. (Whether or not one keeps the information as part of a permanent record is another matter entirely.)

Planning cue traffic is essential for at least two reasons: first, that it helps to organize and simplify the process that the sound operator must follow, and second, that it is really the only way to anticipate potential problems involved in the execution of a cue. For example, look at the group of cues that opens act I, scene 2 (cues 8, 8a, and 8b). Initially, all of these sounds (crickets, wind in pines, and a gust of wind) were to have been components of a single piece of tape, but to incorporate these sounds on a single tape as a single cue would have created difficulties in timing and in level balancing. Having the three sounds on separate machines is easier from the standpoint of adjusting the level of each sound, and it facilitates timing the start of each sound. Although the operator is asked to run three machines instead of one as part of the cue, this solution ultimately makes less demand than if all sounds were on a single tape. (To try to adjust levels within a cue is almost never adequate, and adjustment in timing is impossible.) One can see this kind of problem clearly by setting up a cue traffic list.

Another practice we should emphasize is that of using setup tones to fix recording levels. Because this practice is so very important to producing clean, easily manageable tapes, let's review the procedure.

1. Using the console's setup tone generator, set all output level controls for 0 VU. Do not change the output levels thereafter (or else mark their levels for quick resetting).[4]
2. Set all subsequent equipment (monitor power amps, etc.) that have VU meters for a 0 VU reading.
3. At the head of each reel of tape, record 30 seconds to 1 minute of 1-kHz tone at 0 VU. Use the tone from the console's generator. Switch the meters to "tape" and set the playback level for 0 VU.
4. After recording the tone on tape, replay to check agreement between the 0 VU on tape and 0 VU at the console. Do not change the output level controls thereafter (or else mark their levels for quick resetting).
5. Check each sound to be recorded for maximum level, and adjust the recorder's input level control so that the loudest passage reads at or slightly above 0 VU. When mixing sounds remember that combined sounds read slightly higher than the loudest individual component alone.

[4]Some consoles have no internal tone generator. In such a case, use the tone generator on the tape recorder, if the machine provides one, or else get an external tone generator. If using an external source, set controls on the input module accepting the tone signal for unity gain. That means adjusting line gain trim to pass signal without modification. Set the module output fader to 0 on the index scale. (0 should be approximately 3 dB short of full.) Set output faders to 0 on the index scale. Start the tone. Adjust for 0 VU at the tone source (if meters are provided), then adjust the input module to deliver 0 VU to the outputs.

6. For playback, teach the operator to use the console tone generator to set 0 VU output from the console, then to use the setup tone on tape to set 0 VU output from each tape recorder.

Using setup tones helps ensure predictable levels and satisfactory signal-to-noise ratios and control levels.

Building Composite Cues

Another point to notice is the use of "layering" to build up composite sounds. The layering technique can be used in a number of ways: first, as a means to build up a group of separate tracks that must be played simultaneously but from different loudspeaker sources, as in the environmental sounds in *The Rose Tattoo*. Second, layering can be used as a means of reproducing parallel sound tracks when both sounds use a common group of loudspeakers, but one track must vary in level independently of the other(s). Third, and perhaps most generally useful, layering on a multitrack machine is an excellent way of preparing composite sound effects that must be made up of a number of separate components, but that ultimately will be mixed down to a single one- or two-track tape. Using the multitrack recorder, the various separate components can be put in individually. Layering allows very tight timing of components within the composite effect. Level and even the spectral qualities of each component can be separately controlled.

Let's look at an example of layering for building up a two-track cue. Near the end of act III, scene 1, Alvaro starts his truck, pretending to leave. The cue (44) contains the truck start, an engine rev, air brakes releasing, and then the truck shifting into gear and pulling away. When we originally recorded truck sounds, we didn't really get anything that provided the exact sequence, nor did any of the recorded sounds have the timing needed for the cue.

First, on track 1, we placed the sound of the truck starting. We recorded about ten seconds of engine idle after the start, then made sure that all of the track following (to the length of the cue) was blank. Next, on track 2, the first engine rev was recorded. We did this by cuing a two-track cut of the engine rev very close to its head. Then, with track 1 of the eight-track in synchronous playback mode (i.e., using the record head segment for that track as a pickup head) and track 2 in record mode, we played back track 1. At about 2 seconds into the idle following the start, we rolled the two-track tape of the engine rev, recording that on track 2. On track 3, just after the peak of track 2's engine rev, we dubbed the sound of air brakes. Track 4 received another engine rev, with the sound of gears shifting on track 5. Using track 6, we dubbed in the sound of the truck pulling away.

With all of the sounds on the multitrack tape, we went back through, trying several different variations of control settings on the mixing console until we found satisfactory balances among all of the elements of the cue. These settings included the panning of tracks to the left and right channels of the two-track mix. We made careful notes, detailing all of the actions that we would have to perform during the mix-down, such as fading down track 1 (the engine idle) after the first rev (on track 2) started. Once we set all levels and rehearsed through the mix once or twice, we performed the actual mix-down to two-track tape. Once dubbed, all of the components are on a single cut, correctly balanced and in proper sequence. The process of building the cue was far less onerous than if we had had to make six separate tapes and mix them live, starting and stopping machines and fading controls in precise timing. Live mixing of this cue could have taken many trials before getting the balance and timing just right.

Planning for Operational Simplicity

Although we didn't really discuss this aspect of construction of the tapes in the previous chapters, all of the cues were built with the end in mind of simplifying the operator's work. We try to make the execution of a cue as easy for the operator as reasonably possible. The aim is not to make each cue a matter of simply pushing a "go" button, letting a computer do the real work, but we do want to make it possible for the operator to be able to handle all tape starts and all fades and preset changes comfortably. Keeping actions within a comfortable range for the operator allows him or her to keep attention focused on the production, without distraction from internal timing of fades and other operations.

The purpose of the sound designer in the theatre is to serve the needs of the drama—more particularly, of the director's interpretation of the drama. This is not to say that the designer is merely a servant-executor of someone else's ideas. As we noted earlier, the value of a sound designer lies in that artist's ability to understand intuitively the kinds of sounds required for the drama and to be able to produce them. The director's job is not to plan in detail all of the elements of his or her production but, rather, to provide the overall direction and interpretation and to lead the other artists' efforts into the particular direction he or she needs for those efforts to go. The designers, therefore, must work to realize the intentions of the director. In doing their work, however, designers are going to bring in ideas of their own; that is inescapable. The trick of the craft lies in being able to adapt those ideas to the service of the director's interpretation. In spite of the various statements about theatre as collaborative effort, the director remains the primary interpreter. Never should a designer use collaboration as a pretext to insinuate his or her own ideas into a production at the expense of the director's concept and stated purpose. After all, the show is ultimately the director's conception. The director and only the director can and should have final say over the components of the production.

The planning and execution of a good sound design is a blend of the following qualities: a good intuition for the evocative qualities of sound, an understanding of the relationship between sound and dramatic impact, conceptualization, research, and technical craft. Good theatre is always a blend of many ideas that work together to communicate a single, unified, and emotionally powerful message for the audience. We should never forget that sound is one of the most powerful vehicles of communication that we possess, and its fullest use in dramatic presentation still remains largely unexplored.

Chapter 27

Notes on The Rose Tattoo *in the Computer Age*

The actual production of *The Rose Tattoo* that served as the model for the production example took place in the late 1970s. Techniques used in that production were somewhat more primitive than those described. The example was updated to take account of improvements in equipment through the decade of the 1980s. Now, as this book goes into publication, a new and very different set of circumstances for theatre sound production is emerging: the availability of microprocessor-based devices. Sampling synthesizers, sequencers, DATs, direct-to-disk recording, and computer-based control and editing programs are unquestionably going to change sound design in every possible form of dramatic production. The *Rose Tattoo* production could have been updated to place it entirely in the digital age. However, the complete conversion of sound in all segments of theatre will not take place for several years to come. Consequently, the production example remains in a tape-based format, using processing and editing techniques that belong to the analog age. In this way, those with older technology are served, and theatres with access to new devices still have the basic outlines of a design model. In order to help those who have or will acquire samplers, sequencers, computerized sound scoring consoles, and DAT recorders, we should, however, state what procedures might be used if one were executing the same design for *The Rose Tattoo* using newer systems.

First, the importance of tape recorders would decrease significantly, especially the stereo, reel-to-reel machines. In place of those devices, we would use one or more sampling synthesizers, a sequencer, a MIDI patch bay, and (if possible) a direct-to-disk recorder. Most cues would begin as samples recorded into memory, looped if necessary, and adjusted for optimum utility. Each sample would be assigned as a voice on one segment of the synthesizer keyboard. A small range of keys might be assigned to each segment so that the pitch of each sample could be varied slightly.

For the long, involved environmental cues, patterns of events would be recorded as steps in a "song" held in the sequencer's memory. Settings for synthesized wind, if used, could be stored in the patch bay and activated as needed. Depending on requirements for operational simplicity, the actual performance could either use the samples under control of the sequencer or all of the environmental cues could be recorded to the hard disk system.

Whether the environmental cues are played directly from the sampler or are replayed from disk, use of the sequencer as the device to construct the long environmental sounds has distinct advantages. For example, if, in rehearsal, some individual element of an environmental cue takes on unexpected significance but would serve the purpose better at an earlier or later time in the sequence, moving that one element of the total sound complex is quite easy. Using the sampler, tiny adjustments can be made, perhaps during rehearsal, until the precise effect is achieved. In an analog recording, the entire segment would have to be rebuilt almost from the beginning.

Note that a possible alternative might be to make up the environmental cues using sampling and sequencing, record the cues on hard disk, then make analog cartridge loops just as we did for the production as described in previous chapters. Having new equipment doesn't always mean that that equipment will provide the only solution, or even the best solution. (Even if we chose to use a cartridge, a reworking of the environmental cue to optimize particular sounds would still be relatively easy using sampling and sequencing.) Reasons for using cartridge loops instead of sequenced samples or sounds replayed directly from the hard disk might be that the sampler could be used to better purpose for spot cues or that not enough storage space was available to hold all cues on the hard disk. (Disk storage space is still expensive.)

As to better use of a sampler, various animal and automotive sounds, for instance, could be triggered directly from the sampler keyboard by the operator. Also, having spot cues in sample memory means more flexible playback. Using tape, each instance of a sound required a cut of tape bracketed by leader marked with the cue name and number. To vary the sound, each cut had to be recorded separately with the modifications built in. Using a sampler, the sound can be varied in a number of ways, from immediate repeat (press the key again) to variation in pitch (press another key in the same voice range). Using a sampler, cues such as the parrot squawk and the train whistles could have been varied slightly at each repetition without the difficulty of building several processed copies of the same sound.

Certainly, one cue that was somewhat problematic in the original example would have been absolutely no problem whatsoever using a sampler. That cue is the goat duet, where Alvaro's *Che be-e-ella*, imitates the goat bleat. The operator would simply have hit a key on the sampler's keyboard to activate the goat sample; the actor would say the line, and the operator would hit the key again. A few minutes rehearsal between Alvaro and the operator would have taken care of any timing difficulties.

The sessions to record the original music would certainly involve the direct-to-disk recorder rather than the eight-track or the stereo tape recorders. The one possible reason not to use a hard disk recorder would be if the disk system is a stereo-only unit and not a multitrack recorder. Some of the music would be best recorded onto a multitrack machine. In that case, the eight-track tape recorder would still be the machine of choice. (Again, multitrack hard disk recorders are presently much more expensive than multitrack tape recorders.)

If the hard disk recorder included a waveform editing facility, or if a separate, full-featured waveform editor were used, then all of the processing for auditory perspective would occur using that piece of equipment rather than using discrete EQ, delay, and compression. Certainly, recordings to be used as samples would have to be processed through the waveform editor to set up optimal loops, where needed.

A possible alternative for the stereo tape recorders might be one or more DAT recorders. Sound quality from DATs can be superb, and individual segments of program recorded on DAT tape can be marked for rapid location. DAT machines would be considered as storage for environmental cues as well.

Also useful if available would be a computerized sound scoring console instead of the analog sound scoring console used in our example. The computerized playback console described in Chapter 9, for instance, could handle all of the inputs required, with no reason to insert the mixing console into the playback system, and could flexibly distribute any sounds to any loudspeaker location. Moreover, latest versions of software for such systems include a reasonably full MIDI implementation, so that the console itself could serve as a sequencer, locking specific cue sounds to the console actions required to distribute them. Further, the sound scoring console has the capability to start any or all of the playback equipment, whether the device be a multitrack tape player, DAT, or hard disk recorder.

A clear benefit of using new technology would appear in the rehearsal, in those situations where the sound operator, for various reasons, cued a tape incorrectly. In any of the newer systems, use of time code could make the repositioning of the program much more precise, whether the source were tape, sequencer, or disk.

Although not completely a matter of sound, one possible extension of control that new technology can afford is the precise interlocking of cuing in another production area, such as lighting, to the sound stream, or vice versa. For example, in act 3, scene 2, the dialogue between Rosa and Jack after their night at the beach, a very specific sequence of lighting events could be synchronized to an environmental sound track, bringing in increasing brightness and perhaps the first small predawn glimmers of warm light, after which the first sounds of birds waking could be heard. Ways of giving the operator sufficient control to compensate for variation in actor timing are easily possible, so that the execution of cues, including sound, is not exclusively and mechanically slaved to the time track.

One possibility, which might be useful if no sequencer were available, is that, in executing a complex show from a sampling synthesizer, several operators could be used. Each one would control some few voices on the keyboard, taking cues from a master operator or directly from the stage manager.

The possibilities and advantages of using the newer technology are by no means exhausted with these few suggestions of how *The Rose Tattoo* production might be revised. One fact, however, is very clear, and that is that the new technology is finally going to place the sound designer on the same footing as the lighting designer, as an artist who is free to concentrate more on aesthetic concerns and less on technological matters. While sound technology will probably never be completely transparent, most sound people in the years to come will not have to be any more distracted by problems of ground loops and impedance matching than lighting designers need be with ghost-loading or choosing the right lens focal length. Hardware constraints and system limitations will almost cease to be a factor in design, and the aesthetic capabilities that the new technology provides will make possible the full realization of what sound can really mean as a tool of dramatic production.

Appendix I

Reference Tables

A. Logarithms of Numbers and Decibel Values

Number	Log (Base 10)	dB Power	dB Voltage or Current
0.000000001	-9.0000	-90.00	-180.00
0.000000010	-8.0000	-80.00	-160.00
0.000000100	-7.0000	-70.00	-140.00
0.000001000	-6.0000	-60.00	-120.00
0.000010000	-5.0000	-50.00	-100.00
0.000100000	-4.0000	-40.00	-80.00
0.001000000	-3.0000	-30.00	-60.00
0.010000000	-2.0000	-20.00	-40.00
0.100000000	-1.0000	-10.00	-20.00
0.010000000	-2.0000	-20.00	-40.00
0.020000000	-1.6990	-16.99	-33.98
0.030000000	-1.5229	-15.23	-30.46
0.040000000	-1.3979	-13.98	-27.96
0.050000000	-1.3010	-13.01	-26.02
0.060000000	-1.2218	-12.22	-24.44
0.070000000	-1.1549	-11.55	-23.10
0.080000000	-1.0969	-10.97	-21.94
0.090000000	-1.0458	-10.46	-20.92
0.100000000	-1.0000	-10.00	-20.00
0.200000000	-0.6990	-6.99	-13.98
0.300000000	-0.5229	-5.23	-10.46
0.400000000	-0.3979	-3.98	-7.96
0.500000000	-0.3010	-3.01	-6.02
0.600000000	-0.2218	-2.22	-4.44
0.700000000	-0.1549	-1.55	-3.10
0.800000000	-0.0969	-0.97	-1.94
0.900000000	-0.0458	-0.46	-0.92
1.000000000	0.0000	0.00	0.00
1.100000000	0.0414	0.41	0.83
1.200000000	0.0792	0.79	1.58
1.300000000	0.1139	1.14	2.28
1.400000000	0.1461	1.46	2.92
1.500000000	0.1761	1.76	3.52
1.600000000	0.2041	2.04	4.08
1.700000000	0.2304	2.30	4.61
1.800000000	0.2553	2.55	5.11
1.900000000	0.2788	2.79	5.58
2.000000000	0.3010	3.01	6.02
2.100000000	0.3222	3.22	6.44
2.200000000	0.3424	3.42	6.85
2.300000000	0.3617	3.62	7.23
2.400000000	0.3802	3.80	7.60
2.500000000	0.3979	3.98	7.96
2.600000000	0.4150	4.15	8.30

Number	Log (Base 10)	dB Power	dB Voltage or Current
2.700000000	0.4314	4.31	8.63
2.800000000	0.4472	4.47	8.94
2.900000000	0.4624	4.62	9.25
3.000000000	0.4771	4.77	9.54
3.100000000	0.4914	4.91	9.83
3.200000000	0.5051	5.05	10.10
3.300000000	0.5185	5.19	10.37
3.400000000	0.5315	5.31	10.63
3.500000000	0.5441	5.44	10.88
3.600000000	0.5563	5.56	11.13
3.700000000	0.5682	5.68	11.36
3.800000000	0.5798	5.80	11.60
3.900000000	0.5911	5.91	11.82
4.000000000	0.6021	6.02	12.04
4.100000000	0.6128	6.13	12.26
4.200000000	0.6232	6.23	12.46
4.300000000	0.6335	6.33	12.67
4.400000000	0.6435	6.43	12.87
4.500000000	0.6532	6.53	13.06
4.600000000	0.6628	6.63	13.26
4.700000000	0.6721	6.72	13.44
4.800000000	0.6812	6.81	13.62
4.900000000	0.6902	6.90	13.80
5.000000000	0.6990	6.99	13.98
5.100000000	0.7076	7.08	14.15
5.200000000	0.7160	7.16	14.32
5.300000000	0.7243	7.24	14.49
5.400000000	0.7324	7.32	14.65
5.500000000	0.7404	7.40	14.81
5.600000000	0.7482	7.48	14.96
5.700000000	0.7559	7.56	15.12
5.800000000	0.7634	7.63	15.27
5.900000000	0.7709	7.71	15.42
6.000000000	0.7782	7.78	15.56
6.100000000	0.7853	7.85	15.71
6.200000000	0.7924	7.92	15.85
6.300000000	0.7993	7.99	15.99
6.400000000	0.8062	8.06	16.12
6.500000000	0.8129	8.13	16.26
6.600000000	0.8195	8.20	16.39
6.700000000	0.8261	8.26	16.52
6.800000000	0.8325	8.33	16.65
6.900000000	0.8388	8.39	16.78
7.000000000	0.8451	8.45	16.90
7.100000000	0.8513	8.51	17.03
7.200000000	0.8573	8.57	17.15
7.300000000	0.8633	8.63	17.27
7.400000000	0.8692	8.69	17.38
7.500000000	0.8751	8.75	17.50
7.600000000	0.8808	8.81	17.62
7.700000000	0.8865	8.86	17.73
7.800000000	0.8921	8.92	17.84
7.900000000	0.8976	8.98	17.95
8.000000000	0.9031	9.03	18.06
8.100000000	0.9085	9.08	18.17
8.200000000	0.9138	9.14	18.28
8.300000000	0.9191	9.19	18.38
8.400000000	0.9243	9.24	18.49
8.500000000	0.9294	9.29	18.59
8.600000000	0.9345	9.34	18.69
8.700000000	0.9395	9.40	18.79
8.800000000	0.9445	9.44	18.89
8.900000000	0.9494	9.49	18.99

Number	Log (Base 10)	dB Power	dB Voltage or Current
9.000000000	0.9542	9.54	19.08
9.100000000	0.9590	9.59	19.18
9.200000000	0.9638	9.64	19.28
9.300000000	0.9685	9.68	19.37
9.400000000	0.9731	9.73	19.46
9.500000000	0.9777	9.78	19.55
9.600000000	0.9823	9.82	19.65
9.700000000	0.9868	9.87	19.74
9.800000000	0.9912	9.91	19.82
9.900000000	0.9956	9.96	19.91
10.000000000	1.0000	10.00	20.00
20.000000000	1.3010	13.01	26.02
30.000000000	1.4771	14.77	29.54
40.000000000	1.6021	16.02	32.04
50.000000000	1.6990	16.99	33.98
60.000000000	1.7782	17.78	35.56
70.000000000	1.8451	18.45	36.90
80.000000000	1.9031	19.03	38.06
90.000000000	1.9542	19.54	39.08
100.000000000	2.0000	20.00	40.00
200.000000000	2.3010	23.01	46.02
300.000000000	2.4771	24.77	49.54
400.000000000	2.6021	26.02	52.04
500.000000000	2.6990	26.99	53.98
600.000000000	2.7782	27.78	55.56
700.000000000	2.8451	28.45	56.90
800.000000000	2.9031	29.03	58.06
900.000000000	2.9542	29.54	59.08
1,000.000000000	3.0000	30.00	60.00
10,000.000000000	4.0000	40.00	80.00
100,000.000000000	5.0000	50.00	100.00
1,000,000.000000000	6.0000	60.00	120.00
10,000,000.000000000	7.0000	70.00	140.00
100,000,000.000000000	8.0000	80.00	160.00
1,000,000,000.000000000	9.0000	90.00	180.00

B. International Standards Organization Center Frequency for Equalizers

International Standards Organization Center Frequencies

Octave Centers	2/3 Octave Centers	1/3 Octave Centers
	25.00	25.00
31.50		31.50
	40.00	40.00
		50.00
63.00	63.00	63.00
		80.00
	100.00	100.00
125.00		125.00
	156.00	156.00
		200.00
250.00	250.00	250.00
		315.00
	400.00	400.00
500.00		500.00
	630.00	630.00
		800.00
1,000.00	1,000.00	1,000.00
		1,250.00
	1,600.00	1,600.00
2,000.00		2,000.00
	2,500.00	2,500.00
		3,150.00
4,000.00	4,000.00	4,000.00
		5,000.00
	6,300.00	6,300.00
8,000.00		8,000.00
	10,000.00	10,000.00
		12,500.00
16,000.00	16,000.00	16,000.00
		20,000.00

Appendix II

Production Intercommunication Systems

In most contemporary theatres, the job of providing the system that allows the various production departments to talk to each other during rehearsal and performance usually falls to the sound technicians. Use and operation of production intercommunication equipment, therefore, becomes an additional matter that theatre sound personnel must learn.

Two-Way Communications Fundamentals

A **simplex** system, the most basic form of intercommunication hookup, consists of two stations connected by one audio channel. The system can operate only in a single direction at a time; that is, the party at one station can talk while the other party listens. In order to be useful, the stations must be able to switch functions. Some device, such as a push-to-talk switch, must disable the loudspeaker when the station transmits—switch is pushed—and disable the microphone when the station listens—switch is released. The standby condition is listen. Simplex systems are not useful in theatre, because messages can only go one way at a time, the number of parties to a conversation is limited, and use of at least one hand is required to send a message.

Simultaneous two-way communication takes a pair of audio lines—one for sending and one for receiving. One party transmits on line A and listens on line B while the other party does exactly the reverse. A 2-line, 2-way transmission hookup is called a **duplex** system. The 2-line principle is typical of older two-way intercom systems in which each station has a transmit line that runs to the other station's receive line. Because both microphone and loudspeaker can be active at the same time, the possibility of feedback exists. Therefore, 2-way intercoms use headsets with boom-mounted microphones or must incorporate acoustic separation between microphone and loudspeaker. A traditional duplex system is adequate only if no more than two parties need to communicate. Inclusion of more than two parties, however, entails complicated switching and channel wiring.

Modern intercommunication hook-ups are modelled on the telephone system. Electrically, a telephone system is **network** utilizing a single transmission line. A networked intercommunication system is a modified duplex system, so transmit and receive circuits still must be separated. In a telephone-style network design, each individual station handles the separation locally. Internally, each station is a complete duplex system, containing its own talk and listen amplifiers. The external audio line simply acts as a summing buss, carrying all the voltage variations generated by the individual stations. To the talk amplifiers, the audio line is a very high impedance; to the listen amplifiers, the audio line is a very low impedance.

The audio line in a production intercommunication system is called a **channel**. Like all electrical systems, a communications channel has a generator—usually a microphone preamplifier—that drives a **terminating load**—usually a line amplifier. Successful operation requires one and only one termination. More than one terminating impedance drastically reduces the circuit's operating load impedance, causing degradation in signal quality and limiting the ability of the system to transmit satisfactorily over long cable runs. The terminating station is called the **base station**. All other stations are **bridged** across the audio line. Bridging means that the effective load impedance of the individual station is very high. Each station is connected across the circuit in parallel, therefore each added station lowers the overall load impedance of the circuit. But because impedance is high, each additional station adds only a very small amount to the load. (Review the explanation of parallel resistances in Chapter 2.) Although a network does have practical limits to the number of stations that can be connected to the system, that limit usually exceeds the number that any particular application is likely to need.

Because of the proximity of talk and listen circuits in individual stations, loudness of local transmission creates a problem: The local microphone generates more power at the local listen amplifier than remote signals do. Audibility of the local microphone is called **sidetone**. Some amount of sidetone is necessary because most of us are very uncomfortable using an intercom system if we cannot hear our own voices on the line, so the local station needs a way to regulate sidetone level. Telephone systems handle sidetone by shifting the local microphone signal approximately 90 degrees as it is delivered to the line. Line signal is then combined with local signal inside a mixing device. As a result, the local signal reaching the local earpiece is attenated by partial phase cancellation, but remote signal levels are unaffected. A sidetone adjust control is present in most intercommunication stations, but access to the control is restricted.

Basics of Production Intercommunication

In almost all respects, production intercom equipment bears very little resemblance to any other form of audio equipment. It is an audio-based technology, however, using amplifiers, headphones, microphones, and cabling as basic elements. Almost all modern production intercommunication systems are modular employing several basic stock units: a system power supply, a master (base) station, and several kinds of remote stations. Usually, the power supply and the master station are combined as a single module. The most common form of remote station is a beltpack worn by an individual stagehand or operator. Other kinds of remote stations include wall-mounted boxes, "biscuits" (stations that include a small loudspeaker—often portable), and wireless units. Production intercommunication requires simultaneous, multi-party, hands-free operation. A block diagram of a basic intercom system is shown in Fig. A2-1.

All stations contain a similar set of controls: talk switch, call switch, listen level control, and sidetone adjust. The talk switch activates the local microphone, permitting the station to transmit to the audio channel. The call switch activates signal lights at other stations in the system. The listen level adjusts the loudness of the line for the local station only. Sidetone adjust—a recessed control that can be turned only with a small screwdriver—regulates the amount of the local microphone signal that is heard at the local station. In addition to these standard controls, biscuits and some wall-mounted stations also contain loudspeakers and loudspeaker muting switches.

Basic intercommunications systems use a single channel. More complex hookups involve two or more channels. Master stations are made for applications requiring up to 24 channels. Remote belt-pack units come in single- and dual-channel versions. Biscuits and wall-mounted stations are usually single channel units, but they may incorporate a selector switch to choose between two channels.

FIG. A2-1: BASIC PRODUCTION INTERCOM SYSTEM

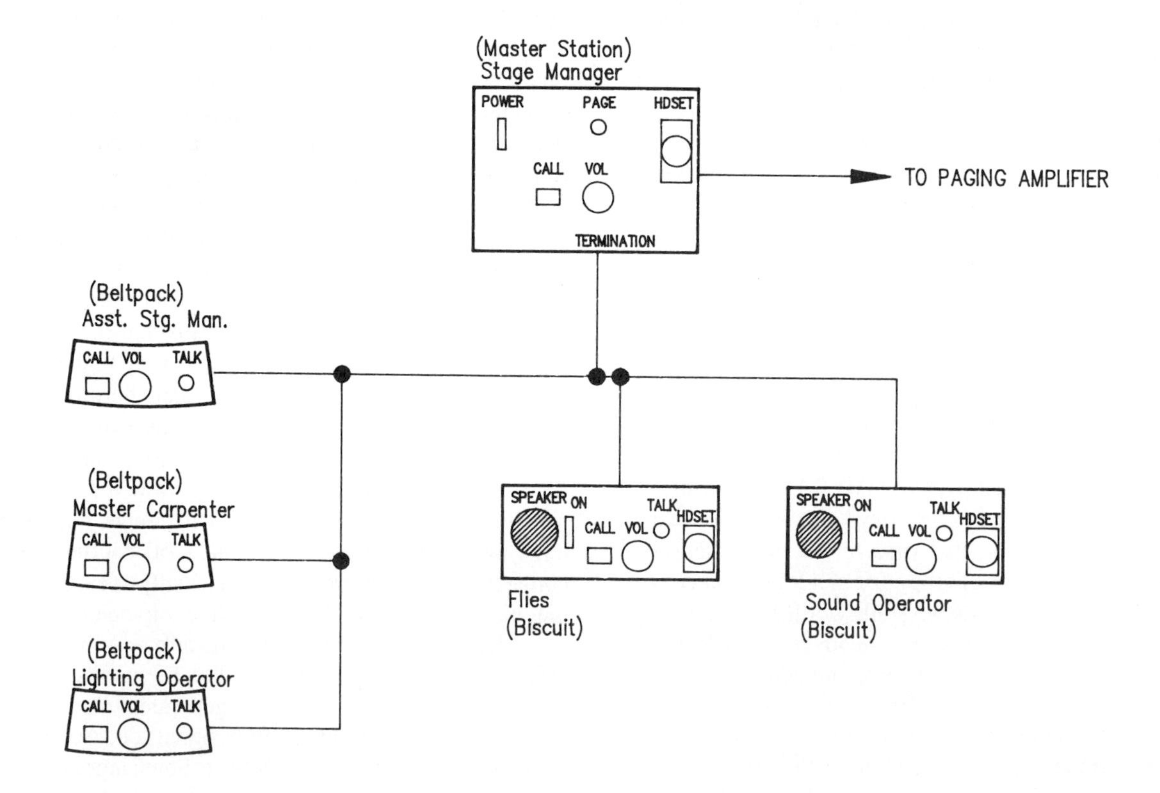

Master Station

A master station is the central operating position of the intercom system. The master station is usually a portable, metal enclosure that contains a power supply, system on/off switch, amplifiers, headset connectors, mic switch, listen-level control, signalling light and switch, and a terminating resistor for each audio channel. The power supply provides operating current and voltage for the master station and places voltage on one of the external lines to power all remote stations connected to the master. The master station's internal talk and listen electronics are no different than for any other station, but they are usually mounted on a larger circuit board that includes electronics for such functions as coupling channels and controlling external audio feeds. Controls are mounted on the station's face panel. The master station's rear panel provides multiple line connectors to branch each channel to various locations in the production complex. Sidetone and channel termination controls are located behind the faceplate and are accessible only by removing a small panel.

Beltpacks

Beltpacks are small, compact units that can be clipped to a belt or a pocket. They are designed for hands-free operation. A remote station includes talk and listen amplifiers, headset connectors, a mike on/off switch, listen level control, call light and switch, and a feed-through connector to allow individual stations to be daisy-chained. Remote stations depend on the system power supply (usually located in the master station) for operating voltage and current. The controls on a remote station operate in the same way that those on the master station do.

Biscuits and Wall-Mounted Stations

Biscuits and wall-mounted units are electrically identical to beltpacks, but they are mounted in metal boxes and often contain a small loudspeaker. Most biscuits and some wall-mounted units can accommodate older, carbon-granule telephone handsets. Biscuits are nominally portable; wall-mounted units are for permanent positions. Biscuits and wall-mounted remote units are used in such locations as front-of-house operations centers, fly rails, orchestra pits, or at any position where beltpacks are not suitable.

Advanced System Configurations

The basic system shown in Fig. A2-1 is perfectly suitable for the run of most productions. Such a system allows the stage manager to communicate with all operating positions and front-of-house and to deliver page announcements to dressing-room and greenroom areas. Technical rehearsals often require a much more elaborate system.

Consider the intercommunication needs of the production staff during technical rehearsals. The stage manager needs a clear channel to all operating positions. Lighting and sound designers also need a clear channel to their operators. But the lighting designer or an assistant may also need to talk to followspot operators on a channel that is completely independent of the designer's channel to the lightboard operator. With the advent of programmable, movable lighting units, the lighting designer or an assistant may need an independent channel to a movable-instrument programmer/operator. Both lighting and sound designers need private channels that enable them to talk to the stage manager apart from the channel that the stage manager uses to call the show. In shows with complex winch and motor activity, a master operator for motorized effects may need a private subsystem to communicate with individual unit operators. All in all, a rehearal intercom hookup for a large show can be quite complex.

Figure A2-2 shows an intercommunication hookup with 3 subsystems. Each subsystem has its own master controller and several remote stations straddle two subsystems: the stage manager's call channel and the designer's or crewhead's instruction channel.

In a system with multiple master stations problems of power and termination always arise. Power is the least difficult to resolve. As long as the output from each power supply is similar in voltage level and polarity, multiple master stations simply provide the reserve power necessary to drive a large number of remote stations. Some system designers, however, dislike mixing power among subsystems, so they use a **pin-lift adapter**. A pin-lift adapter is usually a gender-changer (male-to-male or female-to-female) connector in which the wiring carrying power supply voltage has been cut or removed.

Proper termination requires careful organization of channels. A few rules can help to reduce confusion.

1. Consider the stage manager's station the top position in the hierarchy. Any channel that connects to the stage manager's station is terminated there, including channels originating at the master station of a subsystem. The termination for such a channel should be lifted at the subordinate master unit.

2. All other channels are terminated at the master station controlling their subsystem.

3. Dual-channel remote units may have one channel terminated at one master station while the second channel is terminated at another master station. Under no circumstances should one channel of a dual-channel unit be left unterminated.

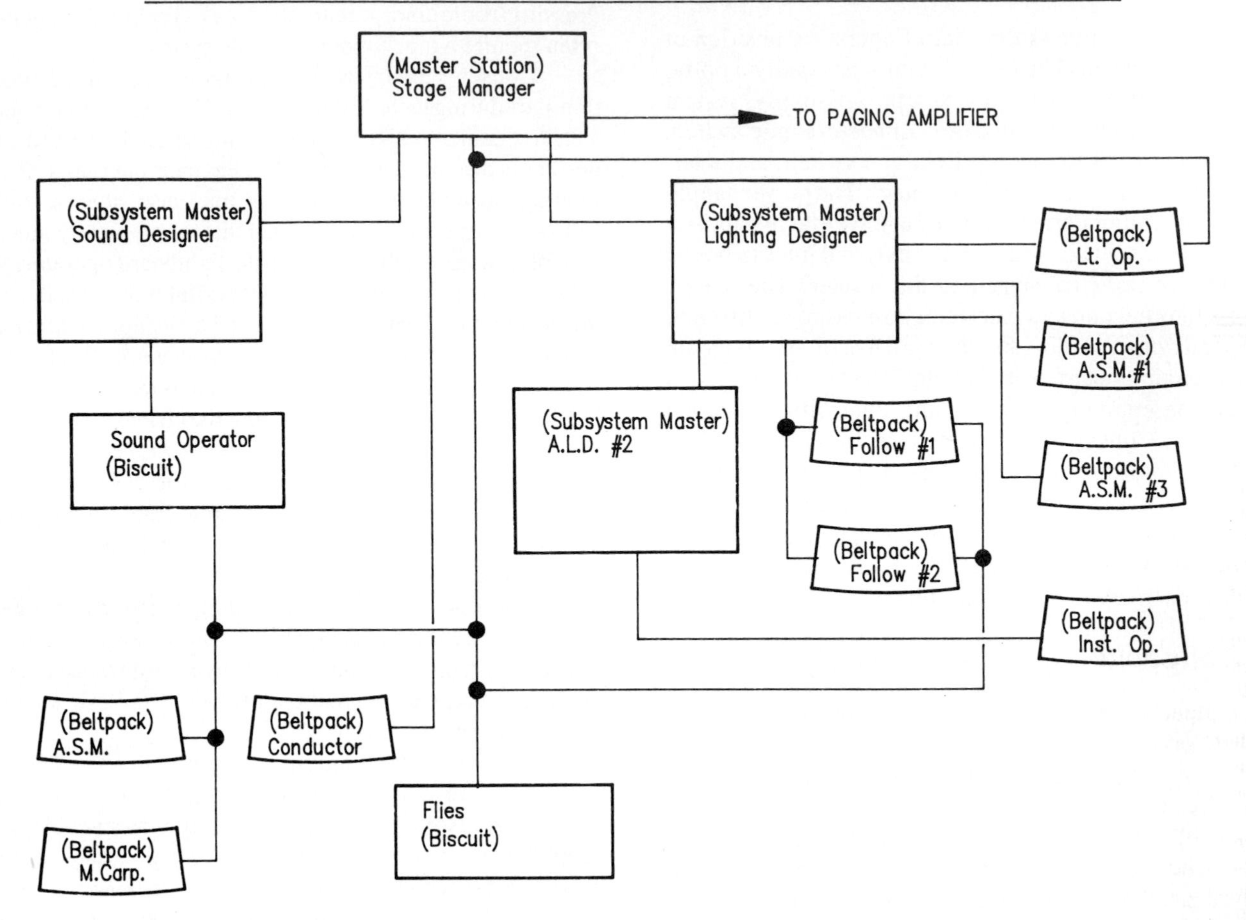

4. Plan interchannel linking. Some master stations have the capability to link their separate channels into a single system. If linking is needed, then the linking switches should be set; otherwise they should be disabled. If people on one channel interfere with those on a different channel, a linkage switch is set. Find it and set it off unless the linkage is desired.
5. Dual-channel stations should be set so that, during rehearsal, the designer's voice is slightly louder than the stage manager's voice. At the remote station, the talk switch on the channel monitoring the stage manager should be left off. The operator should activate it only long enough to acknowledge warns and to report completion of cues. The talk switch to the designer's channel is usually left on.

In Fig. A2-2, both the lighting and sound designers have separate subsystems. That for the lighting designer is quite elaborate, having 6 separate channels, one of which involves a third-level subsystem. The sound designer's subsystem is a simple 2-channel hookup: one channel to the operator and one to the stage manager. The stage manager's station has 4 channels: one for the operators, one to each designer, and one to the conductor.

Theatres that produce only small shows do not need such large intercommunication systems, but all theatres can profit from hookups that provide at least a separate channel for the lighting designer during technical rehearsals. A private channel for the lighting designer significantly improves ability to adjust levels during rehearsals by allowing the lighting designer to talk to the board operator without disturbing the stage manager's call channel. A separate channel for the sound designer is a convenience but not always a necessity.

Index